Praise for Alan Lightman's

THE DISCOVERIES

"Readers are given a rare glimpse into the minds of top researchers."
—*Science News*

"Lightman has a poet's affection for metaphor underpinned with a specificity that makes for meaty, well-observed essays."
—*Time Out New York*

"Careful and wonderfully lucid discussions. . . . A brilliantly guided tour through the human and scientific processes of unveiling nature at a remove from the constraints of our immediate senses." —*Santa Barbara News-Press*

"Lightman vividly explains inaccessible published scientific masterpieces that document each finding. . . . Startlingly comprehensible." —*Library Journal*

"*The Discoveries* is an essential and appealing tour of the high points of twentieth-century science; a brilliant idea, brilliantly executed." —Andrea Barrett, author of *Voyage of the Narwhal*

"Interesting, graceful, and easily understandable. . . . [Will] surprise and stimulate." —*Science Books and Films*

"Vivid, indispensable and deeply moving in its revelation of the orderly complexity of the natural world—and the penetrative power of science."
—Richard Rhodes, author of *The Making of the Atomic Bomb*

"Lightman . . . puts his formidable storytelling powers to best use when exploring the personal lives of scientists that run parallel to their discoveries." —*The Globe and Mail* (Toronto)

ALAN LIGHTMAN

THE DISCOVERIES

Alan Lightman was born in Memphis, Tennessee, and educated at Princeton and at the California Institute of Technology, where he received a Ph.D. in theoretical physics. An active research scientist in astronomy and physics for two decades, he has also taught both subjects on the faculties of Harvard and MIT. Lightman's novels include *Einstein's Dreams*, which was an international bestseller; *Good Benito*; *The Diagnosis*, which was a finalist for the National Book Award; and *Reunion*. His essays have appeared in *The New York Review of Books*, *The New York Times*, *Nature*, *The Atlantic Monthly*, and *The New Yorker*, among other publications. He lives in Massachusetts, where he is adjunct professor of humanities at MIT.

ALSO BY ALAN LIGHTMAN

Origins
Ancient Lights
Time for the Stars
Great Ideas in Physics
Einstein's Dreams
Good Benito
Dance for Two: Selected Essays
The Diagnosis
Reunion
A Sense of the Mysterious

THE DISCOVERIES

The

DISCOVERIES

Great Breakthroughs in 20th-Century Science, Including the Original Papers

ALAN LIGHTMAN

VINTAGE BOOKS
A DIVISION OF RANDOM HOUSE, INC.
NEW YORK

FIRST VINTAGE BOOKS EDITION, NOVEMBER 2006

The Library of Congress has cataloged the Pantheon edition as follows:
Lightman, Alan P., [date]
The discoveries / Alan Lightman.
p. cm.
Includes index.
1. Discoveries in science—Historiography. 2. Discoveries in science—History—20th century—Sources. 3. Science—Historiography. 4. Science—History—20th century—Sources. I. Title.
Q180.55.D57L54 2005
509—dc22 2005040854

Vintage ISBN-10: 0-375-71345-X
Vintage ISBN-13: 978-0-375-71345-3

Book design by M. Kristen Bearse

www.vintagebooks.com

Printed in the United States of America
10 9 8 7 6 5 4

CONTENTS

INTRODUCTION

ON THE NORTHWESTERN SHORE OF AFRICA, some 150 miles south of the Canary Islands, the coastline slightly bulges in a pimple known as Cape Bojador. For Europeans in the early fifteenth century, Cape Bojador marked the boundary between the known and the unknown. North of the cape was civilization and the cities of light. South were the mystical lands of Africa and the Mare Tenebrosum, the Sea of Darkness. No sailor since the ancient Carthaginians had ventured south of Cape Bojador and returned. As Gomes Eanes de Zurara, the Portuguese royal reporter, wrote at the time, "ancient rumors about this Cape . . . have been cherished by the mariners of Spain from generation to generation . . . beyond this Cape there is no race of men nor place of inhabitants . . . and the sea so shallow that a whole league from land it is only a fathom deep, while the currents are so terrible that no ship having once passed the Cape will ever be able to return." Between 1424 and 1434, Prince Henry of Portugal sent fourteen expeditions of ships to round the perilous cape, with its deadly shallows, whirlpools, and violent storms. None succeeded. Yet the unfathomed beckoned. Undeterred, Prince Henry dispatched explorer Gils Eannes for a fifteenth attempt. On this voyage, Eannes gave Cape Bojador a wide berth, steering far to the west, into the Sea of Darkness. As he turned south, he looked back over his shoulder and was astonished to realize that he had left the dreaded cape behind. On his next trip, in 1435, Eannes again rounded Bojador and landed in a bay 150 miles to the south. There, he saw footprints of humans, camels . . .

In the view of historians, Prince Henry did not send his ships south to Africa to colonize or to open fresh trade routes. Instead, he simply wanted to discover what was to be discovered. As Zurara explained, "he had a wish to know the land."

The urge to discover, to invent, to know the unknown, seems so deeply human that we cannot imagine our history without it. Eventually, that passionate urge conquers the fear of the foreign, the fear of the gods, even the fear of personal danger and death. What remains is the pure exhilaration of discovery. We feel that exhilaration in Pablo Picasso's new cubism, and in the "stream of consciousness" of James Joyce and Virginia Woolf, and in the experiments with pentatonic scales in the jazz of Chick Corea and John Coltrane—just as we feel it in the great geographical discoveries of new lands and new seas.

And we feel that exhilaration in the great breakthroughs in science. Werner Heisenberg, one of the founders of quantum physics, describes the transcendent moment in May 1925 when he realized what he had discovered: "it was almost three o'clock in the morning before the final result of my computations lay before me. I had the feeling that, through the surface of atomic phenomena, I was looking at a strangely beautiful interior, and felt almost giddy at the thought that I now had to probe this wealth of mathematical structures that nature had so generously spread out before me. I was far too excited to sleep."

Scientific discoveries tremble with big ideas, not just about science but about human existence as well. Albert Einstein revised our notion of time. Hans Krebs found the universal chemical cycle that provides energy in every cell of every plant and animal on earth—strong evidence for a common origin of life. Jerome Friedman helped discover the quark, believed to be one of the indivisible units of matter. Paul Berg developed the first technique to modify genes and thus to create altered forms of life. Alexander Fleming discovered the first antibiotic, advancing humankind's eternal struggle against mortality and disease. Heisenberg outlined his famous uncertainty principle, showing that the future can never be fully predicted from the past.

Several years ago, I decided to explore some of the great discoveries of science in the twentieth century—to embark on a discovery of discoveries. Were there common patterns of discovery? How did styles of working and thinking vary from one science to the next, and from one scientist to the next? How did the discoverers compare with each other as people? Were they aware of the significance of their work at the time? As part of my undertaking, I collected the actual papers in which the discoveries were first announced, like the first reports of Gomes Eanes de Zurara. Together, these papers would form an unusual kind of history of twentieth-century science.

There came a moment, in the spring of 2002, when I had finally gathered together the twenty-five papers that I would include in this book. I was at home, in my house in Concord, and the golden forsythia were just starting to bloom. For six months I had been badgering astronomers, physicists, chemists, and biologists for nominations of the greatest discoveries in their fields in the twentieth century. The original publication of the theory of relativity. The first quantum model of the atom. The discovery of how nerves communicate with each other, the discovery of the first human hormone, the discovery of the expansion of the universe, the discovery of the structure and secret code of DNA. Some of the discovery papers had been published in obscure journals. Some required translation into English. Some were smudged and faint after being Xeroxed in distant libraries and sent through the mail. Partly using my own judgment, I winnowed down a list of over a hundred landmark papers to twenty-five. Each of these papers had profoundly changed the way that we fathom the world and our place in it. Here were Einstein, Fleming, Bohr, McClintock, Pauling, Watson and Crick, Heisenberg—in their own words—inventing, creating, discovering. Here were the great novels and symphonies of science. On that day in May, there came a moment when I had finished finding and collecting these original writings. I held the stack of twenty-five papers in my arms, a century of scientific thought. My eyes filled with tears.

When I was a graduate student in physics thirty years ago, I had the simple notion of a monolithic mind and method of the scientist. In fact, there is a great variety of minds and patterns of discovery in science. Sometimes the scientist knows pretty much where he or she is going even when the results are revolutionary, as in the findings of Einstein, Planck, and Krebs. By contrast, sometimes the discoveries are completely unexpected or accidental, as shown in the experiments of Bayliss and Starling, Rutherford, Fleming, Hahn and Strassmann, and in the observation of Leavitt. Even theoretical scientists can bc surprised by the conclusions of their pencil-and-paper adventures, as with Steven Weinberg's work. Sometimes the scientists appreciate the significance of their discoveries at the time, as did Einstein, Dicke, Watson and Crick, and Loewi. Other times, the significance is only dimly understood, as with the work of Hubble. In some cases, sheer brilliance leads to discovery. In others, the required ingredients include circumstance and luck.

An equally broad range holds for the scientists themselves. There is no single scientific personality. Great scientists can be bold and self-confident revolutionaries, like Rutherford or Einstein or Watson. Great scientists can also be modest and diffident, like Krebs or Fleming or Meitner. Some scientists, like William Bayliss, are by temperament cautious, meticulous, in love with the details, while others, like Ernest Starling, are brisk, impatient, engaged mainly by the broad sweep of things.

What all of these men and women share is a passion to know, a pure pleasure in solving puzzles, an independence of mind. The American biologist Barbara McClintock recalled that in high school science classes, "I would solve some of the problems in ways that weren't the answers the instructor expected . . . It was a tremendous joy, the whole process of finding that answer, just pure joy." When the German nuclear physicist Lise Meitner was a child, her grandmother cautioned her that she should not sew on the Sabbath or else the heavens would come tumbling down. The little girl decided to do an experiment. She lightly touched her knitting needle to some embroidery and looked up. Nothing happened. Then she took a stitch, waited, looked up. Again, nothing. Finally, Lise was satisfied that her grandmother had been mistaken and went happily about her sewing.

Different kinds of scientists conceive of problems differently. I was trained as a physicist, and I understood well how the physicist thinks about the world. The physicist is a reductionist. The physicist chisels away at a towering building until it is broken down to a heap of individual bricks and cement. What are the fundamental forces and particles of nature? asks the physicist. What are the eternal laws? Physicists make simplifications and idealizations and abstractions until the final problem is so simple that it can be solved by a mathematical law. For example, physicists reduce many phenomena to the model problem of a weight bobbing up and down on a spring, called the harmonic oscillator. The vibrations of atoms in molecules, the sloshing of water in a bowl, the quantum nature of empty space, can all be reduced to weights bobbing up and down on springs, which obey simple equations.

Biologists think differently. Because biology deals with living things, biologists hardly ever go below the level where life is relevant. (A notable exception is modern molecular biology, where physics, biology, and chemistry have merged.) And life requires the interaction between elements in a system. Thus, biology usually deals with *systems*. What is the system by which a living creature regulates and controls its internal processes? What is the system by which a living creature reproduces itself? What is the system by which a living creature gains and utilizes the energy required for life? Where physics might ponder the electrical force between two electrons, biology would be concerned with how the electrical charges on both sides of a cell membrane regulate the passage of substances across the membrane and thus connect the cell to the rest of the organism. Roughly speaking, physics has laws, while biology has concepts. Biology is more of an empirical science than physics because its concepts are close to the observed facts. There are many purely theoretical physicists but few purely theoretical biologists. Chemists lie somewhere between, sometimes acting much like biologists, sometimes like physicists.

All of these differences and similarities are reflected in the essays I have written about the discoveries and in the original discovery papers themselves. In my essays, I have attempted to paint the intellectual and emotional landscape of the discoveries and of the men and women who made them. Each discovery has its own story. Each has its own characters and personalities, its own human drama, its own failures and triumphs and personal ambitions. The essays are structured in ever-deepening layers, so that readers may get a general sense of the discovery and its significance at the beginning, gradually learning more about the lives of the scientists and the details of the discovery, and finally receiving a guided tour through the original discovery paper itself.

A great wealth lies in the discovery papers themselves, and these are included, one or more original papers at the end of each chapter. I've often been struck that philosophy students read Kant's *Critique of Pure Reason*, political science majors read the U.S. Constitution, and literature classes read *Hamlet* and *Moby-Dick*, but students of science hardly ever read the original works of Mendeleyev or Curie or Einstein. Even professional scientists rarely read the original literature in their field, if it is older than a decade or two. The mythology seems to be that in science, as opposed to all other human activities, only the final results matter. According to this belief, a summary or distillation of ideas eliminates the need for the original papers. Moreover, as the subject advances, new mathematical methods are brought to bear, new technologies and instruments become available, ideas and results are recast into an improved form. Would it not be a decided burden to relive much of the outmoded history of science, to struggle through the awkward notations and often half-formed ideas of the original papers?

I believe such a mythology is mistaken. In my view, the first reports of the great discoveries of science are works of art. Like poetry, these papers have their internal rhythms, their images, their beautiful crystallizations, their sometimes fleeting truths. In the original choice of words and metaphors, in the often simple but profound arguments, in the uncertainties and speculations, we can gaze into the mind of a great scientist in a way that no summaries or commentaries can ever provide. In these papers, we see enormously gifted human beings grappling with the nature of the world. Some of the jargon and mathematics are too technical for all but the professional scientist, and the level of genius is sometimes practically incomprehensible. But we can follow the line of thought. And in the discussion of ideas and in the deep questioning, we can recognize fellow thinkers at work.

A few final notes. For length reasons, the original papers have been somewhat abridged. All of the shorter papers have been included in full. For the longer papers, I have omitted the less important or more detailed material. Places in the texts where material has been omitted are indicated by a series of dots. The average omission for all the original papers is about 20 percent.

There are many great scientific discoveries of the twentieth century not included in this book. For obvious reasons, I have had to make a selection. Here, I used my own judgment and the judgment of colleagues to choose those discoveries in pure science with the deepest conceptual significance, those discoveries that most changed thinking and progress in their fields. Discoveries in applied science and technology, such as cloning or television, were not considered. When there were several papers announcing a similar discovery at the same time, I have chosen to focus on only one paper but have mentioned the others. Likewise, when several scientists coauthored a landmark paper, I have usually profiled only one of them in my introductory essay. The twenty-two discoveries are represented by twenty-five papers. In three cases—the discovery of nuclear fission, the discovery of the structure of DNA, and the discovery of the radiation left over from the Big Bang—a theoretical paper by one group of scientists and an experimental paper by another group were so closely connected that I have included both papers for each discovery.

Because many scientific ideas are amplified by later ideas, the discoveries build on each other. For that reason, I have arranged the discoveries in chronological order, beginning with Max Planck's discovery of the quantum in 1900 and ending with Paul Berg's discovery of recombinant DNA in 1972. For example, Planck's quantum (1900) and Rutherford's discovery of the nucleus of the atom (1911) were combined in Bohr's first quantum model of the atom (1913). Von Laue's discovery of the powerful technique of X-ray diffraction (1912) was used by Franklin, Watson, and Crick to discover the structure of DNA (1953), which in turn was used by Paul Berg in his experiments to create new DNA (1972).

Reading the great works of a few scientists might convey the impression that science is carried out mainly by a small number of heroic figures. This impression is false. While the scientists considered here are certainly extraordinary individuals, the enterprise of science is in fact a result of the efforts of many people, all making contributions. Jerome Friedman's experiments that helped discover the quark depended on the previous invention of the magnetic spectrometer. The work of Edwin Hubble in measuring the expansion of the universe rested on the earlier observations of Vasco Melvin Slipher. And so on.

With only twenty-five papers, one must be cautious about generalizing from the demographics, but here are a few statistics: Eighteen of the papers were originally published in English, seven in German (of these seven, five were previously translated into English and two were translated for this book). All but one of the papers originally in German date from the period 1900 to 1927, reflecting the domination of the German-speaking world in science in the early years of the century and the English-speaking world thereafter. Of the individuals or groups authoring each paper, nine were German, nine American, four British, one Austrian/British (Perutz), one from New Zealand (Rutherford), and one Danish (Bohr). Women were involved with four of the twenty-two discoveries considered in this book. Without a doubt, women almost always face greater difficulties than men in pursuing a scientific career. I have discussed these difficulties in detail in the chapters on Leavitt, Meitner, and McClintock. During World War II, a number of the European scientists, including Otto Loewi, Lise Meitner, and Max Perutz, faced special hardships because they were Jewish. (Einstein's political antagonism to the Third Reich was a greater factor in his treatment than was his religion.)

Finally, some of the later chapters involve scientists I have known personally. In these cases, I have permitted myself personal assessments and often more substantial profiles.

For numbers smaller than 1, we use negative superscript numbers to the right of the 10, as in the following examples:

$$10^{-1} = 0.1$$

$$10^{-6} = 0.000001.$$

The negative superscript number tells how many places the decimal point must be moved to the right to make 1. Alternatively, the negative superscript number is one greater than the number of zeros preceeding the 1 before the decimal point.

Using the above notation, the diameter of the earth can be written as about 10^9 centimeters, and the diameter of an atom of hydrogen can be written as about 10^{-8} centimeters. (A centimeter is about 0.39 inches.) The number of atoms in a poppy seed is about 10^{21}.

Another mathematical convention is used for letter symbols representing numbers. If a and b are two numbers, then ab is shorthand for $a \times b$. In other words, when two symbols representing numbers appear side by side, they are meant to be multiplied.

THE DISCOVERIES

1

THE QUANTUM

IN HIS FAMOUS AUTOBIOGRAPHY *The Education of Henry Adams*, published only a few years into the twentieth century, the historian Henry Adams shouted alarm that the sacred atom had been split. Since the ancient Greeks, the atom had been the smallest particle of matter, the irreducible and indestructible element, the metaphor for unity and permanence in all things. Then, in 1897, the British physicist J. J. Thomson found electrons, particles far lighter and presumably smaller than atoms. The next year, Marie Sklodovska (Madame Curie) and her husband Pierre Curie discovered that the atoms of a new element, called radium, continuously hurled out tiny pieces of themselves, losing weight in the process. Now, nothing was permanent—nature no more than human civilizations. The solid had become fragile. Unity had given way to complexity. The indivisible had been divided.

As Adams was summing up the nineteenth century, he was evidently unaware of another scientific bombshell that had just exploded, ultimately as earthshaking and profound as the fracturing of the atom. On December 14, 1900, in a lecture to the stodgy German Physical Society in Berlin, Max Planck proposed the astounding idea of the quantum: energy does not exist as a continuous stream, which can be subdivided indefinitely into smaller and smaller amounts. Rather, he suggested, there is a smallest amount of energy that can be divided no further, an elemental drop of energy, called a quantum. Light is an example of energy. The seemingly smooth flood of light pouring through a window is, in reality, a pitter-patter of individual quanta, each far too tiny and weak to discern with the eye. Thus began quantum physics.

At the time of his lecture, Planck was bald from the middle of his head forward, with a sharp aquiline nose, a mustache, a pair of spectacles fastened to his face, and the overall look of a dull office clerk. He was forty-two years old, almost elderly for a theoretical physicist. Newton had been a youth in his early twenties when he worked out his law of gravity. Maxwell had polished off electromagnetic theory and retired to the country by age thirty-five. Einstein and Heisenberg would be in their mid-twenties when they erected their great monuments.

In 1900, Planck was already established as one of the leading theoretical physicists in Europe. Planck himself had helped legitimize the discipline. Fifteen years earlier, when he secured the rare position of professor of theoretical physics at the University of Kiel, theoretical science was considered an impotent profession, inferior to laboratory experiments. Few students clamored to hear Planck's mathematical lectures. Then, in 1888, after his studies of heat—in which he clarified the Second Law of Thermodynamics and the concept of irreversibility—Planck was appointed professor at the University of Berlin. At the same time, he was made director of the new Institute for Theoretical Physics, founded mainly for him.

At the end of the nineteenth century, physics basked in the glow of extraordinary achievement. Newton's precise laws of mechanics, which described how particles respond to forces, together with Newton's law for gravity had been successfully applied to a large range of terrestrial and cosmic phenomena, from the bouncing of balls to the orbits of planets. The theory of heat, called thermodynamics, had reached its climax with the melancholy but deep Second Law of Thermodynamics: an isolated system moves inexorably and irreversibly to a state of greater disorder. Or, equivalently, every machine inevitably runs down. All electrical and magnetic phenomena had been unified by a single set of equations, called Maxwell's equations after the Scottish physicist James Clerk Maxwell, who completed them. Among other things, these laws demonstrated that light, that most primary of natural phenomena, is an oscillating wave of electromagnetic energy, traveling through space at a speed of 186,282 miles per second. The new areas of physics known as statistical physics and kinetic theory had shown that the behavior of gases and fluids could be understood on the basis of collisions between large numbers of tiny objects, assumed to be the long-hypothesized but invisible atoms and molecules. In short, as Planck scribbled his equations at the dawn of the new century, physics might survey its vast kingdom and be pleased.

Some cracks, however, were starting to show in the marble facade. Aside from the philosophical dismay expressed by Mr. Adams, Thomson's electron was clearly a new type of matter that demanded explanation and raised other questions about the innards of atoms. The "radioactive" disintegrations observed by the Curies involved the unleashing of huge quantities of energy. What was the nature of this energy and where did it come from? Other emissions of electromagnetic radiation from atoms, the so-called atomic spectra, exhibited surprising patterns and regularities but with no theoretical understanding. Equally perplexing were the repeating patterns in the properties of the chemical elements, a phenomenon that scientists suspected was caused by the structure of atoms.

Finally, physicists had observed that a unique kind of light, called black-body light or black-body radiation, emerged from all hot, blackened boxes held at constant temperature. (Set a kitchen oven at some temperature, leave the oven door closed for a long time, and black-body radiation will develop inside—although at any practical cooking temperature this light will be below the frequencies visible to the human eye.) It was already well known to scientists that all hot objects emit light—that is, electromagnetic radiation. In general, the nature of such light varies with the properties of the hot object. But if the radiating object is additionally enclosed within a box and held at constant temperature, its light assumes a special and unvarying form, the so-called black-body radiation.

A particularly mysterious aspect of black-body light was that its intensity and colors were completely independent of the size, shape, or composition of the container—as surprising as if human beings all over the world, upon being asked a question, uttered the same sentence in reply. A heated black box made of charcoal and shaped like a cigar produces precisely the same light as a black box made of dark tin and shaped like a beach ball, provided that the two boxes have the same temperature. The known laws of physics could not explain black-body light. Even worse, the standard working theories of light and of heat actually predicted that a blackened box held at constant temperature should create an *infinite* amount of luminous energy! It was the puzzle of black-body radiation that Max Karl Ernst Ludwig Planck had solved for his lecture of December 14, 1900.

A great deal was already known of the subject. With the use of colored filters and other devices, scientists had measured how much energy there was in each frequency range of black-body light. A colored filter allows light of only a narrow range of frequencies to pass through it. (The frequency of light is the number of oscillations per second. Each frequency of light corresponds to a particular color, just as each frequency of sound corresponds to a particular tone.) The amount of energy in a given frequency range of light is measured by a device called a photometer. Photometers gauge the intensity of light falling on a surface—a glass plate, for example—by comparing that light to another beam of light of known intensity. The comparison can be accomplished, for example, by the relative penetrating power of light through a liquid. More intense light beams have greater penetrating power. (Several decades into the twentieth century, light intensities could be measured more accurately by their electrical effects, with photoelectric detectors.)

The breakdown of a light source into the amount of energy in each range of frequency is called a light spectrum. When the light is black-body light, its spectrum is called a black-body spectrum. Figure 1.1 illustrates two black-body spectra, one for a temperature of 50 K and another for a temperature of 65 K. Here the K stands for Kelvin, the unit

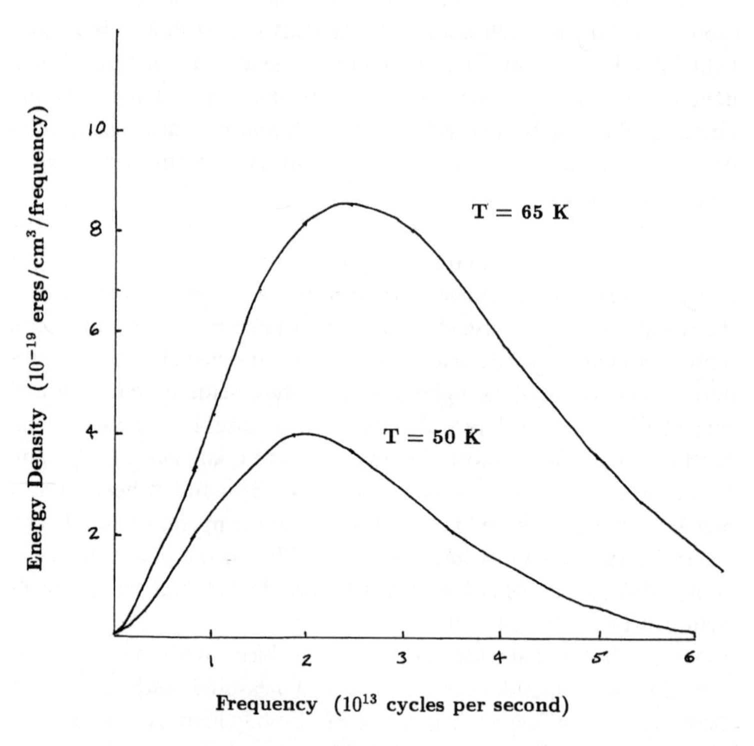

Figure 1.1

of temperature on the absolute temperature scale, which is a form of the Celsius scale with the zero point shifted. The coldest possible temperature lies at 0 K and –273 C.

A more familiar example of a spectrum is the graph that shows how many adults there are in each range of heights. Such a spectrum is usually a bell-shaped curve, with few people at very small heights and few people at very tall heights. As one would expect, the height spectrum varies from one country to the next, since human heights are determined by a large number of variables such as genetics and diet. So, it was remarkable when Planck's predecessor to the Berlin chair, Gustav Kirchhoff, and others, discovered that the black-body spectrum does not vary at all with the details of the container. *The black-body spectrum depends only on a single parameter, the temperature.*

Planck was much impressed by the uniqueness and universality of the black-body spectrum, reasoning that such a universality must be the result of some fundamental new law of nature. A few weeks prior to his December lecture, the German physicist had in fact *guessed* a formula for the spectrum of black-body light. Planck's formula was a mathematical expression for the amount of energy in each range of frequency of black-body light, and it agreed with all experimental measurements. Embracing the aesthetic criteria common to most physicists, Planck found pleasure in the simplicity of his formula, using the word "simple" (*einfach* in German) twice in the first paragraph of his paper.

But a mathematical formula, in itself, is only a tidy summary of quantitative results, like a sun calendar, which tells the number of daylight hours on each day of the year. Such a calendar is useful for making plans, but it does not explain *why* the numbers come out as they do. To know why, we need to know what causes day and night, we need to know that the earth spins on its axis at a certain rate, that the earth also orbits the sun at a certain rate, that the earth's axis is tilted at a particular angle. When we know all of these things, we understand why. With such understanding, we could then *predict* the sun calendar for any planet anywhere in the universe, given the corresponding astronomical facts.

Planck was not satisfied with merely guessing the right formula for black-body light. What compelled him and haunted him was to answer the deeper question: Why? What fundamental, inviolable principles led to that formula, made it a logical necessity, required it and it alone out of all the possible *einfachen* formulas that one could imagine? Why was that same formula observed to be true over and over again, from one experiment to the next, even for experiments that had never been done?

To understand the *why* of his formula, Planck discovered that he had to reject centuries of physical thought that you could chop energy into smaller and smaller pieces indefinitely. Surprisingly, the world did not work in that way. Planck could explain his formula for black-body light only by the radical proposal that there was a smallest piece of energy, called the quantum, which could not be chopped any further. Evidently, energy, like matter, came in granular form. The quantum was the grain of sand on the beach, the penny of currency in the subatomic world. The quantum was indivisible.

Planck was a theoretician, someone who works with pencil and paper and imagines experiments in his mind. To arrive at his conclusions, the German physicist imagined lots of atoms enclosed in a black box, all emitting and absorbing light. In such a situation, the atoms are affected by the surrounding light, and the surrounding light is affected by the atoms. Planck then discovered that if the atoms could absorb or emit energy only in whole chunks, quanta, then the resulting light would necessarily become black-body light.

For much of his life thereafter, Planck was amazed by the success of his quantum proposal. Like other theoretical physicists, he had an almost religious faith in the absolute validity of the laws of nature, which would, as he wrote in 1899, "retain their significance for all times and for all cultures including extraterrestrial and nonhuman ones." For Planck, "the search for the absolute" was "the loftiest goal of all scientific activity."

Yet in spite of Planck's lofty views, he himself did not aspire to make great discoveries. As he told Philipp von Jolly, his professor at the University of Munich, he desired only to understand and perhaps deepen the existing foundations of physics. (In 1878, Jolly actually advised the twenty-year-old Planck not to continue with physics, on the grounds that all the fundamental laws had been discovered.) Planck's cautious manner of "understanding" was to study a subject slowly and carefully, until he had mastered it. Such a conservative and modest approach seemed to grow naturally out of his background as the descendant of a long line of pastors, scholars, and jurists—Planck's father, Wilhelm, was a professor of jurisprudence at Kiel and then Munich—and further to resonate with his loyal support of imperial Germany. Planck's natural restraint carried over to his personal relationships. Marga von Hoesslin Planck, his second wife, wrote to another physicist that her husband was quite proper and reserved with anyone other than his family and could enjoy himself only with people of his own rank, with whom he might take a glass of wine and a cigar and even make a quiet joke.

There were two situations in which Planck abandoned his reserve: with his family and in music. As a young man he wrote to a friend, "How wonderful it is to set everything else aside and live entirely within the family." Many years later, Marga confirmed this feeling in a letter to Einstein upon the death of her husband: "He only showed himself fully in all his human qualities in the family." Planck's other liberation was music. While a student at the University of Munich, he composed songs and a whole operetta; he served as second choirmaster in a school singing group; he played the organ at services in the student's church; and he conducted. For the rest of his life, he played the piano superbly at small musical gatherings in his home. Music, according to Planck's nephew-in-law Hans Hartmann, was the "only domain in life in which [Planck] gave his spirit free rein."

Following Planck's line of argument will help us understand how theoretical scientists think, how they use models, imagination, and logical consistency through mathematics. As it turns out, Planck's paper on the quantum is one of the most conceptually difficult and abstract of any in this book, and the reader will need to exercise some patience and good humor. Planck begins his landmark paper by considering the material atoms that make up the inner walls of the blackened box. After all, these atoms are responsible for creating the observed black-body light, by emitting and absorbing electromagnetic radiation. He idealizes each of these atoms as a "monochromatic vibrating resonator," that is, a system that emits and absorbs light at only a single frequency, say pure red or pure green. A concrete example of one of Planck's monochromatic vibrating resonators would be an electron bouncing up and down, or "vibrating," on a spring. As the electron bounces, it emits light of a particular frequency, the precise number of up-and-down bounces each second. Different frequencies correspond to different rates of bouncing, which in turn are determined by different stiffnesses of the springs. Black-body light is then hypothetically produced by a large number of these bouncing electrons at many different frequencies. All of these ideas are in accord with Maxwell's equations of electromagnetism.

In representing the inner walls of the box as a collection of tiny objects, each vibrating at a single frequency, Planck is using a common strategy in theoretical physics: represent the system under study by a simple model that can be easily analyzed. Five years later, Einstein would use the same strategy in studying time with an imaginary clock made of two mirrors and a light beam bouncing between them, one tick of the clock for each bounce. In fact, in his paper Planck never mentions the atoms in the walls of the container. He refers only to his abstract "resonators," which can be any systems that emit and absorb light at a single frequency.

Planck's imagined picture is the following: begin with an empty black box and heat it to some temperature. At first, there is no light in the box. Then the hot, vibrating resonators begin emitting light at various frequencies, and the box gradually fills up with light. From time to time, some of that light will strike a wall of the box and be absorbed by the resonators there. *Thus the resonators both emit and absorb light.* For a while, everything evolves in time. The amount of energy in each resonator changes, as does the amount of energy in each range of frequency of light contained within the box.

Eventually, however, if the box is held at a constant temperature, a steady equilibrium is achieved. The absorption and emission by the resonators come into balance. Each resonator attains, on average, a particular and constant amount of energy. And the spectrum of light in the box settles into the universal black-body spectrum, which Planck calls the "normal spectrum." After this point, the system of resonators and light ceases further change. The system has arrived at what is called thermodynamic equilibrium. And, indeed, a small keyhole punched into the box to sample the radiation within would reveal Kirchhoff's universal black-body light.

It is important that Planck visualized this picture as dynamic, as a continual giving and taking of energy. Even after attaining equilibrium, the resonators in the walls of the box are constantly emitting and absorbing light and thus exchanging energy with the light inside the box. In turn, the surrounding light is constantly being replenished and depleted by the resonators.

One of Planck's heroes, the Austrian physicist Ludwig Boltzmann (1844–1906), had developed a particularly fruitful way of describing such dynamic systems in terms of probabilities. At each frequency, a certain amount of energy must be shared between the resonators of that frequency and the light of that frequency. And there is a probability that can be calculated for each particular apportionment of this energy. Some apportionments (say one-third of the energy in the resonators and two-thirds in the light) are more probable than others. While the system is out of equilibrium, evolving in time, it "tries out" many different divisions of energy and naturally evolves toward those that are more and more likely, until it reaches the particular apportionment that has maximum probability. This condition is also called maximum entropy, or maximum disorder. (The quantitative measure of disorder in a physical system is called entropy.) One way of stating the famous Second Law of Thermodynamics, which Planck mentions on the second page of his paper, is that systems naturally evolve toward a condition with maximum probability, or maximum entropy, and that this evolution is irreversible. After a system has achieved the apportionment of energy that has largest probability, departures from that condition are relatively unlikely.

All the above was well established, in the most general terms, before 1900. Now we come to the guts of Planck's calculation. The black-body spectrum, being a condition of thermodynamic equilibrium, will represent that particular apportionment of energy between the resonators in the walls of the box and the light inside the box that has maximum probability. The Berlin professor thus had the task of calculating the probability of each possible apportionment of energy and finding the maximum.

Let us look at the resonators alone. Here again, Planck borrowed some ideas from Boltzmann: the probability of finding a particular amount of energy in the resonators is proportional to *the number of different ways*, called complexions, that the energy may be distributed among the resonators. The more complexions, the higher the probability of that amount of energy. An analogy helps to understand this idea. Consider rolling a pair of dice. The probability of rolling the sum of 4 with two dice is higher than rolling a 3, because there are three ways of getting a 4 (3 and 1, 2 and 2, 1 and 3) but only two ways of getting a 3 (1 and 2, 2 and 1). The probability of rolling a 5 is even higher, with four different ways to do it (1 and 4, 2 and 3, 3 and 2, 4 and 1). This method of calculating complexions is how casino operators figure the odds in Las Vegas.

So Planck has reduced his problem to counting the number of different ways that a given amount of energy can be distributed among a given number of resonators—just as if he were figuring how many ways three pounds of sand could be poured into four buckets on the beach. As Planck points out, if the energy could be subdivided into smaller and smaller pieces, ad infinitum, there would be an *infinite* number of pieces of energy that he could distribute among his resonators, leading to an infinite number of ways that the exercise could be done. The entire calculation would then sink in a hopeless morass of infinities.

Planck's revolutionary idea, which he calls "the most essential point of the calculation," is to consider the energy at each frequency as not infinitely divisible, but composed of a number of equal, indivisible parts (later called quanta). These are analogous to the individual grains in the three pounds of sand. And although there are a great many grains of sand in three pounds' worth, that number is not infinite. The number of ways it can be divided among four buckets has a definite limit. Planck, being the consummate theoretician, can shuffle and count any number of grains with a little mathematics and a few strokes of his pen.

Guided by the mathematical form of the black-body formula, Planck proposes that each of his "grains" of energy has a size $h \times \nu$, where ν is the frequency under consideration and h is some constant number. (Henceforth, we will use the shorthand notation that $h\nu$ stands for $h \times \nu$. More generally, any two quantities written side by side means that they are to be multiplied.) For the sake of illustration, suppose the frequency under consideration were 2, and h had the value of 3. Then each individual grain, or quantum, of energy would have a magnitude of $3 \times 2 = 6$. If the total energy available at this frequency were 24, then there would be $24/6 = 4$ quanta, each of energy 6, that could be distributed among the resonators.

Before Planck, everyone assumed that energy could be subdivided ad infinitum. To understand the strangeness of Planck's quantum proposal for energy, consider a wooden swing at a playground. When we raise the swing to a height, we give it gravitational energy—the greater the height, the more energy. Upon release, the swing shuttles back and forth with a frequency, ν, determined only by the length of its rope for small angles of release. All of our intuition suggests that we can raise the swing to any height we choose—that is, we can put any amount of energy into the swing. But Planck's quantum proposal decrees that energy can be put into the swing only in distinct "grains" or quanta, each of size $h\nu$. Thus, the swing can have only certain definite energies: $h\nu$, $2h\nu$, $3h\nu$, $4h\nu$, and so on, separated by equal gaps. The swing cannot have fractional energies, say $2.8 \times h\nu$ or $16.2 \times h\nu$. Evidently, we cannot raise the swing to any height we choose. In daily life, the quantum nature of energy is unobservable because everyday energies are so huge compared to $h\nu$ that we aren't aware of the gaps. (For a typical schoolyard swing, the gap between allowed energies corresponds to changes of height of a billionth of a trillionth of a trillionth of an inch.)

What is the unknown number *h*? By comparing his final formula with the experimental measurements of black-body light (note the references to F. Kurlbaum, O. Lummer, and E. Pringsheim), Planck was later able to determine the value of *h*, $h = 6.55 \times 10^{-27}$ erg seconds, now called Planck's constant. The erg is a unit of energy. For example, a penny dropped from a height of five feet strikes the floor with an energy of four hundred thousand (4×10^5) ergs. The meaning of Planck's constant is such that an elemental vibrating resonator, with a frequency of 1 bounce per second, can change its energy in increments of 6.55×10^{-27} ergs. The extreme smallness of such a number is the reason why quantum effects are completely invisible in the everyday world, as discussed in the example of the schoolyard swing. Planck's constant was a new fundamental constant of nature, like the speed of light—an eternal truth, presumably valid at all times and all places, here and on the far side of the galaxy.

Using his quantum of energy and the well-known mathematics of combinatorics, Planck can now calculate the number of complexions for *N* resonators and *P* quanta, the two equations involving *N* and *P*. These formulas give the number of ways that *P* quanta can be distributed among *N* resonators. Other previously established results from thermodynamics and statistics show how to maximize the number of complexions given the previous equations.

The final formula, the next-to-last equation of the paper, gives the energy density of black-body light in each range of frequency, that is, the black-body spectrum. That formula is identical to Planck's earlier guess, but now Planck knows the why of that formula: energy comes in units of $h\nu$. From that assumption, the formula follows by logical necessity. As advertised, Planck's formula depends only on a single variable, the temperature, denoted by the Greek letter θ. It also depends on two constants, *h* and *k*. The constant *k*, called Boltzmann's constant, is a unit of entropy, the quantitative measure of order and disorder. Boltzmann's constant was already well known to physicists and had been approximately measured, although Planck's black-body formula allowed *k* to be determined much more accurately than ever before. The constant *h* was completely new. The constant *h* measured the size of the quantum.

In a sense, Planck achieved two successes with this work, one conceptual and the other quantitative. Conceptually, he proposed that energy is not a continuous, infinitely divisible substance, as it seems from everyday life, but comes in indivisible units. Energy has a granularity. This idea was no less momentous than Democritus's concept of the atom 2,300 years before. Planck clearly recognized that his new work was of major significance, writing near the beginning of his paper that "I have obtained . . . relations which seem to me to be of considerable importance for other branches of physics and also of chemistry." But he could not have foreseen that his quantum of energy would lead to a wholesale reshaping of physics, called quantum mechanics, along with a radically new conception of reality. For example, one of the findings of quantum mechanics is that all material objects behave as if they exist at many places at once. A closely related result is that the physical world does not obey fully predictive laws but rather evolves within bounds of uncertainty. These ideas will be discussed more in the chapters on Einstein's and Heisenberg's work.

Second, Planck had found a quantitative measure, $h = 6.55 \times 10^{-27}$ erg seconds, for the size of the elementary quantum. It is important to point out that h is not just a pure number, but it has units of energy and time. It therefore sets a "scale." Similarly, the average height of a human being, about 5.7 feet, has units of length and sets the scale of sizes of clothes, buildings, and all things made for human beings. Planck's constant sets the scale for the quantum. Later scientists, such as Niels Bohr and Werner Heisenberg, showed that Planck's quantum constant determines all sizes of the atomic and subatomic domain, down to the smallest structures of time and space. The diameter of an atom depends on h. Heisenberg's Uncertainty Principle depends on h. The tiniest possible size of transistors and computers depends on h. The theoretical density of matter at the birth of the universe depends on h. The smallest increment of time in which time has a meaning depends on h. As Einstein said in his eulogy for Planck in 1948, "he showed convincingly that in addition to the atomistic structure of matter there is a kind of atomistic structure of energy . . . This discovery became the basis of all twentieth century research in physics."

In the tone of Planck's writing, we sense a man who is straightforward and clear, aware that he has made a big discovery and yet taking pains to keep a rein on his enthusiasm. It was a great irony of Planck's career that, despite his restrained nature and modest aspirations, he had proposed a hypothesis that was to change all of physics. Even after other scientists recognized the revolutionary nature of his work, Planck himself wrote in 1910 that "the introduction of the quantum [into the rest of physics] should be done as conservatively as possible, i.e. alterations should only be made that have shown themselves to be absolutely necessary."

Planck deeply respected the logic and lawfulness, the absolute reliability, of the physical world. At the same time, he understood the limits of science in human affairs. In later life, he wrote broad, philosophical essays on the unpredictable nature of human imagination and behavior. His own life was filled with tragic accidents. Planck's first wife, Marie Merck, died in 1909. A son, Karl, died during World War I, and his two daughters, Margarete and Emma, both died during childbirth, in 1917 and 1919, darkly bracketing the year that he was awarded the Nobel Prize. During World War II, Planck, a man revered as much for his personal integrity as for his physics, a father figure for Einstein, was torn between conflicting principles. Although he objected strongly to the policies of the Nazis, he decided to remain in Germany out of a sense of duty. In 1944, another of his sons was executed for suspected complicity in a plot to assassinate Hitler. One day that same year, an Allied bomb dropped on Berlin destroyed most of Planck's manuscripts and books.

ON THE THEORY OF THE ENERGY DISTRIBUTION LAW OF THE NORMAL SPECTRUM

Max Planck

Verhandlungen der Deutschen Physikalischen Gesellschaft (1900)

GENTLEMEN: WHEN SOME WEEKS AGO I had the honour to draw your attention to a new formula which seemed to me to be suited to express the law of the distribution of radiation energy over the whole range of the normal spectrum,[1] I mentioned already then that in my opinion the usefulness of this equation was not based only on the apparently close agreement of the few numbers, which I could then communicate to you, with the available experimental data,[2] but mainly on the simple structure of the formula and especially on the fact that it gave a very simple logarithmic expression for the dependence of the entropy of an irradiated monochromatic vibrating resonator on its vibrational energy. This formula seemed to promise in any case the possibility of a general interpretation much better than other equations which have been proposed, apart from Wien's formula which, however, was not confirmed by experiment.

Entropy means disorder, and I thought that one should find this disorder in the irregularity with which even in a completely stationary radiation field the vibrations of the resonator change their amplitude and phase, as long as one considers time intervals long compared to the period of one vibration, but short compared to the duration of a measurement. The constant energy of the stationary vibrating resonator can thus only be considered to be a time average, or, put differently, to be an instantaneous average of the energies of a large number of identical resonators which are in the same stationary radiation field, but far enough from one another not to influence each other directly. Since the entropy of a resonator is thus determined by the way in which the energy is distributed at one time over many resonators, I suspected that one should evaluate this quantity by introducing probability considerations into the electromagnetic theory of radiation, the importance of which for the second law of thermodynamics was originally discovered by Mr. L. Boltzmann.[3] This suspicion has been confirmed; I have been able to derive deductively an expression for the entropy of a monochromatically vibrating resonator and thus for the energy distribution in a stationary radiation state, that is, in the normal spectrum. To do this it was only necessary to extend somewhat the interpretation of the hypothesis of "natural radiation" which has been introduced by me into electromagnetic theory. Apart from this I have obtained other relations which seem to me to be of considerable importance for other branches of physics and also of chemistry.

I do not wish to give today this deduction—which is based on the laws of electromagnetic radiation, thermodynamics and probability calculus—systematically in all details, but rather to explain to you as clearly as possible the real core of the theory. This can probably be done most easily by describing to you a new, completely elementary treatment through which one can evaluate—without knowing anything about a spectral formula or about any theory—the distribution of a given amount of energy over the different colours of the normal spectrum using one constant of nature only and after that also the value of the temperature of this energy radiation using a second constant of nature. You will find many points in the treatment to be presented arbitrary and complicated, but as I said a moment ago I do not want to pay attention to a proof of the necessity and the possibility to perform it easily and practically, but to the clarity and uniqueness of the given prescriptions for the solution of the problem.

Let us consider a large number of linear, monochromatically vibrating resonators—N of frequency ν (per second), N' of frequency ν', N'' of frequency ν'', . . . , with all N large numbers—which are properly separated and are enclosed in a diathermic medium with light velocity c and bounded by reflecting walls. Let the system contain a certain amount of energy, the total energy E_t(erg) which is present partly in the medium as travelling radiation and partly in the resonators as vibrational energy. The question is how in a stationary state this energy is distributed over the vibrations of the resonators and over the various colours of the radiation present in the medium, and what will be the temperature of the total system.

To answer this question we first of all consider the vibrations of the resonators and try to assign to them certain arbitrary energies, for instance, an energy E to the N resonators ν, E' to the N' resonators ν', ... The sum

$$E + E' + E'' + \ldots = E_0$$

must, of course, be less than E_t. The remainder $E_t - E_0$ pertains then to the radiation present in the medium. We must now give the distribution of the energy over the separate resonators of each group, first of all the distribution of the energy E over the N resonators of frequency ν. If E is considered to be a continuously divisible quantity, this distribution is possible in infinitely many ways. We consider, however—this is the most essential point of the whole calculation—E to be composed of a well-defined number of equal parts and use thereto the constant of nature $h = 6.55 \times 10^{-27}$ erg sec. This constant multiplied by the common frequency ν of the resonators gives us the energy element ε in erg, and dividing E by ε we get the number P of energy elements which must be divided over the N resonators. If the ratio thus calculated is not an integer, we take for P an integer in the neighbourhood.

It is clear that the distribution of P energy elements over N resonators can only take place in a finite, well-defined number of ways. Each of these ways of distribution we call a "complexion," using an expression introduced by Mr. Boltzmann for a similar concept. If we denote the resonators by the numbers 1, 2, 3, . . . , N, and write these in a row, and if we under each resonator put the number of its energy elements, we get for each complexion a symbol of the following form

1	2	3	4	5	6	7	8	9	10
7	38	11	0	9	2	20	4	4	5

We have taken here $N = 10$, $P = 100$. The number of all possible complexions is clearly equal to the number of all possible sets of numbers which one can obtain in this way for the lower sequence for given N and P. To exclude all misunderstandings, we remark that two complexions must be considered to be different if the corresponding sequences contain the same numbers, but in different order. From the theory of permutations we get for the number of all possible complexions

$$\frac{N(N+1)\,.\,(N+2)\,.\,.\,.\,(N+P-1)}{1\,.\,2\,.\,3\,.\,.\,.\,P} = \frac{(N+P-1)!}{(N-1)!P!}$$

or to a sufficient approximation,

$$= \frac{(N+P)^{N+P}}{N^N P^P}$$

We perform the same calculation for the resonators of the other groups, by determining for each group of resonators the number of possible complexions for the energy given to the group. The multiplication of all numbers obtained in this way gives us then the total number R of all possible complexions for the arbitrarily assigned energy distribution over all resonators.

In the same way any other arbitrarily chosen energy distribution E, E', E'', . . . will correspond to the number R of all possible complexions which must be evaluated in the above manner. Among all energy distributions which are possible for a constant $E_0 = E + E' + E'' + \ldots$ there is one well-defined one for which the number of possible complexions R_0 is larger than for any other distribution. We then look for this energy distribution, if necessary by trial, since this will just be the distribution taken up by the resonators in the stationary radiation field, if they together possess the energy E_0. The quantities E, E', E'', . . . can then be expressed in terms of one single quantity E_0. Dividing E by N, E' by N', . . . we obtain the stationary value of the energy U_ν, $U'_{\nu'}$, $U''_{\nu''}$, . . . of a single resonator of each group, and thus also the spatial density of the corresponding radiation energy in a diathermic medium in the spectral range ν to $\nu + d\nu$,

$$u_\nu \, d\nu = \frac{8\pi\nu^2}{c^3} \cdot U_\nu \, d\nu,$$

so that the energy of the medium is also determined.

Of all quantities which occur only E_0 seems now still to be arbitrary. One sees easily, however, how one can finally evaluate E_0 from the given total energy E_t, since if the chosen value of E_0 leads, for instance, to too large a value of E_t, we must decrease it appropriately, and the other way round.

After the stationary energy distribution is thus determined using a constant h, we can find the corresponding temperature θ in degrees absolute[4] using a second constant of nature $k = 1.346 \times 10^{-16}$ erg degree^{-1} through the equation

$$\frac{1}{\theta} = k \frac{d \ln R_0}{dE_0}.$$

The product $k \ln R_0$ is the entropy of the system of resonators; it is the sum of the entropy of all separate resonators.

It would, to be sure, be very complicated to perform explicitly the above-mentioned calculations, although it would not be without some interest to test the truth of the attainable degree of approximation in a simple case. A more general calculation which is performed very simply, using exactly the above prescriptions, shows much more directly that the normal energy distribution determined in this way for a medium containing radiation is given by the expression

$$u_\nu \, d\nu = \frac{8\pi h\nu^3}{c^3} \frac{d\nu}{e^{h\nu/k\theta} - 1},$$

which corresponds exactly to the spectral formula which I gave earlier

$$E_\lambda \, d\lambda = \frac{c_1 \lambda^{-5}}{e^{c_2/\lambda\theta} - 1} \, d\lambda.$$

The formal differences are due to the differences in the definitions of u_ν and E_λ. The first formula is somewhat more general inasfar as it is valid for an entirely arbitrary diathermic medium with light velocity c. I calculated the numerical values of h and k which I mentioned from that formula using the measurements by F. Kurlbaum and by O. Lummer and E. Pringsheim.[5]

I shall now make a few short remarks about the question of the necessity of the above given deduction. The fact that the chosen energy element ε for a given group of resonators must be proportional to the frequency ν follows immediately from the extremely important so-called Wien displacement law. The relation between u and U is one of the basic equations of the electromagnetic theory of radiation. Apart from that, the whole deduction is based upon the single theorem that the entropy of a system of resonators with given energy is proportional to the logarithm of the total number of possible complexions for the given energy. This theorem can be split into two other theorems: (1) The entropy of the system in a given state is proportional to the logarithm of the probability of that state, and (2) The probability of any state is proportional to the number of corresponding complexions, or, in other words, any given complexion is equally probable as any other given complexion. The first theorem is, as far as radiative phenomena are concerned, just a definition of the probability of the state, insofar as we have for energy radiation no other *a priori* way to define the probability than the determination of its entropy. We have here one of the distinctions from the corresponding situation in the kinetic theory of gases. The second theorem is the core of the whole of the theory presented here: in the last resort its proof can only be given empirically. It can also be understood as a more detailed definition of the hypothesis of natural radiation which I have introduced. This hypothesis I have expressed before only in the form that the energy of the radiation is completely "randomly" distributed over the various partial vibrations present in the radiation.[6] I plan to communicate elsewhere in detail the considerations, which have only been sketched here, with all calculations and with a survey of the development of the theory up to the present . . .

1. M. Planck, *Verh. D. Phys. Ges.* **2,** 202 (1900).

2. In the meantime Mr. H. Rubens and Mr. F. Kurlbaum have given a direct confirmation for very long wave lengths. (*S.B. Königl. Preuss. Akad. Wiss*. of 25 October, p. 929 [1900].)

3. L. Boltzmann, especially *S.B. Kais. Ak. Wiss. Wien II*, **76,** p. 373 (1877 [= 1878]).

4. The original states "degrees centigrade" which is clearly a slip [D. t. H.].

5. F. Kurlbaum (*Ann. Phys.* **65** [= **301**], 759 [1898]) gives $S_{100} - S_0 = 0.0731$ Watt cm^{-2}, while O. Lummer and E. Pringsheim (*Verh. Deutsch. Physik Ges.* **2,** 176 [1900]) give $\lambda_m \theta = 2940$ μ degree.

6. M. Planck, *Ann. Phys.* **1** [= 306], 73 (1900). When Mr. W. Wien in his Paris report (Rapports II, p. 38, 1900) about the theoretical radiation laws did not find my theory on the irreversible radiation phenomena satisfactory since it did not give the proof that the hypothesis of natural radiation is the only one which leads to irreversibility, he surely demanded, in my opinion, too much of this hypothesis. If one could prove the hypothesis, it would no longer be a hypothesis, and one did not have to formulate it at all. However, one could then not derive anything essentially new from it. From the same point of view one should also declare the kinetic theory of gases to be unsatisfactory since nobody has yet proved that the atomistic hypothesis is the only one which explains irreversibility. A similar objection could with more or less justice be raised against all inductively obtained theories.

2

HORMONES

THE KLYMOGRAPH, INVENTED BY Carl Friedrich Wilhelm Ludwig in the mid-nineteenth century, is a revolving drum wrapped with a long sheath of smoky graph paper. While the drum slowly rotates, a writing pen traces out blood pressure, bodily secretions, and other vital functions as they change through time—in effect creating a visible motion picture of invisible activities inside a living being.

On January 16, 1902, in a small laboratory at University College in London, two scientists were surprised by the moving picture they saw on their klymograph. Earlier in the day, they had injected a thirteen-pound dog with morphine. Now, they cut open the dog's abdomen and inserted a thin metallic tube, called a cannula, into its pancreas. While the animal was immersed in a saline solution and pumped continuously with oxygen to keep it alive, the two surgeons poured a weak solution of hydrochloric acid into its small intestine. After two minutes, pancreatic juice began slowly dripping out of the cannula, one drop every twenty seconds. Each drop fell upon the flattened end of a delicate lever, which in turn lifted the slender pen of the klymograph.

Pancreatic juice flows from the pancreas to the small intestine and helps with digestion. Normally, the secretion of the juice is triggered by the soupy mush of partly digested food after it leaves the stomach and enters the intestine. In this case, hydrochloric acid served as the trigger.

So far, the two operating scientists, William Maddock Bayliss and Ernest Henry Starling, had merely repeated an experiment done several years earlier by the great Russian physiologist Ivan Petrovich Pavlov. Pavlov, like all biologists, believed in the sacred dogma that parts of the body sent signals to other parts exclusively by means of the nervous system, a system discovered by the ancient Greeks. Substantial evidence supported this belief. In the eighteenth century, the Italian biologist Luigi Galvani had demonstrated that electrical stimulation, later shown to be conducted by nerves, was the cause of muscle contraction. In 1850 Ludwig had shown it was also the nerves that initiated secretion of the salivary glands. Pavlov's own work suggested a major role of the nervous system for intestinal digestion and other functions. The nervous system, with its miles of winding tendrils and pathways, was regarded as *the* communication system by which the body regulated itself, activated organs and muscles, and reacted to changes in the external world. After duplicating Pavlov's results, Bayliss and Starling intended to discover which particular nerves running from the intestine to the pancreas carried the message to begin secreting juice.

The two scientists conducted their experiment that day in Bayliss's small laboratory at the college. The working space was so crammed with odds and ends and various equipment hung from the rafters that a friend said the room "wanted only a stuffed crocodile to make it a complete alchemist's den." At the time, Bayliss was forty-one years old, Starling thirty-five. Bayliss was trained in biology, Starling in medicine. They had begun their highly fruitful collaboration in 1890, the year that Starling received his medical degree from Guy's Hospital Medical School in London. Three years later, in 1893, Bayliss married Starling's sister Gertrude.

The two men complemented each other perfectly. Bayliss, the son of a successful manufacturer, descended from a well-to-do family; Starling, son of a barrister, from middle-class stock. In the words of Charles Lovatt Evans, a younger colleague, Bayliss was "gentle, retiring, patient, over-modest." He was also erudite, careful, and cautious, almost embarrassed when he won a much-publicized court case against an antivivisectionist. By contrast, Starling was "brisk, ambitious, a bit quixotic, and high-strung . . . he had panache, enjoyed the limelight and the use of power for good ends but was too forthright and impatient to be diplomatic." Bayliss lived in a large house, named St. Cuthbert's, attached to a four-acre garden, and issued a standing invitation to colleagues for tea and tennis on Saturdays. Starling, blessed with a beautiful baritone voice and the "panache" to use it, was known to sing German lieder at parties until the outbreak of World War I. Bayliss's style was to work slowly and deliberately, while Starling barreled ahead on the all-or-nothing principle. Bayliss preferred to conduct his researches alone, except in his collaboration with Starling. Starling always worked with others and left the details to his associates. Bayliss had the greater knowledge, Starling the greater daring and vision. Filled with mutual admiration, they disagreed over many things, including the admission of women to the Physiological Society, with Bayliss in favor and Starling opposed on the grounds that it would be improper to dine with ladies while the men "smelled of dog."

After verifying Pavlov's results, the two scientists proceeded to isolate a short loop of the dog's jejunum, the middle portion of its small intestine. (In humans, the jejunum is about nine feet long, comprising a little less than half of the small intestine.) The scientists tied off the loop at both ends and then expertly dissected and removed all of its nerves, leaving it attached to the animal only through arteries and veins. In previous experiments, Bayliss and Starling had systematically destroyed various nerve centers and found continued pancreatic secretion after the introduction of acid into the upper intestine. Now, with all nerves cut out, they expected to find an end to the secretions.

The klymograph told a different story. After a small quantity of acid was poured into the nerveless loop of jejunum, pancreatic juice began flowing at the same rate as before. Evidently, the intestine was signaling the pancreas by some completely new mechanism. After their initial shock, the brothers-in-law scraped some mucus from the jejunum, injected it directly into the bloodstream, and again produced secretion of the pancreas. They had found some kind of chemical messenger in the mucous lining of the small intestine. Furthermore, this chemical was apparently unique in its location and effect. It could not be found in other parts of the body, as other scrapings showed. Furthermore, other substances injected into the bloodstream had no effect on the pancreas. And the chemical messenger was universal. Later tests showed that it could trigger the flow of pancreatic juice in rabbits, monkeys, and humans.

We can sense a bit of the scientists' excitement in the introductory section of their landmark paper of September 12, 1902: "We soon found . . . that we were dealing with an entirely different order of phenomena, and that the secretion of the pancreas is normally called into play not by nervous channels at all, but by a chemical substance which is formed in the mucous membrane of the upper parts of the small intestine under the influence of acid, and is carried thence by the bloodstream to the gland-cells of the pancreas." An observer of the experiment, Sir Charles Martin, wrote later, in classic British understatement, "It was a great afternoon."

Bayliss and Starling had discovered the first hormone. This particular hormone, produced by the upper intestine, they named secretin. (Among the hundreds of other hormones later discovered are insulin, which is secreted by the pancreas and controls blood sugar; follicle-stimulating hormone, secreted from the pituitary to stimulate egg production in the ovaries; growth hormone, which causes the production of protein in muscle cells and the release of energy in the breakdown of fats; and vasopressin, produced in the hypothalamus and acting on the kidneys to restrict the output of urine.) It was Starling, in fact, who later coined the word "hormone," from the Greek *hormon*, meaning to excite or set into motion.

Following the discovery of nerves two thousand years earlier, Bayliss and Starling had found a second mechanism for communication and control in the body. Like Bayliss and Starling themselves, the two mechanisms complement each other. Nerves act and respond in a few thousandths of a second, and they work locally, nerve to adjoining nerve. Hormones take minutes or hours to take effect, they travel over longer distances before reaching their targets, and their actions last longer. If nerves are the sprinters of biology, Bayliss and Starling had discovered the marathon runners. In doing so, they also founded the science of hormones, called endocrinology.

Unlike Max Planck, who was consciously seeking a new law of physics to explain the universal black-body radiation, William Bayliss and Ernest Starling made their discovery by accident. After previous work on electrical and mechanical phenomena in the heart, they had recently turned their attention to the wavelike movements (called peristalsis) and nerve stimulations of the intestines. Here, they were following a well-marked trail. Claude Bernard (1813–1878), who with Ludwig had created the field of modern physiology, had discovered digestive enzymes and gastric juices, including the secretions of the pancreas. Pavlov had shown that the secretion of pancreatic juice is stimulated by acid in the upper intestine. And other researchers had even begun investigating which particular nerves caused the secretion. So it would seem that Bayliss and Starling were carrying out almost routine work, far from the frontiers of biology.

Those frontiers, in the year 1900, were much broader than the frontiers of physics. They included the origin of life; bacteriology and the study of how certain microorganisms transmit and cause disease; the evolution of species; embryology and the question of how dividing germ cells know how to specialize into liver cells and heart cells and other organs in adult creatures; inheritance and the mechanism of passing traits from parent to child; and the organization and function of internal organs. The latter field is called physiology. Bayliss and Starling were physiologists.

As in physics, a great deal of progress had occurred in the previous fifty years. The study of bacteriology and disease had made enormous strides with the "germ theory" of Louis Pasteur and the work of Robert Koch in isolating the bacteria that cause cholera and tuberculosis. One by one, other ancient diseases such as diphtheria, typhoid, gonorrhea, tetanus, and pneumonia were being identified with particular bacteria, and antitoxins were sought. In the early nineteenth century, the cell had been proposed as the basic structural unit of living organisms, and biologists were busy elucidating its parts. By 1890, Theodor Boveri had suggested that the discrete elements of inheritance hypothesized earlier by Gregor Mendel were probably located on chromosomes, the long, rigid bodies in the cell nucleus. But little more was then known about these elements, or genes. Charles Darwin and Alfred Russell Wallace had advanced the theory of evolution, but the detailed workings of that theory were not known. The origin of life remained, and still remains, an open question, although Pasteur had definitively ruled out the old idea of "spontaneous generation." Also unsolved, and still largely unsolved today, was the mechanism by which embryonic cells specialize.

As far as physiology is concerned, a number of scientists suspected that some of the internal organs produced secretions needed for the well-being of the body. As early as 1775, the French physician Théophile Bordeu vaguely suggested that each organ gave off "emanations" used by the body as a whole. In the mid-nineteenth century, Bernard invented the term "internal secretion," applied to the flow of glucose out of the liver. Later in that century, physiologists found that malfunctioning of the adrenal glands, the thyroid, and the pancreas produced known diseases. Thus, by 1900 it was believed that organs could produce essential secretions. Until the work of Bayliss and Starling, however, these secretions were viewed only as assistants to the nerves. The nerves were still believed to be the primary communication and governing system in the body.

The serendipitous discovery of hormones by Bayliss and Starling touched upon a grand theme in biology that embraced every frontier and indeed had haunted the whole discipline since its beginning: the question of whether living matter obeys different laws from nonliving matter. The question is often framed in terms of the debate between vitalism and mechanism. The vitalists argued that there was a special quality of life—some immaterial or spiritual or transcendent force—that enabled a jumble of tissues and chemicals to vibrate with life. That transcendent force was beyond physical explanation. The mechanists, on the other hand, believed that all the workings of a living animal or plant could be ultimately understood in terms of the laws of physics and chemistry.

The opposition of vitalism and mechanism is what one might call a master motif. It extends far beyond biology and resonates with other great dualisms in human thought: mind versus body, spirit versus matter, heaven versus earth, intuition versus reason. Such oppositions, and the deeply felt notion that living matter is fundamentally different from nonliving matter, have always separated biology from other sciences. Biology has always had a special mystique. Indeed, nothing is more mysterious than our own human consciousness. The extent to which that consciousness can be reduced to chemistry and physics remains a much-debated question today.

Many of the vitalist ideas weave back and forth through the history of biology. Plato and Aristotle believed that an idealized "final cause," which was more spirit than matter, impelled a germ cell to develop toward its adult form. René Descartes (1596–1650), who famously articulated the separation between the intangible mind and the tangible body, proposed that the soul interacts with the body in the pineal gland. The specificity of Descartes's proposal suggests how the vitalism versus mechanism debate moved between the realms of philosophy and theology and the more concrete dominion of science. Another theory was that the *élan vital*, or vital spirit, resided in the heart. According to this idea, food changed to blood in the liver, and the blood then went to the heart to be charged with vital spirit. The wondrous activity of nerves was a vitalist phenomenon, baffling until the nature of electricity and nervous electrical conduction was understood. And so on.

As biology advanced through the centuries, the vitalists rarely surrendered but instead retreated to smaller castles and less specific decrees. In the last edition of his *Lärbok i kemien*, considered the most authoritative chemical text of the early-to-mid-nineteenth century, the distinguished Swedish chemist and vitalist Jöns Jacob Berzelius (1779–1848) wrote simply: "In living nature the elements seem to obey entirely different laws than they do in the dead." In almost precise opposition, the mechanist Claude Bernard proclaimed in 1865: "a vital phenomenon has—like any other phenomenon—a rigorous determination, and such determinism could only be a physico-chemical determinism."

By the end of the nineteenth century, most biologists had come to a vague compromise position on this difficult and ongoing debate. The question became: How do chemistry and physics merge with biology? Communications were easily reduced to physics. Digestion and respiration were clearly chemical. But what about regulation and control? What about response to the outside world?

While Bayliss and Starling were scrubbing down in their cluttered laboratory, the revised vitalist argument went something like this: a living organism was mostly a machine that additionally could respond to its environment. Since no one knew of a machine that could respond to changes outside of itself, a living organism was not *merely* a machine. Furthermore, no one knew of a machine that could regulate itself. An internal combustion engine could not alter the timing of its strokes when one of the cylinders blew out, or cool itself down when it wasn't given enough oil. By contrast, living things somehow could move toward the sun, alter their digestive juices according to the food they were eating, sweat when they became hot. These were matters of response and control. And the mechanical model didn't seem to apply.

With the discovery of hormones, Bayliss and Starling had found the internal command and control centers—and in this, their discovery was much larger than a new communication system. The mechanism of response and control was chemical: atoms and molecules. Now, with hormones, there was a mechanism for a living thing to regulate itself. Furthermore, with hormones, an organism could not only be studied but also controlled from the outside. In principle, and later in practice, the hormones discovered by Bayliss and Starling could be manufactured in the lab and injected into living beings, evoking everything from the slow drip of the pancreas to increased sexual response to growth spurts to hunger to changes in mood. Never had the living body come closer to a machine, a self-regulating machine governed not only by physics but also by chemistry. And not only a machine, but a machine that we humans could willfully control. At the start of a new century, we still have not come to terms with the implications of this idea.

In contrast to Planck's paper on the quantum, and indeed most papers in modern physics, Bayliss and Starling's landmark paper of 1902 is conceptually simple and easy to follow. They begin with a succinct history of research on the secretion of pancreatic juice. Here, the scientists show deep reverence for Pavlov (spelled "Pawlow" in their paper). This section ends with the passage quoted earlier, showing a clear awareness of the revolutionary nature of their discovery, if not its full philosophical significance.

Some anatomical terms that may be unfamiliar to all but medical students: the duodenum is the upper ten inches of the small intestine, so named because its length is approximately the width of twelve fingers. The middle section of the intestine, already discussed, is the jejunum. The bottom and longest portion of the small intestine is called the ileum. Chyme is a soupy and acidic mixture of partly digested food and digestive juices. Ferments are acids used in the breakdown of food. Ganglia are central masses of nerves and nerve tissues, with the solar plexus being a large network of nerves in the abdominal cavity that sends nerve impulses to abdominal organs. One of the longest nerves of the body, the vagus nerve runs from the brain through the neck to the stomach, where it innervates the digestive tract. The epithelium is the layer of tissue that lines the inside of an organ. Here, it is the location of the new hormone, secretin, in the upper intestine.

Let us skip to section IV, "The Crucial Experiment," which I have already described. After identifying secretin and its action on the pancreas, Bayliss and Starling now describe their attempts to understand more about the hormone, running one by one through a drill of standard tests. Secretin is not diminished by boiling, distinguishing it from an enzyme. Secretin is produced only in a limited region of the small intestine. It sluggishly passes through parchment paper. It is destroyed by some digestive juices. It is soluble in alcohol. The scientists have done their best to determine the composition of secretin and conclude that "we cannot as yet give any definite suggestion as to the chemical nature of secretin." In fact, secretin, being a protein, is a complex structure of amino acids, and it would not be until the 1920s that such structures could be analyzed with any regularity.

The discussion of the action of secretin in section V is a subtle but important effort to clarify cause and effect. That is, since the blood vessels are dilated at the same time that the pancreas secretes (causing the drop in blood pressure observed in Figure 2), it is possible that secretin directly causes only a dilation of vessels, which in turn causes the pancreatic secretion. However, Bayliss and Starling rule out this possibility. They are able to make some nonacidic extracts of secretin that cause pancreatic secretion without lowering the blood pressure.

In the next sections, the two investigators show that even though they don't know the chemical composition of secretin, it is a universal substance, having the same pancreatic action in dogs, rabbits, men, and monkeys. Secretin is made in only one part of the body, the upper portion of the small intestine. Furthermore, secretin is highly specific in its action. It does not stimulate other parts of the body, such as the salivary glands or the stomach. Like many proteins, secretin is a uniquely shaped key, designed to fit a uniquely shaped lock, although Bayliss and Starling didn't have this understanding at the time. The paper ends with a drab but precise recitation of its results.

The importance of Bayliss and Starling's discovery was immediately recognized. By the time of Starling's prestigious Croonian Lecture at the Royal Society in 1905, he could speak confidently about the "chemical control of functions of the body," about how hormones were part of a system that maintained equilibrium. A new field of biology and medicine had been born. Twenty years later, the knowledge of hormones was given its first important medical application. Canadian scientists Frederick Banting and Charles Best were able to isolate insulin and thus produce a treatment for diabetes.

After the work on secretin, the two scientists continued their illustrious careers. Bayliss went on to study how electricity affects transport of substances across a cell membrane, and later the detailed actions of enzymes, which are proteins that promote biochemical reactions. During World War I, Bayliss worked on wound shock. He found that gum mixed with saline solution could stem the loss of blood and other bodily fluids. Bayliss also completed his lifelong academic labor, *Principles of General Physiology* (1914), considered a landmark textbook in its field.

Starling turned to his celebrated work on the heart, its circulatory functions and its muscular mechanics. Even today, every medical student knows Starling's "Law of the Heart," which states that the force of the muscular contraction is proportional to the extent to which the heart muscle is stretched. Starling wrote his own thick book, *Principles of Human Physiology* (1912), which, with constant revisions, remains a standard international text. That these two great physiologists, who worked closely with each other for fifteen years, chose to write separate textbooks rather than a single coauthored text shows the force of individual personality, ambition, and taste in the enterprise of science.

Starling was also interested in education and sharply critical of contemporary education in England. He argued for "educational reform, or even revolution, for the maintenance of our place in the world . . . in matters of urgent necessity it is unprofitable to count the cost." Such a matter of urgent necessity was World War I. After the war, Starling noted that Germany, unlike England, had long recognized the importance of education as a means for increasing national power. In particular, Starling wrote that the "ignorance of science displayed by members of the [British] government was astounding and disastrous."

THE MECHANISM OF PANCREATIC SECRETION

William Bayliss and Ernest Starling

From the Physiological Laboratory of University College, London

Journal of Physiology (1902)

CONTENTS

I. HISTORICAL

IT HAS LONG BEEN KNOWN THAT the activity of the pancreas is normally called into play by events occurring in the alimentary canal. Bernard[1] found that the pancreatic secretion could be evoked by the introduction of ether into the stomach or duodenum, and Heidenhain[2] studied the relation of the time-course of the secretion to the processes of digestion going on in the stomach and intestines.

Our exact knowledge of many of the factors determining pancreatic secretion we owe to the work of Pawlow and his pupils,[3] who have shown that the flow of pancreatic juice begins with the entry of the chyme into the duodenum and is not excited directly by the presence of food in the stomach itself. The exciting influence of the chyme is due chiefly to its acidity, and a large secretion can be brought about by the introduction of 0.4% hydrochloric acid into the stomach, whence it is rapidly transferred to the duodenum. Pawlow found, however, that other substances, *e.g.* water, oil, introduced into the stomach had a similar, though less pronounced, effect. In each case the effect was produced only when the substances had passed into the duodenum. Pawlow has, moreover, drawn attention to a remarkable power of adaptation presented by the pancreas, the juice which is secreted varying in composition according to the nature of the food which has passed into the duodenum. Thus, with a diet of meat the tryptic ferment is present in relatively largest amount, while a diet of bread causes the preponderance of the amylolytic ferment, and a diet of milk or fat that of the fat-splitting ferment.

Pawlow regards the secretion evoked by the presence of acid in the duodenum as reflex in origin, and ascribes the varying composition of the juice in different diets to a marvellous sensibility of the duodenal mucous membrane, so that different constituents of the chyme excite different nerve-endings, or produce correspondingly different kinds of nerve-impulses, which travel to the gland, or its nerve-centres, and determine the varying activity of the gland-cells.

In searching for the channels of this reflex, Pawlow has shown that, if proper precautions be taken, it is possible to excite a secretion of pancreatic juice by excitation of the divided vagus or splanchnic nerves. The vagus nerves, also, according to him, contain inhibitory fibres.

The question as to the mechanism by which a pancreatic secretion is evoked by the introduction of acid into the duodenum has been narrowed still further by the independent researches of Popielski[4] and of Wertheimer and Lepage.[5] These observers have shown that the introduction of acid into the duodenum still excites pancreatic secretion after section of both vagi and splanchnic nerves, or destruction of the spinal cord, or even after complete extirpation of the solar plexus. Popielski concludes, therefore, that the secretion is due to a peripheral reflex action, the centres of which are situated in the scattered ganglia found throughout the pancreas, and ascribes special importance to a large collection of ganglion cells in the head of the pancreas close to the pylorus. Wertheimer and Lepage, while accepting Popielski's explanation of the secretion excited from the duodenum, found that secretion could also be induced by injection of acid into the lower portion of the small intestine, the effect, however, gradually diminishing as the injection was made nearer the lower end of the small intestine, so that no effect at all was produced from the lower two feet or so of the ileum. Secretion could be excited from a loop of jejunum entirely isolated from the duodenum. They conclude that, in this latter case, the reflex centres are situated in the ganglia of the solar plexus, but they did not perform the obvious control experiment of injecting acid into an isolated loop of jejunum after extirpation of these ganglia. They showed that the effect was not abolished by injection of large doses of atropin, but compared with this the well-known insusceptibility to this drug of the sympathetic fibres of the salivary glands.

The apparent local character of this reaction interested us to make further experiments on the subject, in the idea that we might have here to do with an extension of the local reflexes whose action on the movements of the intestines we have already investigated.[6] We soon found, however, that we were dealing with an entirely different order of phenomena, and that the secretion of the pancreas is normally called into play not by nervous channels at all, but by a chemical substance which is formed in the mucous membrane of the upper parts of the small intestine under the influence of acid, and is carried thence by the bloodstream to the gland-cells of the pancreas.[7]

II. EXPERIMENTAL METHODS

All our experiments were made on dogs which had received a previous injection of morphia, and were anæsthetized with A.C.E. mixture during the course of the experiment. In order to keep the animals' condition constant, artificial respiration was usually employed, a procedure which is especially necessary when both vagi are divided, the anæsthetic bottle being introduced in the course of the blast of air from the pump. The animals had received no food for a period of 18 to 24 hours previously. In the earlier experiments, where a considerable degree of preliminary operative manipulation was required in the abdominal cavity, the animals were placed during the remainder of the experiment in a bath of warm physiological saline, the level of the fluid being above that of the abdominal wound. This method was found to keep them in such good condition throughout a long experiment that it was adopted as a routine practice in all cases. The arterial pressure was always recorded by means of a mercurial manometer connected with the carotid artery in the usual way. The pancreatic juice was obtained by placing a cannula in the larger duct which enters the duodenum on a level with the lower border of the pancreas. To the cannula was connected a long glass tube filled at first with physiological saline; the end of this tube projected over the edge of the bath so that the drops of the fluid as they were secreted fell upon a mica disc cemented to the lever of a Marey's tambour, which was in connection, by means of rubber tubing, with another tambour which marked each drop upon the smoked paper of the kymograph. . . .

IV. THE CRUCIAL EXPERIMENT

On January 16th, 1902, a bitch of about 6 kilos weight, which had been fed about 18 hours previously, was given a hypodermic injection of morphia some 3 hours before the experiment, and during the experiment itself received A.C.E. in addition. The nervous masses around the superior mesenteric artery and cœliac axis were completely removed and both vagi cut. A loop of jejunum was tied at both ends and the mesenteric nerves supplying it were carefully dissected out and divided, so that the piece of intestine was connected to the body of the animal merely by its arteries and veins. A cannula was inserted in the large pancreatic duct and the drops of secretion recorded. The blood-pressure in the carotid was also recorded in the usual way. The animal was in the warm saline bath and under artificial respiration.

The introduction of 20 c.c. of 0.4% HCl into the duodenum produced a well-marked secretion of 1 drop every 20 secs. lasting for some 6 minutes; this result merely confirms previous work.

But, and this is the important point of the experiment, and the turning-point of the whole research, the introduction of 10 c.c. of the same acid into the enervated loop of jejunum produced a similar and equally well-marked effect.

Now, since this part of the intestine was completely cut off from nervous connection with the pancreas, the conclusion was inevitable that the effect was produced by some chemical substance finding its way into the veins of the loop of jejunum in question and being carried in the blood-stream to the pancreatic cells. Wertheimer and Lepage have shown,[8] however, that acid introduced into the circulation has no effect on the pancreatic secretion, so that the body of which we were in search could not be the acid itself. But there is, between the lumen of the gut and the absorbent vessels, a layer of epithelium, whose cells are as we know endowed with numerous important functions. It seemed therefore possible that the action of acid on these cells would produce a body capable of exciting the pancreas to activity. The next step in our experiment was plain, viz. to cut out the loop of jejunum, scrape off the mucous membrane, rub it up with sand and 0.4% HCl in a mortar, filter through cotton-wool to get rid of lumps and sand, and inject the extract into a vein. The result is shown in Fig. 2. The first effect is a considerable fall of blood-pressure, due, as we shall show later, to a body distinct from that acting on the pancreas, and, after a latent period of about 70 secs. a flow of pancreatic juice at more than twice the rate produced at the beginning of the experiment by introduction of acid into the duodenum. We have already suggested the name "secretin" for this body, and as it has been accepted and made use of by subsequent workers it is as well to adhere to it.

In the same experiment we were able to make two further steps in the elucidation of the subject. In the first place the acid extract was boiled and found undiminished in activity, secretin is therefore not of the nature of an enzyme. In the second place, since Wertheimer and Lepage have shown that the effect of acid in the small intestine diminishes in proportion as the place where it is introduced approaches the lower end, so that from the last 6 inches or so of the ileum no secretion of the pancreas is excited, it was of interest to see whether the distribution of the substance from which secretin is split by acids is similar in extent. Fig. 3 shows the result of injecting an extract from the lower 6 inches of the

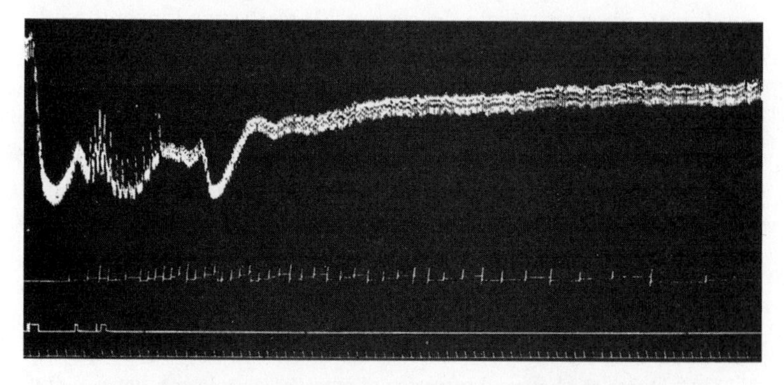

Figure 2. *Effect of injecting acid extract of jejunal mucous membrane into vein*. Explanation as Figure 1. The steps on the drop-tracing are due to a gradual accumulation of secretion on the lever of the drop-recorder, which fluid falls off at intervals. Blood-pressure zero = level of drop recorder.

Upper curve—blood pressure.
Uppermost of three lines—drops of pancreatic juice.
Middle line—signal marking injection of acid extract.
Bottom line—time in 10 second intervals.

ileum made in the same way as the jejunum extract. The fall of blood-pressure is present, but there is no effect on the pancreas. Another preparation from the ileum just above this one also had no effect on the pancreas. A preparation from the jejunum below the previous one had a marked effect, but less than that of the loop above. The distribution of "prosecretin," as we have proposed to call the mother-substance, corresponds therefore precisely with the region from which acid introduced into the lumen excites secretion from the pancreas.

In reply to the objection of Pflüger to this experiment,[9] we admit that it is difficult to be certain that all nerve-channels were absolutely excluded, since the walls of the blood vessels were intact, but we submit that since the result of the experiment was such as has been described it does not in the least matter whether the nerves were all cut or not; the only fact of importance is that it was the belief that all the nerves were cut that caused us to try the experiment of making an acid extract of the mucous membrane and that led to the discovery of secretin.

As to the further objection of Pflüger that it is in no way extraordinary that a body should be extracted from intestinal mucous membrane capable of acting as a stimulant to gland activity since there are many such bodies known, our reply is that secretin, as will be shown more fully later on, is of an entirely specific nature; the experiment described above shows that even from the ileum no such substance can be obtained, and subsequent experiments showed that from no other part of the body

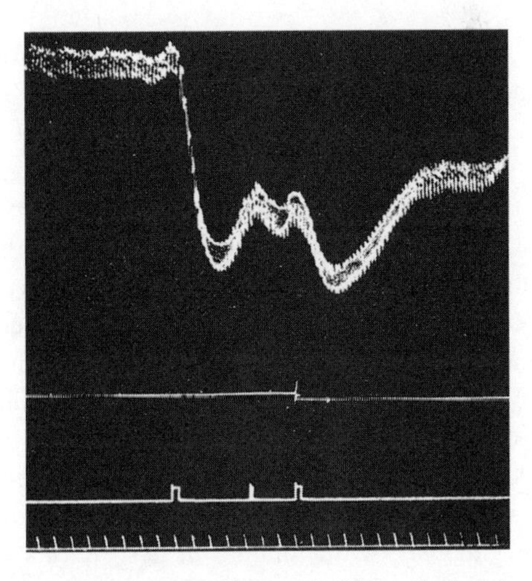

Figure 3. *Effect of acid extract of lower end of ileum.* Explanation as before.

could any body be extracted which caused secretion in the pancreas. And further no other substance known to us, even pilocarpin, which acts so powerfully on most glands, has any effect on the pancreas at all comparable with that of secretin; nor has secretin any action on glands other than the pancreas, except perhaps to a small degree on the secretion of bile.

V. PROPERTIES AND ACTION OF SECRETIN

. . . Secretin is non-volatile, it does not appear in the distillate obtained by passing steam through its solution.

It dialyses through parchment paper, but not readily.

The action of its solutions does not depend on their inorganic constituents, since the ash, prepared for us by Dr Osborne, has no effect on the pancreas.

It is destroyed by digestion with active tryptic solutions for one hour; the effect of peptic solutions is somewhat uncertain, digestion with gastric juice from the dog for one hour did not destroy the activity of a solution of secretin, but considerably diminished it.

Taking these facts together it will be seen that we cannot as yet give any definite suggestion as to the chemical nature of secretin, its solubility in alcohol and diffusibility point to its being a body of low molecular weight; since it is not destroyed by boiling it is not a ferment; and that it is not of the nature of an alkaloid or diamino-acid is shown by the fact of its not being precipitated by tannin.

THE MODE OF ACTION OF SECRETIN. In attempting to form some idea as to the mode of action of this body, the first necessity is to decide whether the substance causing fall of blood-pressure, which we shall in future, for the sake of brevity, call the depressor substance, is the same as that exciting the pancreas to secretion; the importance of this being because the vascular dilatation producing the fall of blood-pressure might be thought to be the cause of the increased activity of the pancreas, which organ no doubt would share in the general splanchnic dilatation. That this is not so, however, can be shown in several ways. Albumoses (Witte's pepton) cause considerable vaso-dilatation in the abdominal viscera, but no secretion of the pancreas, at least not in the doses by which a large fall of blood-pressure is produced (5 c.c. of 5% solution). But we have been able to obtain secretin solutions acting powerfully on the pancreas, with minimal or zero effects on the blood-pressure. The mucous membrane of duodenum and jejunum is rubbed up with sand in a mortar without the addition of acids, this mass is then folded in filter-paper and extracted with absolute alcohol in a Soxhlet apparatus for 24 hours, the mass removed, and boiled with 0.4% HCl, neutralized and filtered in the usual way. The effect of injecting this preparation is shown in Fig. 6. There is no fall of blood-pressure, but a powerful effect on the pancreas. . . .

GENERAL CONDITIONS OF ACTION OF SECRETIN. So far as we have as yet made out this action is wonderfully independent of the state of the animal. It is shown equally well in a dog in the stage of digestion or 20 hours after a meal. The height of the blood-pressure has very little effect. Anæsthetics in all usual doses have no appreciable influence; in this connection we may refer to a statement made by Camus,[10] who, having found the effect of secretin diminished by a previous dose of chloroform and abolished by a strong dose, concludes that the nervous system plays a considerable part in the excitation of the pancreas by this substance. So far as the central nervous system is concerned its influence is excluded by the first experiments in which we injected secretin solutions, since in these the pancreas was cut off from the central nervous system, as well as from the peripheral ganglia, except those in its own substance. With regard to these latter it is plain that we cannot absolutely deny the possibility of secretin acting indirectly through nerve-cells, but we submit that the well-known effect of chloroform as a general protoplasmic poison is quite sufficient to explain the results of Camus. . . .

THE RELATIONS IN SOME OTHER ANIMALS. All the experiments described so far were performed on dogs, and it is of obvious importance to discover whether there is a similar mechanism in other animals.

We find that secretin preparations made from the duodenum of the cat, rabbit, ox, monkey, man and frog are all active as regards the pancreatic secretion of the dog. On the pancreas of the rabbit and monkey we also tested secretin of the dog, rabbit, monkey and man and obtained positive results. We conclude, therefore, that the secretin of all these animals is one and the same body; we hope to extend the list of animals tested at a subsequent date. . . .

VIII. THE FATE OF SECRETIN IN THE ORGANISM

The flow of pancreatic juice produced by an injection of secretin does not last more than about 10 minutes, and the rapid flow not more than

5 minutes, so that the secretin introduced must disappear from the blood.

Now if we make repeated injections of secretin it is possible to produce a continuous flow of juice, without apparent fatigue,[11] for as long as we have made the attempt, that is for eight hours. We thought it possible, therefore, that secretin might be the source of the ferment or ferments of the juice, and that the pancreatic cell might not have very much work to do to convert it into these ferments. If this were so, however, it ought to be possible to obtain secretin back again, either from the pancreatic tissue or the juice. This we have been unable to do. It does not seem, therefore, that secretin disappears in this way. Nor is it to be found in the lymph, or urine, even after repeated doses.

We think that it disappears by oxidation; it has already been mentioned that it is easily destroyed in this way, and it is an obvious means by which the necessary destruction can be provided for. . . .

X. THE ACTION OF SECRETIN ON SOME OTHER GLANDS

SALIVARY GLANDS. Secretin does not excite secretion in these. In one experiment we noted a slow thick secretion from the cannula in the submaxillary duct, but this was abolished at once on cutting the sympathetic on the same side of the neck. This result shows the caution necessary in these experiments and the absolute necessity of taking a blood-pressure tracing, since the explanation of the phenomena was given at once by the blood-pressure curve. The secretin preparation used contained a considerable admixture of depressor substance, and there can be no doubt that the fall of blood-pressure was sufficient to excite the nerve-centre by anæmia, and so produce the flow of sympathetic saliva.

STOMACH. No effect, so far as we were able to make out from the absence of gastric juice after a number of injections of secretin.

SUCCUS ENTERICUS. Also no effect. . . .

XIII. SUMMARY OF CONCLUSIONS

1. The secretion of the pancreatic juice is normally evoked by the entrance of acid chyme into the duodenum, and is proportional to the amount of acid entering (Pawlow). This secretion does not depend on a

nervous reflex, and occurs when all the nervous connections of the intestine are destroyed.

2. The contact of the acid with the epithelial cells of the duodenum causes in them the production of a body (secretin), which is absorbed from the cells by the blood-current, and is carried to the pancreas, where it acts as a specific stimulus to the pancreatic cells, exciting a secretion of pancreatic juice proportional to the amount of secretin present.

3. This substance, secretin, is produced probably by a process of hydrolysis from a precursor present in the cells, which is insoluble in water and alkalis and is not destroyed by boiling alcohol.

4. Secretin is not a ferment. It withstands boiling in acid, neutral or alkaline solutions, but is easily destroyed by active pancreatic juice or by oxidising agents. It is not precipitated from its watery solution by tannic acid, or alcohol and ether. It is destroyed by most metallic salts. It is slightly diffusible through parchment paper.

5. The pancreatic juice obtained by secretin injection has no action on proteids until "enterokinase" is added. It acts on starch and to some extent on fats, the action on fats being increased by the addition of succus entericus. It is, in fact, normal pancreatic juice.

6. Secretin rapidly disappears from the tissues, but cannot be detected in any of the secretions. It is apparently not absorbed from the lumen of the intestine.

7. It is not possible to obtain a body resembling secretin from any tissues of the body other than the mucous membrane of the duodenum and jejunum.

8. Secretin solutions, free from bile-salts, cause some increase in the secretion of bile. They have no action on any other glands.

9. Acid extracts of the mucous membrane normally contain a body which causes a fall of blood-pressure. This body is not secretin, and the latter may be prepared free from the depressor substance by acting on desquamated epithelial cells with acid.

10. There is some evidence of a specific localized action of the vasodilator substances which may be extracted from various tissues.

1. *Physiologie expérimentale*, II. p. 226. Paris, 1856.

2. Hermann's *Handbuch d. Physiologie*, V. p. 183. 1883.

3. *Die Arbeit der Verdauungsdrilsen*. Trans. from Russian, Wiesbaden. 1898. Also *Le travail des glandes digestives*. Paris, 1901.

4. *Gazette clinique de Botkin* (Russ.), 1900.

5. *Journal de Physiologie*, III. p. 335. 1901.

6. This *Journal*, XXIV. p. 99. 1899.

7. A preliminary abstract of the main results of this work was published in the *Proc. Roy. Soc.* LXIX. p. 352. 1902. The experiments, of which an account is given in the present paper, were completed in March, 1902, their publication being delayed by extraneous circumstances.

8. *Journal de Physiologie*, III. p. 695. 1901.

9. *Pflüger's Archiv*, XC. p. 32. 1902.

10. *C. R. Soc. de Biologie*, 25 Avril, 1902, p. 443.

11. There are histological changes in the cells, and Mr Dale, who is now working at the subject, has been able to detect signs of exhaustion.

3

THE PARTICLE NATURE OF LIGHT

ALBERT EINSTEIN DIDN'T TALK until he was four years old. His father Hermann, who ran a failing electrochemical plant in Munich, and his mother Pauline worried that their son might be mentally retarded and considered sending him off for psychiatric help. When young Albert finally did begin speaking, he had the habit of saying everything twice—the first time muttered softly, just for himself, and the second out loud, for everyone else. One cannot help but interpret the boy's doubled speech as an early sign of his deep inner world, a world of silence and solitude where his creative imagination could take flight.

In his need for aloneness, in his fierce independence, and in the originality and beauty of his work, Albert Einstein was an artist as much as a scientist. Indeed, the two endeavors come together in an essay he wrote in midlife: "the aim of science is, on the one hand, a comprehension, as complete as possible, of the connection between the sense experiences in their totality, and, on the other hand, the accomplishment of this aim by the use of the minimum of primary concepts and relations." How similar is this last phrase to Picasso's dictum that the artist should compose with as few elements as possible.

But what happens when the aesthetics of unity and simplicity suggest a conception of nature that contradicts "sense experiences"? Such was the case with Copernicus's proposal that the earth flies through space on an invisible wheel about the sun, or Pasteur's notion that diseases are spread by microscopic living germs, or Einstein's "heuristic view" of light as granular rather than smooth. In these instances of perceptual dissonance, Einstein's inclination was to sharply question the accuracy of the senses rather than to doubt his theoretical principles. Almost always, he was proved right.

Consider the nature of light. Our bodily senses suggest that light is a continuous fluid of energy, completely filling the space that it occupies.

This conception also agrees with the so-called wave theory of light, which for centuries has been successful in explaining many observed properties of light, such as the interference of overlapping streams of light and the bending of light through a prism. But, as Einstein cautions in his 1905 paper, "One should keep in mind, however, that optical observations refer to time averages rather than instantaneous values." Here, Einstein is hinting that if light were in fact composed of large numbers of tiny "particles," most experiments—and certainly the human eye—would not have detected that fact. Analogously, a meteorologist who measures daily rainfall by the rise of water in a cup would not know that the rain actually arrives in individual drops.

Five years earlier, Max Planck had proposed that an atom of matter in contact with light can increase or decrease its energy only in multiples of an indivisible unit of energy. Einstein took Planck's idea a logical but heretical step further and argued that light itself exists in such individual and indivisible units, called quanta (or photons). Quanta are the elementary particles of light. So heretical was Einstein's hypothesis of the quantum nature of light that it was not fully accepted by physicists for almost two decades, even after experimental confirmation. Eventually, however, the quantum nature of light became the centerpiece of a new conception of nature called the wave-particle duality, in which any element of matter or energy behaves partly like a wave, smeared over an extended region of space, and partly like a particle, located at a single position in space. Einstein, with Planck, was the father of quantum physics.

In 1905, Einstein was a poor, twenty-six-year-old clerk working in a patent office in Bern, Switzerland. He and his wife Mileva Maric had secretly given away a daughter named Lieserl, born before their marriage in 1903, and now lived with their infant son Hans Albert in a two-room rented apartment on 49 Kramgasse, reached only by a steep staircase. At this time, the brilliant young physicist felt estranged from the world. He had already renounced his German citizenship at the age of sixteen, in contempt for the authoritarian German military service and his impending draft. In addition, he suffered under his parents' disdain for his Serbian wife. (His mother wrote to him: "she is a book like you—but you ought to have a wife . . . When you'll be 30, she'll be an old hag.") And, since graduating from the Federal Institute of Technology in Zurich in 1900, Einstein had repeatedly been refused jobs by

Europe's academic establishment, many of whose eminences he considered self-satisfied men far below him in scientific ability.

One gets a sense of the young Einstein's bitterness and sharp tongue in a letter he wrote to Maric in December 1901. A month earlier, he had submitted a doctoral thesis to Professor Alfred Kleiner at the University of Zurich, criticizing some of the work of the great Ludwig Boltzmann, a colleague of Kleiner's. To his sweetheart Maric, the twenty-two-year-old Einstein writes:

> Since that bore Kleiner hasn't answered yet, I am going to drop in on him on Thursday . . . To think of all the obstacles that these old philistines put in the way of a person who is not of their ilk, it's really ghastly! This sort instinctively considers every intelligent young person as a danger to his frail dignity, this is how it seems to me by now. But if he has the gall to reject my doctoral thesis, then I'll publish his rejection in cold print together with the thesis and he will have made a fool of himself.

The young Einstein was an embattled loner. Yet, between 1900 and 1905, unemployed much of that time, he managed to publish several scientific papers on his own.

Then, in 1905, still working alone in obscurity, the patent clerk produced five articles that changed physics for all time. Any one of these papers would have brought him lasting recognition. Two provided definitive new evidence for the existence and sizes of atoms and molecules; two proposed a radical new conception of time and space (the theory of special relativity) and tossed out as a by-product the famous formula $E = mc^2$. The fifth paper, for which he later won the Nobel Prize, proposed the quantum nature of light. Evidently, Einstein considered this last paper to be his most radical. In a letter to his friend Conrad Habicht in late May 1905, the young physicist wrote that his granular theory of light was "very revolutionary," the only one of his creations that he ever described in such terms.

Unlike Planck, who set off in his paper to explain a particular phenomenon of light, Einstein begins his article with a sweeping statement of principle. Why, Einstein asks, should there be "a profound formal difference" in the way that physicists view matter and light—the former granular, composed of a finite number of individual atoms, and the latter continuous and infinitely divisible? Here, Einstein subtly reveals his strong philosophical preference for unity. He would like *both* matter and energy to be of a similar nature—either discontinuous or continuous, granular or smooth, composed of a finite number of indivisible units or composed of an infinite number of infinitely divisible fluid elements. Taking his cue from several recent experiments, Einstein proposes the first possibility.

Einstein's emphasis on *unity* in natural phenomena is part of his conception of beauty. And beauty is a powerful guiding principle in his physics. As one of his collaborators, Banesh Hoffman, wrote, "Einstein was motivated not by logic in the narrow meaning of this concept, but by a sense of beauty. In his work he was always looking for beauty." In later chapters, we will learn more about the meaning of beauty in science. (See, especially, Chapter 20.)

Like Planck, Einstein considers the behavior of light in thermodynamic equilibrium with matter—that is, black-body radiation—to be a critical guide to understanding the fundamental nature of matter and energy. However, despite having been influenced by Planck's earlier paper, including Planck's idea of the quantum of energy, Einstein is suspicious of his predecessor's methods and does not start with Planck's theoretically derived law for black-body radiation. Rather, Einstein begins much closer to the experiments. (Even with his philosophical and theoretical principles, Einstein paid attention to experimental results.) He starts with an approximate formula for the observed black-body spectrum, valid at high frequencies where the density of radiation (light) is lowest. As it turns out, the low-density end of the black-body spectrum is where Einstein's intuition tells him that the granular nature of light should be most pronounced, just as the grainy nature of sand is most apparent when the individual grains are loose and well separated. Einstein employs different notation than Planck. His α and β are constants, determined by the quantitative measurements of black-body radiation, and his ρ is what Planck calls u_ν, the energy density of black-body radiation in each range of frequency.

Einstein's strategy is this: he knows the properties of *particles* in thermodynamic equilibrium, such as how the temperature of a gas of air molecules determines its pressure. By comparing the properties of black-body radiation to the properties of particles, he hopes to find similarities. In which case, he can reasonably propose that light is particle-like in nature.

Einstein begins with already established formulas relating entropy to energy and temperature. Recall from Chapter 1 on Planck that entropy is a quantitative measure of the order of a system. Highly disordered systems, like a well-shuffled deck of cards, have high entropy. Well-ordered systems, like a deck with all the cards of each suit together and arranged in ascending numerical order, have low entropy. The First Law of Thermodynamics, established in the nineteenth century, relates the entropy of a system to its energy, temperature, and volume. Einstein easily computes the entropy S of black-body radiation of energy E in a volume v and in a frequency range between ν and $\nu + d\nu$.

But Einstein is not concerned merely with the general formula for entropy. Seeking analogies with phenomena involving particles, he wants to see how the entropy of black-body light depends specifically on the volume of the radiation. That dependence is given in the last equation of section 4. As he then observes: the particular mathematical way that the black-body entropy varies with volume (i.e., with the logarithmic $ln[v/v_0]$) is exactly the same as for the well-established entropy of a dilute gas of matter, known to consist of a finite number of particles. *In other words, Einstein has found a mathematical similarity in the entropy of black-body radiation and the entropy of a gas of particles—suggesting that the radiation may indeed exist in the form of particles*. Evidently, Einstein was the first person to think of this simple but ingenious argument.

Next, Einstein appeals to "Mr. Boltzmann," as did Planck, for a mathematical formula relating the entropy of a state to the *probability* of that state, the latter denoted by W. That formula is the first equation in section 5. Here, R/N stands for Boltzmann's constant, denoted more simply by k in Planck's paper and previously known.

Einstein is, of course, interested in getting back to his earlier formula for the entropy of black-body radiation and how that entropy depends on volume. And he is thinking about light in terms of particles. With Boltzmann's law for entropy in hand, everything must be expressed in terms of probabilities. So Einstein asks the simple mathematical question: What is the probability that a group of n particles, which initially may lie anywhere in a total volume v_0, can later all be found in a smaller volume of size v? This situation is analogous to asking the question: What is the probability that n successive throws of a die will come up with a 3 every time? Since there are a total of six possible numbers one can get with each throw, the probability of getting a 3 on any particular throw is $\frac{1}{6}$. (Equivalently, the chance of finding any one particular particle in volume v is v/v_0.) The probability of getting 3 on the *first* throw is $\frac{1}{6}$. The probability of getting a 3 on the first throw *and* the second throw is $\frac{1}{6} \times \frac{1}{6} = (\frac{1}{6})^2$. The probability of getting a 3 on the first throw *and* the second throw *and* the third throw is $\frac{1}{6} \times \frac{1}{6} \times \frac{1}{6} = (\frac{1}{6})^3$. And so on. The probability of getting a 3 on n throws in a row is $(\frac{1}{6})^n$. Likewise, the probability of finding n particles in a volume v when they could possibly be anywhere in a total volume v_0 is $(v/v_0)^n$, which is the second formula that Einstein gives in section 5. Substituting this formula into the previous equation, Einstein has now computed how the entropy of a gas of n particles depends on volume v.

Next, Einstein does some simple mathematical manipulations of his first formula for the entropy of black-body radiation to bring it into a mathematical form identical to his new formula for the entropy of a gas of particles. By comparison of the two formulas, he is able to deduce not only that black-body radiation behaves like a gas of particles, but also the energy of each "particle" of light, or quantum, of frequency ν. That single quantum energy is $R\beta\nu/N$. This last result is obtained by requiring that the first equation for W, in section 5, equal the second equation for W in section 6. In turn, the two exponents of (v/v_0) must be equal, or $n = EN/(R\beta\nu)$. As a matter of definition, the number of quanta, n, must be equal to the total energy in quanta, E, divided by the energy of a single quantum. From the above equation, the energy of a single quantum is evidently $R\beta\nu/N$.

The constant number $R\beta/N$, multiplying the frequency ν, is exactly what Planck earlier called h, Planck's constant. Einstein can easily evaluate Planck's constant, just as Planck did, by determining β from the observations of black-body radiation. As mentioned earlier, the factor R/N, called Boltzmann's constant, was already known.

It is important to stress that Einstein has not duplicated Planck's work. Using his own original arguments, Einstein has independently derived Planck's idea that energy comes in units of $h\nu$. Furthermore, Einstein has independently determined the numerical value of the constant h in terms of the measurements of black-body radiation. Most crucially, Einstein has gone beyond Planck in applying the quantum idea to light itself. Light, like matter, comes in granular form.

Einstein's paper, like most of his work, is deceptively simple. The language is clear. His arguments are straightforward. The mathematics is not difficult. But the thinking, the intuition about nature, is profound.

It is the mark of a great physicist that after deriving his theoretical results, Einstein then wants to test those results against some actual observed phenomena. He wants a practical application. (Recall that at this time Einstein is employed as a patent clerk, where he spends hours each day staring at the drawings for new drilling gears and electrical transformers and improved typewriters to decide if they will work.) One of those observed phenomena is the so-called photoelectric effect, discovered by the great Hungarian experimental physicist Philip Lenard and utterly baffling in terms of the continuous wave theory of light.

In 1902, Lenard had found that when a metal is illuminated with ultraviolet light, it emits electrons (also called cathode rays). Evidently, the incoming light is absorbed by electrons in the metal, and those energized electrons wriggle loose from their atomic bonds and escape. A continous wave of incoming energy could have produced the same effect. But Lenard further found that the energy of each escaping electron did not vary at all with the intensity of the incident light. In the wave theory of light, one would expect that a greater intensity of incoming light would impart greater force to each electron—just as a heavier tide pounds each rock on the shore with more force—and thus eject each electron with greater energy. Not so, according to Lenard's results. However, Lenard did find the puzzling result that the energy of each escaping electron increased as the *frequency* of the incoming light increased. Even very low-intensity light could eject electrons of high energy if the frequency of that light were high.

Einstein finds a compelling explanation for Lenard's results in terms of his quantum idea. If light comes in a particle-like quantum form, then each electron can absorb only one quantum of light at a time. After an electron absorbs a quantum of light, it escapes with the energy of that quantum—assuming the energy is sufficient to liberate the electron—minus the energy expended to break free of its atomic bonds. Increase the intensity of incoming light and you increase only the number of quanta per second but not the energy of each quantum. Increase the *frequency* of the light, however, and you increase the energy of each quantum, according to Einstein's formula, and thus the energy of each liberated electron. Einstein goes on to work out a definite quantitative prediction for the energy of escaping electrons in a metal in terms of the frequency of the incoming light. It would be a decade before his detailed predictions could be confirmed by the experiments of American physicist Robert Millikan.

But the matter was not finished with Millikan. To the continuing amazement of scientists, further experiments have shown that light behaves *both* as a wave and as a particle. Light, and indeed all matter and

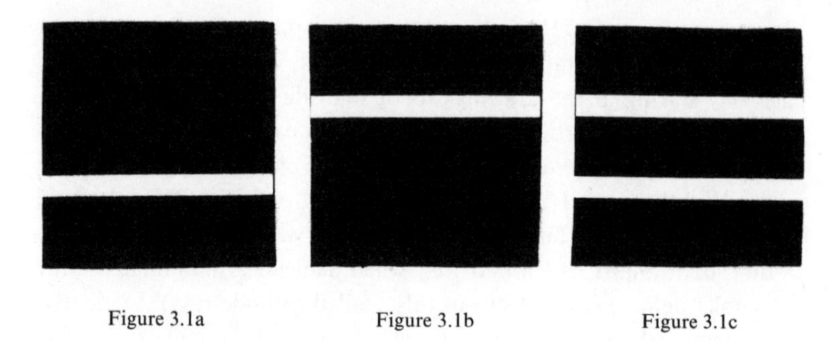

Figure 3.1a Figure 3.1b Figure 3.1c

energy, has a double existence, a dual personality. There is a classic experiment, called the double-slit experiment, that illustrates this wave-particle duality in its most disturbing form. In a darkened room, place a shade between a lightbulb and a screen and make a thin horizontal opening, or slit, in the shade. Make the lightbulb extremely weak, so that it emits only a few photons of light per second. (An extremely sensitive detector at the light source can count each single photon as it leaves the bulb, making a click for each one—click, click, click.) Measure the pattern of light on the screen. It will look something like Figure 3.1a. Next, cover up the first slit in the shade and make a second slit above it. Repeat the experiment and measure the pattern of light on the screen. It will look like Figure 3.1b.

For the third experiment, uncover both slits in the shade. If light consists of individual particles—as indicated by the individual clicks of the detector at the light source, one click and one photon at a time—then each photon must pass through *either* the top slit *or* the bottom slit. This statement would seem obvious. A particle can't be in two places at the same time. Furthermore, because of the high speed of light and the low rate of emission of photons, each photon leaves the light source, passes through the shade, and strikes the screen long before the next photon is emitted. Therefore, the pattern of light on the screen in the third experiment, with both slits open, should be a sum of the patterns seen in the first two experiments. In particular, regions of the screen that were lit up in *either* of the first two experiments should be lit up in the third. Regions of the screen that were dark in *both* of the first two experiments should be dark in the third. The *expected* pattern is shown in Figure 3.1c.

Contrary to these expectations, the pattern of light on the screen is that from two waves emanating from the two slits *simultaneously* and overlapping with each other on the journey to the screen. The wavelike interpretation is shown in Figure 3.2a and the *observed* pattern of light

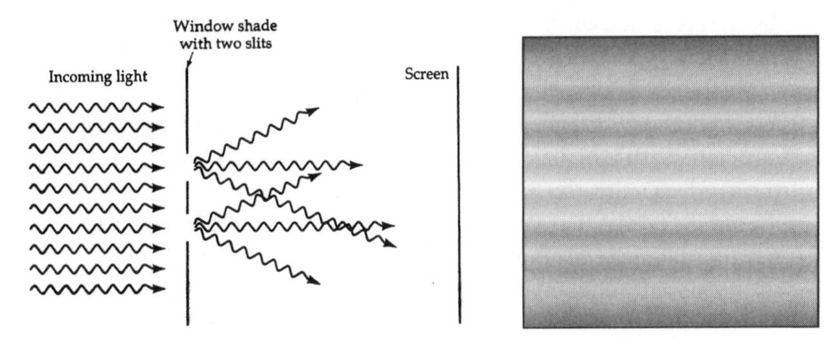

Figure 3.2a

Figure 3.2b

on the screen is shown in Figure 3.2b. Now, there are areas of light where there was darkness for both slits individually. And there are areas of darkness where one or the other slit individually produced light. But we have adjusted our experiment so that only a single photon of light is passing from the bulb to the screen at a time. It is as if each indivisible photon of light *somehow passes through both slits simultaneously* and interferes with itself in route to the screen.

The double-slit experiment has been repeated with electrons. As for photons, each electron behaves as if it travels through both slits simultaneously. The wave-particle duality evidently applies to all matter and energy.

The double-slit experiment suggests to us that our conception of nature as being either wavelike or particle-like, of each particle as existing at only one place at one time, is in error. The "either/or" conception of reality fails. Opposite, and even contradictory, properties seem to coexist in the world of photons and atoms, just as they always have in the world of the mind and the heart. One thinks of Robert Louis Stevenson's Dr. Jekyll and Mr. Hyde, or Caravaggio's famous juxtaposition of bright lights and deep shadows. Or, far more embracing, the ancient Chinese dichotomy of yin and yang. Could the physical universe be so ambiguous and strange? Despite an avalanche of confirming experiments, despite a powerful theory of quantum mechanics, physicists are still troubled by the wave-particle duality of nature.

Einstein could have won the Nobel Prize several times over, for half a dozen major discoveries. But it was his work on the quantum of light that garnered him the prize. He, and all physicists of the world, was so certain that he would win the Nobel that he promised Maric the putative Nobel money in their divorce settlement, two years before he received the prize. The Nobel arrived in 1921. At the time, a journalist gave the following description of Einstein:

> Einstein is tall, with broad shoulders and a scarcely bent back. His head—the head in which the science of the world was newly created—instantly attracts lasting attention . . . A little mustache, dark and very short, adorns a sensuous mouth, very red, rather big, with its corner betraying a permanent slight smile. But the strongest impression is that of stunning youthfulness, very romantic and at certain moments irresistibly reminiscent of the young Beethoven, who, already marked by life, had once been handsome. And then suddenly laughter erupts and one is faced with a student.

In 1905, Einstein could not have foreseen the full wave-particle duality implied by his work. But we do know that he considered his paper on the quantum nature of light to be "revolutionary." In fact, he did not consider his quantum proposal to be on solid ground, derived from first principles. This hesitation is conveyed in the very title of the paper—"On a Heuristic Point of View . . ."—and in the restrained expression of his conclusions: "As far as I can tell, this conception of the photoelectric effect does not contradict . . ." Yet the opening sections of the paper suggest that Einstein feels compelled by his results. They unify the conception of matter and energy.

It is hard to imagine the quiet confidence of the disheveled young man in the Swiss patent office. In the space of a year, he suggested a new way to understand light, laying the foundations of quantum mechanics; he gave firm theoretical evidence for the existence of atoms and molecules; and, as we will see in the next chapter, he radically changed our conception of time.

Although the many ideas of his papers of 1905 were only slowly accepted, Einstein immediately began receiving fan letters from such leading scientific figures as Planck and Lenard. Some of these letters were addressed to "Herr Professor Einstein," even though Einstein had barely completed his doctoral dissertation, and their senders were startled to discover that "A. Einstein" was only a twenty-six-year-old clerk in a patent office. However, by May 1909 Einstein was appointed "extraordinary professor of theoretical physics" at Zurich University, and two months later he received his first of many honorary degrees, from the University of Geneva. In 1913, after two more professorships, Kaiser Wilhelm II confirmed Einstein's appointment to the Prussian Academy of Sciences in Berlin, where he would live and work until 1933.

As Einstein grew older, he never had another year with the same ferocity of intellectual upheaval. Perhaps he needed a kind of gymnastic agility of mind for the efforts of 1905. Perhaps the cataclysm of thought in the Swiss patent office was facilitated by his isolation from the academic establishment and his general sense of alienation from the world. Or perhaps, in the following years, he felt that he had exhausted most accessible topics of fundamental importance, until his great work on gravity a decade later. Even at a young age, Einstein had a genius for knowing what problems to dodge as well as what problems to tackle. In another letter to his friend Habicht in the summer of 1905, he wrote: "There aren't always subjects that are ripe for rumination. At least none that are really exciting."

ON A HEURISTIC POINT OF VIEW CONCERNING THE PRODUCTION AND TRANSFORMATION OF LIGHT

Albert Einstein

Annalen der Physik (1905)

A PROFOUND FORMAL DIFFERENCE exists between the theoretical concepts that physicists have formed about gases and other ponderable bodies, and Maxwell's theory of electromagnetic processes in so-called empty space. While we consider the state of a body to be completely determined by the positions and velocities of an indeed very large yet finite number of atoms and electrons, we make use of continuous spatial functions to determine the electromagnetic state of a volume of space, so that a finite number of quantities cannot be considered as sufficient for the complete determination of the electromagnetic state of space. According to Maxwell's theory, energy is considered to be a continuous spatial function for all purely electromagnetic phenomena, hence also for light, whereas according to the present view of physicists, the energy of a ponderable body should be represented as a sum over the atoms and electrons. The energy of a ponderable body cannot be broken up into arbitrarily many, arbitrarily small parts, but according to Maxwell's theory (or, more generally, according to any wave theory) the energy of a light ray emitted from a point source continuously spreads out over an ever-increasing volume.

The wave theory of light, which operates with continuous spatial functions, has proved itself superbly in describing purely optical phenomena and will probably never be replaced by another theory. One should keep in mind, however, that optical observations refer to time averages rather than instantaneous values; and it is quite conceivable, despite the complete confirmation of the theory of diffraction, reflection, refraction, dispersion, etc., by experiment, that the theory of light, operating with continuous spatial functions, leads to contradictions when applied to the phenomena of emission and transformation of light.

Indeed, it seems to me that the observations of "black-body radiation," photoluminescence, production of cathode rays by ultraviolet light, and other related phenomena associated with the emission or transformation of light appear more readily understood if one assumes that the energy of light is discontinuously distributed in space. According to the assumption considered here, in the propagation of a light ray emitted from a point source, the energy is not distributed continuously over ever-increasing volumes of space, but consists of a finite number of energy quanta localized at points of space that move without dividing, and can be absorbed or generated only as complete units.

In this paper I wish to present the train of thought and cite the facts that led me onto this path, in the hope that the approach to be presented will prove of use to some researchers in their investigations. . . .

4. LIMITING LAW FOR THE ENTROPY OF MONOCHROMATIC RADIATION AT LOW RADIATION DENSITY

The existing observations of "black-body radiation" show that the law

$$\rho = \alpha\nu^3 e^{-\beta\nu/T}$$

originally postulated by Mr. W. Wien for "black-body radiation" is not strictly valid. However, it has been fully confirmed by experiment for large values of ν/T. We shall base our calculations on this formula, bearing in mind, however, that our results are valid only within certain limits.

First of all, this formula yields

$$\frac{1}{T} = -\frac{1}{\beta\nu} \ln \frac{\rho}{\alpha\nu^3},$$

and next, using the relation obtained in the preceding section [omitted here; φ is the entropy/volume/frequency].

$$\varphi(\rho, \nu) = -\frac{\rho}{\beta\nu} \left\{ \ln \frac{\rho}{\alpha\nu^3} - 1 \right\}.$$

Suppose now that we have radiation of energy E, with frequency between ν and $\nu + d\nu$, occupying a volume v. The entropy of this radiation is

$$S = \text{v}\varphi(\rho, \nu)\, d\nu = -\frac{E}{\beta\nu} \left\{ \ln \frac{E}{\text{v}\alpha\nu^3 d\nu} - 1 \right\}.$$

If we restrict ourselves to investigating the dependence of the entropy on the volume occupied by the radiation, and denote the entropy of radiation by S_0 at volume v_0, we obtain

$$S - S_0 = \frac{E}{\beta \nu} \ln \left[\frac{\mathrm{v}}{\mathrm{v}_0} \right].$$

This equation shows that the entropy of monochromatic radiation of sufficiently low density varies with the volume according to the same law as the entropy of an ideal gas or a dilute solution. In the following, we shall interpret this equation on the basis of the principle introduced into physics by Mr. Boltzmann, according to which the entropy of a system is a function of the probability of its state.

5. MOLECULAR-THEORETICAL INVESTIGATION OF THE DEPENDENCE ON VOLUME OF THE ENTROPY OF GASES AND DILUTE SOLUTIONS

. . . If it is meaningful to talk about the probability of a state of a system, and if, furthermore, each increase of entropy can be conceived as a transition to a state of higher probability, then the entropy S_1 of a system is a function of the probability W_1 of its instantaneous state. . . .

If S_0 denotes the entropy of a certain initial state and W is the relative probability of a state of entropy S, we obtain in general:

$$S - S_0 = \frac{R}{N} \ln W.$$

We now treat the following special case. Let a volume v_0 contain a number (n) of moving points (e.g., molecules) to which our discussion will apply. The volume may also contain any arbitrary number of other moving points of any kind. No assumption is made about the law governing the motion of the points under discussion in the volume except that, as concerns this motion, no part of the space (and no direction within it) is preferred over the others. Further, let the number of (aforementioned) moving points under discussion be small enough that we can disregard interactions between them.

This system, which, for example, can be an ideal gas or a dilute solution, possesses a certain entropy S_0. Let us imagine that all n moving points are assembled in a part of the volume v_0 of size v without any other changes in the system. It is obvious that this state has a different value of entropy (S), and we now wish to determine the difference in entropy with the help of Boltzmann's principle.

We ask: How great is the probability of the last-mentioned state relative to the original one? Or: How great is the probability that at a randomly chosen instant of time, all n independently moving points in a given volume v_0 will be found (by chance) in the volume v?

Obviously, this probability, which is a "statistical probability," has the value

$$W = \left(\frac{v}{v_0}\right)^n;$$

from this, by applying Boltzmann's principle, one obtains

$$S - S_0 = R\left(\frac{n}{N}\right) \ln \left(\frac{v}{v_0}\right).$$

6. INTERPRETATION ACCORDING TO BOLTZMANN'S PRINCIPLE OF THE EXPRESSION FOR THE DEPENDENCE OF THE ENTROPY OF MONOCHROMATIC RADIATION ON VOLUME

In section 4 we found the following expression for the dependence of the entropy of monochromatic radiation on volume:

$$S - S_0 = \frac{E}{\beta \nu} \ln \left(\frac{v}{v_0}\right).$$

If we write this formula in the form

$$S - S_0 = \frac{R}{N} \ln \left[\left(\frac{v}{v_0}\right)^{\frac{N}{R}\frac{E}{\beta \nu}}\right]$$

and compare it with the general formula expressing the Boltzmann principle,

$$S - S_0 = \frac{R}{N} \ln W,$$

we arrive at the following conclusion: If monochromatic radiation of frequency ν and energy E is enclosed (by reflecting walls) in the volume v_0, the probability that at a randomly chosen instant the total radiation energy will be found in the portion v of the v_0 is

$$W = \left(\frac{v}{v_0}\right)^{\frac{N}{R}\frac{E}{\beta \nu}}.$$

From this we further conclude that monochromatic radiation of low density (within the range of validity of Wien's radiation formula) behaves thermodynamically as if it consisted of mutually independent energy quanta of magnitude $R\beta\nu/N$. . . .

If monochromatic radiation (of sufficiently low density) behaves, as concerns the dependence of its entropy on volume, as though the radiation were a discontinuous medium consisting of energy quanta of magnitude $R\beta\nu/N$, then it seems reasonable to investigate whether the laws governing the emission and transformation of light are also constructed as if light consisted of such energy quanta. We will consider this question in the following sections. . . .

8. ON THE GENERATION OF CATHODE RAYS BY ILLUMINATION OF SOLID BODIES

The usual view that the energy of light is continuously distributed over the space through which it travels faces especially great difficulties when one attempts to explain photoelectric phenomena, which are expounded in a pioneering work by Mr. Lenard.[1]

According to the view that the incident light consists of energy quanta of energy $(R/N)\beta\nu$, the production of cathode rays by light can be conceived in the following way. The body's surface layer is penetrated by energy quanta whose energy is converted at least partially into kinetic energy of the electrons. The simplest conception is that a light quantum transfers its entire energy to a single electron; we will assume that this can occur. However, we will not exclude the possibility that electrons absorb only a part of the energy of the light quanta.

An electron in the interior of the body having kinetic energy will have lost part of its kinetic energy by the time it reaches the surface. In addition, we will assume that, in leaving the body, each electron must perform some work, P (characteristic for the body). The electrons leaving the body with the greatest perpendicular velocity will be those located right on the surface and ejected normally to it. The kinetic energy of such electrons is

$$\frac{R}{N}\beta\nu - P.$$

If the body is charged to a positive potential Π and surrounded by conductors at zero potential, and if Π is just sufficient to prevent a loss of electrical charge by the body, it follows that

$$\Pi\varepsilon = \frac{R}{N}\beta\nu - P,$$

where ε denotes the charge of the electron; or

$$\Pi E = R\beta\nu - P',$$

where E is the charge of one gram-equivalent of a monovalent ion and P' is the potential of this quantity of negative charge relative to the body.[2]

If one sets $E = 9.6 \times 10^3$, then $\Pi \cdot 10^{-8}$ is the potential in volts that the body acquires when irradiated in vacuum.

To see whether the derived relation agrees in order of magnitude with experiment, set $P' = 0$, $\nu = 1.03 \times 10^{15}$ (which corresponds to the ultraviolet limit of the solar spectrum) and $\beta = 4.866 \times 10^{-11}$. We obtain $\Pi \cdot 10^7 = 4.3$ volts, a result that agrees in order of magnitude with the results of Mr. Lenard.[3]

If the formula derived is correct, then Π, when plotted in Cartesian coordinates as a function of the frequency of the incident light, must give a straight line whose slope is independent of the nature of the substance under study.

As far as I can tell, this conception of the photoelectric effect does contradict its properties as observed by Mr. Lenard. If each energy quantum of the incident light transmits its energy to electrons, independently of all others, then the velocity distribution of the electrons, i.e., the nature of cathode rays produced, will be independent of the intensity of the incident light; on the other hand, under otherwise identical circumstances, the number of electrons leaving the body will be proportional to the intensity of the incident light.[4] . . .

1. P. Lenard, *Ann. d. Phys.* 8 (1902): 169, 170.

2. If one assumes that the individual electron can only be detached by light from a neutral molecule by the expenditure of some work, one does not have to change anything in the relation just derived; one need only consider P' as the sum of two terms.

3. P. Lenard, *Ann. d. Phys.* 8 (1902): 165, 184, table I, fig. 2.

4. P. Lenard, *loc. cit.*, pp. 150, 166–168.

4

SPECIAL RELATIVITY

NOTHING IS MORE BASIC THAN TIME. Time marks all change. Waking and sleeping, sunrise and sunset, the ebb and the flow of the tides, the menstrual cycles of women, the graying of hair, the space between breaths. Although time seems to move at a fidgety rate in our minds, we know that there are timekeeping devices outside of our bodies, ticking off seconds at precise marks and intervals. Clocks, wristwatches, church bells all divide years into months, months into days, days into hours, and hours into seconds, each increment of time marching after the other in perfect succession. And beyond any particular clock, we have faith in a vast scaffold of time, stretched taut through the cosmos, laying down the law of time equally for electrons and people: a second is a second is a second.

In 1905, at the age of twenty-six, Albert Einstein proposed a different law of time: a second is not a second. A second as measured by one clock corresponds to less than a second as measured by another clock in flight with respect to the first. Against all common sense, time is not absolute as it seems. Time is relative to the observer. Astoundingly, Einstein's proposal has been confirmed in the lab.

Einstein's ideas about time had their origin in a different phenomenon. While employed as a lowly patent clerk in Bern, Switzerland, Einstein was struggling to understand why the laws for electricity and magnetism, known as Maxwell's equations, were apparently contradicted by experiments when applied to objects in motion.

Of course, motion involves time. An object cannot move through space without the passage of time. In thinking carefully about motion, Einstein began on a quiet river picnic, like Alice, and found himself chasing the White Rabbit into a fantastic domain. Like Alice, Einstein embarked on his journey with the innocence and daring of a child. Let us try to follow him, one step of logic after the other.

To begin with, motion is a more subtle concept than it seems. When we say that we're driving at forty miles per hour, we actually mean that our car is traveling at forty miles per hour *relative to the road*. The road, of course, is attached to the earth. And the earth spins about on its axis. At the same time, it orbits the sun. The sun, in turn, wheels about the center of the galaxy. And so on. Thus, it would be most difficult to say how fast we are moving in any *absolute* sense. What we can say without ambiguity is that our Chevrolet is passing forty miles of pavement in every hour. For earthly travel, the road, or any landmark stuck to the ground, is our fixed frame of rest.

Maxwell's equations for electricity and magnetism seemed to have built into them a *cosmic* frame of rest, against which all motions could be measured—namely, an almost weightless, invisible substance filling all space. That gossamer substance was called the ether. Among other things, the ether had the job of holding and carrying electric and magnetic forces, as paper is needed to hold paint. More importantly, Maxwell's equations took on their simplest mathematical form for an observer at rest in the ether. For observers not at rest in the ether, the equations were more complex and varied with the particular speed of the observer through the ether. Thus, by doing experiments with magnets and wires and comparing the results to Maxwell's equations, an observer anywhere in the universe could tell whether she was at rest or in motion relative to the ether. (An "observer" in physics is anyone who has rulers, clocks, and other equipment for making measurements. An observer need not be human—a set of instruments that automatically record the results of experiments would suffice.)

One crucial electromagnetic phenomenon was the speed of light. Maxwell's equations demonstrated that light was an undulating wave of electric and magnetic forces traveling through space. All other known waves, such as water waves and sound waves, required a material medium to carry them along. The presumed medium for light was the ether. (In his paper, Einstein calls the ether the "light medium.") Furthermore, Maxwell's equations predicted that light would have a definite speed, 186,282 miles per second, *relative to the ether*. That is, an observer at rest in the ether would measure the speed of light as 186,282 miles per second. An observer moving through the ether would measure a different speed for light, a speed varying with the direction of the light ray.

This latter expectation can be easily understood by a familiar analogy. A water wave has a definite speed relative to still water. When a person sitting at rest in a lake makes a splash, the resulting waves travel outward from her at the same speed in every direction. But if our observer is motoring through the lake when she strikes the water, the waves traveling in the same direction as her motion will move away from her at a lesser speed than the waves traveling in the opposite direction. Indeed, as our observer motors faster and faster, she gains on the waves traveling in her direction and loses on the waves going in the opposite direction. If she moves fast enough, she can actually keep up with the waves traveling in her direction, so that they remain beside her boat.

Physicists reasoned that as the earth orbits the sun, it moves through the ether like a boat through a lake. Thus, the speeds of light rays traveling in different directions should be different. In 1887, in one of the most important scientific experiments of all time, the American physicists Albert Michelson and William Morley carefully measured the speed of light in different directions. They were hoping to determine how fast the earth moved through the ether and thus identify the cosmic frame of rest. To their extreme surprise and disappointment, they found the speed of light to be the same in every direction, 186,282 miles per second. Michelson was so certain that he'd made a mistake that he kept repeating his experiment, with the same discouraging result. Finally, in 1907, he was awarded the Nobel Prize for his "failure," the first American to win the prize since its inception in 1901.

So convinced were physicists of the existence of the ether, indeed the necessity of the ether, that they invented elaborate mechanisms involving the ether to explain the results of Michelson and Morley. Most notable was the theory of the great Dutch physicist Hendrik Antoon Lorentz, a 1902 Nobel Prize winner himself. Lorentz proposed that the electromagnetic forces inside Michelson and Morley's measuring rulers were altered by their motion through the ether in precisely the right way to make it "appear" that light traveled at the same speed in all directions.

According to his autobiography, Einstein had been thinking about the movement of light since the age of sixteen. In the first paragraphs of his paper of 1905, he summarizes the conflict between theory and experiment with regard to "electrodynamics" for moving bodies and casts doubt on the sacred idea of absolute rest. (For different reasons, Aristotle, Newton, and Kant all professed faith in the condition of absolute rest.) Then Einstein announces two "postulates." The first, that the laws of electricity and magnetism "will be valid for all frames of reference for which the equations of mechanics hold good," means that Maxwell's equations should appear identical for all observers traveling at constant velocities, regardless of the directions and speeds of those velocities. Such observers might be called "constant-velocity observers." You know when you are such an observer because you don't feel the telltale jerks and pushes of acceleration. Given the first postulate, Einstein's second postulate means that the speed of light is always measured to be the same number, which he denotes by c, regardless of the motion of the emitter or observer. (As mentioned earlier, $c = 186{,}282$ miles per second.)

In his two postulates, Einstein has banished the ether. Since no ether is observed, he suggests, the ether doesn't exist. Or, in his more modest words, the ether is "superfluous." But the consequences of that simple dismissal are profound. Without an ether, there is no cosmic frame of absolute rest. Therefore, we cannot say in any absolute sense that one observer is at rest and another is in motion. A constant-velocity observer is like a car with no accelerations, no landscape outside the window, no friction from the road. There is no way the passengers of such a car would know how fast they are moving, or even if they are moving at all. *Only motion relative to some other object is measurable and has meaning.* Hence the origin of the famous word "relativity."

Without a condition of absolute rest, all constant-velocity observers should be completely equivalent to one another. Being equivalent, they must measure identical laws of physics (Einstein's first postulate). In particular, since the speed of light comes out of the laws of electromagnetism, all constant-velocity observers must measure the same speed for a passing ray of light.

Immediately, we can see that Einstein's two postulates will require a drastic revision in our notions about time and space. Consider two constant-velocity observers, a man sitting on a bench and another man running at ten miles per hour past the bench. Now suppose a light ray passes by, traveling in the direction opposite to the running man. According to Einstein's postulates, the two fellows will measure *the same* speed for the passing light ray. Common sense screams at us that the man running toward the light ray will see it come at him faster than will the man sitting on the bench. If Einstein's postulates are true, something must be wrong about our commonsense notion of how speeds add and subtract. But speed is distance divided by the time it takes to travel that distance. If our ideas about speed are in error, then so are our ideas about time and space. We have just fallen into the rabbit hole.

With almost childlike simplicity, Einstein begins questioning the meaning of time. "All our judgments in which time plays a part," Einstein writes, "are always judgments of *simultaneous* events. If, for instance, I say, 'That train arrives here at 7 o'clock,' I mean something like this: 'The pointing of the small hand of my watch to 7 and the arrival of the train are simultaneous events.' " Implicitly, Einstein introduces the notion of an "event," which has a precise location in both time and space. For example, an event would be my meeting a friend at the reception desk of the Pierre Hotel in New York at 11:28 a.m. on the morning of July 28, 2003. It is a simple matter, Einstein says, to determine if two events are simultaneous when they happen at the same place, but not so simple when they happen at different places. For the latter, two clocks are required, and those clocks must be synchronized.

Einstein then offers a concrete plan for synchronizing a network of clocks in different locations but at rest with respect to each other. For a trustworthy messenger of time to set the watches in synchrony, he uses a ray of light, which covers any definite distance between two clocks in a definite amount of time. For example, if a light ray is emitted from one clock at noon, at the moment it reaches another clock 186,282 miles away that second clock must be set at noon plus one second.

Using this method, each "local" clock in a particular place can be synchronized with the entire network of clocks. The "time" of an event can now be defined as the time read by the particular clock at the same location as the event. Two events are simultaneous in a particular network if they occur at the same time on the two appropriate local clocks in that network. Other clocks, in motion with respect to the first clocks, are part of their own networks. In Einstein's imagination, different networks of clocks glide past each other like ships in a black sea, each network in synchrony with itself.

Little by little, Einstein is thinking through what it means to measure time, trying not to make any unnecessary assumptions, trying not to let any hidden biases darken his mind. Marcus Aurelius conceived of time as a river, carrying events downstream. Kant proclaimed that time had no independent existence outside the human perception of incidents. And Shelley wrote of "leaden footed time," slower than thought. But Einstein is a physicist. For him, time is a problem for physics. As he recalled at age sixty-seven: "One had to understand clearly what the spatial co-ordinates and the temporal duration of events meant to physics."

Next, the remarkable but not unanticipated surprise. Two events that are simultaneous to one observer (with his network of clocks) are not so for a second observer in relative motion to the first (with his second network of clocks). Imagine the following experiment: let the first observer sit in a train car and place a screen exactly midway between two lightbulbs. Assume further that the screen has a light detector on each side of it, and a bell rings if light strikes both detectors at once. Now, our observer turns on both lightbulbs simultaneously. A light ray from one bulb travels to the left, a light ray from the other travels to the right. The two light rays travel at the same speed, by Einstein's postulates, so they meet at the same time at the screen midway between them. The bell rings. Nothing could be simpler!

Suppose now that the train car is traveling past a second observer on a bench, moving to her right, and consider everything from that second observer's point of view, as shown in Figure 4.1. The figure shows the view at three successive instants, starting from the top. The observer on the bench *sees the two lightbulbs and the screen all moving to the right.* That observer also witnesses the two light rays meet at the same time at the screen. The bell either rings or it doesn't, and since it rang for the first observer, it must ring for the second, indicating that the two light rays meet at the screen at the same time.

But, as seen by the observer on the bench, one of those light rays had to travel farther than the other to reach the screen because, during the period of transit, the screen has moved. In particular, since the screen is moving to the right, it travels toward the left-traveling light ray and away from the right-traveling light ray. *The right-traveling light ray thus had to travel farther than the left-traveling light ray to reach the screen.* Since both light rays have the same speed, according to Einstein's postulates, the right-traveling light ray took longer to reach the screen. Therefore, the lightbulb that emitted the right-traveling ray had to have been turned on *before* the other bulb. In other words, the turning on of the two bulbs—those two events—did not happen simultaneously for the second observer, although they did for the first. Simultaneity is relative to the observer.

The key fact causing the disagreement in simultaneity is that *both* observers see the left-traveling and right-traveling light rays move at the same speed. Before Einstein's ideas about the absence of the ether and the constancy of the speed of light, physicists would have said that at least one of the observers had to be moving through the ether. For that observer, the left-traveling and right-traveling rays could not have had

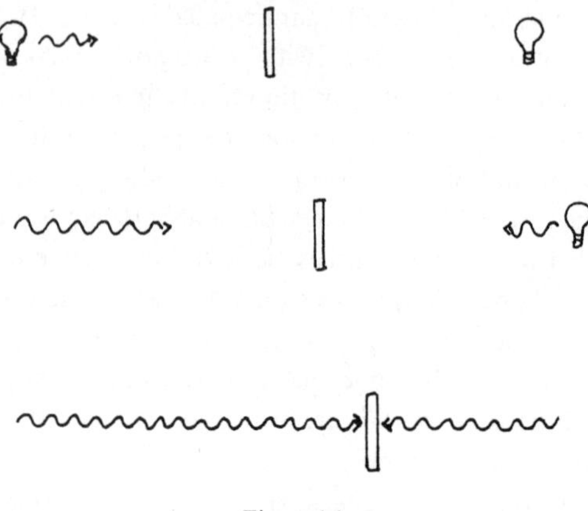

Figure 4.1

the same speed. (One ray is traveling upstream against the ether while the other is traveling downstream with the ether.) The different speeds, in fact, would exactly compensate for the different distances traveled, so that temporal disagreements between the two observers would be eliminated.

One cannot dismiss these results as a peculiarity associated only with light. The turning on of the two lightbulbs could have heralded any two events at opposite ends of the train car. For example, the birth of two babies. In that case, the observer in the train would say that the babies were born at the same time, whereas the observer on the bench would declare that the baby on the left was the elder.

It is striking that Einstein makes so few references to the existing physics literature. One would be tempted to attribute this absence to his isolation from the academic establishment in 1905, but the omission continues throughout his later work. A more plausible explanation might be that Einstein did not seem especially influenced by this or that particular result but instead was guided by the big picture as he saw it.

Here and elsewhere, Einstein used a rare, deductive method of thought, starting from general principles, or "postulates," and then deducing the consequences of those principles. Most other scientists had always worked in an inductive manner, constructing general laws from a range of experimental phenomena, or starting with some simple, elemental processes, and then building up to the more complex. Thus, for example, the early seventeenth-century German astronomer Johannes Kepler pored at great length over the observational data for the orbits of planets before discovering his famous laws of planetary motion. Darwin traveled to Patagonia, Tierra del Fuego, and the Amazon and spent years examining the vital statistics of ostriches and armadillos before "inducing" the principle of natural selection. And Max Planck, in 1900, started with a specific law for a specific observational phenomenon and then worked backward from that law to figure out what it required for the energy allotment in his resonators.

Einstein, a loner in all ways, worked differently. He began with fundamental principles and postulates, creations from his own mind, which were only loosely based on the external world. In the early 1930s, he expressed his philosophy of science in this way: "We now know that science cannot grow out of empiricism alone, that in the constructions of science we need to use free invention which only *a posteriori* [later] can be confronted with experience as to its usefulness." For Einstein, his brain was his eyes, ears, and hands all combined. Experiments were only suggestions. When he needed experiments that had never been done, he imagined them.

Another distinctive and original aspect of Einstein's thinking in this paper was his assumption of a "symmetry principle." A symmetry principle says that a phenomenon should look the same from certain different points of view. For example, a square looks the same when rotated by ninety degrees and is therefore said to have a "four-sided symmetry." In his first postulate, Einstein invokes a powerful symmetry principle: the laws of physics should look the same to different observers as long as they are all moving at constant velocity. In fact, Einstein was the first physicist to invoke a symmetry principle as a *starting point* for investigating the laws of nature. Other physicists would sometimes remark on various symmetries exhibited by their laws and equations *after* those laws and equations had been discovered by other means. But Einstein held symmetry principles to be foundational, and thus clues to discovery. For Einstein, symmetry principles, together with unity, were part of the fabric of nature. Thus, to discover the laws of nature, one started with symmetry principles. (Of course, the laws thus deduced are provisional, as are all human-made theories in science, and must be tested against experiment. Sometimes those proposed laws and their assumed symmetry principles are proved to be false. Nature does not obey all the symmetry principles that we can imagine.) These ideas will be discussed in more detail in Chapter 20.

We have arrived at the nuts and bolts of the paper. Everything follows logically and inevitably from the two postulates. Einstein considers two events and asks how the location in time and space of those two events varies with the perspective of two different observers. The first observer he calls *K*, the second *k*. (Did Kafka, by chance, read Einstein?) Observer *k* moves along the *x* direction at velocity v relative to *K*. Both observers arrange their clocks and rulers so that the first event happens at a time and space position of zero in their respective measuring systems. The second event happens at a space position of *x*, *y*, *z* and time *t* as measured by *K*, and at a space position of ξ, η, ζ, and time τ as measured by *k*. Einstein asks: How do *x*, *y*, *z*, and *t* relate to ξ, η, ζ, and τ? The question is a bit like considering a new system of latitude and longitude—for example, with a zero of latitude at Tokyo instead of the North Pole, and a zero of longitude passing through New York instead of London—and asking how the new and old systems of coordinates compare for the location of Calcutta. The difference, for Einstein's problem, is that one of the coordinates is time. An event has both a time and a place.

Given his postulates, Einstein can work out the shift in the coordinates. He considers two sets of two events: the first set being two ticks of a clock sitting at rest in frame *k*. Each tick happens at a time and a place. The second set is the position of a traveling light ray at two different times. Even though the two observers *K* and *k* will disagree on the time and space positions of the light ray, they must both agree on its speed, as required by the second postulate of relativity. Furthermore, since speed is distance divided by time, if *K* and *k* disagree on the distance traveled by the light ray, as in the example shown in Figure 4.1, they must also disagree on the time taken, so that the ratio of distance divided by time can be the same for both of them.

Using these two sets of events and the complete equivalence of *K* and *k*, Einstein arrives at the "transformation equations" given at the end of the third section. These equations, the foundational equations for the theory of special relativity, quantitatively express how time and space measurements vary between two observers traveling past each other at a relative speed of v. The results depend on v and on the universal velocity of light, *c*. It is important to note that, as in the birth of the babies, the transformation equations apply not just to light. They apply to any events. They signify the relative nature of time and of space. The cool precision of these equations, the detached and measured voice of the young scientist, all belie the tumultuous meaning of Einstein's proposal.

With the transformation equations, Einstein now finds that all moving objects are shortened in the direction of motion by the "relativity factor" $\sqrt{1 - v^2/c^2}$, and all moving clocks are slower by the same factor. "Moving," of course, is a relative term. Each observer has his own network of "stationary" rulers and clocks that are at rest relative to him. All rulers and clocks that move relative to him are shortened and slowed compared to his stationary equipment. For example, if a clock travels by me at half the speed of light, $v = c/2$, then the passage of a second on my stationary clock takes $\sqrt{1 - 1/4} = 0.866$ seconds on the moving clock.

For everyday speeds, which are tiny compared to the speed of light, the relativity factor $\sqrt{1 - v^2/c^2}$ is extremely close to one. (A jet plane flying by me at 600 miles per hour has a relativity factor of .9999999999996.) That is why Einsteinian length contractions and time dilations and discrepancies of simultaneities are completely unnoticable to human sense perceptions. That is why we have such a strong sense of the absolute nature of time. But for high relative speeds or highly sensitive instruments, the relativistic distortions of time and of space become measurable. In fact, the predictions of special relativity have been quantitatively confirmed by many experiments. For example, a subatomic particle called the muon, when at rest in the laboratory, lives an average of 2.2 millionths of a second before disintegrating. Muons traveling at 99.8 percent of the speed of light (relative to the lab) live for 33 millionths of a second (as measured by clocks in the lab) before disintegrating. We know this result because muons have been detected at the earth's surface, after being created at the top of the atmosphere, about seven miles above the ground. Without the effects of relativity, such muons could not last the duration of the trip without disintegrating. The *moving* muon's internal clock ticks more slowly than a *stationary* clock, and exactly by the prescribed amount.

It is impossible to overstate the significance of "On the Electrodynamics of Moving Bodies," perhaps the most important paper in all physics. Unlike almost all other theories of physics, relativity is not a theory about particular forces and particles. Rather it is a theory about the nature of time and space—the stage on which all forces and particles play out their parts. As such, relativity enters every modern theory of physics. Every theory originating before 1905 has been revised according to relativity. Every theory since has incorporated it.

———

The unknown young patent clerk read widely in philosophy and literature, as well as in physics, and was well aware that his relativity theory challenged centuries of thought. Much of his reading he discussed with two other young men, Maurice Solovine and Conrad Habicht. The three friends formed a small society they modestly called the Olympia Academy, which met several nights a week for dinner and conversation. Along with the sausage and the cheese and the fruit and the tea, they discussed Spinoza, Hume, Kant, Sophocles, Racine, Cervantes, Dickens, Mill. Intermingled were such physicists and mathematicians as Mach, Helmholtz, Riemann, Poincaré. As Solovine recalled many years later, "We would read a page or half a page—sometimes only a sentence—and the discussions would continue for several days when the problem was important." The salons were sometimes accompained by one of Einstein's performances on the violin. During the summer, the three friends would end their evenings together by going up to Gurten, south of Bern, where they would wait for the sunrise over the Alps. Such a sight, of course, triggered new discussions of astronomy, after which the tired young men would have dark coffee in a small restaurant and finally start their journey back down the mountain to Bern.

Aside from these dear friends in his youth, Einstein was a loner throughout his life, a man who fiercely guarded his solitude. In later life, he involved himself with many social causes, such as supporting the League for Human Rights, giving numerous lectures around the world on politics and philosophy and education, helping to found the Hebrew University of Jerusalem. Einstein had many romantic relationships in his life. But at the most profound level, he was a solitary man. Unlike almost all other great scientists, he supervised only a single graduate student, and he avoided teaching. In an essay he published in 1931, at the age of fifty-two, Einstein wrote:

> My passionate sense of social justice and social responsibility has always contrasted oddly with my pronounced lack of need for direct contact with other human beings and human communities. I am truly a "lone traveller" and have never belonged to my country, my home, my friends, or even my immediate family with my whole heart.

In 1933, in response to the militarist and anti-Semitic policies of the Third Reich, Einstein left Germany for the Institute for Advanced Study in Princeton. There, he would spend the last twenty-two years of his life.

During those years, the "lone traveller" became even more alone. Part of Einstein's deepening estrangement was simply that he was in a new country with a new language. But, more important, Einstein's intellectual life, the core of his being, disengaged from the rest of physics. First, he could not philosophically accept the fundamental tenet of *uncertainty* present in the new quantum physics, a physics that he had helped to create. (See Chapter 10.) Second, Einstein soon became obsessed with his own nonquantum unified theory of electromagnetism and gravity, which he doggedly pursued for the rest of his life, despite one failure after another. Other physicists regarded these stubborn positions with puzzlement and dismay. While the current of physics swerved in the direction of the powerful new quantum physics, able to explain the behavior of atoms and subatomic particles, Einstein was left in intellectual solitary confinement. The discovery of antimatter, the discovery of a relativistic quantum theory of the electron, the discovery of new fundamental forces in the 1930s all passed by with little comment from Einstein. Even when the great Danish physicist Niels Bohr, with whom he had once had lively discussions, visited Princeton in 1939, Einstein remained cloistered and alone in his paper-strewn office.

In 1953, near the end of his life, Einstein sent a letter to Solovine, the friend of his youth, addressed "To the immortal Olympia Academy." The letter began, "In your short active existence you took a childish delight in all that was clear and reasonable . . . Though somewhat decrepit, we still follow the solitary path of our life by your pure and inspiring light."

ON THE ELECTRODYNAMICS OF MOVING BODIES

Albert Einstein

Annalen der Physik (1905)

IT IS KNOWN THAT Maxwell's electrodynamics—as usually understood at the present time—when applied to moving bodies, leads to asymmetries which do not appear to be inherent in the phenomena. Take, for example, the reciprocal electrodynamic action of a magnet and a conductor. The observable phenomenon here depends only on the relative motion of the conductor and the magnet, whereas the customary view draws a sharp distinction between the two cases in which either the one or the other of these bodies is in motion. For if the magnet is in motion and the conductor at rest, there arises in the neighbourhood of the magnet an electric field with a certain definite energy, producing a current at the places where parts of the conductor are situated. But if the magnet is stationary and the conductor in motion, no electric field arises in the neighbourhood of the magnet. In the conductor, however, we find an electromotive force, to which in itself there is no corresponding energy, but which gives rise—assuming equality of relative motion in the two cases discussed—to electric currents of the same path and intensity as those produced by the electric forces in the former case.

Examples of this sort, together with the unsuccessful attempts to discover any motion of the earth relatively to the "light medium," suggest that the phenomena of electrodynamics as well as of mechanics possess no properties corresponding to the idea of absolute rest. They suggest rather that, as has already been shown to the first order of small quantities, the same laws of electrodynamics and optics will be valid for all frames of reference for which the equations of mechanics hold good.[1] We will raise this conjecture (the purport of which will hereafter be called the "Principle of Relativity") to the status of a postulate, and also introduce another postulate, which is only apparently irreconcilable with the former, namely, that light is always propagated in empty space with a definite velocity c which is independent of the state of motion of the emitting body. These two postulates suffice for the attainment of a simple and consistent theory of the electrodynamics of moving bodies based on Maxwell's theory for stationary bodies. The introduction of a "luminiferous ether" will prove to be superfluous inasmuch as the view here to be developed will not require an "absolutely stationary space" provided with special properties, nor assign a velocity-vector to a point of the empty space in which electromagnetic processes take place.

The theory to be developed is based—like all electrodynamics—on the kinematics of the rigid body, since the assertions of any such theory have to do with the relationships between rigid bodies (systems of co-ordinates), clocks, and electromagnetic processes. Insufficient consideration of this circumstance lies at the root of the difficulties which the electrodynamics of moving bodies at present encounters.

I. KINEMATICAL PART

§ 1. Definition of Simultaneity

Let us take a system of co-ordinates in which the equations of Newtonian mechanics hold good.[2] In order to render our presentation more precise and to distinguish this system of co-ordinates verbally from others which will be introduced hereafter, we call it the "stationary system."

If a material point is at rest relatively to this system of co-ordinates, its position can be defined relatively thereto by the employment of rigid standards of measurement and the methods of Euclidean geometry, and can be expressed in Cartesian co-ordinates.

If we wish to describe the *motion* of a material point, we give the values of its co-ordinates as functions of the time. Now we must bear carefully in mind that a mathematical description of this kind has no physical meaning unless we are quite clear as to what we understand by "time." We have to take into account that all our judgments in which time plays a part are always judgments of *simultaneous events*. If, for instance, I say, "That train arrives here at 7 o'clock," I mean something like this: "The pointing of the small hand of my watch to 7 and the arrival of the train are simultaneous events."[3]

It might appear possible to overcome all the difficulties attending the definition of "time" by substituting "the position of the small hand of my watch" for "time." And in fact such a definition is satisfactory when we are concerned with defining a time exclusively for the place where the watch is located; but it is no longer satisfactory when we have to connect in time series of events occurring at different places, or—what comes to the same thing—to evaluate the times of events occurring at places remote from the watch.

We might, of course, content ourselves with time values determined by an observer stationed together with the watch at the origin of the co-ordinates, and co-ordinating the corresponding positions of the hands with light signals, given out by every event to be timed, and reaching him through empty space. But this co-ordination has the disadvantage that it is not independent of the standpoint of the observer with the watch or clock, as we know from experience. We arrive at a much more practical determination along the following line of thought.

If at the point A of space there is a clock, an observer at A can determine the time values of events in the immediate proximity of A by finding the positions of the hands which are simultaneous with these events. If there is at the point B of space another clock in all respects resembling the one at A, it is possible for an observer at B to determine the time values of events in the immediate neighbourhood of B. But it is not possible without further assumption to compare, in respect of time, an event at A with an event at B. We have so far defined only an "A time" and a "B time." We have not defined a common "time" for A and B, for the latter cannot be defined at all unless we establish *by definition* that the "time" required by light to travel from A to B equals the "time" it requires to travel from B to A. Let a ray of light start at the "A time" t_A from A towards B, let it at the "B time" t_B be reflected at B in the direction of A, and arrive again at A at the "A time" t'_A.

In accordance with definition the two clocks synchronize if

$$t_B - t_A = t'_A - t_B.$$

We assume that this definition of synchronism is free from contradictions, and possible for any number of points; and that the following relations are universally valid:—

1. If the clock at B synchronizes with the clock at A, the clock at A synchronizes with the clock at B.

2. If the clock at A synchronizes with the clock at B and also with the clock at C, the clocks at B and C also synchronize with each other.

Thus with the help of certain imaginary physical experiments we have settled what is to be understood by synchronous stationary clocks located at different places, and have evidently obtained a definition of "simultaneous," or "synchronous," and of "time." The "time" of an event is that which is given simultaneously with the event by a stationary clock located at the place of the event, this clock being synchronous, and indeed synchronous for all time determinations, with a specified stationary clock.

In agreement with experience we further assume the quantity

$$\frac{2\mathrm{AB}}{t'_{\mathrm{A}} - t_{\mathrm{A}}} = c$$

to be a universal constant—the velocity of light in empty space.

It is essential to have time defined by means of stationary clocks in the stationary system, and the time now defined being appropriate to the stationary system we call it "the time of the stationary system."

§ 2. On the Relativity of Lengths and Times

The following reflexions are based on the principle of relativity and on the principle of the constancy of the velocity of light. These two principles we define as follows:—

1. The laws by which the states of physical systems undergo change are not affected, whether these changes of state be referred to the one or the other of two systems of co-ordinates in uniform translatory motion.

2. Any ray of light moves in the "stationary" system of co-ordinates with the determined velocity c, whether the ray be emitted by a stationary or by a moving body. Hence

$$\text{velocity} = \frac{\text{light path}}{\text{time interval}}$$

where time interval is to be taken in the sense of the definition in § 1.

Let there be given a stationary rigid rod; and let its length be l as measured by a measuring-rod which is also stationary. We now imagine the axis of the rod lying along the axis of x of the stationary system of co-ordinates, and that a uniform motion of parallel translation with velocity v along the axis of x in the direction of increasing x is then imparted to the rod. We now inquire as to the length of the moving rod, and imagine its length to be ascertained by the following two operations:—

(*a*) The observer moves together with the given measuring-rod and the rod to be measured, and measures the length of the rod directly by superposing the measuring-rod, in just the same way as if all three were at rest.

(*b*) By means of stationary clocks set up in the stationary system and synchronizing in accordance with § 1, the observer ascertains at what points of the stationary system the two ends of the rod to be measured are located at a definite time. The distance between these two points, measured by the measuring-rod already employed, which in this case is at rest, is also a length which may be designated "the length of the rod."

In accordance with the principle of relativity the length to be discovered by the operation (*a*)—we will call it "the length of the rod in the moving system"—must be equal to the length l of the stationary rod.

The length to be discovered by the operation (*b*) we will call "the length of the (moving) rod in the stationary system." This we shall determine on the basis of our two principles, and we shall find that it differs from l.

Current kinematics tacitly assumes that the lengths determined by these two operations are precisely equal, or in other words, that a moving rigid body at the epoch t may in geometrical respects be perfectly represented by *the same* body *at rest* in a definite position.

We imagine further that at the two ends A and B of the rod, clocks are placed which synchronize with the clocks of the stationary system, that is to say that their indications correspond at any instant to the "time of the stationary system" at the places where they happen to be. These clocks are therefore "synchronous in the stationary system."

We imagine further that with each clock there is a moving observer, and that these observers apply to both clocks the criterion established in § 1 for the synchronization of two clocks. Let a ray of light depart from A at the time[4] t_A, let it be reflected at B at the time t_B, and reach A again at the time t'_A. Taking into consideration the principle of the constancy of the velocity of light we find that

$$t_B - t_A = \frac{r_{AB}}{c - v} \text{ and } t'_A - t_B = \frac{r_{AB}}{c + v}$$

where r_{AB} denotes the length of the moving rod—measured in the stationary system. Observers moving with the moving rod would thus find that the two clocks were not synchronous, while observers in the stationary system would declare the clocks to be synchronous.

So we see that we cannot attach any *absolute* signification to the concept of simultaneity, but that two events which, viewed from a system of co-ordinates, are simultaneous, can no longer be looked upon as simultaneous events when envisaged from a system which is in motion relatively to that system.

§ 3. Theory of the Transformation of Co-ordinates and Times from a Stationary System to another System in Uniform Motion of Translation Relatively to the Former

Let us in "stationary" space take two systems of co-ordinates, i.e. two systems, each of three rigid material lines, perpendicular to one another, and issuing from a point. Let the axes of X of the two systems coincide, and their axes of Y and Z respectively be parallel. Let each system be provided with a rigid measuring-rod and a number of clocks, and let the two measuring-rods, and likewise all the clocks of the two systems, be in all respects alike.

Now to the origin of one of the two systems (k) let a constant velocity v be imparted in the direction of the increasing x of the other stationary system (K), and let this velocity be communicated to the axes of the co-ordinates, the relevant measuring-rod, and the clocks. To any time of the stationary system K there then will correspond a definite position of the axes of the moving system, and from reasons of symmetry we are entitled to assume that the motion of k may be such that the axes of the moving system are at the time t (this "t" always denotes a time of the stationary system) parallel to the axes of the stationary system.

We now imagine space to be measured from the stationary system K by means of the stationary measuring-rod, and also from the moving system k by means of the measuring-rod moving with it; and that we thus obtain the co-ordinates x, y, z, and ξ, η, ζ respectively. Further, let the time t of the stationary system be determined for all points thereof at which there are clocks by means of light signals in the manner indicated in § 1; similarly let the time τ of the moving system be determined for all points of the moving system at which there are clocks at rest relatively to that system by applying the method, given in § 1, of light signals between the points at which the latter clocks are located.

To any system of values x, y, z, t, which completely defines the place and time of an event in the stationary system, there belongs a system of values ξ, η, ζ, τ, determining that event relatively to the system k, and our task is now to find the system of equations connecting these quantities.

In the first place it is clear that the equations must be *linear* on account of the properties of homogeneity which we attribute to space and time.

If we place $x' = x - vt$, it is clear that a point at rest in the system k must have a system of values x', y, z, independent of time. We first define τ as a function of x', y, z, and t. To do this we have to express in equations that τ is nothing else than the summary of the data of clocks at rest in system k, which have been synchronized according to the rule given in § 1.

From the origin of system k let a ray be emitted at the time τ_0 along the X-axis to x', and at the time τ_1 be reflected thence to the origin of the co-ordinates, arriving there at the time τ_2; we then must have

$\frac{1}{2}(\tau_0 + \tau_2) = \tau_1$, or, by inserting the arguments of the function τ and applying the principle of the constancy of the velocity of light in the stationary system:—

$$\frac{1}{2}\left[\tau(0, 0, 0, t) + \tau\left(0, 0, 0, t + \frac{x'}{c - \mathrm{v}} + \frac{x'}{c + \mathrm{v}}\right)\right] = \tau\left(x', 0, 0, t + \frac{x'}{c - \mathrm{v}}\right).$$

Hence, if x' be chosen infinitesimally small,

$$\frac{1}{2}\left(\frac{1}{c - \mathrm{v}} + \frac{1}{c + \mathrm{v}}\right)\frac{\delta\tau}{\delta t} = \frac{\delta\tau}{\delta x'} + \frac{1}{c - \mathrm{v}}\frac{\delta\tau}{\delta t}$$

or

$$\frac{\delta\tau}{\delta x'} + \frac{\mathrm{v}}{c^2 - \mathrm{v}^2}\frac{\delta\tau}{\delta t} = 0.$$

It is to be noted that instead of the origin of the co-ordinates we might have chosen any other point for the point of origin of the ray, and the equation just obtained is therefore valid for all values of x', y, z.

An analogous consideration—applied to the axes of Y and Z—it being borne in mind that light is always propagated along these axes, when viewed from the stationary system, with the velocity $\sqrt{(c^2 - \mathrm{v}^2)}$, gives us

$$\frac{\delta\tau}{\delta y} = 0, \quad \frac{\delta\tau}{\delta z} = 0.$$

Since τ is a *linear* function, it follows from these equations that

$$\tau = a\left(t - \frac{\mathrm{v}}{c^2 - \mathrm{v}^2}x'\right)$$

where a is a function $\phi(\mathrm{v})$ at present unknown, and where for brevity it is assumed that at the origin of k, $\tau = 0$, when $t = 0$.

With the help of this result we easily determine the quantities ξ, η, ζ by expressing in equations that light (as required by the principle of the constancy of the velocity of light, in combination with the principle of relativity) is also propagated with velocity c when measured in the moving system. For a ray of light emitted at the time $\tau = 0$ in the direction of the increasing ξ

$$\xi = c\tau \text{ or } \xi = ac\left(t - \frac{\mathrm{v}}{c^2 - \mathrm{v}^2}x'\right).$$

But the ray moves relatively to the initial point of k, when measured in the stationary system, with the velocity $c - \mathrm{v}$, so that

$$\frac{x'}{c - \mathrm{v}} = t.$$

If we insert this value of t in the equation for ξ, we obtain

$$\xi = a \frac{c^2}{c^2 - \mathrm{v}^2} x'.$$

In an analogous manner we find, by considering rays moving along the two other axes, that

$$\eta = c\tau = ac\left(t - \frac{\mathrm{v}}{c^2 - \mathrm{v}^2} x'\right)$$

when

$$\frac{y}{\sqrt{(c^2 - \mathrm{v}^2)}} = t, \ \ x' = 0.$$

Thus

$$\eta = a \frac{c}{\sqrt{(c^2 - \mathrm{v}^2)}} y \ \text{ and } \ \zeta = a \frac{c}{\sqrt{(c^2 - \mathrm{v}^2)}} z.$$

Substituting for x' its value, we obtain

$$\begin{aligned}
\tau &= \phi(\mathrm{v})\beta(t - \mathrm{v}x/c^2), \\
\xi &= \phi(\mathrm{v})\beta(x - \mathrm{v}t), \\
\eta &= \phi(\mathrm{v})y, \\
\zeta &= \phi(\mathrm{v})z,
\end{aligned}$$

where

$$\beta = \frac{1}{\sqrt{(1 - \mathrm{v}^2/c^2)}},$$

and ϕ is an as yet unknown function of v. If no assumption whatever be made as to the initial position of the moving system and as to the zero point of τ, an additive constant is to be placed on the right side of each of these equations.

We now have to prove that any ray of light, measured in the moving system, is propagated with the velocity c, if, as we have assumed, this is the case in the stationary system; for we have not as yet furnished the proof that the principle of the constancy of the velocity of light is compatible with the principle of relativity.

At the time $t = \tau = 0$, when the origin of the co-ordinates is common to the two systems, let a spherical wave be emitted therefrom, and be propagated with the velocity c in system K. If (x, y, z) be a point just attained by this wave, then

$$x^2 + y^2 + z^2 = c^2t^2.$$

Transforming this equation with the aid of our equations of transformation we obtain after a simple calculation

$$\xi^2 + \eta^2 + \zeta^2 = c^2\tau^2.$$

The wave under consideration is therefore no less a spherical wave with velocity of propagation c when viewed in the moving system. This shows that our two fundamental principles are compatible.[5]

In the equations of transformation which have been developed there enters an unknown function ϕ of v, which we will now determine.

For this purpose we introduce a third system of co-ordinates K′, which relatively to the system k is in a state of parallel translatory motion parallel to the axis of X, such that the origin of co-ordinates of system K′ moves with velocity – v on the axis of X. At the time $t = 0$ let all three origins coincide, and when $t = x = y = z = 0$ let the time t' of the system K′ be zero. We call the co-ordinates, measured in the system K′, x', y', z', and by a twofold application of our equations of transformation we obtain

$$\begin{aligned} t' &= \phi(-\mathrm{v})\beta(-\mathrm{v})(\tau + \mathrm{v}\xi/c^2) = \phi(\mathrm{v})\phi(-\mathrm{v})t, \\ x' &= \phi(-\mathrm{v})\beta(-\mathrm{v})(\xi + \mathrm{v}\tau) = \phi(\mathrm{v})\phi(-\mathrm{v})x, \\ y' &= \phi(-\mathrm{v})\eta = \phi(\mathrm{v})\phi(-\mathrm{v})y, \\ z' &= \phi(-\mathrm{v})\zeta = \phi(\mathrm{v})\phi(-\mathrm{v})z. \end{aligned}$$

Since the relations between x', y', z' and x, y, z do not contain the time t, the systems K and K′ are at rest with respect to one another, and it is clear that the transformation from K to K′ must be the identical transformation. Thus

$$\phi(\mathrm{v})\phi(-\mathrm{v}) = 1.$$

We now inquire into the signification of $\phi(\mathrm{v})$. We give our attention to that part of the axis of Y of system k which lies between $\xi = 0, \eta = 0, \zeta = 0$ and $\xi = 0, \eta = l, \zeta = 0$. This part of the axis of Y is a rod moving perpendicularly to its axis with velocity v relatively to system K. Its ends possess in K the co-ordinates

$$x_1 = \mathrm{v}t,\ y_1 = \frac{l}{\phi(\mathrm{v})},\ z_1 = 0$$

and

$$x_2 = \mathrm{v}t,\ y_2 = 0,\ z_2 = 0.$$

The length of the rod measured in K is therefore $l/\phi(\mathrm{v})$; and this gives us the meaning of the function $\phi(\mathrm{v})$. From reasons of symmetry it is now evident that the length of a given rod moving perpendicularly to its axis, measured in the stationary system, must depend only on the velocity and not on the direction and the sense of the motion. The length of the moving rod measured in the stationary system does not change, therefore, if v and $-\mathrm{v}$ are interchanged. Hence follows that $l/\phi(\mathrm{v}) = l/\phi(-\mathrm{v})$, or

$$\phi(\mathrm{v}) = \phi(-\mathrm{v}).$$

It follows from this relation and the one previously found that $\phi(\mathrm{v}) = 1$, so that the transformation equations which have been found become

$$\begin{aligned} \tau &= \beta(t - \mathrm{v}x/c^2), \\ \xi &= \beta(x - \mathrm{v}t), \\ \eta &= y, \\ \zeta &= z, \end{aligned}$$

where

$$\beta = 1/\sqrt{(1 - \mathrm{v}^2/c^2)}.$$

§ 4. Physical Meaning of the Equations Obtained in Respect to Moving Rigid Bodies and Moving Clocks

We envisage a rigid sphere[6] of radius R, at rest relatively to the moving system k, and with its centre at the origin of co-ordinates of k. The equation of the surface of this sphere moving relatively to the system K with velocity v is

$$\xi^2 + \eta^2 + \zeta^2 = R^2.$$

The equation of this surface expressed in x, y, z at the time $t = 0$ is

$$\frac{x^2}{1 - v^2/c^2} + y^2 + z^2 = R^2.$$

A rigid body which, measured in a state of rest, has the form of a sphere, therefore has in a state of motion—viewed from the stationary system—the form of an ellipsoid of revolution with the axes

$$R\sqrt{(1 - v^2/c^2)},\ R,\ R.$$

Thus, whereas the Y and Z dimensions of the sphere (and therefore of every rigid body of no matter what form) do not appear modified by the motion, the X dimension appears shortened in the ratio $1{:}\sqrt{(1 - v^2/c^2)}$, i.e. the greater the value of v, the greater the shortening. For $v = c$ all moving objects—viewed from the "stationary" system—shrivel up into plain figures. For velocities greater than that of light our deliberations become meaningless; we shall, however, find in what follows, that the velocity of light in our theory plays the part, physically, of an infinitely great velocity.

It is clear that the same results hold good of bodies at rest in the "stationary" system, viewed from a system in uniform motion.

Further, we imagine one of the clocks which are qualified to mark the time t when at rest relatively to the stationary system, and the time τ when at rest relatively to the moving system, to be located at the origin of the co-ordinates of k, and so adjusted that it marks the time τ. What is the rate of this clock, when viewed from the stationary system?

Between the quantities x, t, and τ, which refer to the position of the clock, we have, evidently, $x = vt$ and

$$\tau = \frac{1}{\sqrt{(1 - v^2/c^2)}}(t - vx/c^2).$$

Therefore,

$$\tau = t\sqrt{(1 - v^2/c^2)} = t - (1 - \sqrt{(1 - v^2/c^2)})t$$

whence it follows that the time marked by the clock (viewed in the stationary system) is slow by $1 - \sqrt{(1 - v^2/c^2)}$ seconds per second, or—neglecting magnitudes of fourth and higher order—by $\frac{1}{2}v^2/c^2$.

From this there ensues the following peculiar consequence. If at the points A and B of K there are stationary clocks which, viewed in the stationary system, are synchronous; and if the clock at A is moved with the velocity v along the line AB to B, then on its arrival at B the two clocks no longer synchronize, but the clock moved from A to B lags behind the other which has remained at B by $\frac{1}{2}tv^2/c^2$ (up to magnitudes of fourth and higher order), t being the time occupied in the journey from A to B.

It is at once apparent that this result still holds good if the clock moves from A to B in any polygonal line, and also when the points A and B coincide.

If we assume that the result proved for a polygonal line is also valid for a continuously curved line, we arrive at this result: If one of two synchronous clocks at A is moved in a closed curve with constant velocity until it returns to A, the journey lasting t seconds, then by the clock which has remained at rest the travelled clock on its arrival at A will be $\frac{1}{2}tv^2/c^2$ second slow. Thence we conclude that a balance-clock at the equator must go more slowly, by a very small amount, than a precisely similar clock situated at one of the poles under otherwise identical conditions. . . .

In conclusion I wish to say that in working at the problem here dealt with I have had the loyal assistance of my friend and colleague M. Besso, and that I am indebted to him for several valuable suggestions.

1. The preceding memoir by Lorentz was not at this time known to the author.
2. i.e. to the first approximation.
3. We shall not here discuss the inexactitude which lurks in the concept of simultaneity of two events at approximately the same place, which can only be removed by an abstraction.
4. "Time" here denotes "time of the stationary system" and also "position of hands of the moving clock situated at the place under discussion."
5. The equations of the Lorentz transformation may be more simply deduced directly from the condition that in virtue of those equations the relation $x^2 + y^2 + z^2 = c^2t^2$ shall have as its consequence the second relation $\xi^2 + \eta^2 + \zeta^2 = c^2\tau^2$.
6. That is, a body possessing spherical form when examined at rest.

5

THE NUCLEUS OF THE ATOM

AN OLD PHOTOGRAPH OF THE Cavendish Laboratory at Cambridge University shows a sixtyish Ernest Rutherford, thick, whitened, balding, a bulldog of a man, spectacled and mustached, a cigar growing out of his lips like a permanent part of his face. Despite his three-piece suit, Rutherford somehow looks disheveled. He folds his arms gruffly behind his back, he plants his feet wide apart. With a scowl, he stares off into space, as if pondering some thorny problem, oblivious to the well-groomed young man trying to get his attention. It is the early 1930s, a decade after Rutherford assumed directorship of the most famous laboratory for experimental physics in England, and perhaps the world. In the foreground, a ramshackle cart holds batteries and capacitors, a twisted jumble of wires slithers about in all directions. Above Rutherford's head, a hanging placard reads "Talk Softly Please." Undoubtedly, the sign is aimed at Professor Rutherford himself, whose booming voice could derail any sensitive equipment.

In his day, Rutherford was considered the greatest experimental physicist in the British Commonwealth. No scientist had so dominated the field since Michael Faraday (1791–1867), the discoverer of electromagnetic induction. Rutherford made most of the equipment in his lab with his own hands. Beyond his mechanical ability, he had an unerring intuition about how things worked.

Some of Rutherford's manual dexterity may have come from his father, a small farmer and technologist in New Zealand who also worked as a repairman. As a child, Rutherford took apart clocks and made models of the waterwheels his father built in his mills. The young Rutherford excelled in every subject in school. In 1895, at the age of twenty-four, he won a scholarship for a short-term research position at the Cavendish Laboratory. Although he accomplished some excellent work at this time, Rutherford resented what he considered the snobbish and pretentious airs of Cambridge. After being turned down for an additional fellowship at Cambridge, he accepted a post in 1898 at McGill University in Montreal. There, he became a pioneer in the new field of radioactivity, just discovered by Antoine-Henri Becquerel and Pierre and Marie Curie. Joined in his experiments by British chemist Frederick Soddy, Rutherford studied and described the nature of the mysterious radioactive emanations, measured the rate of disintegration of numerous radioactive elements, and demonstrated what the ancient alchemists had dreamed of for centuries: that in radioactive disintegration, one chemical element is transformed into another. For this work at McGill, Rutherford won the 1908 Nobel Prize in chemistry. Already, he had developed the personal style that C. P. Snow would later call "exuberant, outgoing, and not noticeably modest."

In 1907, after the holder of the chair of physics at the University of Manchester offered to resign so that Rutherford could take his seat, the thirty-six-year-old New Zealander returned to England. It was at Manchester that Rutherford did his greatest work, the discovery of the structure of the atom.

At the time that Rutherford moved to Manchester, the inside of the atom was only poorly understood. The overall masses and sizes of atoms, however, had been known fairly well since the late nineteenth century. An atom of carbon, for example, has a mass of about 2×10^{-23} grams and a radius of about 10^{-8} centimeters. In slightly more familiar terms, it requires about a hundred thousand billion billion atoms of carbon to equal the mass of a U.S. penny, and a hundred million atoms of carbon, back to back, to stretch the width of a penny. It is astounding that scientists were able to determine the sizes and masses of objects so much smaller than anything that could be seen with a microscope. Indeed, even to believe in the existence of the invisible world of the atom required faith in scientific reason.

Let us digress briefly to recount one such chain of reasoning: a determination of the mass of the atom in 1890 by the Danish physicist Ludwig Lorenz (1829–1891). Lorenz applied the theory of electricity and magnetism to calculate the zigzagging deflection of sunlight (an electromagnetic wave) as it passes through the atmosphere. The electrical charges in each molecule of air are set oscillating by the passing wave of light, and those oscillations in turn affect and deflect the light ray. (Light is also deflected, a different amount, in passing through water, leading to the disjointed appearance of a spoon half in water and half in air.) The amount of deflection depends on, among other things, the number of air molecules per cubic centimeter of air. Thus, a measurement of the deflection, plus the theory, gives the number of molecules per cubic centimeter. Weighing the mass of a cubic centimeter of air, in turn, allows an estimate of the mass of a single molecule of air. From chemical experiments, scientists already knew that air is composed of a definite proportion of nitrogen and oxygen molecules, with two atoms per molecule. Thus, with Lorenz's calculations and measurements, one could compute the mass of a single oxygen and nitrogen atom. Also from chemistry, particularly the proportions of weight in which different elements chemically combined, the *relative* masses of different atoms were known. Consequently, with Lorenz's work one could infer the masses of all atoms, from hydrogen to uranium. As it turned out, Lorenz's calculations went unnoticed for some years because he published only in Danish, while most of the well-circulated scientific literature was written in German or English. Other Danish scientists, such as Niels Bohr, were careful to publish in English.

With the knowledge of the mass of an atom, one can estimate its size, in the following way: in solid matter, atoms are packed together closely, nearly side by side. Therefore, the total volume occupied by a million atoms in solid matter is approximately equal to a million times the volume of a single atom, just as the volume of a million marbles crammed together in a box is only slightly larger than a million times the volume of a single marble. We also know that the total mass of a million atoms equals a million times the mass of a single atom. Thus, we arrive at the interesting fact that the total mass of a million atoms in solid matter *divided* by the total volume occupied by those million atoms equals the mass of a single atom divided by the volume of a single atom (the millions cancel each other out). The number of a million was used only as an example and doesn't matter in the end. In other words, the total mass of any solid block of matter divided by the total volume of that block equals the mass of a single atom of the material divided by the volume of a single atom. It is easy to measure the mass and volume of a large group of atoms in solid matter, say a brick of graphite, which is made of carbon atoms. We then know the mass-to-volume ratio of a single atom of carbon. If we know the atom's mass, from previous experiments, we can then deduce its volume. (These simple considerations don't apply to atoms in gases, which are not packed together side by side.)

Now, for the insides of the atom. In 1897, Joseph John (J. J.) Thomson, then director of the Cavendish, discovered a subatomic, electrically charged particle he called the corpuscle, now named the electron. An atom of the lightest element, hydrogen, contained one corpuscle. The heaviest atoms, such as uranium, contained ninety or more. The corpuscle is negatively charged. Since atoms are usually electrically neutral, an equal amount of positive charge must reside somewhere inside the atom to offset the negative corpuscle. Putting these facts and assumptions together, Thomson and others proposed what was called the plum pudding model of the atom: a sphere of smoothly spread positive charge, the "pudding," with the negatively charged corpuscles, the "plums," embedded within it. Like all models in science, the plum pudding model was a simplified mental picture of a real physical object. Scientists have found that mental pictures are helpful in thinking about problems, although models can also be misleading when essential features have been omitted, and they are never as precise as mathematical laws.

Since experiments in deflecting corpuscles with electric and magnetic fields had shown that each corpuscle was many thousands of times lighter than a typical atom, most of the mass of the atom was assumed to lie in the positively charged pudding. Part of that mass was relinquished in the disintegration of radium, polonium, and other radioactive atoms as they ejected pieces of themselves. In particular, one of the pieces spit out by such atoms, the alpha particle, had a mass equal to four hydrogen atoms and a positive charge equal and opposite to two corpuscles. It was Rutherford, in fact, who named the alpha particle. He also named the beta particle, a much lighter particle ejected by radioactive atoms. The beta was eventually found to be a high-speed corpuscle.

Rutherford became fascinated by the alpha particle. It was relatively massive, far more massive than a corpuscle; it was electrically charged, so that it would interact strongly with other electrically charged particles; and it went flying out of radioactive atoms at great speed. Thus, to Rutherford's mind, alpha particles would make perfect projectiles for exploring the insides of atoms. A dab of radioactive material would serve as the cannon. A thick container with a small hole in it would act as the aiming barrel of the cannon and allow passage of only the alpha cannonballs flying in a narrow direction. A sample of the atoms to be probed, the target, was placed a distance from the container in front of the incoming beam of alpha particles. Figure 5.1 below illustrates the experimental setup.

Working with the German physicist Hans Geiger (father of the Geiger counter), Rutherford designed and built several gadgets that could detect and record the impact of single alpha particles. One of these instruments, the so-called scintillation counter, consisted of a zinc sulfide screen that sparkled faintly each time it was struck by an alpha particle. Among other things, scintillation counters could be used to measure the deflections of alpha particle projectiles as they passed through thin foils of target atoms. Where an alpha landed on the screen would tell how much it had been deflected, or scattered.

According to the plum pudding model, with the mass spread out in a thin and diffuse way, an alpha particle would be far more massive and concentrated than anything it was likely to encounter in passing through an atom. In addition, the alpha was hurtling forward at high speed. Thus, its deflection, or change of direction, should be slight. Even a series of consecutive small deflections, as the alpha encountered many corpuscles in each atom and then a number of atoms, should not add up to much more than one degree. Geiger made a quantitative study of these small deflections and found that they increased with the mass of the target atom and the thickness of the foil, as expected, until the foil was so thick that no alpha particles could pass through it.

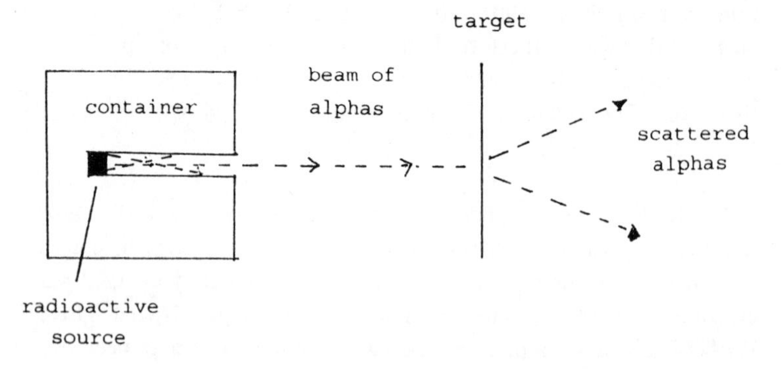

Figure 5.1

In 1909, a student named Ernest Marsden, not yet finished with his bachelor's degree, showed up at Rutherford's lab and asked the professor for an apprentice project. Rutherford gave Marsden the task of measuring *large* deflections of alpha particles passing through thin foils of gold. No such large deflections had been measured, nor were any expected on the basis of the plum pudding model, and no one knows for sure why Rutherford assigned his young student such a "damn fool experiment" (in Rutherford's words).

To everyone's surprise, Marsden and Geiger found large deflections. Indeed, some of the alpha particles bounced back in the direction they had come from, a deflection of 180 degrees. Upon hearing these results, Rutherford's legendary comment was: "It was almost as incredible as if you fired a fifteen-inch shell at a piece of tissue paper and it came back and hit you."

Evidently, there was something heavy and dense in the atom—a peach pit in the pudding. Or perhaps a peach pit and no pudding at all. Furthermore, the fraction of alpha particles that bounced back—about one in 20,000 for alphas with speed 2×10^9 centimeters per second (7 percent of the speed of light) striking a gold foil 0.00004 centimeters thick—was exactly what one would expect if the entire positive charge of each gold atom and most of its mass were concentrated at the center of the atom, in a radius of no larger than 5×10^{-12} centimeters. Such a tiny "nucleus" would be thousands of times smaller than the atom as a whole. (We now know that the actual nuclear radius is even ten times smaller than this upper limit.) As Rutherford correctly inferred from these experiments, most of the atom is empty space. The corpuscles (electrons) orbit at comparatively great distance from the concentrated central nucleus. If the atom were the size of Fenway Park in Boston, then the nucleus at its center, containing all the positive charge and more than 99.9 percent of the mass, would be the size of a pea.

Rutherford mulled over Geiger and Marsden's results for over a year before publicly proposing his new model of the atom, soon called the Rutherford model. In that paper, published in 1911, he also worked out an equation for predicting what fraction of alpha particles should be deflected at each angle.

Unlike Einstein, Rutherford does not begin his paper with lofty principles. Rather, he immediately refers to the specific experimental results of Geiger and Marsden. In particular, he notes that those results are incompatible with the exclusively small deflections expected from the prevailing plum-pudding atomic model. Instead, those findings suggest that the rebounding alphas have experienced a "single atomic encounter" that is "the seat of an intense electric field." These statements are actually summaries of what Rutherford will prove later in his paper.

Rutherford then comes to his calculations. (Surprisingly, for the work of a scientist known mainly as an experimenter, this paper brims with theoretical calculations.) First, Rutherford assumes that the atom contains N corpuscles, each of charge $-e$. He then hypothesizes that the amount of positive charge needed to make the atom electrically neutral, Ne, is concentrated *at the center of the atom.* By assuming this charge to remain fixed and motionless after encounters with the incoming alpha projectiles, he implicitly proposes that all of the atom's mass resides with the concentrated positive charge. The massive central nucleus acts like a Mack truck, which can be struck by a careening Volkswagen (an alpha particle) and barely budge.

The corpuscles are approximated by a total negative charge of $-Ne$ distributed uniformly throughout the entire atom. For values of $N = 100$, roughly correct for gold atoms, and for appropriate masses and velocities of alpha particles, Rutherford then shows that the corpuscular charge outside the center may be neglected for large-angle scatterings of the alphas. In other words, for large deflections, an alpha must pass very close to the center of the atom, where the electric force is the strongest, and the concentrated *positive* charge there is all that matters.

With these assumptions, the physicist from New Zealand calculates the fraction of alpha particles that should be deflected through each range of angle ϕ. (No deflection at all corresponds to $\phi = 0$.) The picture is the following: as an incoming alpha particle hurtles past the nucleus of an atom, it experiences an electrical force from the electrical charge of the nucleus. That force deflects the trajectory of the alpha, much as gravity deflects a pitched baseball. The closer the alpha comes to the concentrated nucleus, the stronger the electrical force and the greater the deflection.

In calculating the expected deflections, Rutherford does not propose any new laws of physics, as did Planck and Einstein. Rather, he manipulates the known laws of mechanics and electricity. Those laws include the constancy of angular momentum and energy of the alpha, and the hyperbolic shape of the trajectory of the alpha—all following from Rutherford's assumptions and the known law for the electrical force between a particle of charge Ne (the central nucleus) and a particle of charge E (the alpha). In qualitative terms, that force increases as the magnitudes of the two charges increases and increases as the distance between them decreases. The closer the two charges, the greater the electrical force between them. What Rutherford is doing is applying known laws to a new situation. This method of working is, in fact, extremely common among scientists, even among theoretical scientists. Many physicists, with any advanced training, could have performed the calculation that Rutherford does in this paper. Rutherford's genius was in conceiving the situation that demanded that calculation.

The final result, equation (5), gives the fraction of scattered alphas that should land on each unit area of a scintillation screen placed a distance r from the target and at an angle ϕ from the initial direction of the alphas. According to that equation, the fraction is proportional to the density of gold atoms in the target, denoted by n, and the thickness of the gold foil, denoted by t. Through the parameter b, the fraction is also proportional to the square of the central charge, $(Ne)^2$, and inversely proportional to the fourth power of the alpha particle velocity, $1/u^4$. Finally, the fraction is proportional to the fourth power of the (trigonometric) cosecant of $\phi/2$, a very specific dependence on angle. For example, the number of alphas scattered at an angle 180 degrees (i.e., straight back) should be about one half of one percent of the number of alphas scattered at an angle of 30 degrees. This last detailed prediction was tested later in further experiments by Geiger and Marsden and confirmed.

In the next section of his paper, "Comparison of Single and Compound Scattering," Rutherford rules out the possibility that the large deflections observed by Geiger and Marsden could have been the result of an accumulation (compounding) of many small-angle deflections. This important calculation demolishes the plum pudding model. The essence of the argument, which also does not go beyond previously known physics and mathematics, is that the probability for an alpha particle to deflect via compounded small scatterings at an angle greater than a few degrees is exponentially small.

Because Geiger and Marsden have not yet measured in detail the fraction of alphas scattered at each angle, Rutherford cannot compare his theoretical equation (5) to their experimental data. However, he can compare it to their results for how the rate of scattering of alphas increases with increasing mass of the target atoms. Those results agree approximately with his theory, if he assumes that the central charge, Ne, is proportional to the atomic weight, A, an assumption not quite true. Rutherford writes: "considering the difficulty of the experiment, the agreement between theory and experiment is reasonably good." As Rutherford is an experimenter himself, he knows firsthand that a theory never agrees completely with experiment, even if the theory is correct. There are usually dozens of sources of experimental error. The lab table might be vibrating with a passing train, the intended vacuum might be spoiled by the faint evaporation of atoms from the container, and so on. A good experimenter tries to identify, understand, and take into account these sources of error.

Rutherford goes on to apply his scattering formula to determine the value of N (number of corpuscles) of gold from observed scattering results at small angles. He obtains a value $N = 97$, in fair agreement with the correct value of 79. One gets the same estimate for N by considering the less frequent scattering at large angles—a significant triumph of the theory, since the large-angle scatterings are produced exclusively by the central charge, while the small-angle scatterings derive partly from the aggregation of the many corpuscles throughout the atom. In this section, Rutherford is systematically testing as many aspects of his formula as he can.

Here and elsewhere, Rutherford's style of thought, like his personality, has a bit of the rough-and-ready character. Aside from the exact and precise calculation leading to equation (5), Rutherford is constantly pulling in experimental results, using approximate numbers, making estimations, comparing one effect against another. If Einstein mainly followed his mind, Rutherford is following his nose. His instinct, from the beginning, led him to consider the alpha particle as a tool to probe the atom. His engineering talent allowed him to construct equipment for detecting alpha particles and making measurements with those particles. His instinct again suggested that he ask Geiger and Marsden to look for *large* deflections of alphas when none were expected. Now, he is convincing himself, and us, of the validity of his revised picture of the atom.

In "General Considerations," Rutherford concludes with a final argument and a suggestion. To play devil's advocate, he asks whether one could imagine something intermediate between his nuclear model and the plum pudding idea—namely that the positively charged mass of the atom might consist of individual particles, like the corpuscles, but lie scattered throughout the entire atom. The answer is no. So many such particles would be needed to cause the observed large deflections that the mass of each particle would be too small to significantly deflect the incoming alphas—in another words, a contradiction.

Rutherford's final suggestion regards the mysterious point of origin of alpha particles. He proposes, in a matter-of-fact tone, that the alphas are expelled from the *nuclei* of radioactive atoms, a suggestion that makes perfect sense in his new model. Furthermore, he points out, the huge repulsive force between a positively charged alpha particle and a positively charged atomic nucleus would naturally explain the high velocities at which alphas are hurled out of radioactive atoms. (Like electrical charges repel each other; positive and negative charge attract.) One can think of the repulsive electrical forces between the positive charges in the atom as compressed springs, poised to jump back. Because the positive charges are crammed so close together in the tiny nucleus, their "springs" are extremely compressed and thus store a great deal of energy. That huge energy of electrical repulsion, as it turns out, is what powers an atomic bomb. (Not nearly so much energy would be stored in a plum pudding form of the atom.)

Rutherford doubted that the enormous energy stored within the nucleus of the atom would ever be harnessed. Others were not so pessimistic. A science fiction writer by the name of H. G. Wells, who paid close attention to the findings of such men as Rutherford, wrote a book called *The World Set Free*. In that novel, published three years after Rutherford's paper, Wells describes a world war in the 1950s in which each of the world's cities is destroyed by a few "atomic bombs" the size of beach balls.

Initially, most of the scientific community ignored Rutherford's new model of the atom. But the Danish theoretical physicist Niels Bohr, who met Rutherford in 1911, was much impressed by the "nuclear" conception of the atom, and it served as the starting point of his quantum model. (As we will see in a later chapter, the very first sentence of Bohr's landmark 1913 paper refers to Rutherford's atom.) While Rutherford had mapped out the basic geography of the atom, Bohr's quantum model showed where the electrons were allowed in that geography as they orbited about Rutherford's nucleus. The locations and energies of the electrons, in turn, determine the manner in which atoms and molecules interact with other atoms, the subject of chemistry. Indeed, it can be argued that all of modern chemistry depends on the Rutherford model of the atom, with negatively charged electrons orbiting a positively charged nucleus, plus the quantum restrictions of Bohr and others.

Some scientists, like Einstein, are loners and do not often collaborate or train students. Others relish having young people around them, relish the master-apprentice relationship, the give-and-take of daily debate. Rutherford was in the second group. A large stable of budding young scientists worked under his guidance—such people as James Chadwick, who discovered the neutron; Peter Kapitza, who pioneered work in low-temperature physics and intense magnetic fields; John Cockcroft, who first split the atomic nucleus by artificial means; Patrick Blackett, who developed the cloud chamber method for detecting subatomic particles—all eventual Nobel Prize winners themselves. Rutherford used to refer to his young apprentices as "my boys." They, in turn, both worshiped and dreaded the bulldog with the loud bark and the infallible nose. In 1921, soon after arriving in Rutherford's lab, young Kapitza wrote to his mother in Petrograd (St. Petersburg): "The Professor is a deceptive character . . . He is a man of immense temperament. He is given to uncontrollable excitement. His moods fluctuate violently. It will need great vigilance if I am going to obtain, and keep, his high opinion."

THE SCATTERING OF α AND β PARTICLES BY MATTER AND THE STRUCTURE OF THE ATOM

Professor Ernest Rutherford, F.R.S.,
University of Manchester[1]

London, Edinburgh and Dublin Philosophical Magazine and Journal of Science, 1911

§ 1.

IT IS WELL KNOWN THAT the α and β particles suffer deflexions from their rectilinear paths by encounters with atoms of matter. This scattering is far more marked for the β than for the α particle on account of the much smaller momentum and energy of the former particle. There seems to be no doubt that such swiftly moving particles pass through the atoms in their path, and that the deflexions observed are due to the strong electric field traversed within the atomic system. It has generally been supposed that the scattering of a pencil of α or β rays in passing through a thin plate of matter is the result of a multitude of small scatterings by the atoms of matter traversed. The observations, however, of Geiger and Marsden[2] on the scattering of α rays indicate that some of the α particles must suffer a deflexion of more than a right angle at a single encounter. They found, for example, that a small fraction of the incident α particles, about 1 in 20,000, were turned through an average angle of 90° in passing through a layer of gold-foil about .00004 cm. thick, which was equivalent in stopping-power of the α particle to 1.6 millimetres of air. Geiger[3] showed later that the most probable angle of deflexion for a pencil of α particles traversing a gold-foil of this thickness was about 0°.87. A simple calculation based on the theory of probability shows that the chance of an α particle being deflected through 90° is vanishingly small. In addition, it will be seen later that the distribution of the α particles for various angles of large deflexion does not follow the probability law to be expected if such large deflexions are made up of a large number of small deviations. It seems reasonable to suppose that the deflexion through a large angle is due to a single atomic encounter, for the chance of a second encounter of a kind to produce a large deflexion must in most cases be exceedingly small. A simple calculation shows that the atom must be a seat of an intense electric field in order to produce such a large deflexion at a single encounter.

Recently Sir J. J. Thomson[4] has put forward a theory to explain the scattering of electrified particles in passing through small thicknesses of matter. The atom is supposed to consist of a number N of negatively charged corpuscles, accompanied by an equal quantity of positive electricity uniformly distributed throughout a sphere. The deflexion of a negatively electrified particle in passing through the atom is ascribed to two causes—(1) the repulsion of the corpuscles distributed through the atom, and (2) the attraction of the positive electricity in the atom. The deflexion of the particle in passing through the atom is supposed to be small, while the average deflexion after a large number m of encounters was taken as $\sqrt{m}\,\theta$, where θ is the average deflexion due to a single atom. It was shown that the number N of the electrons within the atom could be deduced from observations of the scattering of electrified particles. The accuracy of this theory of compound scattering was examined experimentally by Crowther[5] in a later paper. His results apparently confirmed the main conclusions of the theory, and he deduced, on the assumption that the positive electricity was continuous, that the number of electrons in an atom was about three times its atomic weight.

The theory of Sir J. J. Thomson is based on the assumption that the scattering due to a single atomic encounter is small, and the particular structure assumed for the atom does not admit of a very large deflexion of an α particle in traversing a single atom, unless it be supposed that the diameter of the sphere of positive electricity is minute compared with the diameter of the sphere of influence of the atom.

Since the α and β particles traverse the atom, it should be possible from a close study of the nature of the deflexion to form some idea of the constitution of the atom to produce the effects observed. In fact, the scattering of high-speed charged particles by the atoms of matter is one of the most promising methods of attack of this problem. The development of the scintillation method of counting single α particles affords unusual advantages of investigation, and the researches of H. Geiger by this method have already added much to our knowledge of the scattering of α rays by matter.

§ 2.

We shall first examine theoretically the single encounters[6] with an atom of simple structure, which is able to produce large deflexions of an α

particle, and then compare the deductions from the theory with the experimental data available.

Consider an atom which contains a charge $\pm Ne$ at its centre surrounded by a sphere of electrification containing a charge $\mp Ne$ supposed uniformly distributed throughout a sphere of radius R. e is the fundamental unit of charge, which in this paper is taken as 4.65×10^{-10} E.S. unit. We shall suppose that for distances less than 10^{-12} cm. the central charge and also the charge on the α particle may be supposed to be concentrated at a point. It will be shown that the main deductions from the theory are independent of whether the central charge is supposed to be positive or negative. For convenience, the sign will be assumed to be positive. The question of the stability of the atom proposed need not be considered at this stage, for this will obviously depend upon the minute structure of the atom, and on the motion of the constituent charged parts.

In order to form some idea of the forces required to deflect an α particle through a large angle, consider an atom containing a positive charge Ne at its centre, and surrounded by a distribution of negative electricity Ne uniformly distributed within a sphere of radius R. The electric force X and the potential V at a distance r from the centre of an atom for a point inside the atom, are given by

$$X = Ne\left(\frac{1}{r^2} - \frac{r}{R^3}\right)$$

$$V = Ne\left(\frac{1}{r} - \frac{3}{2R} + \frac{r^2}{2R^3}\right).$$

Suppose an α particle of mass m and velocity u and charge E shot directly towards the centre of the atom. It will be brought to rest at a distance b from the centre given by

$$\tfrac{1}{2}mu^2 = NeE\left(\frac{1}{b} - \frac{3}{2R} + \frac{b^2}{2R^3}\right).$$

It will be seen that b is an important quantity in later calculations. Assuming that the central charge is $100e$, it can be calculated that the value of b for an α particle of velocity 2.09×10^9 cms. per second is about 3.4×10^{-12} cm. In this calculation b is supposed to be very small compared with R. Since R is supposed to be of the order of the radius of the atom, viz. 10^{-8} cm., it is obvious that the α particle before being turned back penetrates so close to the central charge, that the field due to the uniform distribution of negative electricity may be neglected. In general, a simple calculation shows that for all deflexions greater than a degree, we may without sensible error suppose the deflexion due to the field of the central charge alone. Possible single deviations due to the negative electricity, if distributed in the form of corpuscles, are not taken into account at this stage of the theory. It will be shown later that its effect is in general small compared with that due to the central field.

Consider the passage of a positive electrified particle close to the centre of an atom. Supposing that the velocity of the particle is not appreciably changed by its passage through the atom, the path of the particle under the influence of a repulsive force varying inversely as the square of the distance will be an hyperbola with the centre of the atom S as the external focus. Suppose the particle to enter the atom in the direction PO (Fig. 1), and that the direction of motion on escaping the atom is OP′. OP and OP′ make equal angles with the line SA, where A is the apse of the hyperbola. $p = \mathrm{SN}$ = perpendicular distance from centre on direction of initial motion of particle.

Let angle POA = θ.

Let V = velocity of particle on entering the atom, v its velocity at A, then from consideration of angular momentum

$$p\mathrm{V} = \mathrm{SA} \,.\, r.$$

From conservation of energy

$$\tfrac{1}{2}m\mathrm{V}^2 = \tfrac{1}{2}m\mathrm{v}^2 + \frac{\mathrm{N}e\mathrm{E}}{\mathrm{SA}},$$

$$\mathrm{v}^2 = \mathrm{V}^2\left(1 - \frac{b}{\mathrm{SA}}\right).$$

Since the eccentricity is sec θ,

$$\begin{aligned} \mathrm{SA} &= \mathrm{SO} + \mathrm{OA} = p \operatorname{cosec} \theta(1 + \cos \theta) \\ &= p \cot \theta/2, \\ p^2 &= \mathrm{SA}(\mathrm{SA} - b) = p \cot \theta/2(p \cot \theta/2 - b), \\ \therefore\ b &= 2p \cot \theta. \end{aligned}$$

The angle of deviation ϕ of the particle is $\pi - 2\theta$ and

$$\cot \phi/2 = \frac{2p}{b}. \qquad (1)$$ [7]

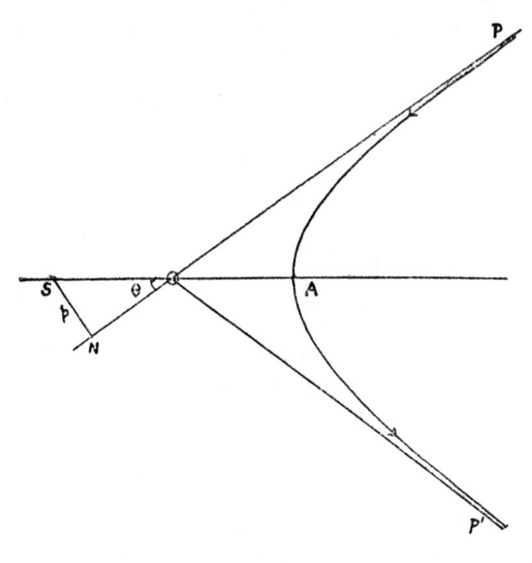

Figure 1

This gives the angle of deviation of the particle in terms of b, and the perpendicular distance of the direction of projection from the centre of the atom.

For illustration, the angle of deviation ϕ for different values of p/b are shown in the following table:—

p/b	10	5	2	1	.5	.25	.125
ϕ	5°.7	11°.4	28°	53°	90°	127°	152°

§ 3. PROBABILITY OF SINGLE DEFLEXION THROUGH ANY ANGLE.

Suppose a pencil of electrified particles to fall normally on a thin screen of matter of thickness t. With the exception of the few particles which are scattered through a large angle, the particles are supposed to pass nearly normally through the plate with only a small change of velocity. Let n = number of atoms in unit volume of material. Then the number of collisions of the particle with the atom of radius R is $\pi R^2 nt$ in the thickness t.

The probability m of entering an atom within a distance p of its centre is given by

$$m = \pi p^2 nt.$$

Chance dm of striking within radii p and $p + dp$ is given by

$$dm = 2\pi pnt.\, dp = \frac{\pi}{4}\, ntb^2 \cot \phi/2 \operatorname{cosec}^2 \phi/2\, d\phi, \tag{2}$$

since

$$\cot \phi/2 = 2p/b.$$

The value of dm gives the *fraction* of the total number of particles which are deviated between the angles ϕ and $\phi + d\phi$.

The fraction ρ of the total number of particles which are deflected through an angle greater than ϕ is given by

$$\rho = \frac{\pi}{4}\, ntb^2 \cot^2 \phi/2. \tag{3}$$

The fraction ρ which is deflected between the angles ϕ_1 and ϕ_2 is given by

$$\rho = \frac{\pi}{4}\, ntb^2 \left(\cot^2 \frac{\phi_1}{2} - \cot^2 \frac{\phi_2}{2}\right). \tag{4}$$

It is convenient to express the equation (2) in another form for comparison with experiment. In the case of the α rays, the number of scintillations appearing on a *constant* area of a zinc sulphide screen are counted for different angles with the direction of incidence of the particles. Let r = distance from point of incidence of α rays on scattering material, then if Q be the total number of particles falling on the scattering material, the number y of α particles falling on unit area which are deflected through an angle ϕ is given by

$$y = \frac{Qdm}{2\pi r^2 \sin \phi.\, d\phi} = \frac{ntb^2.Q. \operatorname{cosec}^4 \phi/2}{16r^2}. \tag{5}$$

Since

$$b = \frac{2NeE}{mu^2},$$

we see from this equation that the number of α particles (scintillations) per unit area of zinc sulphide screen at a given distance r from the point of incidence of the rays is proportional to

(1) $\operatorname{cosec}^4 \phi/2$ or $1/\phi^4$ if ϕ be small;
(2) thickness of scattering material t provided this is small;
(3) magnitude of central charge Ne;
(4) and is inversely proportional to $(mu^2)^2$, or to the fourth power of the velocity if m be constant.

In these calculations, it is assumed that the α particles scattered through a large angle suffer only one large deflexion. For this to hold, it is essential that the thickness of the scattering material should be so small that the chance of a second encounter involving another large deflexion is very small. If, for example, the probability of a single deflexion ϕ in passing through a thickness t is 1/1000, the probability of two successive deflexions each of value ϕ is $1/10^6$, and is negligibly small.

The angular distribution of the α particles scattered from a thin metal sheet affords one of the simplest methods of testing the general correctness of this theory of single scattering. This has been done recently for α rays by Dr. Geiger,[8] who found that the distribution for particles deflected between 30° and 150° from a thin gold-foil was in substantial agreement with the theory. A more detailed account of these and other experiments to test the validity of the theory will be published later. . . .

§ 5. COMPARISON OF SINGLE AND COMPOUND SCATTERING.

Before comparing the results of theory with experiment, it is desirable to consider the relative importance of single and compound scattering in determining the distribution of the scattered particles. Since the atom is supposed to consist of a central charge surrounded by a uniform distribution of the opposite sign through a sphere of radius R, the chance of encounters with the atom involving small deflexions is very great compared with the chance of a single large deflexion.

This question of compound scattering has been examined by Sir J. J. Thomson in the paper previously discussed (§ 1). In the notation of this paper, the average deflexion ϕ_1 due to the field of the sphere of positive electricity of radius R and quantity Ne was found by him to be

$$\phi_1 = \frac{\pi}{4} \cdot \frac{NeE}{mu^2} \cdot \frac{1}{R}.$$

The average deflexion ϕ_2 due to the N negative corpuscles supposed distributed uniformly throughout the sphere was found to be

$$\phi_2 = \frac{16}{5} \frac{eE}{mu^2} \cdot \frac{1}{R} \sqrt{\frac{3N}{2}}.$$

The mean deflexion due to both positive and negative electricity was taken as

$$(\phi_1^2 + \phi_2^2)^{1/2}.$$

In a similar way, it is not difficult to calculate the average deflexion due to the atom with a central charge discussed in this paper.

Since the radial electric field X at any distance r from the centre is given by

$$X = Ne\left(\frac{1}{r^2} - \frac{r}{R^3}\right),$$

it is not difficult to show that the deflexion (supposed small) of an electrified particle due to this field is given by

$$\theta = \frac{b}{p}\left(1 - \frac{p^2}{R^2}\right)^{3/2},$$

where p is the perpendicular from the centre on the path of the particle and b has the same value as before. It is seen that the value of θ increases with diminution of p and becomes great for small values of ϕ.

Since we have already seen that the deflexions become very large for a particle passing near the centre of the atom, it is obviously not correct to find the average value by assuming θ is small.

Taking R of the order 10^{-8} cm., the value of p for a large deflexion is for α and β particles of the order 10^{-11} cm. Since the chance of an encounter involving a large deflexion is small compared with the chance of small deflexions, a simple consideration shows that the average small deflexion is practically unaltered if the large deflexions are omitted. This is equivalent to integrating over that part of the cross section of the atom where the deflexions are small and neglecting the small central area. It can in this way be simply shown that the average small deflexion is given by

$$\phi_1 = \frac{3\pi}{8} \frac{b}{R}.$$

This value of ϕ_1 for the atom with a concentrated central charge is three times the magnitude of the average deflexion for the same value of Ne in the type of atom examined by Sir J. J. Thomson. Combining the deflexions due to the electric field and to the corpuscles, the average deflexion is

$$(\phi_1^2 + \phi_2^2)^2 \text{ or } \frac{b}{2R}\left(5.54 + \frac{15.4}{N}\right)^{1/2}.$$

It will be seen later that the value of N is nearly proportional to the atomic weight, and is about 100 for gold. The effect due to scattering of the individual corpuscles expressed by the second term of the equation is consequently small for heavy atoms compared with that due to the distributed electric field.

Neglecting the second term, the average deflexion per atom is

$$\frac{3\pi b}{8R}.$$

We are now in a position to consider the relative effects on the distribution of particles due to single and to compound scattering. Following J. J. Thomson's argument, the average deflexion θ_t after passing through a thickness t of matter is proportional to the square root of the number of encounters and is given by

$$\theta_t = \frac{3\pi b}{8R}\sqrt{\pi R^2 . n. t} = \frac{3\pi b}{8}\sqrt{\pi n t},$$

where n as before is equal to the number of atoms per unit volume.

The probability p_1 for compound scattering that the deflexion of the particle is greater than ϕ is equal to $e^{-\phi^2/\theta_t^2}$.
Consequently

$$\phi^2 = -\frac{9\pi^3}{64} b^3\, nt \log p_1.$$

Next suppose that single scattering alone is operative. We have seen (§ 3) that the probability p_2 of a deflexion greater than ϕ is given by

$$p_2 = \frac{\pi}{4} b^2 . n. t \cot^2\phi/2.$$

By comparing these two equations

$$p_2 \log p_1 = -.181\phi^2 \cot^2\phi/2,$$

ϕ is sufficiently small that

$$\tan \phi/2 = \phi/2,$$
$$p_2 \log p_1 = -.72.$$

If we suppose

$$p_2 = .5, \text{ then } p_1 = .24.$$

If

$$p_2 = .1, p_1 = .0004.$$

It is evident from this comparison, that the probability for any given deflexion is always greater for single than for compound scattering. The difference is especially marked when only a small fraction of the particles are scattered through any given angle. It follows from this result that the distribution of particles due to encounters with the atoms is for small thicknesses mainly governed by single scattering. No doubt compound scattering produces some effect in equalizing the distribution of the scattered particles; but its effect becomes relatively smaller, the smaller the fraction of the particles scattered through a given angle.

§ 6. COMPARISON OF THEORY WITH EXPERIMENTS.

On the present theory, the value of the central charge N*e* is an important constant, and it is desirable to determine its value for different atoms. This can be most simply done by determining the small fraction of α or β particles of known velocity falling on a thin metal screen, which are scattered between ϕ and $\phi + d\phi$ where ϕ is the angle of deflexion. The influence of compound scattering should be small when this fraction is small.

Experiments in these directions are in progress, but it is desirable at this stage to discuss in the light of the present theory the data already published on scattering of α and β particles.

The following points will be discussed:—

(*a*) The "diffuse reflexion" of α particles, *i.e.* the scattering of α particles through large angles (Geiger and Marsden).

(*b*) The variation of diffuse reflexion with atomic weight of the radiator (Geiger and Marsden).
(*c*) The average scattering of a pencil of α rays transmitted through a thin metal plate (Geiger).
(*d*) The experiments of Crowther on the scattering of β rays of different velocities by various metals.

(*a*) In the paper of Geiger and Marsden (*loc. cit.*) on the diffuse reflexion of α particles falling on various substances it was shown that about 1/8000 of the α particles from radium C falling on a thick plate of platinum are scattered back in the direction of the incidence. This fraction is deduced on the assumption that the α particles are uniformly scattered in all directions, the observations being made for a deflexion of about 90°. The form of experiment is not very suited for accurate calculation, but from the data available it can be shown that the scattering observed is about that to be expected on the theory if the atom of platinum has a central charge of about 100 *e*.

(*b*) In their experiments on this subject, Geiger and Marsden gave the relative number of α particles diffusely reflected from thick layers of different metals, under similar conditions. The numbers obtained by them are given in the table below, where *z* represents the relative number of scattered particles, measured by the number of scintillations per minute on a zinc sulphide screen.

Metal.	Atomic weight.	*z*.	$z/A^{3/2}$.
Lead.............	207	62	208
Gold.............	197	67	212
Platinum	195	63	232
Tin	119	34	226
Silver	108	27	241
Copper..........	64	14.5	225
Iron	56	10.2	250
Aluminum	27	3.4	243
			Average 233

On the theory of single scattering, the fraction of the total number of α particles scattered through any given angle in passing through a thickness t is proportional to $n.A^2t$, assuming that the central charge is proportional to the atomic weight A. In the present case, the thickness of matter from which the scattered α particles are able to emerge and affect the zinc sulphide screen depends on the metal. Since Bragg has shown that the stopping power of an atom for an α particle is proportional to the square root of its atomic weight, the value of *nt* for different elements is proportional to $1/\sqrt{A}$. In this case *t* represents the greatest depth from which the scattered α particles emerge. The number *z* of α particles scattered back from a thick layer is consequently proportional to $A^{3/2}$ or $z/A^{3/2}$ should be a constant.

To compare this deduction with experiment, the relative values of the latter quotient are given in the last column. Considering the difficulty of the experiments, the agreement between theory and experiment is reasonably good.

The single large scattering of α particles will obviously affect to some extent the shape of the Bragg ionization curve for a pencil of α rays. This effect of large scattering should be marked when the α rays have traversed screens of metals of high atomic weight, but should be small for atoms of light atomic weight.

(*c*) Geiger made a careful determination of the scattering of α particles passing through thin metal foils, by the scintillation method, and deduced the most probable angle through which the α particles are deflected in passing through known thicknesses of different kinds of matter.

A narrow pencil of homogeneous α rays was used as a source. After passing through the scattering foil, the total number of α particles deflected through different angles was directly measured. The angle for which the number of scattered particles was a maximum was taken as the most probable angle. The variation of the most probable angle with thickness of matter was determined, but calculation from these data is somewhat complicated by the variation of velocity of the α particles in their passage through the scattering material. A consideration of the curve of distribution of the α particles given in the paper (*loc. cit.* p. 496) shows that the angle through which half the particles are scattered is about 20 per cent greater than the most probable angle.

We have already seen that compound scattering may become important when about half the particles are scattered through a given angle, and it is difficult to disentangle in such cases the relative effects due to the two kinds of scattering. An approximate estimate can be made in the following way:—From (§ 5) the relation between the probabilities p_1 and p_2 for compound and single scattering respectively is given by

$$p_2 \log p_1 = -.721.$$

The probability q of the combined effects may as a first approximation be taken as

$$q = (p_1^2 + p_2^2)^{1/2}.$$

If $q = .5$, it follows that

$$p_1 = .2 \text{ and } p_2 = .46.$$

We have seen that the probability p_2 of a single deflexion greater than ϕ is given by

$$p_2 = \frac{\pi}{4}\, n.\, t.\, b^2 \cot^2 \phi/2.$$

Since in the experiments considered ϕ is comparatively small

$$\frac{\phi \sqrt{p_2}}{\sqrt{\pi n t}} = b = \frac{2\text{N}e\text{E}}{mu^2}.$$

Geiger found that the most probable angle of scattering of the α rays in passing through a thickness of gold equivalent in stopping power to about .76 cm. of air was 1° 40′. The angle ϕ through which half the α particles are turned thus corresponds to 2° nearly.

$$t = .00017 \text{ cm.};\; n = 6.07 \times 10^{22};$$
$$u \text{ (average value)} = 1.8 \times 10^9.$$
$$\text{E}/m = 1.5 \times 10^{14}. \text{ E.S. units};\; e = 4.65 \times 10^{-10}.$$

Taking the probability of single scattering = .46 and substituting the above values in the formula, the value of N for gold comes out to be 97.

For a thickness of gold equivalent in stopping power to 2.12 cms. of air, Geiger found the most probable angle to be 3° 40′. In this case $t = .00047$, $\phi = 4°.4$, and average $u = 1.7 \times 10^9$, and N comes out to be 114.

Geiger showed that the most probable angle of deflexion for an atom was nearly proportional to its atomic weight. It consequently follows that the value of N for different atoms should be nearly proportional to their atomic weights, at any rate for atomic weights between gold and aluminum.

Since the atomic weight of platinum is nearly equal to that of gold, it follows from these considerations that the magnitude of the diffuse reflexion of α particles through more than 90° from gold and the magnitude of the average small angle scattering of a pencil of rays in passing through gold-foil are both explained on the hypothesis of single scattering by supposing the atom of gold has a central charge of about 100 *e*. . . .

§ 7. GENERAL CONSIDERATIONS.

In comparing the theory outlined in this paper with the experimental results, it has been supposed that the atom consists of a central charge supposed concentrated at a point, and that the large single deflexions of the α and β particles are mainly due to their passage through the strong central field. The effect of the equal and opposite compensating charge supposed distributed uniformly throughout a sphere has been neglected. Some of the evidence in support of these assumptions will now be briefly considered. For concreteness, consider the passage of a high speed α particle through an atom having a positive central charge N*e*, and surrounded by a compensating charge of N electrons. Remembering that the mass, momentum, and kinetic energy of the α particle are very large compared with the corresponding values for an electron in rapid motion, it does not seem possible from dynamic considerations that an α particle can be deflected through a large angle by a close approach to an electron, even if the latter be in rapid motion and constrained by strong electrical forces. It seems reasonable to suppose that the chance of single deflexions through a large angle due to this cause, if not zero, must be exceedingly small compared with that due to the central charge.

It is of interest to examine how far the experimental evidence throws light on the question of the extent of the distribution of the central charge. Suppose, for example, the central charge to be composed of N unit charges distributed over such a volume that the large single deflexions are mainly due to the constituent charges and not to the external field produced by the distribution. It has been shown (§ 3) that the fraction of the α particles scattered through a large angle is proportional to $(\mathrm{N}e\mathrm{E})^2$, where N*e* is the central charge concentrated at a point and E the charge on the deflected particle. If, however, this charge is distributed in single units, the fraction of the α particles scattered through a given angle is proportional to $\mathrm{N}e^2$ instead of N^2e^2. In this calculation, the influence of mass of the constituent particle has been neglected, and account has only been taken of its electric field. Since it has been shown that the value of the central point charge for gold must be about 100, the value of the distributed charge required to produce the same proportion of single deflexions through a large angle should be at least 10,000. Under these conditions the mass of the constituent particle would be small compared with that of the α particle, and the difficulty arises of the production of large single deflexions at all. In addition, with such a large distributed charge, the effect of compound scattering is relatively more important than that of single scattering. For example, the probable small angle of deflexion of a pencil of α particles passing through a thin gold foil would be much greater than that experimentally observed by Geiger (§ *b–c*). The large and small angle scattering could not then be explained by the assumption of a central charge of the same value. Considering the evidence as a whole, it seems simplest to suppose that the atom contains a central charge distributed through a very small volume, and that the large single deflexions are due to the central charge as a whole, and not to its constituents. At the same time, the experimental evidence is not precise enough to negate the possibility that a small fraction of the positive charge may be carried by satellites extending some distance from the centre. Evidence on this point could be obtained by examining whether the same central charge is required to explain the large single deflexions of α and β particles; for the α particle must approach much closer to the centre of the atom than the β particle of average speed to suffer the same large deflexion.

The general data available indicate that the value of this central charge for different atoms is approximately proportional to their atomic weights, at any rate for atoms heavier than aluminum. It will be of great interest to examine experimentally whether such a simple relation holds also for the lighter atoms. In cases where the mass of the deflecting atom (for example, hydrogen, helium, lithium) is not very different from that of the α particle, the general theory of single scattering will require modification, for it is necessary to take into account the movements of the atom itself (see § 4).

It is of interest to note that Nagaoka[9] has mathematically considered the properties of a "Saturnian" atom which he supposed to consist of a central attracting mass surrounded by rings of rotating electrons. He showed that such a system was stable if the attractive force was large. From the point of view considered in this paper, the chance of large deflexion would practically be unaltered, whether the atom is considered to be a disk or a sphere. It may be remarked that the approximate value found for the central charge of the atom of gold (100 e) is about that to be expected if the atom of gold consisted of 49 atoms of helium, each carrying a charge 2 e. This may be only a coincidence, but it is certainly suggestive in view of the expulsion of helium atoms carrying two unit charges from radioactive matter.

The deductions from the theory so far considered are independent of the sign of the central charge, and it has not so far been found possible to obtain definite evidence to determine whether it be positive or negative. It may be possible to settle the question of sign by consideration of the difference of the laws of absorption of the β particle to be expected on the two hypotheses, for the effect of radiation in reducing the velocity of the β particle should be far more marked with a positive than with a negative centre. If the central charge be positive, it is easily seen that a positively charged mass, if released from the centre of a heavy atom, would acquire a great velocity in moving through the electric field. It may be possible in this way to account for the high velocity of expulsion of α particles without supposing that they are initially in rapid motion within the atom.

Further consideration of the application of this theory to these and other questions will be reserved for a later paper, when the main deductions of the theory have been tested experimentally. Experiments in this direction are already in progress by Geiger and Marsden.

University of Manchester,
April 1911.

1. Communicated by the Author. A brief account of this paper was communicated to the Manchester Literary and Philosophical Society in February, 1911.
2. Proc. Roy. Soc. lxxxii. p. 495 (1909).
3. Proc. Roy. Soc. lxxxiii. p. 402 (1910).
4. Camb. Lit. & Phil. Soc. xv. pt. 5 (1910).
5. Crowther, Proc. Roy. Soc. lxxxiv. p. 226 (1910).
6. The deviation of a particle throughout a considerable angle from an encounter with a single atom will in this paper be called "single" scattering. The deviation of a particle resulting from a multitude of small deviations will be termed "compound" scattering.
7. A simple consideration shows that the deflexion is unaltered if the forces are attractive instead of repulsive.
8. Manch. Lit. & Phil. Soc. (1910).
9. Nagaoka, Phil. Mag. vii. p. 445 (1904).

6

THE SIZE OF THE COSMOS

FEW EXPERIENCES PROVOKE cosmic questions so much as gazing up on a clear starry night. What are those tiny bright specks glowing in the blackness of space? How big and how far? What nature of thing is that gauzy white sash that extends across the night sky? Where is the earth in the grand sweep of space? When did the universe begin, if at all? Does the cosmos stretch outward and outward without end? Or is there some ultimate boundary of matter and space—beyond which is what?

Every human civilization has attempted to answer such questions. According to the most ancient cosmology on record, the Babylonian *Enuma Elish*, the sky was a vault extending up to unmeasurable height. Heaven and earth joined at a dike built on the waters surrounding earth. The sprinkled stars formed "heavenly writing." And the gauzy white sash, called the Milky Way in later cultures, was the "river of heaven."

Logic and reasoning appear in the earliest known Greek cosmological thought, that of Anaximander in the sixth century B.C. In Anaximander's worldview, the stars consisted of compressed portions of air, while the sun was shaped like a chariot wheel, twenty-eight times the diameter of earth. The rim of this wheel seethed with fire. In the cosmology of the late-eighth-century Arab scholar Jabir ibn Hayyan, stars were not air but divine living beings.

A major problem in testing all of these notions, fanciful and not, was the difficulty in measuring distance in space. Indeed, throughout history, the determination of distance has been the major obstacle for much of astronomy. Without knowing the distance to a shining dot of light, you don't know whether it is a firefly or a star. You don't know how big it is, you don't know how much energy it pumps into space, you don't know what kind of thing it is.

When we look up at the sky, we see only a two-dimensional image, a flat photograph without depth. There are no reference points to tell us how far away are the stars. If we knew a star's actual diameter, we might infer its distance by how small it appears, just as we can look at a far-away car and estimate its distance by knowing its true size. But stars appear only as points of light, even to large telescopes. Likewise, if we knew the intrinsic luminosities of stars, then we could determine their distances by how bright they appear, just as the distance to a 50-watt bulb can be gauged by its apparent brightness. (A definite mathematical relation exists between the intrinsic luminosity of an object, its distance, and its apparent brightness. Determine any two of these quantities and the third can be inferred.) Unfortunately, stars have a wide range of luminosities, most of them unknown until the 1920s and later. Thus, a stellar image could appear dim either because the star is of low luminosity and relatively nearby or because it is of high luminosity but very far away. Consequently, the distance cannot be determined. And without knowledge of their distances, stars, cosmic nebulae, entire galaxies are incomprehensible.

In 1912, a deeply religious and deaf astronomer named Henrietta Leavitt, working at the Harvard College Observatory, discovered a method to determine cosmic distances. In particular, Leavitt found a law for computing the intrinsic luminosities of a certain type of stars called Cepheids. Since the apparent brightnesses of stars could be directly measured, distances could then be inferred. Leavitt gave astronomy the third dimension. After Leavitt's work, Cepheid stars became cosmic lightbulbs of known wattage, floating distance markers scattered through the vast hallways of space. In the following two decades, other astronomers used Leavitt's distance method to measure the size of our galaxy, the Milky Way, and Earth's position in it; to determine that other galaxies exist outside of our own; and, most incredibly, to show that the universe as a whole is expanding, with a beginning in the calculable past.

The ancient Greeks made attempts to determine distances in space. By measuring the directions to planets at different parts of their orbits, combined with geometrical arguments, the Greeks were able to give fair approximations of the *ratios* of distances to the sun and planets. Centuries later, about 1610, Galileo and the first telescope brought about a revolution in astronomy. Peering through his telescope, Galileo was able to see that the Milky Way—the faint, misty band of light that arcs across the sky—was in fact made out of individual stars. But the size of the Milky Way, and the distances to the stars, remained unknown.

The first accurate determination of the distance to the sun was accomplished by the French astronomer Jean Richer in 1672. Richer and his assistants measured the direction to Mars at two observatories, one in Paris and the other on Cayenne Island, off the coast of French Guyana. (The direction is determined by a planet's position relative to the background of distant stars. The direction to the sun cannot be determined by this method, except in a solar eclipse, because the background stars are rendered invisible by the sun's overpowering light.) How did such measurements work? We are all familiar with the phenomenon that when an object is viewed from two perspectives, its direction shifts. Close one eye and look at a tree across the street. Now look at the tree using only the other eye. The direction of the tree shifts. The shift in angle, called the parallax, and the distance between the two observation points (in this case, the separation between your two eyes), can be used to calculate the distance to the tree. Paris and Cayenne were Richer's two eyes. By measuring the parallax to Mars and knowing the distance between Paris and Cayenne, Richer was able to determine the distance to the red planet. He could then determine the distance to the sun, approximately one hundred million miles from earth, knowing from the ancient Greeks the ratios of distances to sun and planets.

A few years later, Isaac Newton estimated the distance to the nearest stars by *assuming* that they were identical to our sun and equal in intrinsic luminosity. He could then calculate how far away the sun would have to be in order to appear as dim as a typical star. The British scientist correctly concluded that the nearest stars are several hundred thousand times the distance to the sun, or roughly twenty or thirty thousand billion miles away. Because distances in the cosmos are so vast, astronomers have resorted to using light-years rather than miles. A light-year is the distance that a beam of light can travel in a year, roughly ten thousand billion (10^{13}) miles. In these terms, the nearest stars are a few light-years away.

About 1838, the German astronomer Friedrich Bessel and others measured the parallax to nearby stars. To observe the tiny angular shift expected from so far away, the astronomers used two observation points separated by as large a distance as possible: the diameter of the earth's orbit about the sun. One observation would be made in the summer, another in winter, when the earth was on the opposite side of its orbit. Bessel's experiment confirmed Newton's estimate. Unfortunately, the parallax method at that time could be used to measure distances only out to about one hundred light-years. Beyond that distance, the shift in direction to a star, even over the large baseline of the earth's orbit about the sun, was too small to detect. Other methods, such as the "moving cluster" method, make certain assumptions about the motions of stars traveling together in a group and can extend the maximum measurable distance to about five hundred light-years. But even five hundred light-years, in cosmic terms, is not so far. Astronomers were almost certain that our strange universe must extend far beyond that. Such was the state of knowledge, and lack of knowledge, at the end of the nineteenth century.

When she arrived at the Harvard College Observatory at the turn of the century, Henrietta Leavitt had no idea that her work would lead to a powerful new method to gauge distances in astronomy. Indeed, her entire career was one of modesty. Even today, Henrietta Leavitt remains a scientist largely unknown to the public, overshadowed by other astronomers who applied her work. There are a few sentences about her work in most college astronomy books. She received no medals, no prizes, no honorary degrees. Much of her life remains shrouded in mystery. Although Leavitt authored a number of scientific papers and left behind a dozen handwritten notebooks of astronomical data, practically no personal letters remain other than her correspondence with Edward Pickering, director of the Harvard College Observatory. A very recently published book by George Johnson, *Miss Leavitt's Stars*, contains almost all that is known about her.

Photographs reveal that Leavitt had a high forehead and a long visage, with her hair pulled back from her face and held behind her head. In most of the pictures of her, both younger and older, she wears a lacy, white, long-sleeved blouse with a high circular collar. Such attire, conservative even in her time, most likely followed from her strict religious upbringing and her own disposition. As astronomer Solon Bailey wrote in his obituary of Leavitt in 1922,

> Miss Leavitt inherited, in a somewhat chastened form, the stern virtues of her puritan ancestors. She took life seriously. Her sense of duty, justice, and loyalty was strong. For light amusements she appeared to care little. She was a devoted member of her intimate family circle, unselfishly considerate in her friendships, steadfastly loyal to her principles, and deeply conscientious and sincere in her attachment to her religion and church. She had the happy faculty of appreciating all that was worthy

> and lovable in others, and was possessed of a nature so full of sunshine that, to her, all of life became beautiful and full of meaning . . . Miss Leavitt was of an especially quiet and retiring nature, and absorbed in her work to an unusual degree.

Bailey also commented that Leavitt was a scientist whose "work was carried out with unusual originality, skill, and patience."

Henrietta Swan Leavitt was born on July 4, 1868, in Lancaster, Massachusetts, one of the seven children of George Leavitt and Henrietta Swan. George was a Congregationalist minister, and Henrietta remained conscientiously religious for her entire life. Miss Leavitt, as she was called all her adult life, attended Oberlin College from 1885 to 1888. From 1888 to 1892 she studied the classics, languages, and astronomy at the Society for the Collegiate Instruction of Women in Cambridge, later to become Radcliffe College. (To receive her astronomy lectures, she would have walked to the Harvard College Observatory, a few hundred yards from the main campus.) One of her college classmates recalled that Leavitt impressed everyone "with the clarity of her mind and the sweet reasonableness of her nature." After receiving her college degree, Leavitt traveled and then spent a couple of years at home in Lancaster with an undisclosed illness.

In 1895, Leavitt revived her passion for astronomy by becoming a volunteer assistant at the Harvard College Observatory. There she joined a dozen other women who had been hired by the dictatorial director of the Observatory, Edward C. Pickering. Such women, alternately called "computers" or "Pickering's harem," were responsible for the painstaking calculation and recording of positions, brightnesses, and spectral colors of literally hundreds of thousands of stars—most of them tiny black points on glass photographic plates, a thousand or more stars on each plate. (The plates were negatives.)

A family crisis in 1900 called Leavitt away to Beloit, Wisconsin, where her father now served as a minister. After an absence of two years, she wrote to Pickering: "I am more sorry than I can tell you that the work I undertook with such delight, and carried to a certain point, with such pleasure, should be left uncompleted." Soon thereafter, at the age of thirty-four, Leavitt accepted an offer from Pickering for a permanent, full-time position at the observatory, for a salary of thirty cents an hour (equivalent to about eight dollars an hour in 2005). In late August 1902, upon arriving back in Massachusetts, she wrote to the director, "At last I find myself free to take [the work] up, and expect to go to the observatory Wednesday afternoon, arriving there between half-past two and three . . . I hope that this long delay has in no way inconvenienced you." By this time, Leavitt was mildly deaf. Over the next few years, her hearing would further deteriorate.

Leavitt's personal story reflects to a large extent the history of women in science. At the end of the nineteenth century, only a handful of women in America, and even fewer in Europe, had been able to find professional employment in science. The chief barriers were lack of educational opportunities and restrictive cultural attitudes toward women's roles and abilities. Most colleges and universities closed their doors to women until the mid-nineteenth century or later. Vassar College, created specifically for women, was founded in 1861; Wellesley in 1870; Smith in 1871.

The first highly recognized female scientist in America was Maria Mitchell (1816–1888), astronomer and discoverer of a comet in 1847. Mitchell, in fact, did not learn her astronomy at a college but rather by helping her father, an amateur astronomer, and by reading astronomical texts while a librarian at the Nantucket Athenaeum. When Vassar opened, Mitchell was made professor of astronomy and director of the observatory there.

In the mid-1870s a woman named Sarah Whiting established a physics laboratory at Wellesley after auditing physics courses at the Massachusetts Institute of Technology (MIT). Academic biological studies were made available to women in 1873, first at the Anderson School of Natural History, off the coast of Cape Cod. In 1884, MIT began to permit women to enroll as regular students.

Astronomy, more than most other sciences, offered opportunities to women. First, much of the work, such as measuring the positions and magnitudes of stars, was perceived to be routine, without requiring advanced education. Second, near the beginning of the twentieth century, scientists undertook a number of large astronomical surveys. The relatively new practice of attaching cameras to telescopes produced photographs containing a huge amount of data. To analyze, measure, compute, and record so much data required a big labor force. And women, often without access to other employment, were willing to work for much less pay than men. Finally, many scientists of both genders believed that the nature of astronomical work was better suited to the female temperament. Williamina Fleming, a colleague of Leavitt's who made important discoveries about unusual stars, put it this way in 1893:

"While we cannot maintain that in everything woman is man's equal, yet in many things her patience, perserverance and method make her his superior." Historian of science Pamela Mack has concluded that between 1875 and 1920 a total of about 160 women were hired at various astronomical observatories in the United States, including the Dudley Observatory in Albany, New York; the Yerkes Observatory in Wisconsin; Mount Wilson in southern California; the U.S. Naval Observatory in Washington, D.C.; the Lick Observatory of the University of California; Columbia; Yale; and other places. The largest number of female astronomical assistants was hired at the Harvard College Observatory. Pickering alone, during his tenure from 1876 to 1919, granted positions to more than forty women.

Edward Pickering was born in 1846 in the aristocratic part of Boston known as Beacon Hill. At the young age of nineteen, he graduated summa cum laude from the Lawrence Scientific School of Harvard. Then he spent a decade at the newly founded MIT, demonstrating his talent for experimental physics and revolutionizing physics education by creating the first physical laboratory in America designed for the instruction of students. In 1876, Pickering left MIT to take up a position as professor of astronomy at Harvard and director of the Harvard College Observatory (HCO).

During the next couple of decades, the emphasis in astronomy underwent a historical transition from measuring the positions and motions of stars to studying individual stars as physical objects. What were stars made of? How did they produce energy? What determined their luminosities and colors? What were the different kinds and categories of stars? To this end, Pickering began enormous surveys to measure the properties of stars, especially their brightnesses and colors. In 1907, as the technical aspects of photography improved and offered revolutionary new methods of astronomical data taking, Pickering began major projects to photograph much of the sky. One such project, the Harvard Photographic Library, eventually produced 300,000 glass plates. Pickering himself routinely took photographs on every clear night and made more than one and a half *million* measurements of stellar brightnesses. Annie Jump Cannon, one of the most distinguished of the HCO astronomers and a student of Sarah Whiting, accomplished the colossal task of classifying nearly 300,000 stars on the basis of their spectral colors.

Pickering had a complex attitude toward his mostly female assistants. On the one hand, he was gracious, compassionate, and encouraging to women. On the other, he clearly exploited his female assistants. In his

1898 Annual Report of the Harvard College Observatory, Pickering wrote: "Many of the assistants [almost all women] are skilful only in their own particular work, but nevertheless capable of doing as much and as good routine work as astronomers who would receive much larger salaries." Astronomer Celia Payne—who arrived at the Observatory not long after Pickering's tenure and who was to make the extremely important discovery that hydrogen is the most abundant element in the universe—said that "Pickering chose his staff to work, not to think."

The women "computers" worked in two rooms at the observatory. Photographs of the rooms, now remodeled beyond recognition, show that they had a homey atmosphere, with flowered wallpaper, mahogany writing tables, star maps, and photographs of famous astronomers (all men) on the walls. There were about eight occupants to a room. The women were expected to work seven hours a day, five of which had to be at the observatory. A photograph from about 1890 shows the women busy at their labors, some squinting through magnifying glasses at the minute dots and smudges on photographic plates, some recording numbers and figures, some consulting reference works. Pickering, who visited his computers once a day, stands majestically in the corner, plump, bearded, dressed in a three-piece suit, gazing over his dominions.

Around 1907, as Pickering began his massive project of photographing the sky, he assigned Leavitt the task of determining the brightnesses of a group of stars in the direction of the celestial north pole. These stars were to be used as standards against which all other stars could be compared. This project was quite difficult. First of all, an astronomer must decide how to quantitatively determine the brightness of a star from its image on a photographic plate. One technique is to begin with a particular star, photograph it with a telescope of known aperture and a camera of known exposure time and standard film, then assign that star a numerical brightness. One can then reduce the aperture of the telescope so that half as much light comes through and photograph the same star. That second image is assigned a brightness equal to half the first. The process continues, gradually building up a catalogue of images of known brightness. The astonomer then analyzes the photographs of stars near the north pole. By comparing these stars with the first set of standards, one determines each of their brightnesses. These stars now represent a new set of standards.

The problem faced by Leavitt was that she was given 277 photographs taken from thirteen different telescopes, with varying apertures and exposure times. Somehow, she had to calibrate and compare all these different data. She ultimately succeeded, with the famous North Polar Sequence, which contains stars with brightnesses varying over a factor of six million from the brightest to the dimmest.

We should emphasize that all of these standard sequences of images dealt only with apparent brightness. Except for the very closest stars, whose distances could be determined by the parallax or moving cluster method, the distances to these stars were not known. What was needed for distances was a method to determine intrinsic luminosity.

About the time that she started work on the North Polar Sequence, Leavitt began the project that would lead to a landmark discovery in science: the study of variable stars. Variable stars, known for centuries, are stars whose brightness varies in a regular way. The leading theory to explain such variations was that variable stars were members of binary star systems: two stars orbiting each other. Each star in a binary pair would periodically eclipse the other, producing regular reductions in light as seen on Earth. About 1914, well after Leavitt's great paper of 1912, it was realized that the light variations in many variable stars are due not to external eclipses but to internal processes in the stars themselves. Instabilities in these particular stars cause them to contract and expand in a rhythmic way over periods of days to weeks. Unlike these variables, most stars change their luminosity only extremely slowly, over a period of billions of years.

To determine whether or not a star was variable, Leavitt used the positive/negative matching method. In this technique, the astronomer starts with positive and negative photographic plates of the same region of the sky but taken at two different times. The two glass plates, each with a thousand or more tiny stellar images, are placed one on top of the other in precise alignment. A variable star would stand out as having a white halo that was bigger or smaller than the underlying black negative image.

Once a variable star is identified, many more photographs are taken at different times and compared to deduce the stars' full cycle of brightening and dimming. Figure 6.1, for example, shows the light cycles of four different variable stars. Here, the vertical scale is a measure of the star's brightness and the horizontal scale is the time (in days) over which the cycle is measured. The dots represent observations made during many different cycles, all folded together into the same cycle. The time needed for a star to make a complete cycle from its brightest point to its

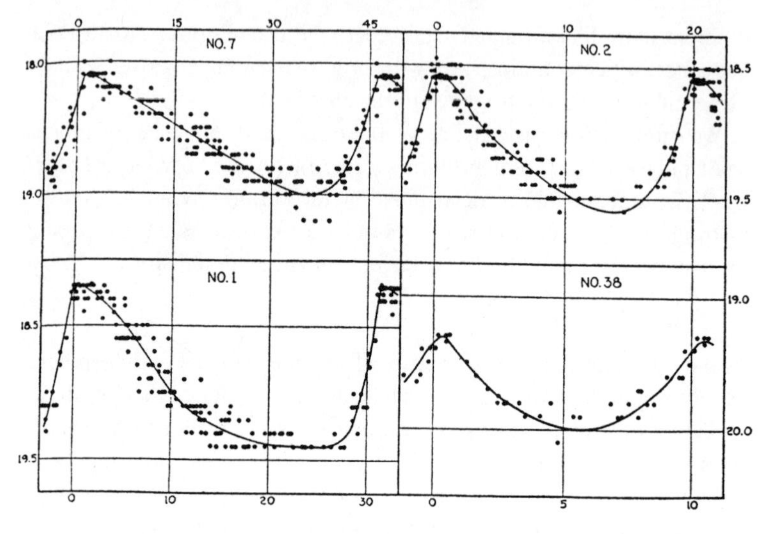

Figure 6.1

dimmest and back to its brightest is called the period. As can be seen from Figure 6.1, star No. 7 has a period of about forty-five days, while star No. 2 has a period of about twenty days.

Over her lifetime, Leavitt identified more than 2,400 variable stars, single-handedly doubling the number of known variable stars. "What a variable star 'fiend' Miss Leavitt is," Princeton professor Charles Young wrote to Pickering. "One can't keep up with the roll of the new discoveries." Leavitt's productivity is even more astounding when considering that her extended illnesses and family affairs took her away from the Observatory for long stretches of time between late 1908 and mid-1911, sometimes for as long as eighteen months at a time. ("Not the least of my trial in being ill is the knowledge of the annoyance it causes you," she wrote to Pickering during one of her absences.)

For her landmark study, Leavitt focused her attention on a particular kind of variable star called a Cepheid. Cepheid stars, first discovered in 1784, are distinctive yellow "supergiant" variable stars that have periods between three and fifty days. In particular, Leavitt studied a group of faint Cepheid stars all located in the same tiny region of space, in a nebula called the Small Magellanic Cloud. (Long after Leavitt's work, it was realized that both the Large and Small Magellanic Clouds are small galaxies, lying outside the Milky Way.)

A critical point is that because all of these stars in the Small Magellanic Cloud are huddled together, it can be assumed that they are close to each other compared to their distance from Earth. That is, these stars are all approximately the same distance from us. Thus, *a Cepheid in this sample that appears twice as bright as another in the sample is twice as intrinsically luminous*. Aside from an overall multiplicative factor, Leavitt is, in effect, measuring the intrinsic luminosities of these stars.

The photographs of the Small Magellanic Cloud were recorded with the twenty-four-inch Bruce telescope at the Harvard southern station in Arequipa, Peru. Because of the faintness of these stars, some of the exposures took up to four hours. Using many photographs and her methods for determining brightnesses, Leavitt measured the apparent brightnesses and periods of twenty-four of these Cepheids.

What Leavitt discovered, to her surprise and delight, was a definite, quantitative relation between the periods of her Cepheids and their apparent brightnesses. (See Figures 1 and 2 of her paper.) The longer the period, the brighter the star. And, since these stars are all about the same distance from Earth, that result translates to: the longer the period, the more intrinsically luminous the star.

No one had predicted such a relationship or any relationship at all. Indeed, based on the prevailing explanation of what caused a Cepheid to vary its light, there should have been no connection between the period and the intrinsic luminosity—the first being an accident of the star's separation from its binary companion and the second an intrinsic property of the individual star. Leavitt described the connection she had discovered as "remarkable." Like Bayliss and Starling's discovery of hormones, or Rutherford and Marsden and Geiger's discovery of the nucleus of the atom, Leavitt's discovery of the period-luminosity relationship for Cepheid stars was completely unexpected.

Leavitt had now found a particular group of stars that could serve as "cosmic beacons," which gauge the enormous distances of space. A cosmic beacon must have three features. (1) It must be easily identifiable. (2) It must have some quality that announces its intrinsic luminosity. (Recall that a star's distance from Earth may be determined from its intrinsic luminosity and its apparent brightness, the latter being directly measurable.) (3) It must be sufficiently plentiful and scattered about the universe.

The Cepheids met all of these requirements. First, they were easily identifiable from their colors and variable light, as had been long known. Second, Leavitt had discovered that their intrinsic luminosities, aside from an overall calibration factor, could be determined from their periods. The periods were directly measurable. And third, Cepheids could be found in lots of places, not just in the Small Magellanic Cloud. In any backwater of space that harbored a Cepheid star, an astronomer could measure the distance to that spot.

One of the first things we notice in looking at Leavitt's paper is that it was published by the Harvard College Observatory. The powerful observatory had its own journal. Second, the paper is signed not by Leavitt but by Pickering, the director of the observatory. In his opening paragraph, Pickering refers to the paper to follow as "prepared by Miss Leavitt." It was customary for the director to sign for the female astronomical assistants.

To read the paper, we must understand a bit of astronomical jargon. For historical reasons, astronomers measure apparent brightness in terms of something called "magnitude," denoted by m. Like apparent brightness, the magnitude m depends on both intrinsic luminosity and distance. (Mathematically, $m = -0.25 - 2.5\ log(L) + 5\ log(d)$, where log denotes the logarithm, L is the intrinsic luminosity in units of our sun's luminosity, and d is the distance to the object in units of parsecs, where a parsec equals 3.3 light-years.) Note that with this odd terminology, the more intrinsically luminous a star, the *smaller* its magnitude, just the opposite of common sense. In these terms, our sun has a magnitude of –26.5, and the naked eye can detect magnitudes of 6.5 and smaller. Astronomers refer to intrinsic luminosity in terms of "absolute magnitude," denoted by M. (Mathematically, $M = 4.75 - 2.5\ log(L)$.) Unlike magnitudes, absolute magnitudes depend only on intrinsic luminosity.

The key results are shown in Figures 1 and 2 of Leavitt's paper, indicating a rather well defined relation between magnitude and period. That is, the magnitudes of stars generally decrease with increasing periods, and *each period corresponds to a fairly well defined magnitude*. Figure 2 presents the same data as Figure 1, the only difference being that the horizontal axis is the logarithm of the period rather than the period itself.

Note the important sentence near the end of the paper: "Since the variables are probably at nearly the same distance from the Earth, their periods are apparently associated with their actual emission of light . . ."

Thus Leavitt did indeed realize that her relation bore upon the intrinsic luminosity of the Cepheids.

It is almost certain that Leavitt understood that her newly found relation between period and magnitude could be used to measure astronomical distances. However, she never says so in her paper. In this regard, her landmark paper differs from the previous papers we have considered. Bayliss and Starling, in their discovery of hormones, and Rutherford, in his discovery of the nucleus of the atom, all found something they were not looking for, but they quickly realized its significance and described that significance in their papers. Why did Leavitt not do the same? One possible reason is that Leavitt and her female colleagues were not in positions where they were encouraged to make ambitious claims. In an atmosphere where the female assistants were chosen "to work, not to think," Leavitt would have faced obstacles in announcing a major conceptual discovery. Certainly, she would not likely have been given access to the instruments she needed to follow up any such claim. Another explanation could be the fear of intellectual theft. The Mount Wilson Observatory in California, in particular, engaged in an intense rivalry with the Harvard College Observatory. Finally, the omission could have been simply a matter of personal style. Considering the overall setting, the first explanation seems the most likely. In any case, there are no surviving letters or journal entries that show Leavitt's opinion about her work or its significance.

A crucial missing piece needed to complete Leavitt's law was a *calibration*. Because Leavitt did not know the distance to the Small Magellanic Cloud, she could determine only the *ratio* of intrinsic luminosities of any two stars in her sample of twenty-four. Just as a person looking at a bunch of lights in a distant office building can easily gauge the *relative* luminosities of the various lightbulbs but not the absolute luminosity of any single bulb, Leavitt knew the *relative* intrinsic luminosities of her Cepheids but not the intrinsic luminosity of any one of them.

In 1913, a year after Leavitt's paper was published, the distinguished Danish astronomer Ejnar Hertzsprung found a Cepheid star much closer to Earth than the Small Magellanic Cloud, close enough so that he could measure its distance by the moving cluster method. From the star's distance and apparent magnitude, he could determine its intrinsic luminosity. After measuring its period, he could then assign a particular intrinsic luminosity to a particular period in Leavitt's graph. Now, everything was calibrated! The resulting law, called the period-luminosity law, is shown in Figure 6.2, where the average luminosity of a Cepheid is graphed against its period.

Cepheid stars, although somewhat rare, can be found in many places throughout the universe. Thus, as mentioned before, they make excellent cosmic beacons. The application of Leavitt's period-luminosity law would be as follows: Find a Cepheid in a cluster of stars whose distance you want to determine. Measure the period and apparent brightness of the Cepheid. From the period and Leavitt's period-luminosity law (Figure 6.2), read off the intrinsic luminosity of the Cepheid. Now, by combining the intrinsic luminosity with the apparent brightness of the star, infer its distance. You now know the distance to the cluster of stars!

Harlow Shapley, a talented young American astronomer who worked at the Mount Wilson Observatory in California and who would eventually succeed Pickering as director of the Harvard College Observatory, was one of the first astronomers to appreciate the importance of Leavitt's work. "Her discovery of the relation of period to brightness is destined to be one of the most significant results of stellar astronomy, I believe," he wrote to Pickering in 1917. That year and the next, Shapley extended and applied Leavitt's work by systematically searching for Cepheids in "globular clusters" of stars at various locations in the Milky Way. He could determine the distance to each from Leavitt's period-luminosity law, and the collection of clusters allowed him for the first time to draw a calibrated map of the galaxy. His result: the Milky Way, a giant pinwheel of stars, was about 300,000 light-years in diameter.

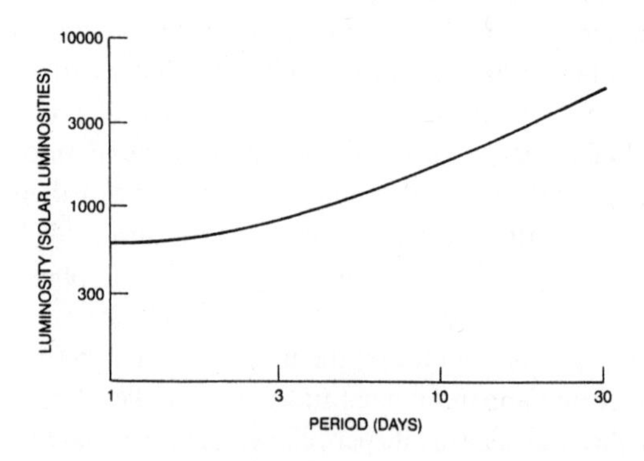

Figure 6.2

Our own solar system languishes about two-thirds of the way out from the center.

For centuries, people had wondered about the nature of the many astronomical nebulae, the cloudy white patches of light in the sky. But no one knew the distances to the nebulae or whether they were part of our galaxy of stars. In 1924, the American astronomer Edwin Hubble found Cepheid stars in the Andromeda nebula and could thus measure its distance, using Leavitt's period-luminosity law. The result was 900,000 light-years, three times the maximum extent of the Milky Way as determined by Shapley. The Andromeda nebula lay outside of our galaxy! (Modern measurements give a diameter of 100,000 light-years for the Milky Way and a distance to the Andromeda galaxy of about two million light-years.)

Andromeda turned out to be the closest large galaxy. Hubble went on to measure the distances to other nebulae, showing that many of them were also entire galaxies far beyond the Milky Way. We now know that, on average, galaxies are separated from each other by about twenty galaxy diameters, and each galaxy contains about a hundred billion stars. The size of the cosmos has become vastly larger. And the galaxy is recognized as the "unit" of matter on the large canvas of the universe.

In 1929, Hubble used Leavitt's law as a foundation to measure the distances to a nearby group of galaxies, all moving away from the Milky Way. He found that their distances were proportional to their outward velocities. From that result, other astronomers concluded that the universe as a whole is expanding. (See Chapter 12 on Hubble and the expansion of the universe.)

With telescopes in space, Cepheid stars can now be seen out to a distance of about 20 million light-years. Thus, simply in quantitative terms, the period-luminosity law extended the distances charted by humankind from 500 light-years to 20 million light-years, increasing the fathomable *volume* of space by the staggering factor of nearly 10^{14}.

None of these truly cosmic advances in knowledge would have been possible without Henrietta Leavitt's discovery of the relation between magnitudes and periods for Cepheid variable stars.

Miss Leavitt spent most of her career at HCO working not on variable stars but on Pickering's project of establishing magnitude standards for stars. She was not allowed to pursue the implications and applications of her most important work. As Celia Payne commented, "It may have been a wise decision to assign the problems of photographic photometry to Miss Leavitt . . . But it was also a harsh decision, which condemned a brilliant scientist to uncongenial work, and probably set back the study of variable stars for several decades." Uncongenial or not, Leavitt seems to have gone about her work cheerfully and with some humor. Antonia Maury, a colleague at Harvard, recalled that Leavitt, upon pondering one particularly perplexing star, commented that "we shall never understand it until we find a way to send up a net and fetch the thing down!"

Leavitt's title at HCO, from the beginning to the end, was "assistant." In 1919, she and her widowed mother took up residence in a brick building on Linnean Street, several blocks from the Observatory. But soon Leavitt was ill again, now with cancer. She died on December 12, 1921, at the age of 53. Shortly before her death, Leavitt wrote out her will, leaving her possessions and assets to her mother: bookcase and books $5, folding screen $1, rug $40, table $5, chair $2, desk $5, table $5, rug $20, bureau $10, bedstead $15, two mattresses $10, two chairs $2, one @ $100 face value First convertible 4% Liberty Bond $96.33, one @ $50 face value Fourth 4¾% Liberty Bond $48.56, one @ $50 face value Victory 4¼% Note $50.02.

One of her very few awards was an honorary membership in the American Association of Variable Star Observers. In 1925, Professor Mittage-Leffler of the Swedish Academy of Science wrote to Leavitt to say that he wished to nominate her for a Nobel Prize. The Swedish scientist didn't know that she had been dead for three years.

PERIODS OF 25 VARIABLE STARS IN THE SMALL MAGELLANIC CLOUD

Circular of the Astronomical Observatory of Harvard College (1912)

THE FOLLOWING STATEMENT REGARDING the periods of 25 variable stars in the Small Magellanic Cloud has been prepared by Miss Leavitt.

A Catalogue of 1777 variable stars in the two Magellanic Clouds is given in H.A. 60, No. 4. The measurement and discussion of these objects present problems of unusual difficulty, on account of the large area covered by the two regions, the extremely crowded distribution of the stars contained in them, the faintness of the variables, and the shortness of their periods. As many of them never become brighter than the fifteenth magnitude, while very few exceed the thirteenth magnitude at maximum, long exposures are necessary, and the number of available photographs is small. The determination of absolute magnitudes for widely separated sequences of comparison stars of this degree of faintness may not be satisfactorily completed for some time to come. With the adoption of an absolute scale of magnitudes for stars in the North Polar Sequence, however, the way is open for such a determination.

Fifty-nine of the variables in the Small Magellanic Cloud were measured in 1904, using a provisional scale of magnitudes, and the periods of seventeen of them were published in H.A. 60, No. 4, Table VI. They resemble the variables found in globular clusters, diminishing slowly in brightness, remaining near minimum for the greater part of the time, and increasing very rapidly to a brief maximum. Table I gives all the periods which have been determined thus far, 25 in number, arranged in the order of their length. The first five columns contain the Harvard Number, the brightness at maximum and at minimum as read from the light curve, the epoch expressed in days following J.D. 2,410,000, and the length of the period expressed in days. The Harvard Numbers in the first column are placed in Italics, when the period has not been published hitherto. A remarkable relation between the brightness of these variables and the length of their periods will be noticed. In H.A. 60, No. 4, attention was called to the fact that the brighter variables have the longer periods, but at that time it was felt that the number was too small to warrant the drawing of general conclusions. The periods of 8 additional variables which have been determined since that time, however, conform to the same law.

TABLE I.

PERIODS OF VARIABLE STARS IN THE SMALL MAGELLANIC CLOUD.

H.	Max.	Min.	Epoch.	Period.	Res. *M.*	Res. *m.*	H.	Max.	Min.	Epoch.	Period.	Res. *M.*	Res. *m.*
			d.	*d.*						*d.*	*d.*		
1505	14.8	16.1	0.02	1.25336	−0.6	−0.5	1400	14.1	14.8	4.0	6.650	+0.2	−0.3
1436	14.8	16.4	0.02	1.6637	−0.3	+0.1	*1355*	14.0	14.8	4.8	7.483	+0.2	−0.2
1446	14.8	16.4	1.38	1.7620	−0.3	+0.1	1374	13.9	15.2	6.0	8.397	+0.2	−0.3
1506	15.1	16.3	1.08	1.87502	+0.1	+0.1	818	13.6	14.7	4.0	10.336	0.0	0.0
1413	14.7	15.6	0.35	2.17352	−0.2	−0.5	*1610*	13.4	14.6	11.0	11.645	0.0	0.0
1460	14.4	15.7	0.00	2.913	−0.3	−0.1	*1365*	13.8	14.8	9.6	12.417	+0.4	+0.2
1422	14.7	15.9	0.6	3.501	+0.2	+0.2	*1351*	13.4	14.4	4.0	13.08	+0.1	−0.1
842	14.6	16.1	2.61	4.2897	+0.3	+0.6	827	13.4	14.3	11.6	13.47	+0.1	−0.2
1425	14.3	15.3	2.8	4.547	0.0	−0.1	*822*	13.0	14.6	13.0	16.75	−0.1	+0.3
1742	14.3	15.5	0.95	4.9866	+0.1	+0.2	823	12.2	14.1	2.9	31.94	−0.3	+0.4
1646	14.4	15.4	4.30	5.311	+0.3	+0.1	824	11.4	12.8	4.	65.8	−0.4	−0.2
1649	14.3	15.2	5.05	5.323	+0.2	−0.1	821	11.2	12.1	97.	127.0	−0.1	−0.4
1492	13.8	14.8	0.6	6.2926	−0.2	−0.4							

The relation is shown graphically in Figure 1, in which the abscissas are equal to the periods, expressed in days, and the ordinates are equal to the corresponding magnitudes at maxima and at minima. The two resulting curves, one for maxima and one for minima, are surprisingly smooth, and of remarkable form. In Figure 2, the abscissas are equal to the logarithms of the periods, and the ordinates to the corresponding magnitudes, as in Figure 1. A straight line can readily be drawn among each of the two series of points corresponding to maxima and minima, thus showing that there is a simple relation between the brightness of the variables and their periods. The logarithm of the period increases by about 0.48 for each increase of one magnitude in brightness. The residuals of the maximum and minimum of each star from the lines in Figure 2 are given in the sixth and seventh columns of Table I. It is possible that the deviations from a straight line may become smaller when an absolute scale of magnitudes is used, and they may even indicate the corrections that need to be applied to the provisional scale. It should be noticed that the average range, for bright and faint variables alike, is about 1.2 magnitudes. Since the variables are probably at nearly the same distance from the Earth, their periods are apparently associated with their actual emission of light, as determined by their mass, density, and surface brightness.

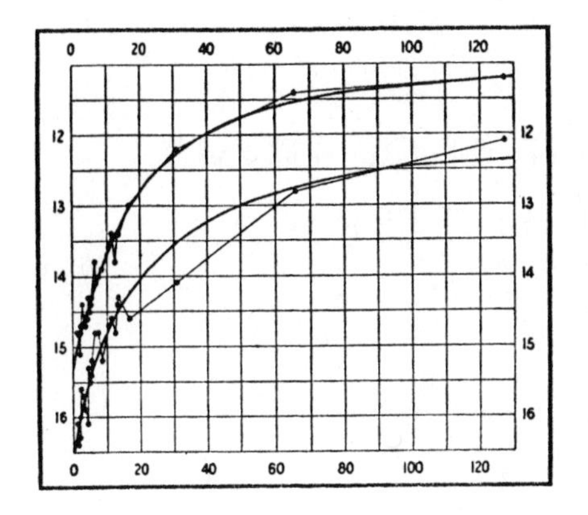

FIGURE 1

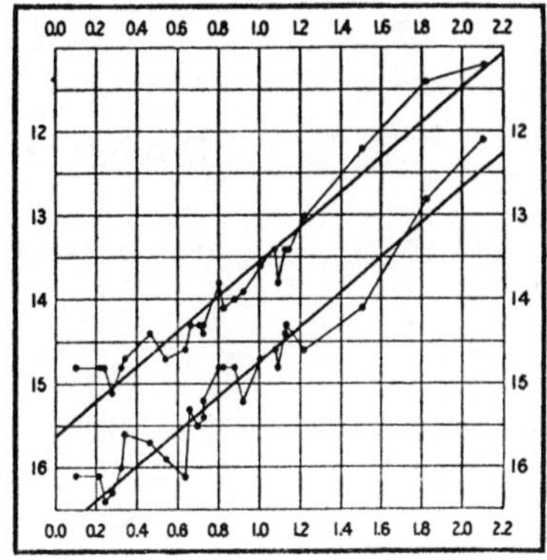

FIGURE 2

The faintness of the variables in the Magellanic Clouds seems to preclude the study of their spectra, with our present facilities. A number of brighter variables have similar light curves, as UY Cygni, and should repay careful study. The class of spectrum ought to be determined for as many such objects as possible. It is to be hoped, also, that the parallaxes of some variables of this type may be measured. Two fundamental questions upon which light may be thrown by such inquiries are whether there are definite limits to the mass of variable stars of the cluster type, and if the spectra of such variables having long periods differ from those of variables whose periods are short.

The facts known with regard to these 25 variables suggest many other questions with regard to distribution, relations to star clusters and nebulae, differences in the forms of the light curves, and the extreme range of the length of the periods. It is hoped that a systematic study of the light changes of all the variables, nearly two thousand in number, in the two Magellanic Clouds may soon be undertaken at this Observatory.

Edward C. Pickering
March 3, 1912.

7

THE ARRANGEMENT OF ATOMS IN SOLID MATTER

EVERY FEW YEARS, my wife ventures down into our dark, grimy basement and returns with a certain cardboard box. After sweeping off the dust, she carefully unloads the precious cargo, piece by piece, onto our dining room table. Here lies the rock collection of her childhood. She is still fascinated by the beautiful crystals, and so am I. The yellowish calcite, whose tiny peaks look like a miniature mountain range. Cut a precipice and you miraculously form a perfectly flat plateau in the shape of a hexagon. Or the dense nugget of galena, lead gray, metallic, heavy to hold. It gives way to the chisel in three different directions, revealing perfect cubes and octahedrons. Cut the stone into even smaller pieces (my wife begins objecting) and you discover still more cubes and octahedrons, octahedrons within octahedrons, and so on. Can this go on? Or the amber-colored topaz, a dozen stubby fingers poking out from a gnarled fist, each finger with a tip shaped like a rhombus. Or the pink halite, which looks like a small cocktail filled with pink, cloudy ice cubes.

How could the inanimate world create such flat surfaces and ordered symmetries? They appear more like human-made buildings than the randomly curving coastlines and clouds and leaf shapes that we associate with nature. Deep, secret forces must be at work.

In 1784, the French mineralogist René-Just Haüy offered the insightful proposal that the visible symmetries of crystals are possibly a consequence of invisible symmetries, an orderly arrangement of the smallest elements of crystals. Those smallest invisible elements would be atoms and molecules, although the idea of the atom was still hypothetical at the time. Certainly no one knew the dimensions of atoms, if they existed at all.

Taking Haüy's good idea, a nineteenth-century botanist and physicist named Auguste Bravais found that only a small number of arrangements of atoms could fit together to make a repeating pattern in space. Bravais discovered fourteen such possible arrangements, called unit cells. Four of Bravais's unit cells are shown in Figure 7.1. The lower left is simply a cube, with an atom at each of its eight corners. Clearly, cubes can be stacked in a repeating pattern that fills up space, as illustrated in Figure 7.2. Here, we have shown only three cubes, to avoid clutter, but the stacking can be continued in all directions indefinitely. The unit cell in the upper left of Figure 7.1 is like a cube, but with skewed angles between the edges at each corner, a kind of Salvador Dalí cube. These arrangements, too, can be stacked together in a periodic pattern. The cell in the upper right has some sides that are rectangular, and an additional atom on the top and bottom faces. At the lower right is a cube with an additional atom in the middle. By mathematics alone, Bravais had discovered the language of crystals. But it was the language of a mythical country. No unit cell had ever been seen.

Skipping ahead for a moment beyond the momentous discovery of 1912, Figure 7.3 shows one of the first crystals to be analyzed at the atomic level, sodium chloride, also known as common table salt. Here, the small, light-colored spheres are sodium atoms and the large, dark spheres are chlorine. The lines connecting the spheres are drawn only to help visualize the arrangement. This Tinkertoy structure is, in fact, one of the fancier cells in Bravais's fourteen-point grammar. The overall shape is a cube, with an atom of chlorine at each of the eight corners, but with an additional atom in the middle of each edge and face. A sin-

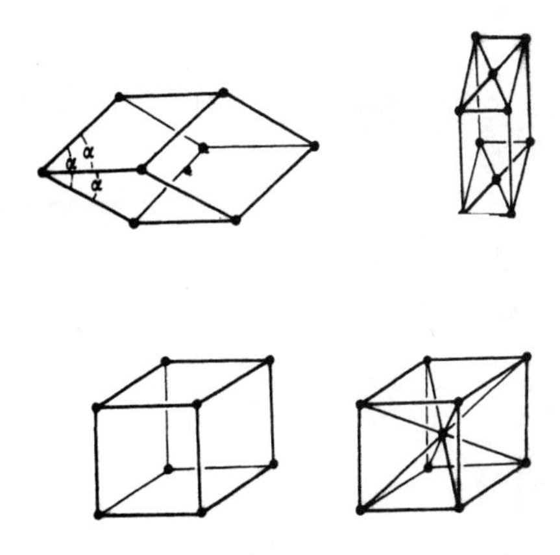

Figure 7.1

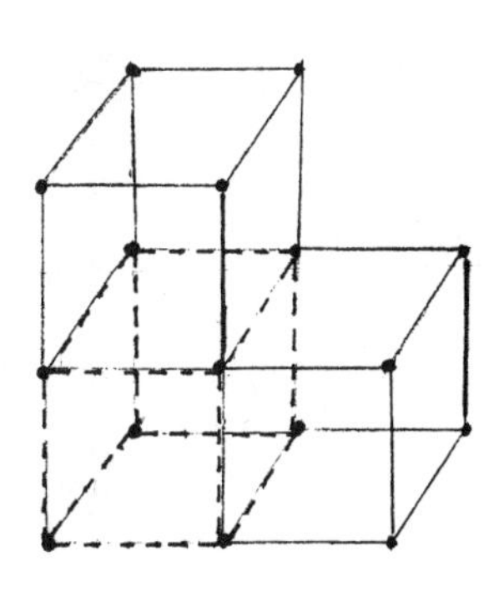

Figure 7.2

gle *molecule* of sodium chloride, that is, the smallest substance that has the chemical properties of sodium chloride, would be a single atom of sodium bound to a single atom of chlorine. But a *unit cell* of sodium chloride requires more than two atoms. The unit cell, shown in Figure 7.3, is the smallest unit from which a crystal of sodium chloride can be built up by repetition in three dimensions—that is, by stacking this model on top of itself vertically, horizontally, and outward, over and over, millions of times—in effect creating a tiny city built of sodium and chlorine bricks. Bravais would have been delighted.

Yet Bravais, with his geometrical diagrams, had only begun. For the manner in which atoms arrange themselves in space reveals much more than an explanation of the beautiful symmetries of rocks. These arrangements are fundamental to the physical properties of all solid matter. These arrangements govern how different substances bend and twist and interact in space. Even more profoundly, these arrangements reveal something about the push and pull between atoms, which cause the orderly structures in the first place. The hexagon that appears when I excavate a small mountain peak of my wife's calcite crystal is quietly whispering about the electrical forces between closely packed atoms of calcium, carbon, and oxygen—a world one hundred million times smaller than the marble-sized piece of rock.

In *Gulliver's Travels*, Jonathan Swift imagines a land where people are six inches tall. To such a Lilliputian, the fine stitching in a piece of silk would look like strands of fetuccine, and a chessboard would be a sculp-

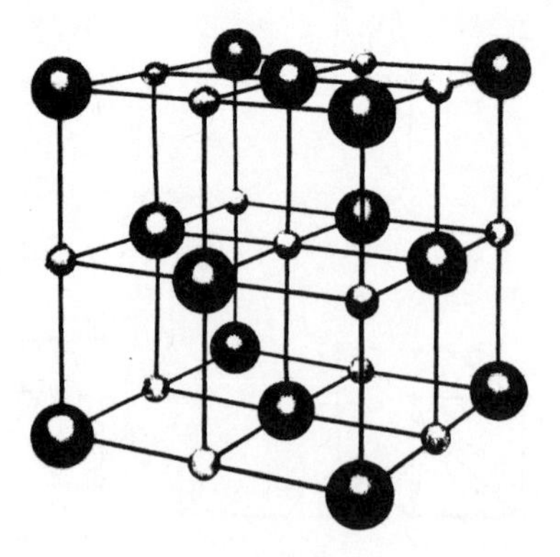

Figure 7.3

ture museum. But Swift's little people would still be millions of times too big to explore a unit cell of sodium chloride. Even the best microscopes of the nineteenth century could not see anything smaller than a couple of ten-thousandths of a centimeter, still thousands of times larger than an element of Bravais's grammar. For well over a century after the speculations of Haüy, scientists held little hope of ever peering into the tiny city of crystals.

Then, one wintry evening in February of 1912, a thirty-two-year-old German physicist named Max von Laue, once the star pupil of Max Planck, was "suddenly struck" by a thought: What happens when a beam of X-rays is aimed at a crystal? X-rays, discovered in the late nineteenth century, were believed by some scientists to be traveling waves of electromagnetic energy, like other forms of light, but with very small wavelengths ranging from 10^{-6} to 10^{-9} centimeters (a millionth to a billionth of a centimeter). In an inspiring vision, von Laue imagined that X-rays would scatter off the neat rows of atoms like water waves scattering off a line of buoys. Downstream, such waves merge, overlap, and produce patterns that reveal the spacings of the buoys. This process, called diffraction, was already known for visible light reflecting off rows of parallel grooves on a flat sheet of glass. Now von Laue hypothesized that waves of X-rays, after flowing through the miniature city of a crystal, would leave periodic patterns on a photographic plate—architectural blueprints of the placement of *individual atoms*. In effect, X-rays could be used to photograph the inside of a crystal.

Von Laue, a theorist himself, talked excitedly about his predicted effect but had trouble persuading colleagues to do the experiment. At the time, he was a relatively unknown privatdozent, or lecturer, at the University of Munich. X-rays were not waves at all, some physicists said, but rather high-speed particles. Other scientists believed that ambient heat causes atoms in solids to constantly jostle, like an enthusiastic crowd at a football game, and that this movement would blur any attempted photograph. As von Laue recounts in his Nobel lecture of 1915, "the acknowledged masters of our science entertained doubts . . . A certain amount of diplomacy was necessary before Friedrich [an assistant to the boss, Professor Arnold Sommerfeld] and Knipping [a doctoral student] were finally permitted to carry out the experiment according to my plan." The experiments began on April 21, 1912. Von Laue worked out the mathematics. Friedrich and Knipping did the experiment, starting with a piece of copper sulfate crystal. They were successful, as can be seen from Figure 2 of their paper.

Wasting little time, Von Laue sent one of his X-ray photographs to Einstein, in Prague, who wrote back to him that "your experiment is one of the finest things to have happened in physics." Two days later, an elated Einstein wrote to physicist Ludwig Hopf: "It is the most wonderful thing I have ever seen. Diffraction on individual molecules, whose arrangement is thus made visible."

Max von Laue was born in 1879, the son of a German military official. Owing to his father's various postings, the young von Laue moved from Pfaffendorf to Brandenburg to Altona to Posen to Berlin to Strassburg. After a year of military service in 1898, he went to the University of Strassburg, and then finished his undergraduate work at Göttingen. In 1902, von Laue made his pilgrimage to the University of Berlin to study with Planck, the greatest European theoretical physicist of the day. There, von Laue applied Planck's ideas of entropy to radiation fields and championed Einstein's new theory of relativity in 1905. A photograph of von Laue from this period shows a very handsome man with strong features and a mustache, a steady gaze, a civilized, almost regal bearing, and a sensitive but determined expression.

From early in his training, von Laue took a special interest in optics and the wavelike behavior of light. As he recalls, "I had finally been able to cultivate what one could almost term a special feeling or intuition for wave processes." What is such a "special feeling" in science? Many of the greatest practitioners, from Einstein to Richard Feynman, have attempted to describe it. In part, this special feeling is a thorough knowledge of the mathematics of a process. It is the ability to comprehend a thing from several points of view. And it is certainly the ability to visualize a phenomenon, even one that is not visible to the eye. In his mind's eye, von Laue was able to imagine a voyage through the tiny city of a crystal.

All of us have some experience with waves. The wave after wave of the ocean, rolling onto a beach. Or the vibrating curve of a plucked violin string. Or the oscillating line on a heart monitor. A critical part of von Laue's "special feeling" for waves was an understanding of how waves overlap, sometimes canceling each other and sometimes reinforcing. This process is called interference. Figure 7.4 illustrates how interference comes about. The waves shown here have the same wavelength, that is,

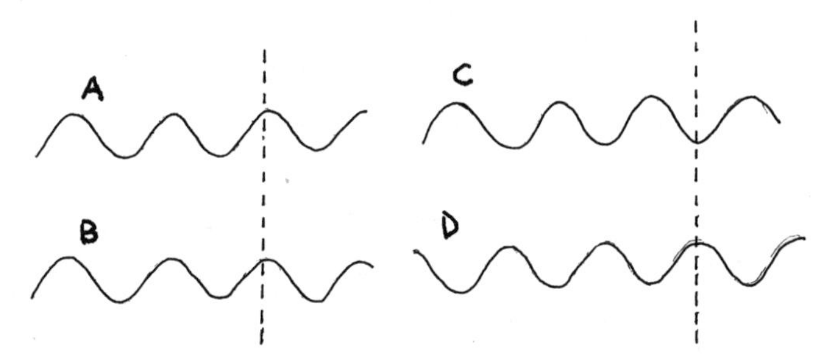

Figure 7.4

the same distance between successive crests and troughs. Waves A and B are positioned relative to each other so that their crests line up and their troughs line up. Such waves are said to be "in phase." When waves A and B overlap, they will reinforce each other, producing a stronger wave than either wave alone. Waves C and D, on the other hand, are "out of phase" with each other. Each trough of wave C lines up with a crest of wave D. When waves C and D overlap, they will cancel each other and leave stillness. Two particles that converge on a point cannot eliminate each other, but two waves can.

An important application of the interference of waves is a diffraction grating, which is a row of parallel grooves that scatter incoming light. The grooves can be carved in glass or in polished metal. The first crude diffraction grating was built in 1801 by the British physicist Thomas Young, who was studying the properties of light. Like Mozart, Young was a child prodigy. He learned to read at age two, had digested the Bible by six, and was one of the first people to translate Egyptian hieroglyphics. With his diffraction grating, Young helped prove the wave nature of light.

Figure 7.5 schematically illustrates one of Young's diffraction gratings. Here the vertical row of dots represents a cross section of the grooves. Parallel beams of light come in from the left, scatter off the grooves, and then emerge to the right in all directions. In general, even if the incoming waves are in phase with each other, they will be out of phase after scattering. However, those waves that scatter in special directions will be in phase and reinforce each other. The special directions of reinforcement are determined by the wavelength of light and the spacing between the grooves. (For example, for a wavelength equal to one-fifth the distance between grooves, the scattered waves would reinforce in the

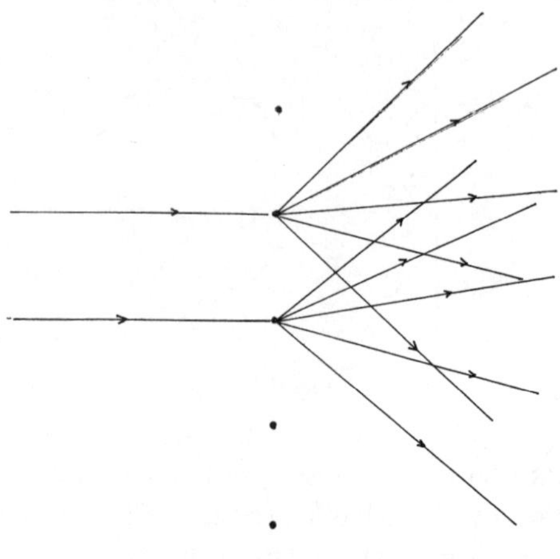

Figure 7.5

directions at angles of 11.5 degrees, 23.6 degrees, and 36.9 degrees relative to the incoming beam.) In these special directions, we would see bright spots of light.

How such magic occurs is revealed in Figure 7.6. Consider waves A and B, initially in phase with each other. These two waves will scatter from the grooves in all directions, as suggested in Figure 7.5, but for purposes of illustration the figure shows just one particular direction of scattering, indicated by the arrow. This direction has been carefully chosen for illustration for a reason: wave B, after it exits the grooves to the right, travels *exactly one whole wavelength* to catch up to wave A. Consequently, by the time the two waves have reached the dashed line, they are back in phase and will reinforce each other at any further point downstream. Wave B is like a watch that has been stopped for exactly twenty-four hours and then started again, finding itself back in synchrony with another watch, wave A. If all of the grooves are the same distance apart, wave A will likewise be exactly one whole wavelength behind the wave above it (in the direction shown), and those waves will also emerge in phase. And on and on, through thousands of grooves, each scattered wave exactly one wavelength behind the wave above it, with crests and troughs lining up. Thus, all waves scattering *in this particular direction* will be in phase.

By comparison, consider the smaller scattering angle shown for waves C and D. In this particular direction, wave D, after it exits the grooves to the right, travels *exactly one half of a wavelength* to catch up to wave C.

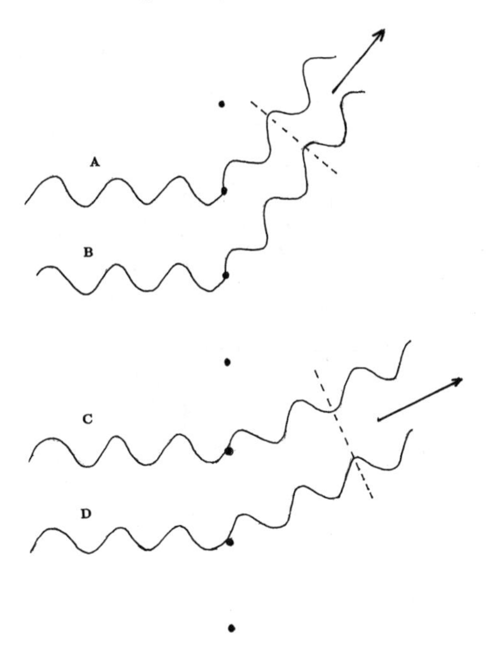

Figure 7.6

Thus, these two waves are out of phase with each other by the time they have reached the dashed line. They will cancel each other out, producing darkness. Indeed, every pair of waves scattering in this direction cancel each other out.

The end result is that the scattered waves will reinforce each other and produce bright spots only at certain angles (directions). As mentioned earlier, these special angles can be calculated simply in terms of the wavelength of incoming light and the distance between grooves. At all other angles (directions), the waves will be more or less out of phase with each other and cancel each other out.

Von Laue's revelation was that the tiny city of a crystal would act as a diffraction grating in three dimensions. For a crystal, individual atoms substitute for grooves and scatter incoming light. Analogously to the parallel grooves of a diffraction grating, the atoms in a crystal of matter are spaced at precisely repeating intervals, a crucial requirement for diffraction to work. As shown in Figure 7.7, parallel rays of X-rays come in

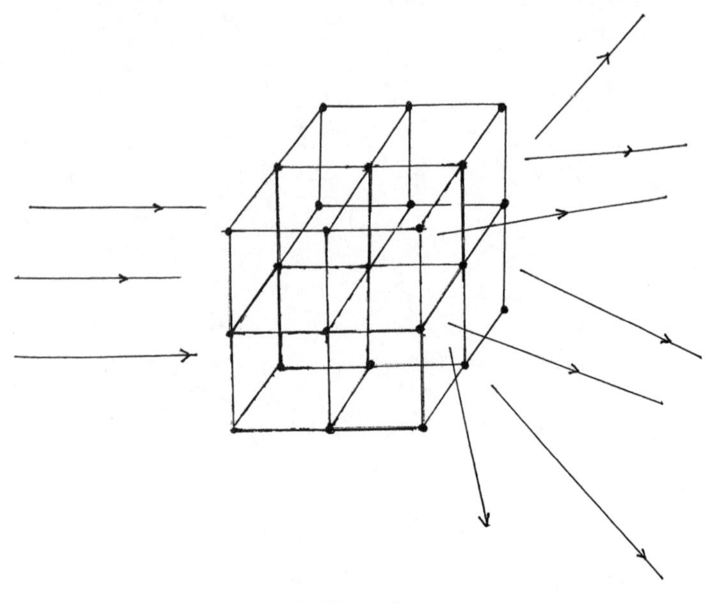

Figure 7.7

from the left, scatter within the three-dimensional crystal, and emerge in many directions on the right.

Three dimensions offer new complications. For the standard diffraction grating, interference between waves occurs only in one dimension, along the single row of grooves. But in a three-dimensional lattice, such as is shown in Figure 7.7, waves can interfere with each other along the three perpendicular rows of atoms. Thus full reinforcement requires that scattered waves arriving from atoms above and below each other, right and left of each other, and in front of and behind each other all be in phase when they emerge from the crystal. Depending on the spacing between atoms and unit cells, and the wavelength and direction of the incoming light, there may be no emergent direction for which all of these conditions are satisfied. When such directions do exist, as evidenced by bright spots of reinforced light on a photographic plate, the particular locations and patterns of those spots reveal much detail about the crystal lattice.

But why X-rays? Because only waves of such short wavelengths can probe the tiny distances between atoms in solid matter. A sewing thread can measure the eye of a needle, but a shoestring cannot. It was well known in 1912 that atoms are roughly 3×10^{-8} centimeters apart in solid matter. Visible light has wavelengths ranging from 4×10^{-5} to 7.7×10^{-5} centimeters, far too large to investigate the tiny spaces between atoms. But X-rays are perfect for the job. As mentioned earlier, X-rays had been measured to have wavelengths ranging from 10^{-6} to 10^{-9} centimeters. The shorter waves in this range would be roughly one-tenth the space between atoms, the best ratio for a diffraction grating to work.

To fully appreciate von Laue's idea, we must remember that in 1912, X-rays, also called Röntgen rays after their discovery by Wilhelm Röntgen in 1895, were not well understood. The X in the word stood for "unknown." All that was known for sure was that X-rays had extraordinary penetrating power. Experiments by the British physicist Charles Glover Barkla suggested that X-rays were electromagnetic waves of very short wavelength. On the other hand, some scientists believed that X-rays were material particles, like the high-speed electrons called β-rays. Thus, the diffraction patterns observed by Friedrich and Knipping, which could be plausibly produced only by the overlap of waves, not only provided a powerful tool for the study of atomic architecture in solid matter but also gave confirming evidence of the wave nature of X-rays.

Now, the paper itself. It is unusual for it to be rigidly divided into theoretical and experimental sections, with the authors Friedrich, Knipping, and von Laue separately apportioned between these two parts. But it is not unusual to have several scientists jointly author a largely experimental paper. Experiments, unlike theoretical ideas and calculations, often require the varied skills and resources of a team of people.

Von Laue begins by referring to the recent research of Barkla and to the older notions of Bravais that atoms in a crystal may be arranged in an orderly lattice. He then quickly outlines his proposal: by analogy to one-dimensional interference patterns with optical (visible) light, X-rays should produce three-dimensional interference patterns after traversing a crystal.

At the end of this brief introductory section, von Laue announces that "Friedrich and Knipping have tested the above hypothesis at my instigation." By habit, von Laue was a man who gave credit where credit was due. For example, in his Nobel address, he graciously acknowledges the influence of physicist Peter Paul Ewald in his thinking about electromagnetic waves in crystals. But here, in his landmark paper of 1912, von Laue clearly lays claim to the idea of X-ray diffraction. The Nobel committee in Sweden was convinced and awarded the prize to von Laue and von Laue alone. In an analogous manner, it is Ernest Rutherford who always gets acknowledged for discovering the nucleus of the atom, although the actual experiment was done by his assistants Geiger and Marsden, at Rutherford's suggestion.

In the next section of his paper, von Laue performs a standard calculation of the interference of electromagnetic waves. What is new about this calculation is that he considers the scattering centers to be distributed in three dimensions rather than one or two. He then specializes to the simple case when the unit cell is a cube, of side a, realizing that this special case will not apply to all crystals but will simplify the analysis. Here and elsewhere, von Laue is more interested in the principles of his ideas than in the details.

The main result is given in the last equation. This equation states that full reinforcement of the scattered waves will occur only if three conditions are satisfied, one for each dimension. The first condition, for example, requires that waves scattering off atoms in a row parallel to the x-axis travel distances differing from each other by a whole number of wavelengths, so that they are in phase with each other. As in the case of one-dimensional scattering, this condition is true only for certain directions α. The second and third conditions apply to atoms in rows parallel to the y and z directions, respectively.

The operational use of these three von Laue conditions is as follows: For each bright spot observed on the photographic film, one can measure its direction from the crystal and obtain the three angles α, β, and γ. (Technically speaking, α, β, and γ are the cosines of the angles the direction makes with respect to the x-axis, y-axis, and z-axis.) One then takes ratios of the angles. These ratios result in the ratios of integers: $\alpha/\beta = h_1/h_2$ and $\alpha/\gamma = h_1/h_3$. Next, one tries to find small integers h_1, h_2, and h_3 that satisfy these requirements. Once the integers are known, one goes back to the original equations and solves for λ/a. If λ is known, one thus derives a. If a is known, one thus derives λ. (Recall that λ is the wavelength of the X-ray and a is the length of a side of the unit cell.) Each individual spot can be the result of reinforcement of only a single wavelength. But if the incoming beam contains different wavelengths, different spots can be the result of different wavelengths. Evidently, von Laue, Friedrich, and Knipping found from their results, shown in Figures 2 and 3 of their paper, that they required five different incoming wavelengths to fit all of their spots. There may have been other wavelengths as well, but not all wavelengths are able to satisfy the three von Laue conditions.

In the last section of his part, "General Summary," von Laue argues that the observed interference phenomena could not have been produced by particle rays. Essentially, an incoming particle can strike only a single atom at a time and thus not stimulate a coordinated emission from rows of atoms, as can a plane of incoming waves all in phase with each other. Significantly, von Laue was at least as concerned with proving the wavelike nature of X-rays as he was with developing a new tool to measure the atomic spacings of solids. In retrospect, the latter has been far more important.

In the experimental part of the paper, Friedrich and Knipping begin by describing their apparatus. It is extremely important that the incoming X-ray beam be narrow, so that all incident waves travel the same distance to the crystal and thus be in phase with each other upon arrival there. Also important is that the various pieces of equipment be carefully lined up. The experimenters must be assured that the beam of X-rays, which is invisible to their eyes, properly makes its lightning-speed journey through all the collimating shutters and pinholes in order to strike the crystal in the bull's-eye.

I find it surprising and quaint that the crystal is attached to the goniometer (a device that can rotate the crystal) with common wax, although this substance is still used. That the exposure time took as long as hours, and sharp diffraction spots were still seen, seems extraordinary. Any shaking of the apparatus during the long exposure, a slight twisting of the crystal relative to the incoming X-ray beam, would have completely destroyed the delicate interference pattern and produced a dim blur on the films.

Several aspects of Figures 2 and 3, the key results, are worth noting. First, many diffraction spots on the actual photographic film were too dim to show up well in the reproduction of Figure 2 but are indicated in the schematic diagram of Figure 3. Second, the spots have a fourfold symmetry. That is, if we rotate the diagram by ninety degrees, it looks like the original. This high degree of symmetry is both characteristic of the diffraction process in general and also revealing of the particular symmetry of the zinc sulfide crystal under examination. Friedrich and Knipping consider their results to be "beautiful proof" of the orderly arrangement of atoms in a crystal. The positions of the various spots, using von Laue's three conditions, yield information about the incoming wavelengths and size of the unit cell. Von Laue, Friedrich, and

Knipping have achieved humankind's first glimpse into the atomic world of a crystal.

A mere two years after his paper of 1912, Max von Laue was awarded the Nobel Prize—one of the shortest intervals between discovery and prize of any Nobel. The importance of the work was instantly recognized. As early as 1913, the British physicist William Lawrence Bragg greatly improved upon von Laue's mathematical method and, with his father William Henry Bragg, used the new calculations and experiments to give the first detailed analysis of crystals. For this work, father and son Bragg also won the Nobel (the only father and son team to do so). In his Nobel Prize lecture, the younger Bragg begins by declaring: "You have already honoured with the Nobel Prize Prof. von Laue, to whom we owe the great discovery which has made possible all progress in a new realm of science, the study of the structure of matter by the diffraction of X-rays."

The year of his great discovery, von Laue was appointed professor of physics at the University of Zurich, and later, in 1919, the same exalted position at the University of Berlin. As a person, von Laue was highly respected for his character and judgment, and for several decades he had great influence in directing German science. However, he detested the National Socialist Party and Hitler's regime. In the middle to late 1930s, when Einstein and his "Jewish physics" were denounced by the Nazis, von Laue was practically the only German physicist who continued to support Einstein. During World War II, rather than support the German war effort, von Laue instead wrote a history of physics, which appeared in four editions and seven translations. The father of X-ray diffraction also loved speed. He routinely traveled at high speed on a motorbike to his lectures and, later, at high speed in cars. At age eighty, while driving at high velocity to his laboratory, he had a fatal collision with a motorcyclist.

Because almost all solid matter exists in crystalline form at sufficiently low temperatures, X-ray diffraction has become a far-ranging method of analysis, hailed by chemists and biologists alike as an essential tool for understanding the structure of matter. For example, a crucial step in the determination of the double-helical structure of DNA in 1953 was the X-ray diffraction work of Rosalind Franklin. In 1960, Max Perutz used X-ray diffraction to determine the structure of hemoglobin, the first protein molecule to be unraveled (along with its smaller cousin myoglobin). Today, almost all biochemical laboratories use X-ray diffraction techniques to uncover the spatial structure of organic molecules.

When I recently visited a laboratory at Brandeis University one morning, a young graduate student rushed up to me, without knowing who I was, and blurted out that he had just decoded the structure of some complex organic molecule with only half of the data. Apparently, he had been working all night. His eyes were bloodshot, his hands trembled. Employing X-ray diffraction and analyzing the interference patterns with a computer, he had an ironclad fit to his molecule after using only a portion of the available information—just as a crossword puzzle can often be finished without the use of all the clues. There is only one way the letters fit together to make words, one inescapable arrangement. "How could this be?" he asked, with confusion, excitement, and awe. And I was equally awed, awed that there is a miniature world underneath our gaze, a hidden world that truly exists, a world of both beauty and logic.

INTERFERENCE PHENOMENA WITH RÖNTGEN RAYS

W. Friedrich, P. Knipping, and M. von Laue

Communicated by A. Sommerfeld in the Proceedings on 8 June 1912

Sitzungsberichte der Kuniglich Bayerischen Akademic der Wissenschaften (1912)

THEORETICAL PART

M. von Laue

Introduction

BARKLA'S RESEARCH IN THE LAST FEW YEARS (C. G. Barkla, *Philosophical Magazine*, z B. 22, pg. 398, 1911) has shown that Röntgen-rays experience diffraction in matter similar to diffraction of light in translucent materials. Röntgen-rays also, however, induce the atoms of the material to send out spectrally homogeneous fluorescences, which are totally characteristic for the substance.

In addition, in 1850 Bravais introduced into crystallography the theory that atoms in a crystal are ordered in a lattice. If Röntgen-rays are really composed of electromagnetic waves, then one could assume that the lattice would produce interference by free or forced vibrations [of the electrons in the atoms], leading, specifically, to interference patterns of the same nature as those that are known in optical interference spectra. The constants of this lattice [distances between neighboring atoms] can be calculated from the molecular weight of the crystallized compound, its density, the number of molecules per gram equivalent, and crystallographic data. One always finds for these constants an order of magnitude of 10^{-8} centimeters, while the wavelength of the Röntgen-rays are of the order of 10^{-9} centimeters, as shown by Walter and Pohl (B. Walter and R. Pohl, *Annalen der Physik*, vol. 25, pg. 715, 1908; vol. 29, pg. 331, 1908), Sommerfeld (A. Sommerfeld, *Annalen der Physik*, vol. 38, pg. 473, 1912), and Koch (P. P. Koch, *Annalen der Physik*, vol. 38, pg. 507, 1912). A complication in these observations is that a 3-fold periodicity appears in these lattices, while with optical [one-dimensional] lattices, there is only one direction, and at most two in a [two-dimensional] cross-lattice.

Friedrich and Knipping have tested the above hypothesis at my instigation. They will describe these experiments and results themselves in the second part of this paper.

The Theory and Its Quantitative Comparison with Experiment

We would like to express the above thought mathematically. The midpoint positions of an atom are determined through the rectangular coordinates, x, y, z. We will consider the most common crystal space group, that is the triclinic lattice [unit-cell]. The edges of the elementary parallelepiped could therefore have any length and any angle between them. By choosing particular values of lengths and angles, one can get to any other type of lattice [unit cell]. Let us denote the length and direction of these edges by the vectors $\mathbf{a}_1$, $\mathbf{a}_2$, and $\mathbf{a}_3$. Then the midpoint of an atom lies at

$$x = ma_{1x} + na_{2x} + pa_{3x},$$

$$y = ma_{1y} + na_{2y} + pa_{3y},$$

$$z = ma_{1z} + na_{2z} + pa_{3z},$$

where m, n, and p are integers.

For a single atom, we will make the assumption that its vibrations are purely sinusoidal. Needless to say, this assumption may not be any more correct than it is in optics. However, just as in optics, one can describe spectral inhomogeneous rays [not all of the same wavelength] through Fourier analysis. The wave coming from any one atom can then be described at any distance from the atom through the expression

$$\Psi \frac{exp\,(-ikr)}{r},$$

where r is the radius vector from the atom to the observation point, Ψ is a function of the direction, and $k = 2\pi/\lambda$, where λ is the wavelength of the Röntgen-ray.

If the direction of the primary [incoming or undeflected] Röntgen-ray is denoted by α_0, β_0, and γ_0, [cosines of the angles made with the x, y, and z axes], then the amplitude of the primary ray is the superposition [summed over all atoms]

$$\Psi \frac{exp\,[-ik(r + x\alpha_0 + y\beta_0 + z\gamma_0)]}{r}.$$

If we then let the scattered ray have direction (α, β, and γ), then

$$r = R - (x\alpha + y\beta + z\gamma),$$

and the scattered ray has the amplitude [summed over all values of m, n, and p],

$$\Psi \frac{exp\,(-ikR)}{R} \Sigma\, exp\, [ik[x(\alpha - \alpha_0) + y(\beta - \beta_0) + z(\gamma - \gamma_0)]].$$

. . . In the special case where all three edges of the lattice [unit-cell] have the same length and are perpendicular, so that we can align the coordinate axes with these directions, and the incident Röntgen-ray travels in the z direction, then the equations simplify to:

$$a_{1x} = a_{2y} = a_{3z} = a,$$

with all the other **a** components zero, and

$$\alpha_0 = 0, \qquad \beta_0 = 0, \qquad \gamma_0 = 1.$$

The condition for maximum intensity [at the observation point] then reads:

$$\alpha = h_1\lambda/a, \qquad \beta = h_2\lambda/a, \qquad 1 - \gamma = h_3\lambda/a,$$

where h_1, h_2, and h_3 are integers.

On a photographic plate with which the [scattered] rays interact, the curves α = constant and β = constant are hyperbolae whose midpoint lies at the point of impact of the primary Röntgen-ray. If only the first two relations are satisfied [for α and β], then one sees the well known [two-dimensional] cross-lattice spectrum, in which in every crossection of two hyperbolae there appears an intensity maximum. The curves γ = constant choose from the cross-lattice spectrum those points that appear in Figures 2 and 3. . . .

One has to be careful to recognize that any lattice is not uniquely divided into elementary parallelepipeds but can be sorted into an infinite number of these shapes. In a regular lattice, for instance, one can dissect such parallelepipeds into many other [unit cells]. In a cubic array, for example, any side is the diagonal of another cube, and thus the principal directions have a two-fold symmetry axis. The intensity maxima must therefore be consistent with different perspectives of a cube. In fact, Figures 2 and 3 show that the spots order themselves in a circle around a two-fold axis. Here, one has irradiated along the axis of a regular crystal and the photographic plate stands perpendicular to the axis. . . .

What we can see is that the theory can be fitted to a four-fold symmetric description under the assumption that there are multiple wavelengths from $0.038a$ to $0.15a$. Since a is equal to 3.3×10^{-8} centimeters for zinc sulfide, the wavelengths must therefore occur in the interval from 1.3×10^{-9} to 5.2×10^{-9} centimeters. . . .

General Summary

Let's discuss at the end, without any reference to the formulae, the question of how these experiments relate to the wave nature of Röntgen-rays. That the diffracted rays have a wave character is shown by the sharpness of the intensity maxima, which are easy to understand from the interference phenomenon but not easy to understand from the corpuscular assumption [the assumption of the finite size and materiality of the atom]. [The wave character of Röntgen-rays is also demontrated by] the ease with which the rays go through the material, only possible for the fastest β-rays.

Nevertheless, one could doubt the wave nature of the primary rays. Imagine, therefore, that the atoms of the crystal are stimulated by a particle ray. The particle nature of Röntgen-rays, which is assumed by many researchers, can be included here under the definition of particle rays. Coherent diffraction can only arise from rows of atoms which are hit by the same particle and are parallel to the z direction. Atoms that have a certain offset in the $x - y$ direction are then stimulated from different particles. For these atoms, specific phase differences could not occur [for particle rays].

The breaking of the circle that appears [in Figure 2], therefore, cannot be understood based on the particle interpretation. In addition, the primary [undeflected] and diffracted rays are so similar that one can deduce the wave nature of the latter from the wave nature of the former. Those diffracted rays that come from the crystal have a spectral homogeneity. The primary ray, instead, as consistent with the Bremsstrahlung experiments of Sommerfeld, must consist of nonperiodic waves [a mixture of wavelengths]. Undetermined at the moment is whether the periodic rays are initially determined by fluorescence in the crystal or whether they are already present in the primary ray and are merely sorted by the crystal.

EXPERIMENTAL PART

W. Friedrich and P. Knipping

For an experimental proof of the previously described theoretical ideas, a definitive experimental protocol was used after a number of attempts with a provisional apparatus, shown in Figure 1. Anticathode A in a Röntgen-ray tube produces Röntgen-rays in a narrow beam of approximately 1 millimeter in diameter passing through the shutters B_1 to B_4. This beam goes through crystal *Kr*, which is set onto goniometer *G*. Around the crystal in various directions and at various distances were photographic plates *P*, on which the intensity distribution was registered from the secondary reflections from the crystal. In order to prevent unwanted diffraction, a lead backstop *S* was used, and the whole detector was enclosed in a lead box *K*. We determined the exact line-up optically. Using a cathode meter whose end had a cross hair, we could determine the alignment of all of the pieces, including the beam, the

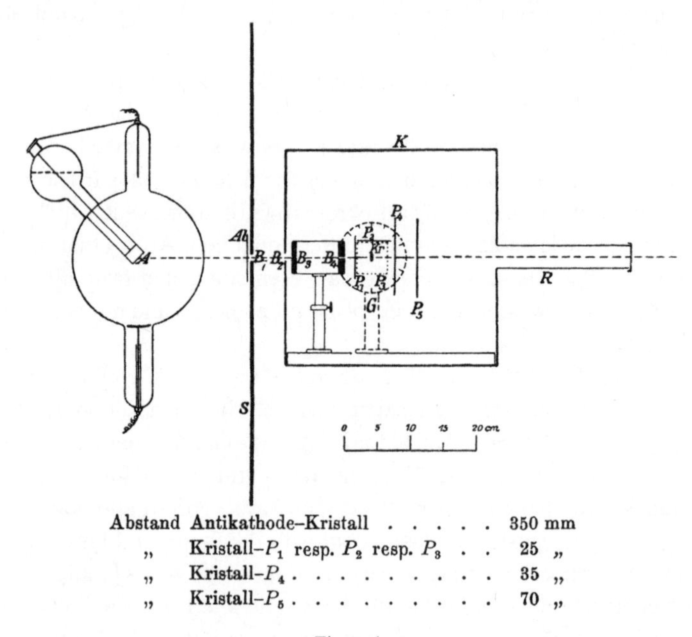

Figure 1

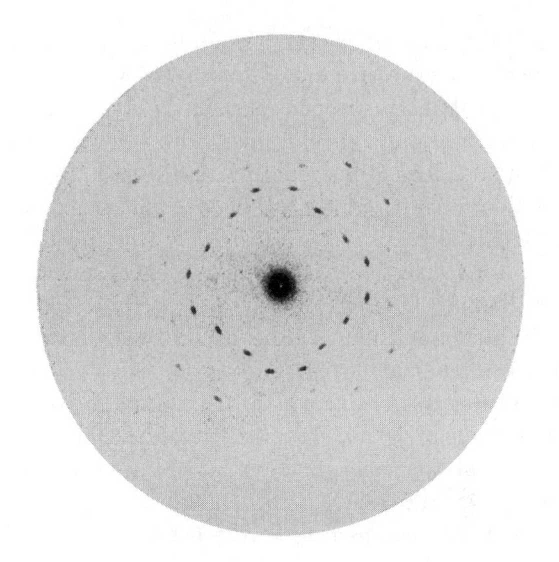

Figure 2

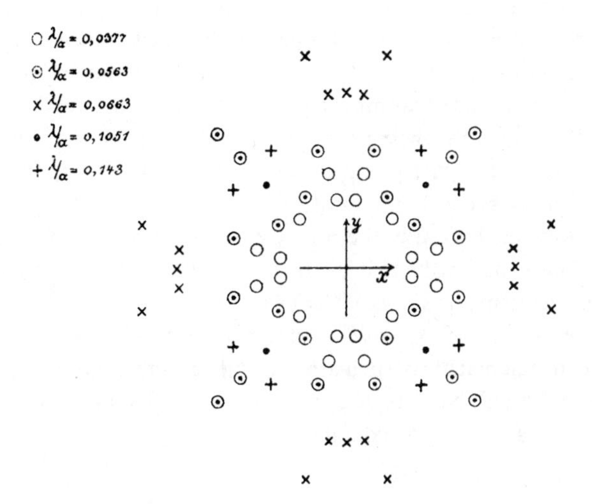

Figure 3

shutters, and the goniometer axis. A control experiment showed that this optical method translated to Röntgen-rays as well. . . .

When the apparatus was properly aligned, the crystal was attached to the goniometer head with a small amount of wax. It was oriented in the beam with the aid of the cathode meter with cross hair. This very important calibration could be obtained within one minute [an angular measure equal to 1⁄60 degree].

After we tried several detectors that were not suitable, we used Schleubner-Röntgen films, which were developed with Rodinal (1:15) and gave the best results. With a power of 2–10 milliampres in the tube, 1–20 hours were required for exposure time. . . .

Copper sulfate crystals were used as the control. In order to be sure that the reflections originated from the crystal structure of the copper sulfate, the crystals were pulverized, put into a small paper box, and the experiment was repeated. . . .

We assumed that the observation of a regular system with a space group more simple than triclinic would be easier to interpret. Therefore, zinc blende, with which we had already made some attempts, was used and produced more intense secondary reflections. . . .

The positions of the secondary spots [in Figure 2] are completely symmetrical with regard to the through [undeflected, or primary] beam. One can draw two pairs of mutually perpendicular symmetry axes in the figure. There are also two mirror planes. The fact that two symmetry axes and two mirror planes appear on the plate is one of the most beautiful proofs for the space group of the crystal. No other aspect beyond the space group needs to be considered. . . .

If we turn the crystal in the primary beam, we must conclude from the previous experiments that the picture on the plate turns with it. In fact, such experiments are totally conclusive. . . .

During the time of exposure, we noticed that the intensity of the primary beam fluctuated, particularly in older Röntgen-ray tubes, by as much as 6–12 fold. Nevertheless, as shown in the diagrams, the secondary spots remained sharp and unmoved. . . .

8

THE QUANTUM ATOM

AT THE END OF 1911, a twenty-six-year-old physicist named Niels Bohr traveled to Manchester to speak to Professor Rutherford. The young man was deeply discouraged. He had been developing some new hypotheses about atoms and, the previous summer, tried to present them to the celebrated J. J. Thomson, discoverer of the electron, father of the plum pudding model, director of the Cavendish Laboratory in Cambridge. But Thomson had shown little interest in young Bohr's ideas and buried his latest paper under a pile of unread manuscripts. Now, Bohr pleaded again for an audience. We might imagine the scene in Rutherford's hodgepodge of a lab. The smell of battery acid floats through the air. Rutherford is barking orders to his assistants while fiddling with the zinc sulfide screen of a misbehaving scintillation counter. Then he looks up to see Bohr, standing timidly with his notebook, dressed in a suit and battered shoes, his pipe hidden away in his pocket, unable at the moment to reveal his big, bucktoothed smile. Could the professor possibly spare a few minutes of his time? Bohr asks, in a voice as faint as the rustle of a leaf.

In many ways, Rutherford and Bohr were opposites. While Rutherford was loud, explosive, confident almost to a fault, Bohr was quiet, shy, self-effacing, sensitive, philosophical. While Rutherford roared directly at a problem full throttle, Bohr tended to circle it slowly, debating all possibilities in his mumbled whispers. (Years later, according to legend, Bohr admonished his students: "You should speak as clearly as you think, but not more so.") Rutherford was an experimentalist, a nuts-and-bolts man. Bohr worked only with pencil and paper.

Despite these differences, Rutherford immediately took to the soft-spoken Danish physicist, recognized his brilliance beneath the diffident manner, and invited him to stay on at Manchester. In the following six months, there developed between the two men an almost father-son relationship, filled with mutual admiration as well as affection. This support may have been just what Bohr needed. Within another year, now back at the University of Copenhagen, he unveiled his revolutionary model of the atom. Many scientists balked at Bohr's proposals because they turned classical physics on its head. But Einstein was delighted. When he heard about the astonishing agreement between Bohr's theory and experimental results, he proclaimed that this was one of the great discoveries of science.

Bohr's model of the atom combined Rutherford's nuclear model with the new quantum physics of Planck and Einstein. In particular, Bohr hypothesized that an electron in circular orbit about the central nucleus of an atom could not have a continuous range of energies—corresponding to a continuous range of average distances from the nucleus—but only certain energies, separated by gaps. As if a planet could circle a central star at 100 million miles or 400 million miles or 900 million miles, but nothing in between.

As Bohr stood before Professor Rutherford that first time in the winter of 1911, he was haunted by two major unsolved problems in physics. Both had to do with the nature of atoms. The first was a paradox. According to the well-established theory of electromagnetism, laid out sixty years earlier, any electrically charged particle that is diverted from a straight-line path should radiate electromagnetic waves and thereby lose energy. In particular, the electrons orbiting the central nucleus in Rutherford's atom should continuously lose energy, causing them to spiral into the tiny nucleus. A simple calculation, based on numbers known at the time, indicated that all atoms should collapse by this process in far less than a second. Yet it was as certain as sunrise that atoms did not behave in this way. Aside from the newly discovered radioactive atoms, atoms clearly maintained their structure and sizes without any change. If Rutherford also worried about these theoretical contradictions, he had kept his worries to himself.

The second conundrum concerned the periodic chemical properties of the elements. In 1869, the Russian chemist Dmitri Ivanovich Mendeleyev published his profound but mysterious Periodic Table, a classification of chemical elements according to increasing atomic weights. In particular, Mendeleyev showed that elements arranged in order of increasing weights showed regular repetitions of chemical properties. By the early 1900s, scientists understood that chemical bonds were formed by electrical forces between atoms. Thus, Thomson and other scientists were struggling to explain Mendeleyev's regular repetitions in terms of the placement of the electrically charged electrons within atoms.

A third mystery in physics—although there had been few attempts to explain it—was the so-called atomic spectra. We have already discussed the general idea of a radiation spectrum in connection with Planck's paper of 1900, in Chapter 1. By the mid-nineteenth century, it had been experimentally shown that each type of atom, when energized by heat, emits light only at *particular* frequencies unique to that atom. For example, hydrogen atoms emit light at frequencies (in cycles per second) of 3.08×10^{15}, 8.19×10^{14}, 6.17×10^{14}, and others; carbon atoms emit light at frequencies of 9.58×10^{15}, 1.21×10^{15}, 3.32×10^{14}, and others. Like fingerprints, the frequencies of light emitted by each type of atom distinguish it from all other atoms. Atomic spectra could be measured to very high accuracy with prismlike devices that fanned out the different frequencies, or colors, in different directions.

While the observed features of atomic spectra did not exactly violate any known laws of physics, neither could they be explained. For example, why were there *gaps* between the emitted frequencies of each atom, as if a singer could sing only the notes C-sharp, G, and A, with no notes in between? The precision with which spectra could be measured, the stubborn insistence of an atom to emit the same frequencies of light over and over, always identical to the emission of other atoms of its kind but different from all other atoms, only served to emphasize the degree of ignorance of science at the time.

In 1885, the Swiss physicist and mathematician Johann Jakob Balmer discerned *patterns* in the string of numbers that were the frequencies of light emitted by hydrogen. (Hydrogen is the simplest of all atoms, consisting of a single electron orbiting a single proton.) Furthermore, Balmer managed to construct a remarkable organizing formula for these patterns. Balmer's formula was:

$$\nu = 3.29 \times 10^{15} \times \left(\frac{1}{n^2} - \frac{1}{m^2} \right),$$

where ν stands for the emitted frequency in cycles per second, $n = 2$, and m can be any whole number ranging from 3 and higher. For example, for $m = 3$, the above formula gives

$$\nu = 3.29 \times 10^{15} \times \left(\frac{1}{4} - \frac{1}{9} \right) = 6.17 \times 10^{14},$$

which is one of the observed frequencies of hydrogen. Other frequencies are obtained by setting $m = 4, 5, 6$, and so on, in turn, like a piano player striking a sequence of ascending keys. (Balmer actually had a slightly different formulation of his findings, which he discussed in terms of wavelengths rather than frequencies. But his results were equivalent to the above equation.)

Balmer had no theory at all for this formula. It was just a mathematical shorthand for expressing the patterns he saw. As we have discussed in Chapter 1 on Planck, such a mathematical formula, by itself, is like a sun calendar that gives the number of daylight hours on each day of the year. A sun calendar is useful for making plans, but it does not offer any explanation for the scientific principles that cause the results.

The Swiss physicist went on to speculate that all other emitted frequencies of hydrogen, even those not yet observed, could be obtained by letting n, as well as m, range over all possible whole numbers. For example, setting $n = 3$ and letting $m = 4$ or $m = 5$ would give two more predicted frequencies of hydrogen light. Balmer's formula was a wondrous magical trick. And nobody, including Balmer, knew why it worked.

Beautiful patterns, as in Balmer's formula, do not happen by chance. Like the periodic ebb and flow of the tides, or the perfect six-sided symmetry of each snowflake, a simple pattern is compelling evidence for a simple mechanism or fundamental principle of nature. But what was the mechanism?

Surprisingly, Bohr had never seen Balmer's formula at the time that he visited Rutherford in Manchester. However, he did know about the first two unsolved problems. And, like many other physicists, he knew about the quantum idea of Planck. Finally, he knew firsthand about the failure of classical (that is, nonquantum) physics in the atomic domain. For his doctoral dissertation, Bohr had studied the behavior of electrons in matter and found unequivocally that classical electromagnetic theory could not account for the observed magnetic properties of metals. Classical physics clearly needed revision.

In 1912, Bohr began postulating atomic models in which certain parameters of the electron's orbit were "quantized"—that is, they could have only certain values. In this way, he hoped eventually to explain the two unsolved problems: why electrons did not spiral into the nucleus and why elements had periodically repeating chemical properties.

Quantum models of the atom were in the air. Other scientists, including Arthur Eric Haas, a doctoral student at the University of Vienna; the Dutch theoretical physicist Hendrik Antoon Lorentz; and the British physicist John William Nicholson, were introducing Planck's quantum constant into atomic models, attempting to explain not only the outstanding problems mentioned above but also the *dimensions* of the atom. The size of the atom was fairly well known from experiment. But theorists wanted to know *why* it had this size. What fundamental principles determined the size of an atom? Rutherford's atomic model gave no clue as to *why* the electrons orbited the central nucleus at distances of about 10^{-8} centimeters. Planck's constant for the quantum of energy, these scientists suspected, might be a key factor.

For Bohr, the competition was closing in. In a letter at the time, he wrote that "I am afraid I must hurry if [my work] is to be new when it appears. The question is indeed such a burning one."

Then, in February 1913, Copenhagen's expert on spectroscopy, H. M. Hansen, asked Bohr to explain Balmer's formula. Apparently, Bohr had never before seen that magical equation. "As soon as I saw Balmer's formula," Bohr later recalled, "the whole thing was immediately clear to me."

Let us try to reconstruct Bohr's thoughts. The young Danish theorist may well have been inspired by two crucial clues from Balmer's formula. First, the emitted frequencies involved one number *subtracted* from another. (As mentioned earlier, Balmer had no inkling of why such a subtraction was involved; he knew only that a subtraction seemed required to match the patterns in the data.) Second, Balmer's formula worked—that is, fit the observed frequencies of hydrogen—only if each of the numbers n and m could change only in whole steps.

Suppose, Bohr might have reasoned, that each of the two whole numbers in Balmer's formula represented an allowed energy level of the electron as it orbited the atomic nucleus. That these levels could change only in steps was reminiscent of Planck's quantum idea that the energy of atomic resonators could change only by unit increments or steps.

Suppose further that, for some strange quantum reason, these orbits were stable, meaning that electrons could not emit radiation (and energy) while circling in an allowed orbit. Unable to emit radiation, a circling electron would hoard its energy forever, like a miser who stashes his money under the bed and never spends a cent. But if an electron could *jump* from one of these orbits to another orbit of lower energy, then it would necessarily have to spend some of its energy—by an amount equal to the energy of the second orbit subtracted from the energy of the first. (Aha, we're now getting a glimmer of the true significance of subtraction in Balmer's formula.) Since *total* energy cannot change, the energy the electron loses in making such a jump would be given to the creation and emission of light, which carries its own energy. The miser's stash under his bed has finally decreased, and the missing coins have appeared in the grocer's hands in the form of light.

For the coup de grâce, recall Einstein's proposal of 1905. Light energy comes in individual quanta, with the *frequency of a quantum proportional to its energy*. Applied here, an electron jumping from one allowed orbit to another emits a quantum of light whose frequency is proportional to its energy. The energy in that light quantum, in turn, is the difference in energy of two allowed orbits of a circling electron. And those individual orbital energies can only be represented by whole numbers. Putting all of this together, finally, the emitted atomic frequencies would always be proportional to the difference of two numbers, each of which could change only by whole steps.

Bohr begins his paper by referring to "Prof. Rutherford's" recent nuclear model of the atom and stating that the Rutherford atom is unstable, that is, it should collapse. Bohr seems to imply that the earlier, Thomson atomic model was stable. But in fact it too should have collapsed under the classical laws of physics, both by radiation of the electrons and by mechanical instabilities. (Thomson originally tried to stabilize his model by assuming that there were so many electrons that they formed almost a solid ring of charge within the "pudding" of positive charge. In such a situation, the electrons would not radiate and spiral inward. However, by 1910, experiments at the Cavendish had shown that such an assumption of thousands of electrons violated the experimental evidence.) In any case, Bohr remarks on the "general acknowledgement of the inadequacy of the classical electrodynamics" in the tiny world of the atom and suggests that Planck's idea of the quantum may offer a solution.

Guided by the above reasoning, Bohr then makes some critical assumptions. First, the possible energy W of an orbiting electron cannot be any value, as would be the case under classical theory, but is restricted to being a whole multiple, τ, of one-half the orbital frequency, ω, times Planck's constant, h. (Bohr calls a whole number an "entire number.") This assumption is written mathematically in equation (2). Note that there are two frequencies in Bohr's paper: the orbital frequency, denoted by ω, which is the number of orbits per second that an electron makes about the central nucleus of an atom; and the frequency of light emitted when an electron changes from one orbit to another, denoted by ν.

The proposal here is very similar to Planck's 1900 assumption that the energy of a "resonator" in contact with black-body radiation can be only a multiple of the frequency of that resonator. However, Bohr's factor of ½ seems somewhat arbitrary. In physics, volumes can be written on such seemingly inconsequential numbers. Why not a factor of 1, as Planck used, or a factor of ⅔, or 2? At the very end of Bohr's paper, in §3, he shows that the precise factor must be ½ so that for large values of τ, where classical physics should be increasingly valid, the emission frequencies predicted by his model agree with those predicted by classical electromagnetic theory. Even this argument is not really a justification of the factor of ½ but rather a necessity for the results to come out right. A full justification of this factor would have to await the full quantum theory of Werner Heisenberg and Erwin Schrödinger, fifteen years later. Indeed, throughout his paper, despite the beautiful agreements between theory and experiment, Bohr is keenly aware that his theory hinges on unproved assumptions, that it is incomplete and provisional, just as Einstein called his theory of the light quantum "heuristic."

Next, *reverting to classical physics* for the balance of electrical force and acceleration, Bohr obtains the various formulas of equation (3). Notice now that all of the properties of the electron's orbit—its energy, W, its orbital frequency, ω, and its distance, a, from the central nucleus—are completely determined in terms of known constants: the electrical charge, e, and mass, m, of the electron, Planck's constant, h, plus the single quantum number τ. (Don't confuse this m, for the electron mass, with the m in Balmer's formula.) Thus the quantum number τ determines everything. As τ increases from 1 to 2 to 3 and so on, different allowed orbits are mapped out, at larger and larger radii.

For $\tau = 1$, that is, for the *first* quantum level, and for a nucleus with a single positive charge, $E = e$, Bohr puts in the known numbers for e, m, and h and *deduces* the diameter of the hydrogen atom, $2a = 1.1 \times 10^{-8}$ centimeters. This number agrees well with experimental results and is a major triumph of the theory.

Bohr's next assumption is that the electron cannot radiate energy while in one of these quantum configurations. Bohr calls these stable configurations "stationary states."

Now, Bohr digresses to mention the very recent work of John Nicholson, who also proposed a quantum model for the atom. By the time Nicholson's model had been published, Bohr had already developed most of his own theory. Nicholson almost got it right, and thus might have won the Nobel instead of Bohr, but he was missing the critical idea of stationary states (in which electrons cannot radiate) and the quantum jumping between states (during which electrons emit a single gasp of energy). In Nicholson's model, electrons radiate *while* they are in particular quantum states of definite energy, continuously changing their energy and thus contradicting the starting assumption that they had only particular, quantized energy levels. Furthermore, Nicholson's predicted emission frequencies do not have the mathematical form of Balmer's formula.

Next, Bohr augments and clarifies his assumptions, laid out as (1) and (2). While ordinary (nonquantum) mehanics can account for the force balance of an electron in a stationary orbit, that mechanics cannot describe how an electron "passes" from one orbit to another. Furthermore, when an electron does change orbit, it emits light of a single frequency, which Bohr calls "homogeneous radiation."

At this point, still limbering up in the first section of his paper, Bohr has already proposed to violate classical physics in several ways: (1) electrons in atoms can have only certain energies and orbits, (2) electrons cannot radiate while in those allowed orbits, (3) somehow, electrons can "pass" from one allowed orbit to another, and when they do so they emit a single quantum of energy, with frequency given by Einstein's 1905 relation, $E = h\nu$.

It is most interesting that Bohr describes electrons as "passing" from one orbit to another. Yet he can offer no physical picture for what he means by this verb. In classical physics, the electron can accurately be described as "spiraling" from one orbit to another as it slowly loses energy. We can visualize such a spiral, as we have all seen water winding down a drain, eagles wheeling in an updraft, and other spiraling things.

But the work of Balmer and Planck, as interpreted by Bohr, suggests that the electron cannot occupy the space between orbits in any previously known way—otherwise it would radiate energy in a continuous manner. Somehow, it is possible for an electron to begin at one energy level, corresponding to one orbit, and suddenly reappear at another energy level and orbit. I have just now used the word "reappear." Bohr uses the word "pass." Some scientists use "jump." But, in fact, we have no adequate vocabulary to describe such phenomena because all of our vocabulary comes from our human experience with the world. And we humans have no experience, no intuition, no direct sensory connection with the atomic world of the quantum. In that world, our language fails us.

Bohr himself was aware of these difficulties with language. In 1928, as quantum mechanics was being further developed, he wrote:

> we find ourselves here on the very path taken by Einstein in adapting our modes of perception borrowed from the sensations to the gradually deepening knowledge of the laws of Nature. The hindrances met on this path originate above all in the fact that, so to say, every word in the language refers to our ordinary perception.

Finally, Bohr is ready to get paid for his sacrificial and heretical assumptions. He specializes his considerations to hydrogen, the simplest atom, with a single electron orbiting a nucleus of a single positive charge. If the electron begins in a stationary state represented by $\tau = \tau_1$ and ends in a state with $\tau = \tau_2$, the difference in energy is easily calculated by Bohr's previous formulas. By the law of the conservation of energy, the change in energy of the electron must equal the energy in light emitted. Then, using the Einstein relation that the energy of the light quantum equals $h\nu$, the emitted frequency is given by equation (4) in Bohr's paper. By fixing one of the τs and letting the other vary, Bohr can reproduce the various "series" of spectral emissions observed by Balmer, Paschen, and other physicists.

Equation (4) is the glowing triumph of Bohr's atomic theory. Notice that it has exactly the same mathematical form as the experimentally observed frequencies, summarized by Balmer's formula (with τ_1 standing for m and τ_2 for n). When Bohr substitutes in known values for the "constants" m, e, and h, he arrives at the overall numerical coefficient outside the parentheses:

$$\frac{2\pi^2 me^4}{h^3} = 3.1 \times 10^{15} \textit{ cycles per second},$$

which is fairly close to the experimentally observed number of 3.29×10^{15}, given uncertainties in the measured values of the constants. More recent values of these constants show agreement to several decimal places. Thus, given the assumptions of his model, Bohr has theoretically calculated *both* the size of atoms and their energy emissions in terms of fundamental constants of nature—and found agreement with known experimental results! Numbers that previously had to be accepted as givens, without explanation, have now been explained in terms of fundamental principles.

In another critical success of his theory, Bohr applies it to the case of a "singly ionized" helium atom, that is, one electron orbiting a nucleus of charge $2e$, twice the charge of the hydrogen nucleus. The resulting numbers agree with the frequencies of certain emissions measured by the spectroscopists Alfred Fowler, E. C. Pickering, and others. Formerly, these emissions had been attributed to hydrogen, without any theoretical justification. Here is a concrete prediction. Bohr is not just proposing a theory to explain things already known. He is *predicting* that the material substance responsible for the emissions observed by Fowler and Pickering is helium, not hydrogen. Later experiments showed that Bohr was correct. With the success of helium, Fowler and other physicists, who were almost scandalized by Bohr's drastic assumptions, decided that there must be something to his ideas.

Niels Bohr, with this paper, became widely regarded as the father of modern atomic physics. Subsequently, he would delve even more deeply into the atom and become a pioneer in nuclear physics, the behavior not of the outer electrons but of the inner nucleus of protons and neutrons.

One cannot help wondering why it was Bohr, and not other theoreticians, who conceptualized the first quantum model of the atom. Nicholson was close. For example, why not Einstein, who was familiar with the problem of atomic spectra and the radiative instability of the Rutherford atom, who clearly understood the idea of the quantum, and who was only a few years older than Bohr? After seeing Bohr's paper, Einstein told the Hungarian physicist György Hevesy that he also had been thinking about the problem of atomic spectra and a quantum model of the atom but, in Hevesy's recollection, "had no pluck to develop it." This comment is somewhat puzzling, as Einstein was not particularly modest about his scientific abilities.

Perhaps it was the complexity of emission spectra for every element except hydrogen that discouraged Einstein, who liked things simple. Or perhaps it was that Einstein, in 1912, was preoccupied with his new theory of gravity, general relativity. Maybe Bohr's success and priority of discovery derived from his special fascination with paradoxes. In this case, the paradox was that classical physics could apparently live side by side with quantum physics. The former correctly described the mechanical balance of forces of an electron in a stationary orbit, while the latter decreed, without pictures, that the electron could jump or pass or reappear somewhere else without making the trip.

Indeed, Bohr's way of thinking in terms of paradoxes and contradictions formed the philosophical foundation of his Institute for Theoretical Physics, a house-sized building on Blegdamsvej 15 in Copenhagen. Here, Bohr created a scientific "school," in the vein of Arnold Sommerfeld's group in Munich and Werner Heisenberg's group in Leipzig. At Bohr's institute, daily stimulation from brilliant seminars and disturbing new ideas could dismast slow thinkers. The American physicist John Wheeler, who studied with Bohr in the mid-1930s (and who trained my own thesis adviser, Kip Thorne, making me a great-grandstudent of Bohr) recalls Bohr's usual method of explanation as a "one-man tennis match." Each hit of the ball would be some telling contradiction to previous results, raised by a new experiment or theory. After each hit Bohr would run around to the other side of the court quickly enough to return his own shot. "No progress without a paradox," was Bohr's mantra. The worst thing that could happen in a visitor's seminar was the absence of surprises, after which Bohr would politely mutter those dreaded words, "That was interesting."

ON THE CONSTITUTION OF ATOMS AND MOLECULES

Niels Bohr, Dr. phil. Copenhagen[1]

Philosophical Magazine (1913)

INTRODUCTION

IN ORDER TO EXPLAIN the results of experiments on scattering of α rays by matter Prof. Rutherford[2] has given a theory of the structure of atoms. According to this theory, the atoms consist of a positively charged nucleus surrounded by a system of electrons kept together by attractive forces from the nucleus; the total negative charge of the electrons is equal to the positive charge of the nucleus. Further, the nucleus is assumed to be the seat of the essential part of the mass of the atom, and to have linear dimensions exceedingly small compared with the linear dimensions of the whole atom. The number of electrons in an atom is deduced to be approximately equal to half the atomic weight. Great interest is to be attributed to this atom-model; for, as Rutherford has shown, the assumption of the existence of nuclei, as those in question, seems to be necessary in order to account for the results of the experiments on large angle scattering of the α rays.[3]

In an attempt to explain some of the properties of matter on the basis of this atom-model we meet, however, with difficulties of a serious nature arising from the apparent instability of the system of electrons: difficulties purposely avoided in atom-models previously considered, for instance, in the one proposed by Sir J. J. Thomson.[4] According to the theory of the latter the atom consists of a sphere of uniform positive electrification, inside which the electrons move in circular orbits.

The principal difference between the atom-models proposed by Thomson and Rutherford consists in the circumstance that the forces acting on the electrons in the atom-model of Thomson allow of certain configurations and motions of the electrons for which the system is in a stable equilibrium; such configurations, however, apparently do not exist for the second atom-model. The nature of the difference in question will perhaps be most clearly seen by noticing that among the quantities characterizing the first atom a quantity appears—the radius of the positive sphere—of dimensions of a length and of the same order of magnitude as the linear extension of the atom, while such a length does not appear among the quantities characterizing the second atom, viz. the charges and masses of the electrons and the positive nucleus; nor can it be determined solely by help of the latter quantities.

The way of considering a problem of this kind has, however, undergone essential alterations in recent years owing to the development of the theory of the energy radiation, and the direct affirmation of the new assumptions introduced in this theory, found by experiments on very different phenomena such as specific heats, photoelectric effect, Röntgen-rays, &c. The result of the discussion of these questions seems to be a general acknowledgment of the inadequacy of the classical electrodynamics in describing the behaviour of systems of atomic size.[5] Whatever the alteration in the laws of motion of the electrons may be, it seems necessary to introduce in the laws in question a quantity foreign to the classical electrodynamics, *i. e.* Planck's constant, or as it often is called the elementary quantum of action. By the introduction of this quantity the question of the stable configuration of the electrons in the atoms is essentially changed, as this constant is of such dimensions and magnitude that it, together with the mass and charge of the particles, can determine a length of the order of magnitude required.

This paper is an attempt to show that the application of the above ideas to Rutherford's atom-model affords a basis for a theory of the constitution of atoms. It will further be shown that from this theory we are led to a theory of the constitution of molecules.

In the present first part of the paper the mechanism of the binding of electrons by a positive nucleus is discussed in relation to Planck's theory. It will be shown that it is possible from the point of view taken to account in a simple way for the law of the line spectrum of hydrogen. Further, reasons are given for a principal hypothesis on which the considerations contained in the following parts are based.

I wish here to express my thanks to Prof. Rutherford for his kind and encouraging interest in this work.

PART I.–BINDING OF ELECTRONS BY POSITIVE NUCLEI

§ 1. General Considerations

The inadequacy of the classical electrodynamics in accounting for the properties of atoms from an atom-model as Rutherford's, will appear very clearly if we consider a simple system consisting of a positively charged nucleus of very small dimensions and an electron describing closed orbits around it. For simplicity, let us assume that the mass of the electron is negligibly small in comparison with that of the nucleus, and further, that the velocity of the electron is small compared with that of light.

Let us at first assume that there is no energy radiation. In this case the electron will describe stationary elliptical orbits. The frequency of revolution ω and the major-axis of the orbit $2a$ will depend on the amount of energy W which must be transferred to the system in order to remove the electron to an infinitely great distance apart from the nucleus. Denoting the charge of the electron and of the nucleus by $-e$ and E respectively and the mass of the electron by m, we thus get

$$\omega = \frac{\sqrt{2}}{\pi} \frac{W^{3/2}}{eE\sqrt{m}}, \quad 2a = \frac{eE}{W} \ldots \quad (1)$$

Further, it can easily be shown that the mean value of the kinetic energy of the electron taken for a whole revolution is equal to W. We see that if the value of W is not given, there will be no values of ω and a characteristic for the system in question.

Let us now, however, take the effect of the energy radiation into account, calculated in the ordinary way from the acceleration of the electron. In this case the electron will no longer describe stationary orbits. W will continuously increase, and the electron will approach the nucleus describing orbits of smaller and smaller dimensions, and with greater and greater frequency; the electron on the average gaining in kinetic energy at the same time as the whole system loses energy. This process will go on until the dimensions of the orbit are of the same order of magnitude as the dimensions of the electron or those of the nucleus. A simple calculation shows that the energy radiated out during the process considered will be enormously great compared with that radiated out by ordinary molecular processes.

It is obvious that the behaviour of such a system will be very different from that of an atomic system occurring in nature. In the first place, the actual atoms in their permanent state seem to have absolutely fixed dimensions and frequencies. Further, if we consider any molecular process, the result seems always to be that after a certain amount of energy characteristic for the systems in question is radiated out, the systems will again settle down in a stable state of equilibrium, in which the distances apart of the particles are of the same order of magnitude as before the process.

Now the essential point in Planck's theory of radiation is that the energy radiation from an atomic system does not take place in the continuous way assumed in the ordinary electrodynamics, but that it, on the contrary, takes place in distinctly separated emissions, the amount of energy radiated out from an atomic vibrator of frequency ν in a single emission being equal to $\tau h\nu$, where τ is an entire number, and h is a universal constant.[6]

Returning to the simple case of an electron and a positive nucleus considered above, let us assume that the electron at the beginning of the interaction with the nucleus was at a great distance apart from the nucleus, and had no sensible velocity relative to the latter. Let us further assume that the electron after the interaction has taken place has settled down in a stationary orbit around the nucleus. We shall, for reasons referred to later, assume that the orbit in question is circular; this assumption will, however, make no alteration in the calculations for systems containing only a single electron.

Let us now assume that, during the binding of the electron, a homogeneous radiation is emitted of a frequency ν, equal to half the frequency of revolution of the electron in its final orbit; then, from Planck's theory, we might expect that the amount of energy emitted by the process considered is equal to $\tau h\nu$, where h is Planck's constant and τ an entire number. If we assume that the radiation emitted is homogeneous, the second assumption concerning the frequency of the radiation suggests itself, since the frequency of revolution of the electron at the beginning of the emission is 0. The question, however, of the rigorous validity of both assumptions, and also of the application made of Planck's theory, will be more closely discussed in § 3.

Putting

$$W = \tau h \frac{\omega}{2}, \ldots\ldots \qquad (2)$$

we get by help of the formula (1)

$$W = \frac{2\pi^2 me^2 E^2}{\tau^2 h^2}, \quad \omega = \frac{4\pi^2 me^2 E^2}{\tau^3 h^3}, \quad 2a = \frac{\tau^2 h^2}{2\pi^2 meE}. . \qquad (3)$$

If in these expressions we give τ different values, we get a series of values for W, ω, and a corresponding to a series of configurations of the system. According to the above considerations, we are led to assume that these configurations will correspond to states of the system in which there is no radiation of energy; states which consequently will be stationary as long as the system is not disturbed from outside. We see that the value of W is greatest if τ has its smallest value 1. This case will therefore correspond to the most stable state of the system, *i. e.* will correspond to the binding of the electron for the breaking up of which the greatest amount of energy is required.

Putting in the above expressions $\tau = 1$ and $E = e$, and introducing the experimental values

$$e = 4.7 \times 10^{-10}, \quad \frac{e}{m} = 5.31 \times 10^{17}, \quad h = 6.5 \times 10^{-27},$$

we get

$$2a = 1.1 \times 10^{-8} \text{ cm.}, \quad \omega = 6.2 \times 10^{15} \frac{1}{\text{sec.}}, \quad \frac{W}{e} = 13 \text{ volt.}$$

We see that these values are of the same order of magnitude as the linear dimensions of the atoms, the optical frequencies, and the ionization-potentials.

The general importance of Planck's theory for the discussion of the behaviour of atomic systems was originally pointed out by Einstein.[7] The considerations of Einstein have been developed and applied on a number of different phenomena, especially by Stark, Nernst, and Sommerfield. The agreement as to the order of magnitude between values observed for the frequencies and dimensions of the atoms, and values for these quantities calculated by considerations similar to those given above, has been the subject of much discussion. It was first pointed out by Haas,[8] in an attempt to explain the meaning and the value of Planck's constant on the basis of J. J. Thomson's atom-model, by help of the linear dimensions and frequency of an hydrogen atom.

Systems of the kind considered in this paper, in which the forces between the particles vary inversely as the square of the distance, are discussed in relation to Planck's theory by J. W. Nicholson.[9] In a series of papers this author has shown that it seems to be possible to account for lines of hitherto unknown origin in the spectra of the stellar nebulæ and that of the solar corona, by assuming the presence in these bodies of certain hypothetical elements of exactly indicated constitution. The atoms of these elements are supposed to consist simply of a ring of a few electrons surrounding a positive nucleus of negligibly small dimensions. The ratios between the frequencies corresponding to the lines in question are compared with the ratios between the frequencies corresponding to different modes of vibration of the ring of electrons. Nicholson has obtained a relation to Planck's theory showing that the ratios between the wave-length of different sets of lines of the coronal spectrum can be accounted for with great accuracy by assuming that the ratio between the energy of the system and the frequency of rotation of the ring is equal to an entire multiple of Planck's constant. The quantity Nicholson refers to as the energy is equal to twice the quantity which we have denoted above by W. In the latest paper cited Nicholson has found it necessary to give the theory a more complicated form, still, however, representing the ratio of energy to frequency by a simple function of whole numbers.

The excellent agreement between the calculated and observed values of the ratios between the wave-lengths in question seems a strong argument in favour of the validity of the foundation of Nicholson's calculations. Serious objections, however, may be raised against the theory. These objections are intimately connected with the problem of the homogeneity of the radiation emitted. In Nicholson's calculations the frequency of lines in a line-spectrum is identified with the frequency of vibration of a mechanical system in a distinctly indicated state of equilibrium. As a relation from Planck's theory is used, we might expect that the radiation is sent out in quanta; but systems like those considered, in which the frequency is a function of the energy, cannot emit a finite amount of a homogeneous radiation; for, as soon as the emission of radiation is started, the energy and also the frequency of the system are altered. Further, according to the calculation of Nicholson, the systems are unstable for some modes of vibration. Apart from such objections—which may be only formal—it must be remarked, that the theory in the form given does not seem to be able to account for the well-known laws of Balmer and Rydberg connecting the frequencies of the lines in the line-spectra of the ordinary elements.

It will now be attempted to show that the difficulties in question disappear if we consider the problems from the point of view taken in this paper. Before proceeding it may be useful to restate briefly the ideas characterizing the calculations on p. 166. The principal assumptions used are:

(1) That the dynamical equilibrium of the systems in the stationary states can be discussed by help of the ordinary mechanics, while the passing of the systems between different stationary states cannot be treated on that basis.

(2) That the latter process is followed by the emission of a *homogeneous* radiation, for which the relation between the frequency and the amount of energy emitted is the one given by Planck's theory.

The first assumption seems to present itself; for it is known that the ordinary mechanics cannot have an absolute validity, but will only hold in calculations of certain mean values of the motion of the electrons. On the other hand, in the calculations of the dynamical equilibrium in a stationary state in which there is no relative displacement of the particles, we need not distinguish between the actual motions and their mean values. The second assumption is in obvious contrast to the ordinary ideas of electrodynamics, but appears to be necessary in order to account for experimental facts.

In the calculations on p. 166 we have further made use of the more special assumptions, viz. that the different stationary states correspond to the emission of a different number of Planck's energy-quanta, and that the frequency of the radiation emitted during the passing of the system from a state in which no energy is yet radiated out to one of the stationary states, is equal to half the frequency of revolution of the electron in the latter state. We can, however (see § 3), also arrive at the expressions (3) for the stationary states by using assumptions of somewhat different form. We shall, therefore, postpone the discussion of the special assumptions, and first show how by the help of the above principal assumptions, and of the expressions (3) for the stationary states, we can account for the line-spectrum of hydrogen.

§ 2. Emission of Line-spectra

SPECTRUM OF HYDROGEN. General evidence indicates that an atom of hydrogen consists simply of a single electron rotating round a positive nucleus of charge e.[10] The reformation of a hydrogen atom, when the electron has been removed to great distances away from the nucleus—*e. g.* by the effect of electrical discharge in a vacuum tube—will accordingly correspond to the binding of an electron by a positive nucleus considered on p. 166. If in (3) we put $\mathrm{E} = e$, we get for the total amount of energy radiated out by the formation of one of the stationary states,

$$\mathrm{W}_\tau = \frac{2\pi^2 m e^4}{h^2 \tau^2}.$$

The amount of energy emitted by the passing of the system from a state corresponding to $\tau = \tau_1$ to one corresponding to $\tau = \tau_2$, is consequently

$$\mathrm{W}_{\tau_2} - \mathrm{W}_{\tau_1} = \frac{2\pi^2 m e^4}{h^2}\left(\frac{1}{\tau_2^2} - \frac{1}{\tau_1^2}\right).$$

If now we suppose that the radiation in question is homogeneous, and that the amount of energy emitted is equal to $h\nu$, where ν is the frequency of the radiation, we get

$$W_{\tau_2} - W_{\tau_1} = h\nu,$$

and from this

$$\nu = \frac{2\pi^2 me^4}{h^3}\left(\frac{1}{\tau_2^2} - \frac{1}{\tau_1^2}\right). \ldots\ldots \quad (4)$$

We see that this expression accounts for the law connecting the lines in the spectrum of hydrogen. If we put $\tau_2 = 2$ and let τ_1 vary, we get the ordinary Balmer series. If we put $\tau_2 = 3$, we get the series in the ultra-red observed by Paschen[11] and previously suspected by Ritz. If we put $\tau_2 = 1$ and $\tau_2 = 4, 5, \ldots$, we get series respectively in the extreme ultra-violet and the extreme ultra-red, which are not observed, but the existence of which may be expected.

The agreement in question is quantitative as well as qualitative. Putting

$$e = 4.7 \times 10^{-10}, \quad \frac{e}{m} = 5.31 \times 10^{17}, \quad \text{and} \quad h = 6.5 \times 10^{-27},$$

we get

$$\frac{2\pi^2 me^4}{h^3} = 3.1 \times 10^{15}.$$

The observed value for the factor outside the bracket in the formula (4) is

$$3.290 \times 10^{15}.$$

The agreement between the theoretical and observed values is inside the uncertainty due to experimental errors in the constants entering in the expression for the theoretical value. We shall in § 3 return to consider the possible importance of the agreement in question.

It may be remarked that the fact, that it has not been possible to observe more than 12 lines of the Balmer series in experiments with vacuum tubes, while 33 lines are observed in the spectra of some celestial bodies, is just what we should expect from the above theory. According to the equation (3) the diameter of the orbit of the electron in the different stationary states is proportional to τ^2. For $\tau = 12$ the diameter is equal to 1.6×10^{-6} cm., or equal to the mean distance between the molecules in a gas at a pressure of about 7 mm. mercury; for $\tau = 33$ the diameter is equal to 1.2×10^{-5} cm., corresponding to the mean distance of the molecules at a pressure of about 0.02 mm. mercury. According to the theory the necessary condition for the appearance of a great number of lines is therefore a very small density of the gas; for simultaneously to obtain an intensity sufficient for observation the space filled with the gas must be very great. If the theory is right, we may therefore never expect to be able in experiments with vacuum tubes to observe the lines corresponding to high numbers of the Balmer series of the emission spectrum of hydrogen; it might, however, be possible to observe the lines by investigation of the absorption spectrum of this gas (see § 4).

It will be observed that we in the above way do not obtain other series of lines, generally ascribed to hydrogen; for instance, the series first observed by Pickering[12] in the spectrum of the star ζ Puppis, and the set of series recently found by Fowler[13] by experiments with vacuum tubes containing a mixture of hydrogen and helium. We shall, however, see that, by help of the above theory, we can account naturally for these series of lines if we ascribe them to helium.

A neutral atom of the latter element consists, according to Rutherford's theory, of a positive nucleus of charge $2e$ and two electrons. Now considering the binding of a single electron by a helium nucleus, we get, putting $E = 2e$ in the expressions (3) on p. 166, and proceeding in exactly the same way as above,

$$\nu = \frac{8\pi^2 me^4}{h^3}\left(\frac{1}{\tau_2^2} - \frac{1}{\tau_1^2}\right) = \frac{2\pi^2 me^4}{h^3}\left(\frac{1}{\left(\frac{\tau_2}{2}\right)^2} - \frac{1}{\left(\frac{\tau_1}{2}\right)^2}\right).$$

If we in this formula put $\tau_2 = 1$ or $\tau_2 = 2$, we get series of lines in the extreme ultra-violet. If we put $\tau_2 = 3$, and let τ_1 vary, we get a series which includes 2 of the series observed by Fowler, and denoted by him as the first and second principal series of the hydrogen spectrum. If we put $\tau_2 = 4$, we get the series observed by Pickering in the spectrum of ζ Puppis. Every second of the lines in this series is identical with a line in the Balmer series of the hydrogen spectrum; the presence of hydrogen in the star in question may therefore account for the fact that these lines are of a greater intensity than the rest of the lines in the series. The series is also observed in the experiments of Fowler, and denoted in his paper as the Sharp series of the hydrogen spectrum. If we finally in the above formula put $\tau_2 = 5, 6, \ldots,$ we get series, the strong lines of which are to be expected in the ultra-red.

The reason why the spectrum considered is not observed in ordinary helium tubes may be that in such tubes the ionization of helium is not so complete as in the star considered or in the experiments of Fowler, where a strong discharge was sent through a mixture of hydrogen and helium. The condition for the appearance of the spectrum is, according to the above theory, that helium atoms are present in a state in which they have lost both their electrons. Now we must assume that the amount of energy to be used in removing the second electron from a helium atom is much greater than that to be used in removing the first. Further, it is known from experiments on positive rays, that hydrogen atoms can acquire a negative charge; therefore the presence of hydrogen in the experiments of Fowler may effect that more electrons are removed from some of the helium atoms than would be the case if only helium were present.

SPECTRA OF OTHER SUBSTANCES. In case of systems containing more electrons we must—in conformity with the result of experiments—expect more complicated laws for the line-spectra than those considered. I shall try to show that the point of view taken above allows, at any rate, a certain understanding of the laws observed.

According to Rydberg's theory—with the generalization given by Ritz[14]—the frequency corresponding to the lines of the spectrum of an element can be expressed by

$$\nu = F_r(\tau_1) - F_s(\tau_2),$$

where τ_1 and τ_2 are entire numbers, and $F_1, F_2, F_3, \ldots.$ are functions of τ which approximately are equal to

$$\frac{K}{(\tau + a_1)^2}, \frac{K}{(\tau + a_2)^2}, \ldots$$

K is a universal constant, equal to the factor outside the bracket in the formula (4) for the spectrum of hydrogen. The different series appear if we put τ_1 or τ_2 equal to a fixed number and let the other vary.

The circumstance that the frequency can be written as a difference between two functions of entire numbers suggests an origin of the lines in the spectra in question similar to the one we have assumed for hydrogen; *i. e.* that the lines correspond to a radiation emitted during the passing of the system between two different stationary states. For systems containing more than one electron the detailed discussion may be very complicated, as there will be many different configurations of the electrons which can be taken into consideration as stationary states. This may account for the different sets of series in the line spectra emitted from the substances in question. Here I shall only try to show how, by help of the theory, it can be simply explained that the constant K entering in Rydberg's formula is the same for all substances.

Let us assume that the spectrum in question corresponds to the radiation emitted during the binding of an electron; and let us further assume that the system including the electron considered is neutral. The force on the electron, when at a great distance apart from the nucleus and the electrons previously bound, will be very nearly the same as in the above case of the binding of an electron by a hydrogen nucleus. The energy corresponding to one of the stationary states will therefore for τ great be very nearly equal to that given by the expression (3) on p. 166, if we put $E = e$. For τ great we consequently get

$$\lim (\tau^2 . F_1(\tau)) = \lim (\tau^2 . F_2(\tau)) = \ldots . = \frac{2\pi^2 me^4}{h^3},$$

in conformity with Rydberg's theory.

§ 3. General Considerations continued

We shall now return to the discussion (see p. 168) of the special assumptions used in deducing the expressions (3) on p. 166 for the stationary states of a system consisting of an electron rotating round a nucleus.

For one, we have assumed that the different stationary states correspond to an emission of a different number of energy-quanta. Considering systems in which the frequency is a function of the energy, this assumption, however, may be regarded as improbable; for as soon as one quantum is sent out the frequency is altered. We shall now see that we can leave the assumption used and still retain the equation (2) on p. 166, and thereby the formal analogy with Planck's theory.

Firstly, it will be observed that it has not been necessary, in order to account for the law of the spectra by help of the expressions (3) for the stationary states, to assume that in any case a radiation is sent out corresponding to more than a single energy-quantum, $h\nu$. Further information on the frequency of the radiation may be obtained by comparing calculations of the energy radiation in the region of slow vibrations based on the above assumptions with calculations based on the ordinary mechanics. As is known, calculations on the latter basis are in agreement with experiments on the energy radiation in the named region.

Let us assume that the ratio between the total amount of energy emitted and the frequency of revolution of the electron for the different stationary states is given by the equation $W = f(\tau) \cdot h\omega$, instead of by the equation (2). Proceeding in the same way as above, we get in this case instead of (3)

$$W = \frac{\pi^2 m e^2 E^2}{2h^2 f^2(\tau)}, \quad \omega = \frac{\pi^2 m e^2 E^2}{2h^3 f^3(\tau)}.$$

Assuming as above that the amount of energy emitted during the passing of the system from a state corresponding to $\tau = \tau_1$ to one for which $\tau = \tau_2$ is equal to $h\nu$, we get instead of (4)

$$\nu = \frac{\pi^2 m e^2 E^2}{2h^3} \left(\frac{1}{f^2(\tau_2)} - \frac{1}{f^2(\tau_1)} \right).$$

We see that in order to get an expression of the same form as the Balmer series we must put $f(\tau) = c\tau$.

In order to determine c let us now consider the passing of the system between two successive stationary states corresponding to $\tau = N$ and $\tau = N - 1$; introducing $f(\tau) = c\tau$, we get for the frequency of the radiation emitted

$$\nu = \frac{\pi^2 m e^2 E^2}{2c^2 h^3} \cdot \frac{2N - 1}{N^2 (N - 1)^2}.$$

For the frequency of revolution of the electron before and after the emission we have

$$\omega_N = \frac{\pi^2 m e^2 E^2}{2c^3 h^3 N^3} \text{ and } \omega_{N-1} = \frac{\pi^2 m e^2 E^2}{2c^3 h^3 (N - 1)^3}.$$

If N is great the ratio between the frequency before and after the emission will be very near equal to 1; and according to the ordinary electrodynamics we should therefore expect that the ratio between the frequency of radiation and the frequency of revolution also is very nearly equal to 1. This condition will only be satisfied if $c = \frac{1}{2}$. Putting $f(\tau) = \tau/2$, we, however, again arrive at the equation (2) and consequently at the expression (3) for the stationary states. . . .

April 5, 1913.

1. Communicated by Prof. E. Rutherford, F.R.S.

2. E. Rutherford, Phil. Mag. xxi. p. 669 (1911).

3. See also Geiger and Marsden, Phil. Mag. April 1913.

4. J. J. Thomson, Phil. Mag. vii. p. 237 (1904).

5. See f. inst., 'Théorie du ravonnement et les quanta.' Rapports de la réunion à Bruxelles, Nov. 1911. Paris, 1912.

6. See f. inst., M. Planck, *Ann. d. Phys.* xxxi. p. 758 (1910); xxxvii. p. 642 (1912); *Verh. deutsch. Phys. Ges.* 1911, p. 138.

7. A. Einstein, *Ann. d. Phys.* xvii. p. 132 (1905); xx. p. 199 (1906); xxii. p. 180 (1907).

8. A. E. Haas, *Jahrb. d. Rad. u. El.* vii. p. 261 (1910). See further, A. Schidlof, *Ann. d. Phys.* xxxv. p. 90 (1911); E. Wertheimer, *Phys. Zeitschr.* xii. p. 409 (1911), *Verh. deutsch. Phys. Ges.* 1912, p. 431; F. A. Lindemann, *Verh. deutsch. Phys. Ges.* 1911, pp. 482, 1107; F. Haber, *Verh. deutsch. Phys. Ges.* 1911, p. 1117.

9. J. W. Nicholson, Month. Not. Roy. Astr. Soc. lxxii. pp. 49, 139, 677, 693, 729 (1912).

10. See f. inst. N. Bohr, Phil. Mag. xxv. p. 24 (1913). The conclusion drawn in the paper cited is strongly supported by the fact that hydrogen, in the experiments on positive rays of Sir J. J. Thomson, is the only element which never occurs with a positive charge corresponding to the loss of more than one electron (comp. Phil. Mag. xxiv. p. 672 (1912)).

11. F. Paschen, *Ann. d. Phys.* xxvii. p. 565 (1908).

12. E. C. Pickering, Astrophys. J. iv. p. 369 (1896); v. p. 92 (1897).

13. A. Fowler, Month. Not. Roy. Astr. Soc. lxxiii. Dec. 1912.

14. W. Ritz, *Phys. Zeitschr.* ix. p. 521 (1908).

9

THE MEANS OF COMMUNICATION BETWEEN NERVES

In one of the most remarkable narratives of scientific discovery, Otto Loewi, at age eighty-seven, recalled how the idea came to him for testing the way nerves communicate. The thought arrived in a dream:

> The night before Easter Sunday of [1921] I awoke, turned on the light, and jotted down a few notes on a tiny slip of paper. Then I fell asleep again. It occurred to me at six o'clock in the morning that during the night I had written down something most important, but I was unable to decipher the scrawl. The next night, at three o'clock, the idea returned. It was the design of an experiment to determine whether or not the hypothesis of chemical transmission [of the nervous impulse from nerves to their respective organs] that I had uttered seventeen years ago was correct. I got up immediately, went to the laboratory, and performed a simple experiment on a frog heart according to the nocturnal design . . .

Loewi clearly misremembered some details. The first experiment reported in his 1921 paper was carried out in February, well before Easter. However, the point of origin of the experiment, in a dream, and the scribbled slip of paper seem to be true. That story he told to friends at the time. Einstein had his "thought experiments," Rutherford had his wildly intuitive "damn fool" ideas, but very few scientists have reported receiving their great ideas in a dream. Years later, Loewi wrote that the process of his discovery of 1921 "shows that an idea may sleep for decades in the unconscious mind and then suddenly return. Further, it indicates that we should sometimes trust a sudden intuition without too much skepticism."

In 1936, on his way back to Graz from receiving the Nobel Prize,

Loewi stopped in Vienna to meet Sigmund Freud. Unfortunately, nothing has been recorded about their conversations.

At the time of Loewi's dream, in 1921, it had long been known that the nervous system is a means of internal communication within living creatures. It was also well known that signals travel down the spindly filaments of a nerve in the form of electricity. What was not known was how nerves conveyed their electrical impulses to muscles, organs, and other nerves—in the process activating motion, respiration, digestion, circulation, reproduction, even thought. In short, how did nerves talk to the rest of the body? Most biologists believed that such internerve communication was also electrical. In that view, tiny electrical currents flowed from nerves to heart muscles or thyroid glands or the waiting antennae of other nerves.

Loewi's late-night experiment was not only simple but elegant. First, he isolated the hearts of two frogs and removed the nerves from the second heart. (One is reminded of the procedure used by Bayliss and Starling.) Into both hearts Loewi inserted a tube filled with Ringer solution, a liquid that matches the concentration of salts in the body and keeps isolated organs alive. Loewi then stimulated the vagus nerve of the first heart. The vagus slows down the functions of organs, and the heart's rate of throbbing decreased as expected. After a few minutes, he poured the Ringer solution from the first heart into the second, nerveless heart. *It also slowed.* Its beating diminished just as if its own vagus nerve had been roused. Similarly, when Loewi stimulated the accelerator nerve of the first heart and then transferred its liquid to the second heart, that heart speeded up. The results showed without doubt that a stimulated nerve releases some chemical substance, which then activates organs. The transmission of nerve impulses to organs is chemical, not electrical.

Like the landmark experiments of Bayliss and Starling in 1902 and of Rutherford in 1911, Loewi's experiment was beautiful in design, far-reaching and inescapable in its conclusions. Unlike these previous scientists, Loewi knew exactly what he was looking for. After his landmark experiment of 1921, Loewi and his collaborators identified not only the chemical agents of nerves, the so-called neurotransmitters such as adrenaline and dopamine and serotonin, but also other natural substances that both inhibit and enhance those agents. Living communication takes place through an intricate network of checks and balances and bodily controls. Loewi had unmasked the secret messengers of the body. In doing so, he had discovered both the messengers themselves and the world they lived in. Over the next decades, Loewi's work would impact everything from knowledge of brain function to treatment of neurological diseases to drugs that affect the receptors of Loewi's neurotransmitters and cure illnesses ranging from high blood pressure to stomach ulcers. As Loewi's friend and fellow Nobelist Henry Dale remarked in Loewi's 1962 obituary, Loewi's discoveries "opened a new vista" in biology.

On that late night of 1921, Loewi would have been working in his laboratory at the University of Graz, where he had been head of pharmacology since 1909. At the time he was forty-seven years old. Photographs show a stocky man with a full face, bushy eyebrows, and a faint mustache. Loewi loved good food, good wine, and especially good conversation. In Dale's recollections, Loewi was "a ready and uninhibited talker," whose "enthusiasms and dislikes were easily aroused, and given prompt and effective expression." Those enthusiasms sometimes expressed themselves as impatience in new ventures. On a trip to England in 1902, Loewi learned English at breakneck speed so that he could talk to the great British physiologists, and he did not like his hasty grammar reproved. "I have not time to learn English correctly," he barked at Dale on the occasion. "I wish to speak it fast."

Loewi also loved the arts. As he bent over his lab table, alone in the room, pressing the frog hearts with his gloved hands, he might have been humming portions of *Tristan und Isolde* or *Die Walküre*. From an early age, he had adored the music of Wagner. The teenaged Loewi had also appreciated painting, especially the early Flemish painters, and wished to study the history of art.

Otto Loewi was born in June 1873 in Frankfurt am Main. He was the first child and only son of Jacob Loewi, a Jewish wine merchant, and Anna Willstädter. Loewi recalls that at the Frankfurt Gymnasium he did poorly in science and mathematics but well in the humanities. Following his parents' insistence that he pursue a practical career, young Loewi enrolled as a medical student at Strassburg University in 1891. But his heart was not there. For much of his medical training, he skipped the medical courses to hear lectures in philosophy and in the history of German architecture. He barely passed the *Physicum*, his first medical school examination. Finally, his interest in medicine was ignited by the lectures of Professor Bernard Naunyn, an experimental pathologist and head of internal medicine.

A subtle but influential step in Loewi's scientific development may have been receiving his thesis topic at Strassburg from Professor Oswald Schmiedeberg, widely regarded as the father of modern pharmacology. Schmiedeberg and his colleagues undoubtedly introduced Loewi to the study of the role of chemicals in the inner workings of living organisms. In his autobiography, Loewi devotes little space to Schmiedeberg. Yet Schmiedeberg was already quite famous for having shown that minute quantities of the drug muscarine had the same effect on a frog's heart as electrical stimulation of the vagus nerve—a result that partly foreshadowed Loewi's future landmark experiment. At the time of Schmiedeberg's early work, however, the understanding of nerve architecture was far more primitive than it was in 1900, or even in 1921.

The modern conception of nerves began with the work of Italian physicist and physiologist Luigi Galvani. In the 1790s, Galvani demonstrated that electrical currents caused muscle contractions in frogs. (Galvani deemed such electricity "animal electricity," distinct from the "natural electricity" of lightning and the "artificial electricity" obtained by rubbing cats' fur.) Frogs make particularly good subjects for such experiments because their muscle action is so responsive and pronounced.

Of course, muscles are not nerves. In 1842, another Italian scientist, Carlo Matteucci, showed that an injured animal muscle produced an electrical current. Thus, electricity not only stimulated the body but also occurred naturally within the body. The work of Matteucci and other scientists created the new field of "electrophysiology." In 1849, the German electrophysiologist Emil Heinrich du Bois-Reymond found Matteucci's "injury current" also in nerves, demonstrating the electrical nature of the nervous system. Although the details were not known, scientists now understood that signals and information are carried along nerves by electrical currents.

The first detailed picture of a nerve came from Italian pathologist Camillo Golgi (1843–1926). Golgi invented a method of staining nerve cells and fibers by hardening them with potassium or ammonium bichromate and then immersing them in a weak solution of silver nitrate. Nerve cells stained in this way clearly show their main elements. In publications beginning in 1873, Golgi concluded that nerves consist of several parts: fine tendrils, called dendrites, leading to a roundish nerve body, called the soma, leading to a single, thin extension called an axon. (See Figure 9.1.) Nerve axons vary in length from 10^{-3} centimeters to 100 centimeters and are typically 10^{-4} centimeters wide.

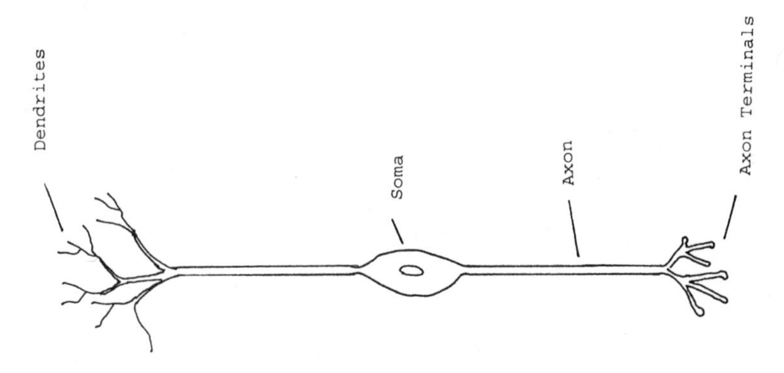

Figure 9.1

In the 1890s, the Spanish neuroanatomist Santiago Ramón y Cajal (1852–1934) showed that the electrical signal coursing through a nerve begins in the dendrites and moves to the axon. Most significantly, Ramón y Cajal found that nerves do not touch other nerves. There is a microscopic gap, called the synaptic cleft or simply the synapse, between the ending axon of one nerve and the beginning dendrites of another. The synapse is about 2×10^{-6} centimeters wide, or fifty times smaller than the thickness of an axon.

Ramón y Cajal's identification of the synapse, the empty space between connecting nerves, was extremely important. It meant that the nervous system is not a continuous wire. Each nerve has a beginning and ending. Each nerve is a self-contained unit, a concept later called the neuron doctrine. One can thus think of the nervous system as a complex network of individual nerves, or neurons, with each neuron receiving signals from thousands of other neurons and then sending on its own signal to yet other neurons. Figure 9.2 illustrates a small portion of such a network.

An electrical signal passes through one neuron in about one one-thousandth of a second. Then, somehow, that signal is transmitted to another neuron or organ across the synaptic gap. A critical question was the means of that transmission. Until Loewi's work of 1921, the prevailing theory was electrical transmission.

In 1896, his fresh medical degree in hand, Loewi went to work at the city hospital in Frankfurt. After witnessing a high rate of death among patients with pneumonia, he decided that clinical medicine was not for

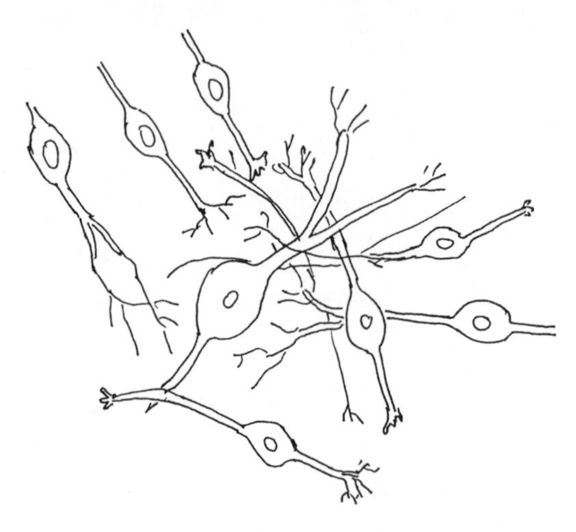

Figure 9.2

him. The young scientist preferred research in physiology. Fortunately, an excellent opportunity presented itself, in the form of an assistantship with Professor Hans Horst Meyer, a leading pharmacologist at the University of Marburg. At age twenty-five, Loewi thus began what would be perhaps the most influential learning period of his life. He would work with Meyer for more than ten years—in Marburg for six years, from 1898 to 1904, and then five more years in Vienna, from 1904 to 1909.

In his *Autobiographic Sketch* of 1960, Loewi remembered Meyer as "a great scientist and person." Indeed, Meyer became a powerful mentor in Loewi's life, both in substance and in style, dispelling the notion that the master-apprentice relationship is not important in the sciences. We will see a similar influential relationship between Hans Krebs and his mentor Otto Warburg.

One of the repeating themes during his years with Meyer was Loewi's work on metabolism. In 1902, he achieved his first truly important discovery: animals do not need to eat proteins whole but can internally build proteins from their constituents, the amino acids. To arrive at this conclusion, the twenty-nine-year-old scientist had to feed dogs an unappetizing meal made of the "degradation products" of the pancreas. "For a long time I encountered great difficulties," Loewi later wrote, "particularly because most of the dogs did not relish the unusual fare. I persisted, however, because I had not the slightest doubt that I would succeed in the end. My perserverance was rewarded" after several years' work. One can sense in these comments Loewi's strong self-confidence.

The year 1902 was eventful for Loewi for another reason. In that year he traveled to England to visit the British physiologists. The great German scientist Carl Ludwig had ruled physiology for much of the nineteenth century, but when Ludwig died in 1895, the center of gravity of physiological research shifted to England. In 1902, the ambitious young Loewi went to London to meet William Bayliss and Ernest Starling, who had just discovered the first hormone and, further, the entire hormonal system as a second means of communication (see Chapter 2). Loewi was "charmed by Starling's appearance, his expressive features, his shining eyes." Loewi then went on to Cambridge, where he learned of John Langley's work in dividing the nervous system into two parts: those nerves that slowed down organs and functions, called the parasympathetic nerves, and those nerves that speeded things up, the sympathetic nerves.

In Cambridge, Loewi also met T. R. Elliott, then a young graduate, who was beginning a series of brilliant experiments demonstrating that the actions of the sympathetic nerve system could be reproduced by injecting the chemical adrenaline. But reproducing the actions of nerves with a chemical injected from outside the organism does not necessarily mean that that chemical is internally produced by the nerves themselves, or that it carries the impulses of nerves. Two years later, Elliott ventured the second, more significant statement. He hypothesized that adrenaline, in fact, was the means by which sympathetic nerves transmitted their impulses to other nerves within the organism. In 1903, Loewi himself made a similar conjecture: the chemical muscarine might transmit the nerve impulses of the vagus nerve to the heart. Thus, the idea of chemical transmission of nerve impulses was in the air.

Furthermore, there were problems with the electrical theory. Nerve impulses were known to be one-way signals. A signal that went from nerve A to nerve B never went from nerve B to nerve A. Yet, basic physics decreed that electrical currents could flow in either direction. Another problem was the observed fact that nerves could either inhibit or excite other nerves and organs, a behavior at odds with the all-or-nothing nature of electrical discharge in the electrical theory.

Yet most scientists continued to support the electrical theory. An electrical mechanism for signals *within* single nerves was established. Furthermore, chemical agents for internerve communication might not be stable enough, or might not act swiftly enough. Thus, despite the work of Langley and Elliott, most physiologists subscribed to the electrical theory of nervous transmission. By the late teens, the theory of chemical transmission was largely discredited.

Where do creative ideas come from? The process is as mysterious in the sciences as in the arts. Between 1909, when he moved to Graz on his own, and 1921, Loewi published some twenty scientific papers, on subjects ranging from the effect of inorganic ions on the heart muscle, to modification of diabetes by drugs, to carbohydrate metabolism. He accomplished what was considered excellent work. Yet, as Dale writes in his obituary, Loewi did nothing in this period to suggest that he was about "to step suddenly beyond the normal limits" of his past research and perform his landmark experiment. Indeed, Loewi seems to have forgotten his early suggestion that nerves might transmit their impulses by chemicals rather than by electricity. Much of his time in Graz was taken up with teaching and preparing for his five weekly lectures, during which he "suffered from a kind of stage fright." In the evenings, he often attended chamber concerts, some of them performed in his own home, and socialized with writers, actors, and philosophers.

Loewi begins his landmark paper with a clear statement of the problem to be investigated: "the mechanism of action of nerve stimulation." Next, he briefly discusses his methods, noting that the vagus nerve of his frogs is still attached to the sinus, a dilated blood vessel. Oxygen is continuously perfused, or bubbled, through the Ringer solution, which is sometimes called the perfusate. Loewi reports on following up his first experiment with many experiments on a variety of frogs: esculenta (edible frogs), temporaria (common frogs), and toads (no comment). The large number of experiments over a period of time, all giving the same results, clearly reinforces the certainty of his conclusions.

In the first set of experiments, Loewi describes how stimulation of the vagus nerve produces the well-known negative inotropy, meaning decrease in the force of contraction of the heart, and negative chronotropy, meaning decrease in the beating rate of the heart. As described earlier, he gives evidence that the Ringer solution collected from the period of vagus stimulation can by itself cause these activities. The chemical atropine inhibits the action of whatever chemical has been produced by the vagus nerve and secreted into the Ringer solution. (In later work, scientists showed that atropine works by blocking the receptors of the neurotransmitter on the receiving nerve or organ.) Figure 1 of Loewi's paper shows the changes in heart stroke volume or force of contraction (height of the lines), and Figure 2 shows the changes in beating rate (horizontal spacing of the lines). It is difficult, and gruesome, for the nonbiologist to realize that these are beating, throbbing hearts lying on Loewi's lab table.

Loewi's Figures 1 and 2 are electrocardiograms, which register the electrical current in the heart. Although he does not discuss his equipment, Loewi almost certainly measured such currents with a galvanometer, invented in the early nineteenth century and named after Luigi Galvani. A galvanometer works on the principle that a current running through a coil of wire placed near a magnet will cause the coil to rotate. The force of twist of the coil measures the strength of the current. And the tiny back-and-forth movement of the coil measures changes in the current over time. Loewi would have placed electrodes on his frog hearts and then run the electrodes to his galvanometer. The tiny twisting of the wire coil can be mechanically recorded by a device like Carl Ludwig's klymograph or electrically recorded by a more modern device like an oscilloscope.

In the second set of experiments, Loewi reports that the Ringer solution collected during the period in which the accelerans (accelerator) nerve is stimulated produces an increase in stroke volume, indicating increased cardiac activity. Again, Ringer solution collected from the Normal Period, that is, a period in which nerves are not stimulated, has no effect on the heart.

In the discussion section, Loewi follows a cautious, Socratic line of logical argument. The substances in the Ringer solution that have the same effect on the heart as nerve stimulation are either newly synthesized as a result of nerve stimulation or not. If not, if the substances already existed, then they must be released as a result of nerve stimulation. (In later experiments, Loewi showed that the second possibility is true.) The substances in the Ringer solution may be the agents that activate the heart, or they may be produced by the activity of the heart. Loewi's own experiments here favor the first possibility, since the Ringer solution collected after nerve stimulation can activate a nerveless heart.

The actual behavior of neurotransmitters, demonstrated by Loewi in later years, is shown in Figure 9.3. The neurotransmitter molecules (indicated by dots) are bound up in molecular containers (indicated by circles) in the axon terminals. When a nerve is stimulated, the containers move to the outer membrane of the axon terminal and release the chemical neurotransmitter. The neurotransmitter molecules cross the synaptic gap and are received by receptors (indicated by boxes) in the dendrites of other nerves, or in the outer membranes of organs. The receptor molecules, in turn, initiate an electrical current through the movement of electrically charged atoms, and the nervous signal continues.

In the last paragraph of his paper, Loewi announces that he will follow up these experiments with other investigations to determine the identity of the "substances," and answer other questions. Indeed, those researches would occupy him and his collaborators for the next fifteen years. Loewi's tone throughout is modest, confident in what he has demonstrated but careful not to claim too much from his results. In later papers, Loewi's claims would slowly ascend in volume, until his Nobel Prize lecture of 1936, which resounds with such statements as "so much for the field of activity and the importance of the neurochemical mechanism."

Otto Loewi remained at the University of Graz until March 11, 1938, the day that the Nazis invaded Austria. During that night, a dozen young German soldiers broke into his bedroom and took him to jail, where he was soon joined by his two youngest sons and other Jewish males of the

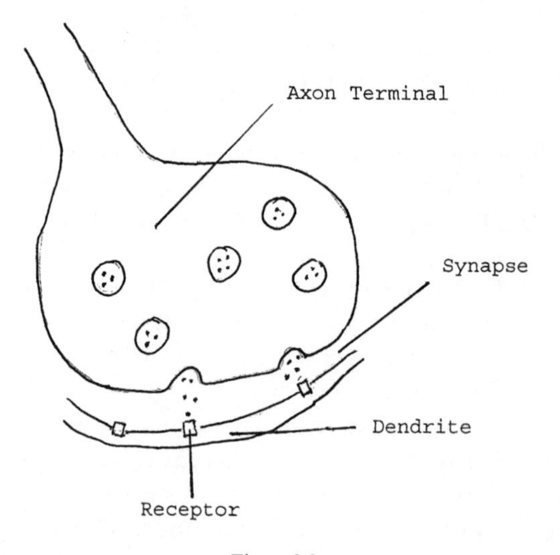

Figure 9.3

city. After a few months he and his children were released. Loewi was soon offered a safe haven in London and later in Brussels. However, he was not allowed to emigrate until he agreed to transfer his Nobel Prize money from a bank in Stockholm to a bank controlled by the Nazis.

In 1940, Loewi took up the position of Research Professor of Pharmacology at the College of Medicine of New York University and remained there for the rest of his life.

After his initial publication in *Pflügers Archiv* in 1921, Loewi wrote more than a dozen more papers in that journal over the next fifteen years, adding more details to the chemical transmission process. Eventually, he identified the principal neurotransmitters, including acetylcholine, the slowing-down transmitter, and adrenaline (also called epinephrine), the speeding-up transmitter. Furthermore, he and his collaborators found a class of enzymes called esterases that metabolized and destroyed the neurotransmitters after their action, and another class, called eserines, that inhibited the esterases. Such chemical controls provide a far more elaborate and finely balanced system than would be achieved by electrical transmission of nervous impulses.

Strictly speaking, Loewi's initial experiments with nerves and heart muscles showed only that nerves communicate with *organs* via the chemical neurotransmitters. However, Dale soon extended Loewi's work to prove that nerves communicate with other nerves as well by the same chemical process. At first, Loewi resisted the idea that his chemical transmission process could be extended from the involuntary nervous system to the voluntary system. Sudden muscle contractions and other aspects of the voluntary system, Loewi argued, required a much faster response than could be provided by chemical messengers between synapses. However, Dale again showed that the voluntary nervous system was also controlled by chemical transmission. Thus, Loewi's work had far more universality than he first imagined. It applies to all nerves in all living organisms. Most neuroscience today rests on the foundation that Loewi first laid in 1921.

In a remarkable passage of his autobiography, Loewi recounts his fearful thoughts while in prison in 1938. He does not mention his two sons or his wife. Instead, he feared the loss of being able to make public his most recent research. "When I was awakened that night and saw the pistols directed at me, I expected, of course, that I would be murdered. From then on during days and sleepless nights I was obsessed by the idea that this might happen to me before I could publish my last experiments."

ON THE HUMORAL TRANSMISSION OF THE ACTION OF THE CARDIAC NERVE

Otto Loewi

From the Pharmacological Institute of the University of Graz. Undertaken with support from Prince Liechtenstein-Spende

Received 20 March 1921

Pflügers Archiv (1921)

THE MECHANISM OF ACTION OF nerve stimulation is not known. Considering that certain chemicals act almost identically to the stimulation of a particular nerve, it is possible that substances are synthesised under the influence of a nerve stimulation, and that these substances are responsible for the effect of the stimulation. It is probably not possible to address this question under the conditions of a whole-animal experiment. The only option would be to use isolated organs. This type of work has in fact only been presented by Howell, who concluded that the action of the vagus nerve is due to secretion of potassium during the stimulation. However, the experimental results were disproved.

METHODS

I chose hearts from cold-blooded animals, because this allows the experimental possibility of rendering chemicals that may be generated as a result of a stimulation detectable through enrichment of the small volume of perfusate.

I chose the well-known cannulated heart method of Straub with the modification that the dissected left vagus was still attached to the sinus, and was bridged over an electrode. If the nerve is maintained moist and the stimulus occasionally, if only briefly, interrupted, it can often remain excitable for several hours.

The Ringer solution contained 0.6% NaCl, 0.01% KCl, 0.02% $CaCl_2$ + 6 H_2O, 0.05% $NaHCO_2$. Oxygen was continuously perfused. The experiments were mostly conducted in February and March on freshly caught esculenta (10 experiments), temporaria (4 experiments) and common toads (4 experiments).

EXPERIMENTS

All experiments showed the same results.

1 Experiments with inhibitory vagal stimulation

After the heart was washed with Ringer solution to remove traces of blood, the Ringer was not changed for a certain period. At the end of this period (Normal Period) it was pipetted out and stored. Then the vagus was electrically stimulated for the same time period, with short interruptions. The outcome in frogs was the well-known negative inotropy and chronotropy. The perfusate collected during the Vagus Stimulation Period was also pipetted and kept. After the heart had recovered from the vagal stimulation it was exposed alternately to the solutions from each period. The perfusate collected during the Normal Period did not act differently from fresh Ringer, i.e., it had no influence (see Fig. 1). During exposure to the Ringer collected during the Vagus Stimulation Period, a clear inotropic effect occurred regularly (Figs. 1 and 2), sometimes accompanied by a negative chronotropic effect (Fig. 2). The latter effect had hardly been expected because under the experimental conditions, the sinus barely came into contact with the perfusate. Figure 1 shows that atropine immediately blocks the response.

2 Experiments with excitatory vagal stimulation

It could be that vagal stimulation of the atropinized frog heart results in the release of an excitatory substance because of the presence of accelerans fibres whose action cannot be blocked by atropine. However this obvious thought could not be checked because of a dearth of frog material. This is why I turned to the use of toads. At this time of year, these react normally to vagal stimulation with steep increases in pulse amplitude and frequency (Fig. 3a) [omitted here] The experiment was carried out as the one described above. Fig. 3b [omitted] shows that the perfusate of the Normal Period is completely without influence, but the perfusate collected during the accelerans stimulation period leads to a significant increase in stroke volume. It is particularly noteworthy that the perfusate collected during the accelerans stimulation period was obtained 3.5 hours after the start of the experiment, i.e., after the heart had been washed out numberless times and the accelerans had already been stimulated for one hour. Considering this long previous history of the heart it appeared desirable to test the activity of the perfusate also on a completely fresh toad heart. Fig. 3c [omitted] shows that in a fresh heart the activity is nearly the same as with the previously treated.

DISCUSSION OF THE RESULTS

These experiments teach us that under the influence of the stimulation of inhibitory and excitatory nerves it is possible to detect substances in the solution of the heart with a similar effect to those which are attributed to a nerve stimulus. So under the influence of nerve stimulation these substances are either newly synthesized or released from precursors, or they had been previously synthesized and were released only when the cells became permeable as a result of the stimulation. There are two possible explanations for the relevance of the substances: one explanation is that they could be synthesized independently of the type of mechanical heart activity, directly evoking the specific response of the heart to the nerve stimulus which would then be only indirectly effective. It is not surprising that the action under these experimental conditions lags behind that of a nerve stimulus, because it can be assumed that only a small part of the substances synthesized or released from a precursor in or at the cell enters the perfusate, and, in addition, the perfusate

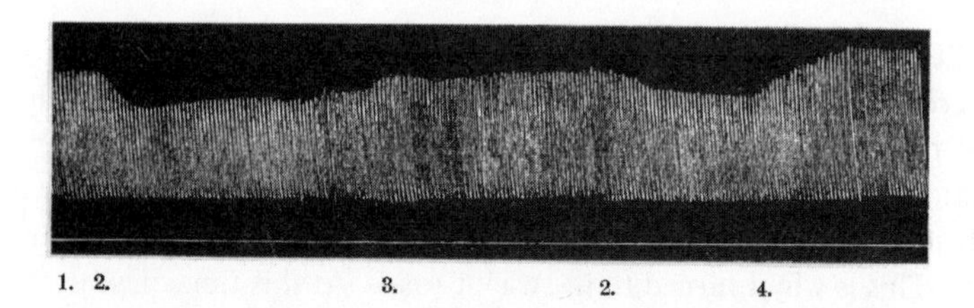

Figure 1. Esculenta. 1 and 3 indicate points of injection of Ringer solution without vagus nerve stimulation. 2 indicates point of injection of Ringer solution after vagus nerve stimulation. 4 indicates injection of atropine.

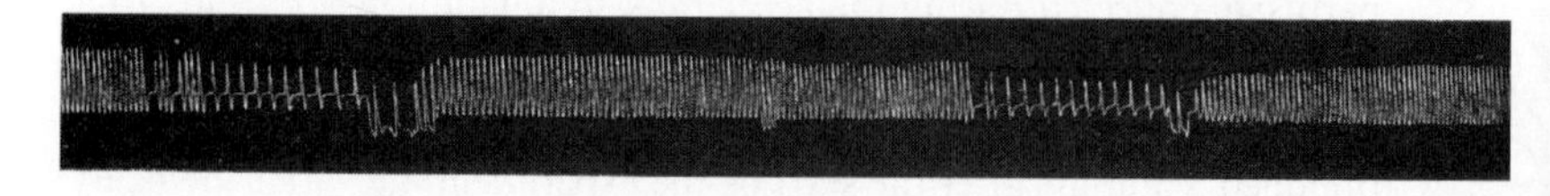

Figure 2. Temporaria.

causes a high degree of dilution. The second possibility is that the substances are only a product of the special type of heart activity caused by the nerve stimulus; in this case they would have an effect that is only by chance identical with the nerve stimulus.

Concerning the question of the identity of the substances, it can at this moment only be excluded that the chemical product of the vagal stimulation is potassium, because an increased effect of potassium cannot be blocked by atropine that we used in our experiments. Once I have the necessary animal material I intend to investigate the nature of the substances as well as many other questions raised by the experiments described.

10

THE UNCERTAINTY PRINCIPLE

IN HIS *MEMOIRS* (2001), Edward Teller describes a moment during his apprenticeship with Werner Heisenberg in the late 1920s. One evening Teller went to dinner at Heisenberg's bachelor apartment, where he was delighted to see "an excellent grand piano." A musician himself, and possibly trying to impress his mentor, Teller mentioned that he had been playing Beethoven and Mozart but was particularly fond of Bach's Prelude in E-flat Minor. At which point, Heisenberg sat down and performed the piece beautifully, even substituting a two-handed mezzo forte for the usual one-handed forte.

Heisenberg's apartment was conveniently located in the same building where he met his students at the University of Leipzig. Like Ernest Rutherford in Cambridge and Niels Bohr in Copenhagen, Heisenberg had created an international "school" of physicists in Leipzig. Teller recalls about twenty young men in the group, including himself (a Hungarian), some Germans, a few Americans, two Japanese, an Italian, an Austrian, a Swiss, and a Russian. Heisenberg, their leader, was twenty-seven years old. Already, he had formulated the theory of quantum mechanics.

Heisenberg's twenty disciples ate and drank physics. But they also met once a week in the evening for jokes, Ping-Pong, and chess. Even here Heisenberg demonstrated his supremacy. After being beaten by his students at Ping-Pong, he proceeded to train intensively on a long boat trip from Shanghai to Europe. On his return, Heisenberg could not be defeated.

Such competitiveness Teller describes as "half serious, half joking." Whatever it was, it began at a young age. As a boy, Heisenberg was physically frail. His wife Elisabeth recalls that to solve this problem he ran a few kilometers around Luitpold Park every evening, constantly checking his time with a stopwatch. After three years of the regimen, he had acquired the strength of an athlete and took up mountain climbing, hiking, and skiing. Ping-Pong was easy.

By the age of twenty-one, Heisenberg had completed his Ph.D. in theoretical physics. The year was 1923. His teacher at the time, Max Born, recalled later that his student "looked like a simple farm boy, with short, light hair, clear bright eyes, and a radiant expression on his face . . . His unbelievable quickness and acuteness of apprehension enabled him to do a colossal amount of work without much effort."

After working with Born at the University of Göttingen, Heisenberg made the pilgrimage to Copenhagen to work with Bohr, the revered father figure of so many young physicists of the day. There, in 1925, Heisenberg did his great work developing a rigorous mathematical framework for Bohr's informal quantum model of the atom.

Two years later, in 1927, Heisenberg published his landmark paper on the Uncertainty Principle, a consequence of his earlier work on quantum mechanics. That paper stated that nature is unknowable beyond certain limits. Just as the angel Raphael tells Adam in *Paradise Lost* that "the great Architect / Did wisely to conceal and not divulge / His secrets to be scann'd by them who ought / Rather admire," Heisenberg announced to the world that a good part of nature is permanently hidden from view. Matter and energy cannot be measured and gauged with complete accuracy. The state of the physical world, or even a single electron, hovers in a cloud of uncertainty. Consequently, and in contradiction to centuries of scientific thought, the future cannot be predicted from the past.

When Heisenberg began working with Bohr in 1924, there were both answers and questions about the new quantum ideas—a beautiful prelude for a breakthrough in science. Almost everything hinged on the strange "wave-particle duality" of nature.

Only one year before, in 1923, the American physicist Arthur Compton had experimentally confirmed Einstein's proposal that light consisted not of a continuous wave of energy but of a swarm of individual particles called photons. When Compton shined X-rays at electrons, each electron rebounded as if it had been hit by a single tiny billiard ball. From the amount and direction of rebound, Compton could infer that each billiard ball of light had its own energy and its own definite momentum, just as did particles like electrons. According to classical physics, the momentum p of a particle of mass m and velocity v is just the product of the two, $p = mv$. In a collision between two particles, momentum may be shifted from one particle to the other, but the total momentum of the two particles remains constant. Compton found the same phenomenon with photons and electrons. The beam of X-rays behaved as if it were billions of individual photons. In the case of a photon, which has no mass but instead a wavelength λ, its effective momentum was experimentally found by Compton to be $p = h/\lambda$, where h is Planck's quantum constant. Such a result accords exactly with Einstein's proposals.

In opposition to this particle view of light were centuries of evidence that light behaved like a wave. For centuries, scientists had documented that when light travels through a small hole, it spreads out in all directions, just as a water wave spreads out in ripples after flowing around a rock. These spreading waves can then overlap with each other, producing ridges and dips and other "interference patterns" as described in Chapter 7 on Max von Laue and X-ray diffraction. There is no problem in thinking of light as a wave. The problem occurs in thinking of light also as a particle. Particles, from all experience, do not spread out over space. They remain highly localized, in one place at a time. How can something behave both as a particle and as a wave?

The "double-slit experiment," discussed in Chapter 3 on Einstein's first paper, illustrates this kind of double existence, or wave-particle duality, of light. Recall the essential feature of that experiment: even when the source of light is so dim that it emits only one photon a second, one tiny billiard ball of definite energy and momentum each second—so that only one photon at a time strikes a screen with two well-separated holes—the light behaves as if it *passes through both holes simultaneously*. That is, each photon seems to take at least two different paths to the screen, one going through one hole and one through the other. Furthermore, downstream of the screen, the light interferes with itself like the overlapping of two waves, a wave emerging from each hole.

Shortly after Compton's work, the French physicist Louis-Victor de Broglie proposed that the wave-particle duality should apply to matter as well as to light. In particular, a particle of momentum p, whether it be an electron or a photon, should have an effective wavelength of $\lambda = h/p$, precisely equivalent to Compton's result. In other words, even a single electron should behave as if it occupies a diffuse region of space. Like a wave, an electron should be able to overlap with itself, disappearing where you might expect it and appearing where you might not. In 1927, the American physicists Clinton Davisson and L. H. Germer experimentally confirmed de Broglie's hypothesis. They fired electrons at a large

Figure 10.1

crystal of nickel and found the same wave interference patterns that von Laue had found for X-rays bombarding crystals.

A critical feature of the wave-particle duality is that the path of a moving particle can be measured and known only in a probabilistic sense. For example, the pattern of light that is found in the double-slit experiment, as seen again in Figure 10.1, is created only after many photons have struck the recording film. One can say, for example, that 20 percent of the photons land at the brightest line in the middle, 5 percent in each of the bright lines on either side of the middle, 2 percent in each of the next pair of lines, and so on. But one cannot predict where each individual photon will land on the film. By contrast, if particles had no wavelike properties, if particles did not spread out and interfere with themselves, then the path of each individual particle from emission to the screen could be determined.

Even when the film is replaced by a large group of individual photon detectors, so that each emitted photon is detected one at a time, indicated perhaps by a loud click, we cannot predict which detector will click after each photon is emitted at the source. All we can do is add up the clicks at different locations and give the fraction of photons that land in each spot. That fraction is called the probability. Analogously, when we throw a pair of dice, we cannot predict with certainty what sum will come up. However, we can accurately give the probability of each sum, that is, the fraction of times we would get each sum if we rolled the dice millions of times.

This probabilistic nature of reality required that physics be revised. In pre-quantum physics, each particle was viewed as having a definite position and velocity at each moment of time. As a particle moved through space, one could map out its continuous trajectory, like a marble rolling across the floor. Such notions seem obvious. But according to quantum physics, one cannot determine the path of a single particle traveling from A to B because each particle behaves as if it takes many different paths to get from A to B. At best, one can determine average paths taken by many particles.

How does one describe such a reality? Bohr attempted it, with his 1913 quantum model of the atom. By postulating that each electron could have only certain energies, corresponding to certain orbits about the atomic nucleus, Bohr was able to explain the particular frequencies of radiation emitted by hydrogen. Bohr's quantized orbits were equivalent to thinking of each electron as a wave, which could wrap evenly around the nucleus only at certain radii. However, Bohr had constructed a model, not a theory. His quantum postulate was ad hoc, without fundamental justification. Furthermore, Bohr could not describe the path taken by an electron when it "jumped" from one allowed quantum orbit to another, nor could he explain the likelihood of that jump. Finally, Bohr's simplified model applied only to atoms of hydrogen, with a single electron orbiting the nucleus.

In 1925, the twenty-three-year-old Heisenberg, a pupil of Bohr, worked out a detailed theory of quantum mechanics. To accomplish this feat, Heisenberg used a branch of mathematics called matrix algebra. In this scheme, each physical object, like a photon or an electron, is represented by an *array* of numbers rather than a single number for position, a single number for momentum, and so on. The array, as opposed to a single number, reflects the multiplicity of possibilities for the object. Another matrix of numbers represents a measurement. When the measurement matrix is multiplied by the object matrix, the result represents a physical measurement of the object, like a photon detector clicking when it is struck by a single photon.

In his autobiography, Heisenberg describes the transcendent creative moment when he realized that his new theory of quantum mechanics would succeed. At the end of May 1925, after months of struggling with his theory, he fell ill with hay fever and took a two-week leave of absence from the University of Göttingen.

> I made straight for Heligoland, where I hoped to recover quickly in the bracing sea air . . . Apart from daily walks and long swims, there was nothing in Heligoland to distract me from my problem . . . When the first terms seemed to accord with the energy principle, I became rather

> excited, and I began to make countless arithmetical errors. As a result, it was almost three o'clock in the morning before the final result of my computations lay before me. The energy principle had held for all the terms, and I could no longer doubt the mathematical consistency and coherence of the kind of quantum mechanics to which my calculations pointed. At first, I was deeply alarmed. I had the feeling that, through the surface of atomic phenomena, I was looking at a strangely beautiful interior, and felt almost giddy at the thought that I now had to probe this wealth of mathematical structures that nature had so generously spread out before me. I was far too excited to sleep.

The following year, the Austrian physicist Erwin Schrödinger proposed an alternative formulation of quantum mechanics, representing objects by continuous waves of probability rather than matrices. Heisenberg and Schrödinger's different formulations were shown to be equivalent, and both men won the Nobel Prize for the development of quantum mechanics.

Heisenberg's guiding principle in his highly mathematical theory was that objects in themselves have no physical meaning. Only *measurements* of objects have physical significance. One can meaningfully speak of the location of an object at A and at B if that object is measured at A and at B—as when we measure a single photon when it is emitted at A and later when it strikes a detector at B. But it is without meaning, according to Heisenberg, to discuss the object between measurements, that is, in its path from A to B. It is almost as if the object did not exist between measurements. Such a notion, of course, rudely violates all common sense. As Heisenberg stated in his Nobel Prize lecture of 1933, "the natural phenomena in which Planck's constant plays an important role can be understood only by largely forgoing a visual description of them." Such a view is as much philosophical as physical.

Heisenberg's philosophical side may have been partly shaped by his broad education and background. He was born in Würzburg into a world of privilege and culture, in December 1901. His father, August Heisenberg, was a scholar of languages and became Professor of Middle and Modern Greek Languages at the University of Munich. His mother, Annie Wecklein, was a poet and the daughter of the headmaster of the Max-Gymnasium in Munich. Extremely protected by his parents, the young Heisenberg would stop speaking to anyone he thought had treated him unfairly, and he refused ever again to look at a certain schoolteacher who once slapped his hands with a rod. Like the young Einstein, Heisenberg developed an inner freedom and independence and a propensity to question all things.

After attending the Maximillian school in Munich, Heisenberg went to the University of Munich to study with Arnold Sommerfeld, and then to Göttingen to work with Born. Undoubtedly, Heisenberg's philosophical disposition was strongly influenced by his time in Copenhagen with Niels Bohr, from 1924 to 1926. Bohr embodied one of the keenest philosophical minds in all science. As Heisenberg later said of his training, "I learned physics, along with a dash of optimism, from Sommerfeld; from Max Born, mathematics; and Niels Bohr introduced me to the philosophical background of scientific problems."

Heisenberg begins his 1927 paper on the Uncertainty Principle with an acknowledgment of the inconsistencies in the "physical interpretation" of quantum mechanics. While the young German physicist feels that "the mathematical scheme of quantum mechanics needs no revision," he questions our understanding of such fundamental concepts as mass, position, and velocity—in other words, the basic ideas of mechanics and motion. The sweep and daring of such a challenge remind us of the opening of Einstein's 1905 paper on relativity.

To support his suspicions, Heisenberg points to one of the fundamental equations of his quantum mechanics: $\mathbf{qp} - \mathbf{pq} = -i\hbar$. Here $\mathbf{q}$ represents a measurement of a particle's position and $\mathbf{p}$ represents a measurement of a particle's momentum. The symbol $\hbar$ stands for the all-important Planck's constant h divided by 2π. And the symbol i is the square root of -1, a strange and beautiful number that we regretfully do not have time here to discuss. The physical meaning of this equation is that if we measure the position of a particle and then its momentum, we get a different answer than if we measured the momentum first and then the position. *Each measurement unavoidably disturbs the particle, in a way that depends on what quantity is measured, so that the results of a subsequent measurement are altered.* Before quantum mechanics, physicists believed that they could simultaneously measure the position and momentum of a particle with as much precision as they wanted, a circumstance represented by the equation $\mathbf{qp} - \mathbf{pq} = 0$. Quantum effects make such precise and simultaneous measurements impossible.

Heisenberg interprets quantum mechanics in terms of discontinuities.

As we have seen in Bohr's paper, the electrons orbiting an atom have discontinuous energies, in that they are allowed only certain energies, with gaps between those energies. Heisenberg illustrates his idea of discontinuity with the graph, called the worldline, of a particle moving through space and time. If the curve is discontinuous, with gaps in certain places, then it becomes impossible to define position and velocity (the tangent to the curve) in those places.

The next section of Heisenberg's paper is the most powerful. Recall that his guiding principle is to concentrate on what can be observed. He asks us to consider exactly what is to be understood by the words "position of the object." (Later, he does the same with the word "velocity.") Those words have no meaning, for Heisenberg, except as defined by an actual physical observation of the object.

Then, in the style of Einstein, he constructs a hypothetical but quite feasible experiment. Suppose we want to determine the position of an electron. We do so by shining light on it. The light, after scattering from the electron, is collected by a lens and focused into an image on a photographic plate or some other detector. From the location of the image on the detector, we deduce the position of the electron. (Our human eye determines positions of objects the same way, with a focusing lens and a retina for a detection screen.)

The key idea is that light, as shown by Einstein and Compton, carries momentum. So when a photon of light strikes the electron, it gives the electron a push and the electron goes flying away. *We have disturbed the electron in order to view it.*

If the electron had been sitting quietly at rest before our observation, it is now moving. Furthermore, it is moving in some unknown direction, because we cannot measure the angle of the deflected photon with complete accuracy. We know only that it passed through the lens. *To determine the position of the electron, we have given it a partly unknown velocity.*

There is a trade-off between the uncertainty in position and the uncertainty in momentum of the electron. (As mentioned earlier, momentum is velocity multiplied by mass. Like velocity, momentum has a direction.) We can measure the position of the electron with greater and greater accuracy by using light of smaller and smaller wavelength. (As discussed in Chapter 7 on von Laue, to probe smaller and smaller details, we need finer and finer probing devices.) But, as can be seen in Compton's relation, $p = h/\lambda$, a smaller wavelength corresponds to a larger momentum for the photon, resulting in a larger (partly unknown) momentum transfer to the electron.

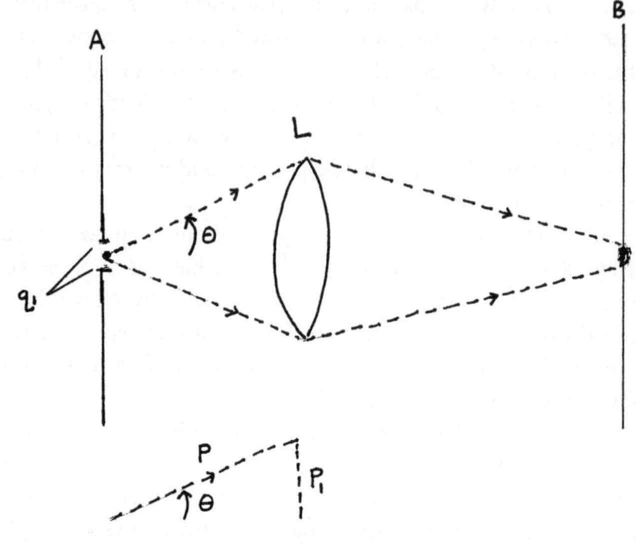

Figure 10.2

Now, we will make these ideas quantitative. The situation is illustrated in Figure 10.2. Here, the electron is somewhere in the plane A on the left. We shine light on it, the light is focused by the lens, denoted by L, and then strikes the photographic plate in plane B on the right.

Because of the wave nature of light and the phenomenon of diffraction of light by the lens, the image of the electron on the photographic plate will not be a perfect point. Instead, the image will be blurred, as if the electron were smeared over a region $q_1 = \lambda / sin\theta$, to use Heisenberg's notation. Here, λ is the wavelength of light used and θ is half the angle of the possible light ray trajectories through the lens (indicated by dashed lines in the illutration). This smearing formula for q_1 comes out of the wave theory of light and has been well known since the early nineteenth century. Because of the light blurring, it is impossible to locate the position of the electron exactly. However, the uncertainty in position, q_1, can be made smaller by decreasing the wavelength λ of light.

Unfortunately, decreasing the wavelength increases the (unknown) momentum given to the electron by the photon. We cannot know the exact path that a photon of light follows in traveling from the electron in plane A to its image in plane B. All we know is that the photon path must be somewhere between the dashed lines. If the initial momentum of the photon is p, then after its deflection by angle θ it will have a sideways momentum $p_1 = p\ sin\theta$, as shown at the bottom of the figure. It is this sideways momentum that is uncertain.

Since θ is the maximum possible deflection, we know that the sideways momentum of the photon after deflection may be anywhere from zero to $p\ sin\theta$. From the law that total momentum is constant, the electron will acquire an equal and opposite sideways momentum (in the plane A). Thus, the sideways momentum of the electron after the scattering with light will be anywhere between zero and $p\ sin\theta$. In other words, the momentum of the electron after the scattering is *uncertain* to this amount. Using Compton's relation between momentum and wavelength, $p = h/\lambda$, we can express this uncertainty in electron momentum as $p_1 = h\ sin\theta/\lambda$.

We can make this uncertainty in momentum smaller by *increasing* the wavelength. However, such an increase will make the uncertainty in position, q_1, *larger*.

A quantitative measure of these two competing effects can be seen by multiplying q_1 and p_1 together. From the above equations, we get

$$q_1 p_1 = h.$$

The above result is the famous Heisenberg Uncertainty Principle. It says that although we can measure position as accurately as we want (q_1 very small), *or* momentum as accurately as we want (p_1 very small), we cannot measure both as accurately as we want. The equation gives the combined uncertainty. The less uncertainty in one, the more uncertainty in the other. As an example, suppose that an electron, which has a mass of 9.1×10^{-28} grams, has an uncertainty in its velocity of 1 centimeter per second. Given that electrons in atoms travel at velocities of a hundred million centimeters per second or so, this is a relatively tiny uncertainty. Then, according to Heisenberg's relation, the uncertainty in the electron's position is about 7 centimeters, huge on the scale of atoms. Or suppose we make the uncertainty in position to be 10^{-9} centimeters, about a tenth the size of an atom. Then the uncertainty in velocity is a whopping 7×10^9 centimeters per second.

But such uncertainties, and indeed all quantum effects, are totally negligible in the everyday world. For example, let's apply Heisenberg's equation to a baseball instead of an electron. A baseball has a mass of about 140 grams. An uncertainty in velocity of 1 centimeter per second corresponds to an uncertainty in position of 4.7×10^{-29} centimeters, far, far smaller than the size of an atomic nucleus.

Heisenberg's Uncertainty Principle is a statement of principle. In pre-quantum physics, it was of course recognized that any practical measurement had uncertainties associated with it—the glass of the lens could have been ground more finely, the lab table might be shaking a little, and so on. But, *in principle*, one could build instruments that made these uncertainties as small as one wished. Heisenberg's uncertainties are different. No matter what instruments we ever build, there is a fundamental limitation on how small we can make the uncertainties. This fundamental limit is caused by the essential and unavoidable wave-particle duality of light.

In the next section of his paper, Heisenberg extends the uncertainty principle to other pairs of observables, such as measurements of energy **E** and time **t,** and measurements of angular momentum **J** and angle **w.** I have omitted the following highly mathematical sections of the paper, but included the final "Addition in Proof." The tone of this section suggests not only Heisenberg's willingness to clarify his discussion but also his deep respect for his mentor "Professor Bohr."

Quantum mechanics, along with relativity, is the cornerstone of all modern physics. The understanding of atoms and subatomic particles, lasers, silicon chips, and so much else in the world of today depends fundamentally on quantum mechanics. If Planck's constant were much smaller than it is, then quantum effects would dwindle to irrelevance even for subatomic particles. If Planck's constant were much larger, as in George Gamow's classic fantasy *Mr. Thompkins Explores the Atom*, then ordinary chairs, along with their occupants, would routinely vanish and reappear across the room. Quantum mechanics has altered our view of the nature of reality. Heisenberg's Uncertainty Principle means, among other things, that the future cannot be determined from the past. The future position of a particle can be determined only if its current position and velocity are known. The Uncertainty Principle decrees that this condition is possible only to within limits. The Uncertainty Principle decrees that the world of certainty envisioned by Galileo and Newton does not exist.

Werner Heisenberg seemed to be suffering personally from his Uncertainty Principle when I went to hear him speak at the California Institute of Technology in the early 1970s. At the time, I was a graduate student in physics and enthralled to meet a living legend in my field. The auditorium was packed. Several hundred faculty and students fidgeted anxiously in our chairs. The speaker was introduced. And then Heisenberg shuffled to the lecturn. The person once described as looking like "a simple farm boy, with clear bright eyes, and a radiant expression" struck me as a weary old man, with a wrinkled face and a dark weight on his back. But that was not my biggest surprise. At the following reception for Heisenberg at the elegant Caltech faculty club, Richard Feynman, a professor at Caltech and a Nobel Prize winner himself, stood up and verbally attacked Heisenberg for the foolishness of his lecture, indeed mocked him to his face. Underneath Feynman's scathing remarks, I detected not only a disagreement with Heisenberg's new scientific work, but also contempt for the man, a deep resentment that the founder of quantum physics had helped the Nazis try to build an atomic bomb. I was stunned by the drama.

Here was a firsthand example of the controversy and bitterness Heisenberg aroused among many fellow scientists when he chose to remain in Germany during World War II. Such controversy haunted the rest of his life, although he continued to receive medals, honorary degrees, and invitations to speak.

For most of the war, Heisenberg was a professor of physics at the University of Berlin and director of the Kaiser Wilhelm Institute for Physics. The majority of Jewish German scientists, such as Einstein, Lise Meitner, and Hans Krebs, had already fled Germany or been expelled. Other scientists, such as Max Planck and Max von Laue, remained in Germany but opposed the Nazi regime and managed to avoid doing war-related research.

Heisenberg's motives and thinking will probably always remain shrouded in a cloud of uncertainty. He did indeed work on the development of a German atomic bomb. But there is evidence that he and his colleagues never thought a bomb would be built by the Germans. Their estimate of the required time and expense of the project far exceeded the resources available. One can wonder what Heisenberg would have done if the needed resources had been given to him.

Heisenberg did not support the brutal regime of the Nazis. As the Austrian-American physicist Victor Weisskopf later said, "It must have driven him to utter despair and depression that his beloved country had fallen so deeply into the abyss of crime, blood, and murder." Heisenberg could have emigrated, he could have joined the anti-Nazi underground movement, or he could have retired from public life. But he was not a hero, and, having won the Nobel, he was too prominent to retire. What of emigration, the first possibility? Here, Heisenberg had a terrible choice. In her biography titled *Inner Exile*, Heisenberg's wife Elisabeth says that her husband felt that leaving Germany would have spared his reputation but nothing else. In emigrating, "he would have abandoned his friends and his students, his family in general, physics . . . just to save himself. It was a thought that he could not bear."

ON THE PHYSICAL CONTENT OF QUANTUM KINEMATICS AND MECHANICS

Werner Heisenberg

Zeitschrift für Physik (1927)

First we define the terms *velocity*, *energy*, etc. (for example, for an electron) which remain valid in quantum mechanics. It is shown that canonically conjugate quantities can be determined simultaneously only with a characteristic indeterminacy (§1). This indeterminacy is the real basis for the occurrence of statistical relations in quantum mechanics. Its mathematical formulation is given by the Dirac-Jordan theory (§2). Starting from the basic principles thus obtained, we show how microscopic processes can be understood by way of quantum mechanics (§3). To illustrate the theory, a few special *gedankenexperiments* are discussed (§4).

We believe we understand the physical content of a theory when we can see its qualitative experimental consequences in all simple cases and when at the same time we have checked that the application of the theory never contains inner contradictions. For example, we believe that we understand the physical content of Einstein's concept of a closed 3-dimensional space because we can visualize consistently the experimental consequences of this concept. Of course these consequences contradict our everyday physical concepts of space and time. However, we can convince ourselves that the possibility of employing usual space-time concepts at cosmological distances can be justified neither by logic nor by observation. The physical interpretation of quantum mechanics is still full of internal discrepancies, which show themselves in arguments about continuity versus discontinuity and particle versus wave. Already from this circumstance one might conclude that no interpretation of quantum mechanics is possible which uses ordinary kinematical and mechanical concepts. Of course, quantum mechanics arose exactly out of the attempt to break with all ordinary kinematic concepts and to put in their place relations between concrete and experimentally determinable numbers. Moreover, as this enterprise seems to have succeeded, the mathematical scheme of quantum mechanics needs no revision. Equally unnecessary is a revision of space-time geometry at small distances, as we can make the quantum-mechanical laws approximate the classical ones arbitrarily closely by choosing sufficiently great masses, even when arbitrarily small distances and times come into question. But that a revision of kinematical and mechanical concepts is necessary seems to follow directly from the basic equations of quantum mechanics. When a definite mass m is given, in our everyday physics it is perfectly understandable to speak of the position and the velocity of the center of gravity of this mass. In quantum mechanics, however, the relation $\mathbf{pq} - \mathbf{qp} = -i\hbar$ between mass, position, and velocity is believed to hold. Therefore we have good reason to become suspicious every time uncritical use is made of the words "position" and "velocity." When one admits that discontinuities are somehow typical of processes that take place in small regions and in short times, then a contradiction between the concepts of "position" and "velocity" is quite plausible. If one considers, for example, the motion of a particle in one dimension, then in continuum theory one will be able to draw (Fig. 1) a worldline $x(t)$ for the track of the particle (more precisely, its center of gravity), the tangent of which gives the velocity at every instant. In contrast, in a theory based on discontinuity there might be in place of this curve a series of points at finite separation (Fig. 2). In this case it is clearly meaningless to speak about one velocity at one position (1) because one velocity can only be defined by two positions and (2), conversely, because any one point is associated with two velocities.

The question therefore arises whether, through a more precise analysis of these kinematic and mechanical concepts, it might be possible to clear up the contradictions evident up to now in the physical interpretations of quantum mechanics and to arrive at a physical understanding

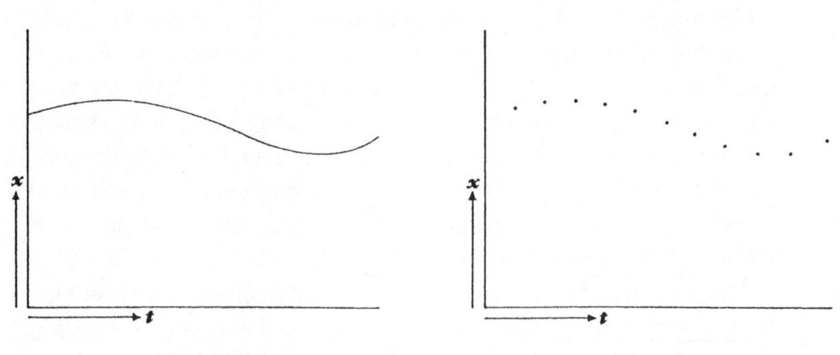

Figure 1

Figure 2

Max Planck, 1921.
Courtesy American Institute of Physics Emilio Segrè Visual Archives

Ernest Starling, 1887.
College Collection, Library Services, University College London

William Bayliss (to the right of the blackboard) and Ernest Starling (to the left of the blackboard), ca. 1905.
College Collection, Library Services, University College London

Albert Einstein at his desk at the Swiss patent office, 1905.

Photograph by Lucien Chavan, courtesy American Institute of Physics Emilio Segrè Visual Archives.

Ernest Rutherford, ca. 1890.
United Kingdom Atomic Energy Authority, courtesy American Institute of Physics Emilio Segrè Visual Archives, Physics Today *Collection*

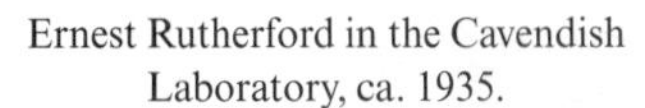

Ernest Rutherford in the Cavendish Laboratory, ca. 1935.
Photograph by C. E. Wynn-Williams, courtesy American Institute of Physics Emilio Segrè Visual Archives. Permission to reproduce granted by the Department of Physics, Cambridge University.

Henrietta Leavitt, ca. 1898.
Courtesy Harvard College Observatory

The "computers" at the Harvard College Observatory, standing in front of the observatory entrance, 1917. Henrietta Leavitt is sixth from the left.
Courtesy Harvard College Observatory

Max von Laue in his motorcar.
Courtesy Archiv zur Geschichte der Max Planck-Gesellschaft, Berlin-Dahlem

Niels Bohr in the early 1920s.

Courtesy American Institute of Physics Emilio Segrè Visual Archives

Otto Loewi.
Courtesy Public Affairs Office, NYU Medical Center

Werner Heisenberg (center), with Enrico Fermi (left)
and Wolfgang Pauli (right), ca. 1928.
Photograph by F. D. Rasetti, courtesy American Institute of Physics Emilio Segrè Visual Archives, Segrè Collection

of the quantum-mechanical formulas.[1] In order to be able to follow the quantum-mechanical behavior of any object one has to know the mass of this object and its interactions with any fields and other objects. Only then can the Hamiltonian function be written down for the quantum-mechanical system. (The following considerations ordinarily refer to nonrelativistic quantum mechanics, as the laws of quantum electrodynamics are still very incompletely known.)[2] About the "Gestalt" (construction) of the object any further assumption is unnecessary; one most usefully employs the word "Gestalt" to designate the totality of these interactions.

When one wants to be clear about what is to be understood by the words "position of the object," for example of the electron (relative to a given frame of reference), then one must specify definite experiments with whose help one plans to measure the "position of the electron"; otherwise this word has no meaning. There is no shortage of such experiments, which in principle even allow one to determine the "position of the electron" with arbitrary accuracy. For example, let one illuminate the electron and observe it under a microscope. Then the highest attainable accuracy in the measurement of position is governed by the wavelength of the light. However, in principle one can build, say, a γ-ray microscope and with it carry out the determination of position with as much accuracy as one wants. In this measurement there is an important feature, the Compton effect. Every observation of scattered light coming from the electron presupposes a photoelectric effect (in the eye, on the photographic plate, in the photocell) and can therefore also be so interpreted that a light quantum hits the electron, is reflected or scattered, and then, once again bent by the lens of the microscope, produces the photoeffect. At the instant when position is determined—therefore, at the moment when the photon is scattered by the electron—the electron undergoes a discontinuous change in momentum. This change is the greater the smaller the wavelength of the light employed—that is, the more exact the determination of the position. At the instant at which the position of the electron is known, its momentum therefore can be known up to magnitudes which correspond to that discontinuous change. Thus, the more precisely the position is determined, the less precisely the momentum is known, and conversely. In this circumstance we see a direct physical interpretation of the equation $\mathbf{pq} - \mathbf{qp} = -i\hbar$. Let q_1 be the precision with which the value q is known (q_1 is, say, the mean error of q), therefore here the wavelength of the light. Let p_1 be the precision with which the value p is determinable; that is, here, the discontinuous change of p in the Compton effect. Then, according to the elementary laws of the Compton effect p_1 and q_1 stand in the relation

$$p_1 q_1 \sim h. \tag{1}$$

That this relation (1) is a straightforward mathematical consequence of the rule $\mathbf{pq} - \mathbf{qp} = -i\hbar$ will be shown below. Here we can note that equation (1) is a precise expression for the facts which one earlier sought to describe by the division of phase space into cells of magnitude h. For the determination of the position of the electron one can also do other experiments—for example, collision experiments. A precise measurement of the position demands collisions with very fast particles, because for slow electrons the diffraction phenomena—which, according to Einstein, are consequences of de Broglie waves (as, for example, in the Ramsauer effect)—prevent a sharp specification of location. In a precise measurement of position the momentum of the electron again changes discontinuously. An elementary estimate of the precision using the formulas for de Broglie waves leads once more to relation (1).

Throughout this discussion the concept of "position of the electron" seems well enough defined, and only a word need be added about the "size" of the electron. When two very fast particles hit the electron one after the other within a very short time interval Δt, then the positions of the electron defined by the two particles lie very close together at a distance Δl. From the regularities which are observed for α-particles we conclude that Δl can be pushed down to a magnitude of the order of 10^{-12} cm if only Δt is sufficiently small and particles are selected with sufficiently great velocity. This is what we mean when we say that the electron is a corpuscle whose radius is not greater than 10^{-12} cm.

We turn now to the concept of "path of the electron." By path we understand a series of points in space (in a given reference system) which the electron takes as "positions" one after the other. As we already know what is to be understood by "position at a definite time," no new difficulties occur here. Nevertheless, it is easy to recognize that, for example, the often used expression, the "1s orbit of the electron in the hydrogen atom," from our point of view has no sense. In order to measure this 1s "path" we have to illuminate the atom with light whose wavelength is considerably shorter than 10^{-8} cm. However, a single photon of such light is enough to eject the electron completely from its "path" (so that only a single point of such a path can be defined).

Therefore here the word "path" has no definable meaning. This conclusion can already be deduced, without knowledge of the recent theories, simply from the experimental possibilities.

In contrast, the contemplated measurement of position can be carried out on many atoms in a 1s state. (In principle, atoms in a given "stationary" state can be selected, for example, by the Stern-Gerlach experiment.) There must therefore exist for a definite state—for example, the 1s state—of the atom a probability function for the location of the electron which corresponds to the mean value for the classical orbit, averaged over all phases, and which can be determined through the measurement with an arbitrary precision. According to Born,[3] this function is given by $\psi_{1s}(q)\overline{\psi}_{1s}(q)$ where $\psi_{1s}(q)$ designates the Schrödinger wave function belonging to the 1s state. With a view to later generalizations I should like to say—with Dirac and Jordan—that the probability is given by $S(1s, q)\overline{S}(1s, q)$, where $S(1s, q)$ designates that column of the matrix $S(E, q)$ of transformation from E to q that belongs to the energy $E = E_{1s}$.

In the fact that in quantum theory only the probability distribution of the position of the electrons can be given for a definite state, such as 1s, one can recognize, with Born and Jordan, a characteristically statistical feature of quantum theory as contrasted to classical theory. However, one can say, if one will, with Dirac, that the statistics are brought in by our experiments. For plainly *even in classical theory* only the probability of a definite position for the electron can be given as long as we do not know the phase of [the motion of the electron in] the atom. The distinction between classical and quantum mechanics consists rather in this: classically we can always think of the phase as determined through suitable experiments. In reality, however, this is impossible, because every experiment for the determination of phase perturbs or changes the atom. In a definite stationary "state" of the atom, the phases are in principle indeterminate, as one can see as a direct consequence of the familiar equations

$$\mathbf{Et} - \mathbf{tE} = -i\hbar \quad \text{or} \quad \mathbf{Jw} - \mathbf{wJ} = -i\hbar,$$

where $\mathbf{J}$ is the action variable and $\mathbf{w}$ is the angle variable.

The word "velocity" can easily be defined for an object by measurements when the motion is free of force. For example, one can illuminate the object with red light and by way of the Doppler effect in the scattered light determine the velocity of the particle. The determination of the velocity is the more exact the longer the wavelength of the light that is used, as then the change in velocity of the particle, per light quantum, by way of the Compton effect is so much less. The determination of position becomes correspondingly inexact, in agreement with equation (1). If one wants to measure the velocity of the electron in the atom at a definite instant, then, for example, one will let the nuclear charge and the forces arising from the other electrons suddenly be taken away, so that the motion from then on is force-free, and one will then carry out the measurement described above. As above, one can again easily convince oneself that a [momentum] function $p(t)$ cannot be defined for a given state—such as the 1s state—of an atom. On the contrary, there is again a probability function for p in this state which according to Dirac and Jordan has the value $S(1s, p)\overline{S}(1s, p)$. Here $S(1s, p)$ again designates that column of the matrix $S(E, p)$—that transforms from **E** to **p**—which belongs to $E = E_{1s}$.

Finally we come to experiments which allow one to measure the energy or the value of the action variable J. Such experiments are especially important because only with their help can we define what we mean when we speak of the discontinuous change of the energy and of J. The Franck-Hertz collision experiments allow one to base the measurement of the energy of the atom on the measurement of the energy of electrons in rectilinear motion, because of the validity of the law of conservation of energy in quantum theory. This measurement in principle can be carried out with arbitrary accuracy if only one forgoes the simultaneous determination of the position of the electron or its phase (see the determination of p, above), corresponding to the relation $\mathbf{Et} - \mathbf{tE} = -i\hbar$. The Stern-Gerlach experiment allows one to determine the magnetic or an average electric moment of the atom, and therefore to measure quantities which depend only on the action variable J. The phases remain undetermined in principle. It makes as little sense to speak of the frequency of the light wave at a definite instant as of the energy of an atom at a definite moment. Correspondingly, in the Stern-Gerlach experiment the accuracy of the energy measurement decreases as we shorten the time during which the atom is under the influence of the deflecting field.[4] Specifically, an upper bound is given for the deviating force through the circumstance that the potential energy of that deflecting force can at most vary inside the beam by an amount which is considerably smaller than the differences in energy of the stationary states. Only then will a determination of the energy of the stationary states be at all possible. Let E_1 be an amount of energy which satisfies this condition (E_1 also fixes the precision of the energy measurement). Then E_1/d specifies the highest allowable value for the deflecting force, if d is the breadth of the beam (measurable through the spacing of the slits employed). The angular deviation of the atomic beam is then $E_1 t_1/dp$, where we designate by t_1 the time during which the atoms are under the influence of the deflecting field, and by p the momentum of the atoms in the direction of the beam. This deflection must be of at least the same order of magnitude as the natural broadening of the beam brought about by the diffraction by the slits, if any measurement is to be possible. The diffraction angle is roughly λ/d if λ denotes the de Broglie wavelength; thus,

$$\lambda/d \sim E_1 t_1/dp,$$

or, as $\lambda = h/p$,

$$E_1 t_1 \sim h. \tag{2}$$

This equation corresponds to equation (1) and shows how a precise determination of energy can only be obtained at the cost of a corresponding uncertainty in the time. . . .

ADDITION IN PROOF

After the conclusion of the foregoing paper, more recent investigations of Bohr have led to a point of view which permits an essential deepening and sharpening of the analysis of quantum-mechanical correlations attempted in this work. In this connection Bohr has brought to my attention that I have overlooked essential points in the course of several discussions in this paper. Above all, the uncertainty in our observation does not arise exclusively from the occurrence of discontinuities, but is tied directly to the demand that we ascribe equal validity to the quite different experiments which show up in the corpuscular theory on one hand, and in the wave theory on the other hand. In the use of an idealized gamma-ray microscope, for example, the necessary divergence of the bundle of rays must be taken into account. This has as one consequence that in the observation of the position of the electron the direction of the Compton recoil is only known with a spread which then leads to relation (1). Furthermore, it is not sufficiently stressed that the simple theory of the Compton effect, strictly speaking, only applies to free electrons. The consequent care needed in employing the uncertainty relation is, as Professor Bohr has explained, essential, among other things, for a comprehensive discussion of the transition from micro- to macromechanics. Finally, the discussion of resonance fluorescence is not entirely correct because the connection between the phase of the light and that of the electronic motion is not so simple as was assumed. I owe great thanks to Professor Bohr for sharing with me at an early stage the results of these more recent investigations of his—to appear soon in a paper on the conceptual structure of quantum theory—and for discussing them with me.

Copenhagen,
Institute for Theoretical Physics of the University.

1. The present work has arisen from efforts and desires to which other investigators have already given clear expression, before the development of quantum mechanics. I call attention here especially to Bohr's papers on the basic postulates of quantum theory (for example, *Zeits. f. Physik, 13*, 117 [1923]) and Einstein's discussions on the connection between wave field and light quanta. The problems dealt with here are discussed most clearly in recent times, and the problems arising are partly answered, by W. Pauli ("Quantentheorie," *Handbuch der Physik*, Vol. XXIII, cited hereafter as *l.c.*); quantum mechanics has changed only slightly the formulation of these problems as given by Pauli. It is also a special pleasure to thank here Herr Pauli for the repeated stimulus I have received from our oral and written discussions, which have contributed decisively to the present work.

2. Quite recently, however, great advances in this domain have been made in the papers of P. Dirac (*Proc. Roy. Soc. A114*, 243 [1927] and papers to appear subsequently).

3. The statistical interpretation of de Broglie waves was first formulated by A. Einstein (*Sitzungsber. d. preussische Akad. d. Wiss.*, p. 3 [1925]). This statistical feature of quantum mechanics then played an essential role in M. Born, W. Heisenberg, and P. Jordan, Quantenmechanik II (*Zeits. f. Physik*, *35*, 557 [1926]), especially chapter 4, §3, and P. Jordan (*Zeits. f. Physik*, *37*, 376 [1926]). It was analyzed mathematically in a seminal paper of M. Born (*Zeits. f. Physik*, *38*, 803 [1926]) and used for the interpretation of collision phenomena. One finds how to base the probability picture on the theory of the transformation of matrices in the following papers: W. Heisenberg (*Zeits. f. Physik*, *40*, 501 [1926]), P. Jordan (*Zeits. f. Physik*, *40*, 661 [1926]), W. Pauli (remark in *Zeits. f. Physik*, *41*, 81 [1927]), P. Dirac (*Proc. Roy. Soc. A113*, 621 [1926]), and P. Jordan (*Zeits. f. Physik*, *40*, 809 [1926]). The statistical side of quantum mechanics is discussed more generally in P. Jordan (*Naturwiss.*, *15*, 105 [1927]) and M. Born (*Naturwiss.*, *15*, 238 [1927]).

4. In this connection see W. Pauli, *l.c.*, p. 61.

11

THE CHEMICAL BOND

LINUS PAULING ONCE RECOUNTED the moment he decided to become a chemist. The year was 1914. Thirteen years old, a sophomore at Washington High School in Portland, Oregon, Pauling was invited by his friend Lloyd Jeffress to see some chemical experiments. Up in a second-floor bedroom, as Linus looked on, the other boy combined sugar and potassium chlorate in a ceramic bowl and then poured in sulfuric acid. Immediately, the mixture sizzled and seethed, producing a gush of steam and a mound of black carbon. Linus already had an interest in science, especially insects and minerals. Now, he was fascinated by the idea that some substances could change into others. "I am going to be a chemist!" he announced on the spot.

When young Linus got home, he began reading the chemistry book left by his father, a pharmacist who had died suddenly of peritonitis four years earlier. Another pharmacist and family friend gave the boy some chemicals to begin his own experiments. A neighbor brought him pieces of glassware. A grandfather who worked as a night watchman at a foundry secured bottles of sulfuric acid, nitric acid, and potassium permanganate.

Linus Pauling's career had begun. By the end of his long life, in 1994, he was widely considered the greatest chemist of the twentieth century. At the center of his many accomplishments, Pauling employed the new quantum theory of the European physicists to study how atoms bond to each other. More than anyone, Pauling pioneered the modern theory of the chemical bond.

What metals should be combined with gold to give it more strength? Why does salt taste salty and sugar sweet? What makes rubber soft and elastic? Why does iron rust? What happens when food is digested? How is penicillin extracted and purified from the *Penicillium* fungus? Is it possible to create a synthetic form of silk? How does one make a lubricating oil that does not get too thin in hot weather, nor too thick in cold weather? These are all questions for chemistry. While physics is concerned with the elementary particles of matter and the forces between them, chemistry is more concerned with the properties of *aggregate* bits of matter and how that matter reacts with other matter. While the physicist would study the structure of an individual atom, the chemist would investigate how one atom interacts with other atoms to form molecules and compounds. Many properties of matter in the world of human experience derive from the way that atoms bond with other atoms. Such is the province of chemistry.

In 1922, when Linus Pauling received his B.S. degree in chemical engineering from Oregon State College and began graduate work at the California Institute of Technology (Caltech), there was no fundamental theory of the chemical bond. Yet chemists knew much about the bond from experiments.

First was the atom. In the early years of the nineteenth century, the British chemist John Dalton had found that chemical elements combined to make compounds in very specific relative weights. For example, a gram of hydrogen would always combine with eight grams of oxygen to make water. Dalton's finding lent support to the ancient Greek notion of atoms and the plausible hypothesis that individual atoms of different elements have different weights. The atom became the basic unit of chemistry. But the main interest of chemistry was how atoms combined with other atoms to form molecules.

Soon after Dalton's work, in 1819, the Swedish chemist Jöns Jakob Berzelius proposed that atoms bonded to other atoms by means of electrical forces. Positive charges attract negative charges. Such basic electrical phenomena were well known at this time, although the particular subatomic elements responsible for charge, such as the electron and proton, were far beyond detection.

As mentioned in Chapter 8 on Bohr's quantum atom, in 1869, Russian chemist Dmitri Ivanovich Mendeleyev published his mysterious Periodic Table, a classification of chemical elements according to increasing atomic weights. Elements arranged in order of increasing weights showed regular repetitions of chemical properties. (Elements in the same columns of the table had the same properties, as Pauling says in his paper.) For example, begin with lithium, the third-lightest element. Skip the next seven elements in the list and come to sodium, eleventh-lightest element, which has the same chemical properties as lithium. Skip another seven elements and come to potassium, nineteenth-lightest element, which has similar properties to lithium and sodium. Or begin with beryllium, fourth-lightest element. Skipping seven elements brings us to magnesium, with similar properties as beryllium. Another seven and we come to calcium, with similar properties. An understanding of these magical patterns, like the patterns of the spectral emissions of atoms, would have to await the development of quantum physics, fifty years into the future. However, even in the nineteenth century, chemists realized that the chemical properties of substances were determined by the way in which one atom bonds to another, that is, the nature of the chemical bond.

By the end of the nineteenth century, it had become apparent that there were two different kinds of chemical bonds. The first, called the polar or ionic bond, occurs when a positively charged atom attracts a negatively charged atom. Since atoms are normally electrically neutral, with their positive charge counterbalanced by an equal amount of negative charge, a polar bond would be formed when a neutral atom transfers much of its negative charge to another neutral atom, rendering the first atom positive and the second negative. The second kind of bond, called the nonpolar bond and later the covalent bond, occurs between two neutral atoms. Here, the two bonding atoms are closer together and some of the electrical charges within the two atoms pull on both atoms. Of the two kinds of bonds, the covalent bond is stronger, more versatile, and generally more important in complex phenomena.

With J. J. Thomson's discovery of the negatively charged electron in 1897 and Ernest Rutherford's discovery of the positively charged atomic nucleus in 1911, these notions of chemical bonds became more precise. The electrons, orbiting the central nucleus in the outer parts of an atom, would be the particles responsible for chemical bonds. In ionic bonds, electrons would be transferred from one atom to another. In covalent bonds, electrons would remain in or near their initial atoms but pull on the positive nuclei of two atoms at once.

In 1916, the eminent American chemist Gilbert Newton Lewis (1875–1946) proposed that covalent bonds are brought about by the *sharing of pairs* of electrons between two atoms. According to Lewis, each of the two atoms in a covalent bond would contribute one electron to the bond. The two electrons would form a pair that would travel together and be shared by the two atoms. Furthermore, the pair of shared electrons would attract the positively charged nuclei of *both* atoms, thus holding them together and creating the bond. Lewis's proposal had a great deal of explanatory power. However, it was only a proposal, without theoretical foundation and without much quantitative detail. That foundation would not be possible until 1925, when a full theory of quantum mechanics was created.

The academic year 1926–1927 was enormously important for Pauling and the history of chemistry. In that year, the twenty-five-year-old Pauling, already a rising star with a fresh Ph.D. from Caltech and twelve published papers under his belt, received a Guggenheim fellowship. The young chemist traveled to Europe to learn the new quantum physics. Much of that physics had been founded by men of his own age. Werner Heisenberg was also twenty-five at the time. Wolfgang Pauli, of whom more will be said later, was only a year older.

Pauling spent the first part of his fellowship year at the Institute of Theoretical Physics in Munich under the direction of Arnold Sommerfeld, then went to Niels Bohr's institute in Copenhagen, then to Zurich. At Sommerfeld's institute, bubbling with the new quantum physics, Pauling found himself in the enviable position of being the lone chemist. He realized immediately that quantum mechanics would provide the basis for understanding the structure of molecules and the nature of the chemical bond.

In Zurich, Pauling was strongly influenced by two other young scientists, Walter Heitler, aged twenty-three, and Fritz London, twenty-seven. Heitler and London were just completing the first quantum mechanical calculation of a shared-electron chemical bond, two electrons shared in the hydrogen molecule, H_2. Pauling's ambition was set on fire. A photograph from this period shows him to be tall and lanky, with a prominent nose and Adam's apple, a square chin, curly brown hair, and, most important, a look of supreme confidence. As he once wrote, even as a boy "I had the feeling that I could understand everything . . . if I try hard enough."

That self-confidence seems to have come at an early age, brought on both by natural ability and by family circumstance. Besides his childhood obsession with insects and minerals, Pauling had been a precocious reader. By the age of nine, he had read the Bible, Darwin's *Origin of Species*, ancient histories, and some of the *Encyclopedia Britannica*, which he often recited to a younger cousin. After his father died that year, Pauling's mother Belle began to depend on him to help support himself and his two younger sisters, Pauline and Lucile. The oldest child accepted his new duties.

When insects did not satisfy young Pauling's budding scientific mind, he began studying minerals. His books discussed white grains of quartz, pink feldspar, and black grains of mica, all of which he was able to recognize in the chips of granite he found in a rock wall near his house. To his delight, the world of theory could be matched with the world of experiment. The world of family responsibility could be balanced with the world of science. Nowhere were Pauling's self-confidence, seriousness, deliberateness of purpose, and self-awareness more evident than in the first entry of his diary, dated August 20, 1917. He was sixteen at the time.

> Today I am beginning to write the history of my life . . . This "history" is not intended to be written in diary form or as a continued narrative—rather it is to be a series of essays on subjects most important to my mind. It will serve to remind me of resolutions made, of promises, and also of good times had, and of important occurrences in my passage through this "vale of tears." . . . Often, I hope, I shall glance over what I have written before, and ponder and meditate on the mistakes that I have made.

Upon his return from Europe in 1927, Pauling was appointed assistant professor of chemistry at Caltech. The next year, at age twenty-seven, he published the first of his pioneering papers on the chemical bond.

Underlying Pauling's seminal paper of 1928 was the realization that the orbits of electrons in isolated, single atoms were not of the right shape for bonding with other atoms. (Quantum mechanics decrees that electron orbits come in very particular shapes, such as spheres, dumbbells, cloverleaves, and so on. Some of these orbits are shown in Figure 11.1.) Instead, Pauling proposed, bonding orbits arise from a combination, or "hybrid," of single-atom orbits. His brilliant idea was that those hybrid orbits that were most elongated and directed, like a pointed finger, would be the ones used by nature in the covalent bond. The most highly directed orbits would reach out the farthest from one atom to another, providing the strongest possible bond between them. In a sense, bonded atoms should point to each other. In arriving at his proposal, Pauling was likely guided as much by deep common sense as by knowledge of theoretical physics, as much by thinking in pictures as by solving mathematical equations.

To understand more of Pauling's paper, it is first necessary to review some basic ideas of quantum physics. Indeed, we will have to discuss more scientific concepts in this chapter than in any previous chapter. Most of these come from physics. At the end, we will have laid the foundations for much of modern chemistry.

A key concept of the new quantum physics was the idea of allowed orbits. As Niels Bohr had proposed with his crude model in 1913, electrons in orbit about the nucleus of an atom could have only certain energies, separated by regularly spaced gaps. (See Chapter 8 on Bohr.) The detailed quantum theory of Werner Heisenberg and Erwin Schrödinger extended Bohr's idea of restricted energies to restricted "quantum states." A quantum state involved not only the energy of the electron but also the shape and orientation of its orbit. Those energies and orbits could have only certain values. Thus, they were "quantized."

Another important concept was the notion of probability rather than certainty. The electron, or any subatomic particle, was represented by a "wave function," which gave the probability that the electron would be at a particular place at a particular time. (A wave function is also sometimes called an eigenfunction, as in Pauling's paper.) Each quantum state of the electron—that is, each specification of its energy and orbital parameters—would correspond to a particular wave function. Because the electron acts like a wave as well as a particle, one must think of it as occupying many positions simultaneously, with some positions more likely than others.

The allowed orbits of quantum mechanics could be divided into "shells" and "subshells." Each value of the energy corresponded to a definite shell. For example, Bohr's $\tau = 1$ quantum state is called the K-shell, the $\tau = 2$ state is called the L-shell, and so on. Roughly speaking, the shell of an electron specifies its average distance from the nucleus. By contrast, the *subshell* of the electron is related to the *shape* of its orbit. (Two electrons can have the same average distance from the nucleus but very different shapes for their orbits.) The subshell of an electron is specified by another quantum number, denoted by l. The $l = 0$ quantum state is called the *s*-subshell, or *s*-level. The $l = 1$ quantum state is called the *p*-subshell, or *p*-level. And so on, to higher l. Just as with the quantum number τ, the l quantum number can change only by integral amounts, leading to the conclusion that both the energy and orbital shape of the electron are "quantized" and can have only certain values.

Figure 11.1a shows the shape of the *s*-subshell orbit. One can imagine the atomic nucleus at the origin of the XYZ coordinates. The way to interpret the illustration is the following: The farther the curve is from the origin, the greater is the probability of finding the electron in that particular direction, so that the shape of the curve gives the variation of probability in different directions. As can be seen, the *s*-subshell is a perfect sphere, at equal distances in all directions, meaning that an electron in the *s*-subshell is equally likely to be found in any direction from the nucleus.

Figure 11.1b shows the shape of the *p*-subshell orbits. Now, the shape is not a perfect sphere about the origin, but looks more like a dumbbell, with two touching spheres. That dumbbell can have three different *orientations*, directed along the X-axis, Y-axis, or Z-axis, leading to three different *p*-subshells, called here p_x, p_y, and p_z. For example, consider the p_x subshell. The spherical lobes extend farthest from the origin at the two points on the X-axis denoted by A and B. Thus an electron in the p_x

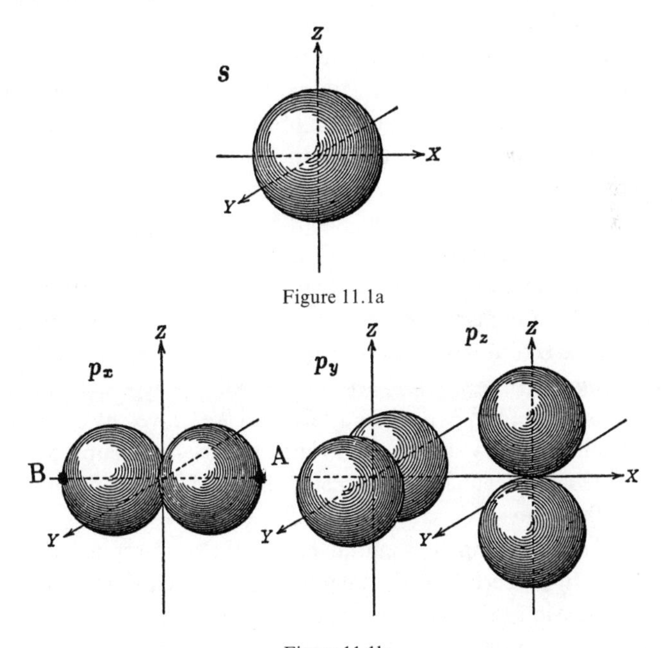

Figure 11.1a

Figure 11.1b

subshell has the greatest probability of being found along the X-axis, defining a particular direction in space. As one moves in a direction toward a point on the Z-axis or the Y-axis, the probability decreases toward zero. The same kind of analysis applies to the p_y and p_z subshells. The subshell of an electron's orbit is specified by the shape (*l* value) and orientation, (e.g., p_x, p_y, or p_z).

Shells and subshells constitute the geography of the quantum world. To an electron, shells and subshells are the continents between oceans.

Two other quantum ideas are critical for understanding Pauling's paper: the intrinsic spin of a subatomic particle and the exclusion principle, both formulated about 1925.

To explain the observed radiations from atoms and the behavior of electrons, the Dutch-American physicists George Uhlenbeck and Samuel Goudsmit proposed that each electron spins about its own invisible axis like a tiny gyroscope. Unlike in classical (pre-quantum) physics, that spin could never slow down or speed up. As for mass and electrical charge, the electron spin was a fixed property of the electron. The Austrian theoretical physicist Wolfgang Pauli further suggested that the electron's spin could have one of two orientations. It could point "up" or "down." The direction of "up" was arbitrary, but, once chosen, the electron could point either along that direction or opposite to that direction, with nothing in between. Thus, the spin orientation, like the orbit orientation, was quantized. The orientation of electron spin became an additional part of the quantum state of the electron. Now, the complete quantum state consisted of *four* properties: the energy, the shape of the orbit, the orientation of the orbit, and the orientation of spin, all specified by quantum numbers.

Pauli, like Einstein and Heisenberg, was a prodigy. In 1919, at age nineteen, he had so mastered Einstein's new theory of general relativity that he was asked to write a book-length article on the subject for the *Encyclopedia of Mathematical Sciences*. By age twenty-one, Pauli had received his Ph.D. from Arnold Sommerfeld at the University of Munich, where Pauling was to visit a few years later. For the next couple of years, Pauli worked in Copenhagen with the great Bohr, father of the quantum model of the atom.

Soon after discussing the quantum nature of electron spin, Pauli used the symmetry properties of the quantum wave function to deduce his famous Exclusion Principle (later called the Pauli Exclusion Principle): Two electrons cannot occupy the same quantum state. In other words, two electrons in orbit about an atomic nucleus cannot have the same quantum numbers for energy, orbital shape, orbital orientation, and orientation of spin. In particular, *at most two electrons can occupy the same shell and subshell*, one with its spin pointing up and one with its spin down. Because of their different spin orientations, these two electrons would have different quantum states. But a third electron in the same shell and subshell is excluded because there are only two possible spin orientations. A third electron would necessarily have the same quantum state as one of the first two and thus be forbidden by Pauli's principle.

The Pauli Exclusion Principle explains how shells and subshells are filled up as one progresses to heavier and heavier elements, with more and more electrons orbiting the central nucleus. The K-shell has only a single subshell, the *s*-subshell. Thus there can be only two electrons in the K-shell (one with spin "up" and one with spin "down.") The lightest atom, hydrogen, has a single electron. The next-lightest, helium, has two electrons. Beyond helium, the K-shell is completely filled and cannot hold any more electrons. With the next-lightest element, lithium, one must begin filling the L-shell. The L-shell has four subshells: one *s*-subshell and three *p*-subshells, p_x, p_y, and p_z. Since each of these four subshells can hold two electrons (one with spin up and one with spin down), the L-shell can hold a maximum of eight electrons. For example, lithium, with a total of three electrons, has two electrons in the K-shell and one electron in the L-shell. Neon, with a total of ten electrons, has two electrons in the K-shell and eight in the L-shell. Beyond neon, the L-shell is completely filled and one must begin filling the next shell, the M-shell. Sodium, for example, with a total of eleven electrons, has one electron in the M-shell. And so on.

Electrons in completely filled shells are "locked up." They cannot be shared with other atoms. Thus, only the electrons in unfilled shells, such as the one electron in the K-shell of hydrogen or the one electron in the L-shell of lithium, are available for sharing with other atoms. Since it is this sharing that creates the covalent chemical bond, *the chemical properties of an element are largely determined by the number of electrons in the outermost, unfilled shell.* Such electrons are called valence electrons, and the number of them is called the valence of the element.

Pauli's Exclusion Principle, at last, explained the mysterious repeating patterns in the Periodic Table. Since valences regularly repeat as shells are filled and one starts over with the next shell, the chemical properties of elements repeat. Lithium has the same chemical properties as sodium because both elements have one electron in their outermost unfilled shell.

The covalent chemical bond also hinges on the Exclusion Principle. Two electrons that are shared by two atoms in a molecule can pull most effectively on both atoms, and thus bond them most strongly, when those electrons spend most of their time between the two atomic nuclei, in the same region of space. In turn, the electrons occupy the same region of space when their spatial quantum numbers, that is, their energy and orbital quantum numbers, are the same. By the Exclusion Principle, their spin orientations must then be opposite. Thus, the shared-electron chemical bond is formed by a pair of electrons with the same spatial quantum state and opposite spins. In such a situation, as Pauling says at the beginning of his paper, "the wave function [of the molecule] is symmetric in the positional [spatial] coordinates of the two electrons . . . so that each electron is partially associated with one nucleus and partially with the other."

Linus Pauling's particular genius was his physical insight in finding solutions to the equations of quantum mechanics. Those equations can be solved exactly only when one electron is involved. As soon as there are two or more electrons, as is always the case with the covalent bond, then only approximate solutions are possible. Strictly speaking, the quantum states previously discussed apply only to single electrons orbiting a single atomic nucleus. Even in a single atom, when more than one electron is present the tidy and well-defined quantum states corresponding to particular values of τ, l, and the other quantum numbers become only approximate. In molecules, where electrons orbit two atomic nuclei and not just one, the situation is even more complex. As Pauling says in his paper, when two atoms bond together in a molecule, "the interchange energy resulting from the formation of shared electron bonds is large enough to change the quantization, destroying the two sub-shells with $l = 0$ and $l = 1$ of the L shell." The single-atom subshells such as the s and p orbitals in Figure 11.1 are no longer appropriate.

How does one proceed? Finding good approximate solutions in any field is as much art as science. And here, physical insight is a premium. Pauling was guided by his understanding that for a pair of electrons to pull strongly on two atomic nuclei, bonding them together, the electrons should overlap as much as possible. Furthermore, Pauling reasoned, *such overlap would be highest for the most pointed orbits.* More pointed orbits provide more physical space for the two bonding electrons to travel together between the two nuclei, pulling on both. More pointed orbits also reach out farther from one atomic nucleus to the next.

To illustrate the idea of pointedness, the *p*-subshell orbits (Figure 11.1b) are more pointed than the *s*-subshell orbit (Figure 11.1a), which is not pointed at all. The orbit of Figure 11.4, to be discussed later, is even more pointed. The most pointed orbits form the strongest bonds. *And atoms in a molecule always adjust their interactions and orientations to form the strongest bond possible*. In a similar way, two nearby magnets without any friction will naturally orient themselves so that their opposite poles come as close together as possible. Or, as yet another example, a marble rolling on a bumpy floor tends to settle at the lowest dip in the floor, where it is closest to the center of the earth. The particular arrangement with the strongest bonds is called "stable." Unstable bonds change and shift to become stable. Stable bonds have the smallest possible energy. Stable bonds, like the marble settling in the dip of the floor, lie at the bottom of the "bowl of energy" and stay where they are.

Given his guiding principle of pointed orbits, Pauling used a technique that he calls the "Heisenberg-Dirac resonance phenomenon." In that method, one seeks a combination of two single-atom orbits, or two approximate quantum states in general, that has stronger bonds than either of the single-atom orbits alone. In the early 1930s, Pauling coined the term "hybrid" for such combined orbits. Figure 11.4 depicts a hybrid orbit.

Pauling applied these ideas to the extremely important chemical bond of the carbon atom. Carbon is the central element in biology. Because of its bonding properties, allowing it to form a great variety of strong bonds with one, two, three, or four other atoms, carbon is ideal for creating the complex molecules required for life. An atom of carbon has two K-shell electrons and four L-shell electrons, the latter available for bonds.

Figure 11.2a Figure 11.2b

One of the simplest carbon molecules is methane, CH_4, consisting of one atom of carbon and four of hydrogen. In methane, each of the four L-shell electrons in a carbon atom pairs with one electron from a hydrogen atom. Methane is often represented by Figure 11.2a, where each line stands for a covalent chemical bond. Another representation, devised by Gilbert Lewis and shown in Figure 11.2b, replaces each line by two dots. The two dots stand for the two electrons shared in the bond.

Now, according to the simplest quantum approximation to the carbon atom, using the single-atom orbits, the four valence electrons of carbon should form four shared-electron bond pairs, with one pair filling the *s*-subshell and one pair in each of the three *p*-subshells. As can be seen in Figure 11.1b, the three *p*-subshells extend at right angles to each other. On the basis of this simple theory, one would expect that three of the carbon bonds in a methane molecule should be at right angles (90°) to each other. The fourth, belonging to the spherically symmetric *s*-subshell, could be in any arbitrary direction.

This picture, however, does not agree with experiment. Since the nineteenth century, it had been known from experiment that the four carbon bonds in a carbon-based molecule point in the four directions of a regular tetrahedron, making angles of 109.47° with each other. The situation is illustrated in Figure 11.3. A regular tetrahedron, as shown in Figure 11.3a, is a four-sided, three-dimensional figure with each side an equilateral triangle. (Here, the dashed line is the invisible edge of the bottom triangle. This line is omitted in Figure 11.3b.) Figure 11.3b shows the observed bonds of the carbon atom. If the carbon atom is placed at the center of the tetrahedron, represented by the large dot, then the four carbon bonds, the dashed lines going from the central dot to the four corners of the tetrahedron, make angles of 109.47° with each other.

For a few years, this contradiction between the right angles of the *p*-subshell orbits and the observed tetrahedral angles of carbon bonds

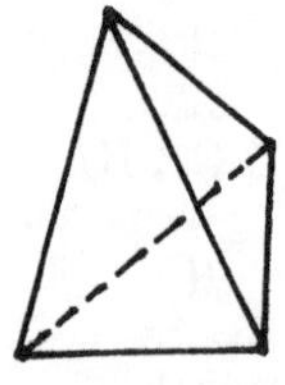

Figure 11.3a

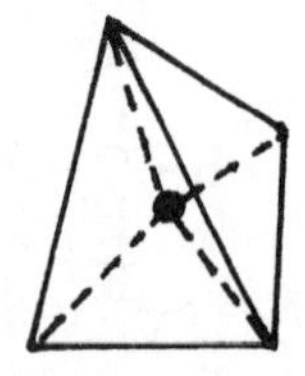

Figure 11.3b

constituted a problem for the application of quantum theory to the chemical bond.

Pauling was able to find a *hybrid* of the *s*- and *p*-subshells, denoted by (*s-p*), that was considerably more pointed than either the *s* or *p* alone and would thus have lower energy than the *s* or *p* orbits alone. Take the X-direction, for example, as the direction of the first bond. Pauling found that the combination $(s\text{-}p)_1 = \frac{1}{2}s + \frac{\sqrt{3}}{2}\,p_x$ had the maximum possible concentration in the X-direction. That hybrid subshell is shown in Figure 11.4. When Pauling found hybrid *s-p* subshells for the other three electron pairs, $(s\text{-}p)_2$, $(s\text{-}p)_3$, and $(s\text{-}p)_4$, he discovered that they were all identical to $(s\text{-}p)_1$, except for a rotation of direction, and that rotated angle was exactly 109.47°. What a triumph! Pauling's identical hybrid orbits for the bonds of carbon indeed pointed in the direction of the corners of a regular tetrahedron. And Pauling knew why. Furthermore, he had developed a powerful new technique for calculating other chemical bonds.

All of these ideas and calculations are summarized in Pauling's 1928 paper by one paltry sentence, matter-of-fact and bland: "It has been further found that as a result of the resonance phenomenon a tetrahedral arrangement of the four bonds of the quadrivalent carbon atom is the stable one." No details are given. However, one can sense what must have been his excitement when he describes the calculations in detail, a decade later, in his landmark textbook *The Nature of the Chemical Bond*: "A surprising result of the calculation, of great chemical significance . . . is that the second best [most pointed] bond orbital is equivalent to the first . . . and that *its bond direction makes the tetrahedral angle of* 109.47° *with that of the first*." The italics are Pauling's.

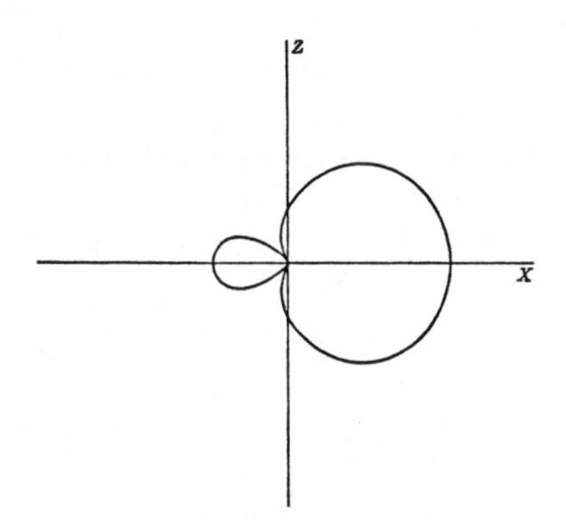

Figure 11.4

Some historians of chemistry consider Pauling's next paper in the series, published in 1931 in the *Journal of the American Chemical Society*, to be the most important paper in chemistry in the twentieth century. However, the paper of 1928 was the first. And it is clear in this paper that Pauling had already conceived of, formulated, and successfully applied his key concept of the hybrid bond, even though he refers to it only briefly.

The discovery that he could explain the bonds of carbon was a "surprising result." Pauling loved surprises. Surprises stirred him up. Surprises provoked his imagination. In a letter to a friend in 1980, Pauling wrote: "I tried to fit knowledge that I acquired into my system of the world . . . When something comes along that I don't understand, that I can't fit in, that bothers me: I think about it, mull over it, and perhaps ultimately do some work with it . . . Often I'm not interested in something new that's [already] been discovered, because even though it's new, it doesn't surprise me and interest me."

Over his career, Pauling was stirred and provoked to many more discoveries. In addition to his fundamental work on the chemical bond, central to most of chemistry, Pauling studied the structure of substances with X-ray crystallography (see Chapter 7 on von Laue), he did pioneering work in molecular biology, he uncovered some of the fundamental structures of proteins. In later years, he was a biologist as much as a chemist. In 1935, he found how oxygen bonds to hemoglobin, a critical factor in how oxygen is carried by the blood. In 1936, he discovered how protein molecules fold up and unfold, now known to be a key factor in their behavior and properties. In 1949, he made the first identification of a molecular disease, sickle-cell anemia, caused when red blood cells assume an elongated shape. In a lecture years later, Pauling said, "The way to have good ideas is to have lots of ideas." His 1954 Nobel Prize in chemistry was earned several times over.

In the last half of the twentieth century, Linus Pauling's stature grew to legendary proportions. No other scientist except Einstein was as well known to the American public. Like Einstein, Pauling projected something of a homespun quality. Pauling had grown up in a family of modest means, chopped wood and mopped kitchen floors to work his way through college, enjoyed love stories and romances in addition to more intellectual reading, became a national champion of vitamin C, loved Doris Day movies, was happily married for fifty-nine years.

In 1945, at the end of World War II, Pauling's career broadened dramatically. Especially because of the extreme power and horror of nuclear weapons, physicists and chemists all over the world were forced to take a stand on war. Earlier, we have seen the positions taken by von Laue, Heisenberg, and others. With his characteristic deliberateness, Pauling decided that for the rest of his life he would devote half his time to peace activism—learning about international relations and international law, treaties, national histories, the peace movement, and other subjects relating to the question of how to abolish war from the world. He gave numerous lectures on the subject. With his wife Ava Helen, he circulated a petition to end nuclear weapons tests and in 1958 presented the petition (with 11,000 signatures from forty-nine countries) to the United Nations. That year he also published his influential antiwar book *No More War!* In 1961 he helped organize a major peace conference in Oslo. For these efforts, Pauling was awarded the 1962 Nobel Peace Prize, the only person ever to win a Nobel in both science and peace.

In an essay in the early 1950s, just at the time the hydrogen bomb was being developed, Pauling wrote:

> How do people of different beliefs, different natures, different ideals, different races, get along together? How does a man get along with a neighbor whom he does not like? Not by preparing to fight him—that is not the civilized method. Instead, different people and different groups of people have learned to live together in peace, to respect one another's qualities, even the differences—they have learned this in every sphere except that of international relations. Now the time has come for nations to learn this lesson . . . I am sure that we may have hope. The stage is now set for a great act—the final abolition of war and the achievement of a permanent peace.

THE SHARED-ELECTRON CHEMICAL BOND

Linus Pauling

Gates Chemical Laboratory, California Institute of Technology

Communicated March 7, 1928

Proceedings of the National Academy of Sciences (1928)

WITH THE DEVELOPMENT OF the quantum mechanics it has become evident that the factors mainly responsible for chemical valence are the Pauli exclusion principle and the Heisenberg-Dirac resonance phenomenon. It has been shown[1,2] that in the case of two hydrogen atoms in the normal state brought near each other the eigenfunction which is symmetric in the positional coördinates of the two electrons corresponds to a potential which causes the two atoms to combine to form a molecule. This potential is due mainly to a resonance effect which may be interpreted as involving an interchange in position of the two electrons forming the bond, so that each electron is partially associated with one nucleus and partially with the other. The so-calculated heat of dissociation, moment of inertia, and oscillational frequency[2] of the hydrogen molecule are in approximate agreement with experiment. London[3] has recently suggested that the interchange energy of two electrons, one belonging to each of two atoms, is the energy of the non-polar bond in general. He has shown that an antisymmetric (and hence allowed) eigenfunction symmetric in the coordinates of two electrons can occur only if originally the spin of each electron were not paired with that of another electron in the same atom. The number of electrons with such unpaired spins in an atom is, in the case of Russell-Saunders coupling, equal to $2s$, where s is the resultant spin quantum number, and is closely connected with the multiplicity, $2s + 1$, of the spectral term. This is also the number of electrons capable of forming non-polar bonds. The spins of the two electrons forming the bond become paired, so that usually these electrons cannot be effective in forming further bonds.

It may be pointed out that this theory is in simple cases entirely equivalent to G. N. Lewis's successful theory of the shared electron pair, advanced in 1916 on the basis of purely chemical evidence. Lewis's electron pair consists now of two electrons which are in identical states except that their spins are opposed. If we define the chemical valence of an atom as the sum of its polar valence and the number of its shared electron pairs, the new theory shows that the valence must be always even for elements in the even columns of the periodic system and odd for those in the odd columns. The shared electron structures assigned by Lewis to molecules such as H_2, F_2, Cl_2, CH_4, etc., are also found for them by London. The quantum mechanics explanation of valence is, moreover, more detailed and correspondingly more powerful than the old picture. For example, it leads to the result that the number of shared bonds possible for an atom of the first row is not greater than four, and for hydrogen not greater than one; for, neglecting spin, there are only four quantum states in the *L*-shell and one in the *K*-shell.

A number of new results have been obtained in extending and refining London's simple theory, taking into consideration quantitative spectral and thermochemical data. Some of these results are described in the following paragraphs.

It has been found that a sensitive test to determine whether a compound is polar or non-polar is this: If the internuclear equilibrium distance calculated for a polar structure with the aid of the known properties of ions agrees with the value found from experiment, the molecule is polar; the equilibrium distance for a shared electron bond would, on the other hand, be smaller than that calculated. Calculated[4] and observed values of the hydrogen-halogen distances in the hydrogen halides are in agreement only for HF, from which it can be concluded that HF is a polar compound formed from H^+ and F^- and that, as London had previously stated, HCl, HBr, and HI are probably non-polar. This conclusion regarding HF is further supported by the existence of the hydrogen bond. The structure

$$\left[:\ddot{\underset{\cdot\cdot}{F}}:H:\ddot{\underset{\cdot\cdot}{F}}:\right]^-$$

for the acid fluoride ion and a similar one for H_6F_6 are ruled out by Pauli's principle, if the shared pairs be of London's type. The ionic structure

$$:\ddot{\underset{\cdot\cdot}{F}}:^-\ H^+\ :\ddot{\underset{\cdot\cdot}{F}}:^-,$$

in which the proton holds the two fluoride ions together by electrostatic forces (including polarization), is, of course, allowed. This conception of the hydrogen bond, requiring the presence of a proton, explains the observation that only atoms of high electron affinity (fluorine, oxygen, and nitrogen) form such bonds.

Compounds of septivalent chlorine probably contain chlorine with a polar valence of +3 and with four shared electron bonds. Thus the perchlorate ion would be formed from

$$\cdot\dot{\underset{\cdot}{\mathrm{Cl}}}\cdot^{+3} \text{ and } 4\cdot\ddot{\underset{\cdot\cdot}{\mathrm{O}}}\!:^{-}$$

and would have the structure

$$\left[\begin{array}{ccc} & :\ddot{\mathrm{O}}: & \\ :\ddot{\underset{\cdot\cdot}{\mathrm{O}}}: & \ddot{\underset{\cdot\cdot}{\mathrm{Cl}}} & :\ddot{\underset{\cdot\cdot}{\mathrm{O}}}: \\ & :\underset{\cdot\cdot}{\mathrm{O}}: & \end{array}\right]^{-}.$$

This structure is, of course, also possible for FO_4^-, inasmuch as there are positions for four (and only four) unpaired electrons in the *L*-shell of fluorine. London assumed that septivalent chlorine had a polar valence of zero and seven shared electron bonds, and pointed out in explanation of the non-existence of valences other than one for fluorine that seven unpaired *L*-electrons cannot occur. This simple explanation, however, does not make impossible fluorine with a polar valence of +3 and with four shared bonds; the non-existence of higher valences of fluorine must accordingly instead be explained on energetic considerations, and is connected with its very large ionization potentials.

In the case of some elements of the first row the interchange energy resulting from the formation of shared electron bonds is large enough to change the quantization, destroying the two sub-shells with $l = 0$ and $l = 1$ of the *L*-shell. Whether this will or will not occur depends largely on the separation of the *s*-level ($l = 0$) and the *p*-level ($l = 1$) of the atom under consideration; this separation is very much smaller for boron, carbon, and nitrogen than for oxygen and fluorine or their ions, and as a result the quantization can be changed for the first three elements but not for the other two. The changed quantization makes possible the very stable shared electron bonds of the saturated carbon compounds and the relatively stable double bonds of carbon, which are very rare in other atoms, and in particular are not formed by oxygen. This rupture of the *l*-quantization also stabilizes structures in which only three electron pairs are attached to one atom, as in molecules containing a triple bond ($N_2 = \vdots N\vdots N\vdots$), the carbonate, nitrate, and borate ions

$$\left[\begin{array}{ccc} & & :\!\ddot{\text{O}}\!: \\ :\!\ddot{\text{O}}\!: & \text{C} & \\ & & :\!\ddot{\text{O}}\!: \end{array}\right]^{-},$$

etc., the carboxyl group,

$$\begin{array}{lll} & & :\!\ddot{\text{O}}\!: \\ \text{R}:\text{C} & & \\ & & :\!\ddot{\text{O}}\!: \\ & & \quad\ \text{H} \end{array},$$

and similar compounds. It has further been found that as a result of the resonance phenomenon a tetrahedral arrangement of the four bonds of the quadrivalent carbon atom is the stable one.

Electron interactions more complicated than those considered by London also result from the quantum mechanics, and in some cases provide explanations for previously anomalous molecular structures.

It is to be especially emphasized that problems relating to choice among various alternative structures are usually not solved directly by the application of the rules resulting from the quantum mechanics; nevertheless, the interpretation of valence in terms of quantities derived from the consideration of simpler phenomena and susceptible to accurate mathematical investigation by known methods now makes it possible to attack them with a fair assurance of success in many cases.

The detailed account of the material mentioned in this note will be submitted for publication to the *Journal of the American Chemical Society*.

1. Heitler, W., and London, F., *Z. Phys.*, **44,** 455 (1927).
2. Y. Sugiura, *Ibid.*, **45,** 484 (1927).
3. F. London, *Ibid.*, **46,** 455 (1928).
4. L. Pauling, *Proc. Roy. Soc.*, **A114,** 181 (1927). In this paper it was stated that the consideration of polarization of the anion by the cation would be expected to reduce the calculated values; a number of facts, however, indicate that with ions with the noble gas structure the effect of polarization on the equilibrium distance is small.

12

THE EXPANSION OF THE UNIVERSE

IT IS A CHILLY EVENING in the late 1920s, on the top of Mount Wilson in southern California. After dinner, Edwin Hubble walks along the dirt path from the little monastery to the great domed building of the hundred-inch Hooker telescope, the largest telescope on earth. The dome alone is ninety-five feet in diameter and a hundred feet high. Wearing his camel-hair coat and black beret, Hubble enters the shedlike door. He climbs the metal stairway to the concrete floor of the telescope mount, climbs another flight of stairs to the platform surrounding the instrument, then one more set of steps to the "observing platform." Except for a single assistant, he is alone. Outside, darkness is falling. At his signal, the massive dome slowly opens with a long rumbling sound. The sky is a deep purple gash flecked with stars. Hubble calls out his orders to the assistant below, the hours and degrees of the pointing angle he wants, and the telescope wheels around to its target, whining metal against metal, clicking and clanking on its seventeen-foot gears.

Tonight, Hubble will make photographs of distant galaxies, so distant that their light has spent millions of years traveling through space to the earth. The astronomer studies the view through the eyepiece. When the position is correct, he strikes a match, lights his pipe, and sits down in his low bentwood chair. From now on, a clock drive will automatically revolve the telescope at precisely the correct rate to compensate for the earth's rotation, so that the galaxies do not slide out of view. But the clock drive is silent. The lights in the observatory are extinguished, leaving the astronomer in darkness except for the faint glow of his pipe and the gauzy light of the stars overhead. Here he will sit for hours through the cold silent night. From time to time, he will call out new orders to his assistant, stand, and stare through the eyepiece. The telescope itself looms above him like a giant squatting bird. Its torso, the light tube, extends for some thirty feet. Its massive hind legs and haunches, the support structure, cling to the floor with a steel grip. This bird weighs one hundred tons. Hubble is a large man himself, six feet two inches tall, barrel-chested, a former heavyweight boxer. But sitting beneath the tail of the giant bird as it peers into space, he is the size of an ant.

On February 4, 1931, in the small library of the Mount Wilson Observatory, a stone's throw from the Hooker telescope, Albert Einstein announced that his original conception of a static universe was no longer valid. Forty-one-year-old Edwin Hubble, as well as a half dozen other leading astronomers, stood nearby. As a result of the discoveries of Edwin Hubble, Einstein went on, the universe must be considered to be in motion. The cosmos was expanding. Space itself was stretching, with distant galaxies flying away from each other like dots painted on the skin of a swelling balloon. According to a reporter from the Associated Press, "a gasp of astonishment swept through the library." And the news flashed over the world.

In 1543, Copernicus proposed that the sun, rather than the earth, was the center of our planetary system. In all of the following centuries, Edwin Hubble's discovery of the expansion of the universe was probably the most important event in astronomy. If the universe is expanding, then it is changing. Indeed, the universe must have evolved through a series of grand epochs, each unimaginably different from the last. In particular, the universe was smaller and denser in the past. At one time, the galaxies were touching. Farther in the past, stars had not yet formed out of the dense clouds of primordial gas. Farther still in the past, electrons would be boiled off the outer parts of atoms. Go far enough into the past, and all of the matter that we see in the cosmos was crushed together into a volume tinier than an atom. That point in time, or just before, was the "beginning," now called the Big Bang. By measuring the rate that the universe is expanding, astronomers can calculate that the Big Bang occurred about fifteen billion years ago. Thus, Hubble's discovery had enormous meaning, not only for science but also for philosophy, theology, and even human psychology.

Edwin Hubble had taken a winding route to the moment in the mountaintop library with Einstein. Born in Marshfield, Missouri, in late 1889 and schooled in Chicago, he could easily have been a professional athlete or a lawyer or a half dozen other occupations instead of an astronomer. By sixteen, Edwin was the star of the Central High

School basketball team in Chicago. In a single track meet his senior year, he won the pole vault, the shot put, the standing high jump, the running high jump, the discus, and the hammer throw, and on May 6, 1906, he established the Illinois state record in the high jump. After compiling a superb academic record at the University of Chicago, he won a Rhodes Scholarship, which had been a personal obsession. Another obsession, astronomy, met strong resistance by his father, who was a lawyer and insurance agent. At Oxford, Edwin read not astronomy or mathematics but jurisprudence. He explained to a friend that on his return home he would have to earn money to support his family. But Edwin still made time for winning track events at Oxford, and he held an exhibition boxing match with a French champion. When Hubble came back to the United States in 1913, he opened a law office in Louisville, Kentucky, where his parents had moved. The law, however, could never fill his stomach. After a year Hubble went back to the University of Chicago as a graduate student. Astronomy it would be.

Edwin Hubble seems to have been an almost superhuman individual: handsome, physically powerful and athletic, bright, restless, ambitious, arrogant, aloof. At an early age he read widely in the classics and history as well as in science. A classmate in high school remembers that Edwin rejected his teachers' authority and questioned them with a smart aleck attitude. Another classmate, Albert Colvin, recalls that Edwin had "a way of acting as though he had all the answers . . . He always seemed to be looking for an audience to which he could expound some theory or other." According to his sister Betsy, Edwin "tried to do things to prove he was capable of doing them," and often retold stories about himself with exaggerated and heroic proportions.

Hubble eventually succeeded in his ambitions beyond any possible exaggeration. Ironically, when he published his famous paper of 1929, claiming a linear relation between distances and receding velocities of the galaxies, he suspected that he'd done something important, but he didn't know what it was.

Before Hubble's discovery, nearly every human culture on earth had believed in a universe without change, a cosmos in stasis. The stars in the sky certainly appear to be fixed and unmoving, aside from the stately revolution caused by the spin of the earth. As Aristotle wrote in *On the Heavens*, "Throughout all past time, according to the records handed down from generation to generation, we find no trace of change either in the whole of the outermost heaven or in any one of it proper parts." Indeed, scientists and poets alike took the fixed heavens as the supreme metaphor for constancy and permanence, in contrast to the ephemeral nature of all phenomena on earth. Copernicus, who was willing to challenge so much of our astronomical thinking, wrote that "the state of immobility is regarded as more noble and godlike than that of change and instability, which for that reason should belong to the Earth rather than to the [Universe]." And in Shakespeare's *Julius Caesar*, Caesar says to Cassius:

> But I am constant as the northern star,
> Of whose true-fix'd and resting quality
> There is no fellow in the firmament.

Einstein, in his 1917 theory of cosmology, simply assumed that the universe did not change on the grand scale. Indeed, he was so certain that the universe had to be static that he was willing to revise and complicate the elegant equations of his 1915 theory of gravity, general relativity, in order to account for the presumed immobility of the heavens. Those equations show how matter and energy generate gravity, and how that gravity in turn affects the geometry of space and time. In his revision, Einstein added a number to his equations, sometimes called the lambda term or the cosmological constant. The lambda term acts as a kind of repulsive force, balancing the attractive force of gravity and thus allowing the stars and nebulae of the universe to hold steady at fixed positions. As the great German-born physicist said at the end of his paper: "That [lambda] term is necessary only for the purpose of making possible a quasi-static distribution of matter, as required by the fact of the small velocities of stars."

Unbeknownst to Einstein, who was not an astronomer, certain recent astronomical data were already suggesting that the material of the universe did not sit quietly in balanced equilibrium. Since 1912, a onetime farm boy from Mulberry, Indiana, named Vasco Melvin Slipher had been amassing evidence that some of the "nebulae" were flying away from the solar system at fantastic speeds. Slipher had been given access to the twenty-four-inch telescope of the Lowell Observatory in Arizona. (The twenty-four inches refers to the diameter of the lens or mirror of the telescope. Larger diameters can gather more light and thus see fainter objects, as well as resolve finer details.)

Nebulae are the permanent, cloudlike, misty patches of light in the sky. Many have been known since antiquity. Galileo, with his first telescope, showed that some of the nebulae are congregations of individual stars, too dim and close together to distinguish with the naked eye. The most dramatic cosmic nebula is the faint band of light that arches across the night sky, called the Milky Way, or sometimes just the Galaxy. The Milky Way is the great spiral-shaped system of stars that is home to our own star, the sun. Today, we know the Milky Way contains about a hundred billion stars. We also know that there are actually three kinds of nebulae: the globular clusters, which are spherical systems of about a million stars located within the Milky Way; the galactic nebulae, which are clouds of dust and gas also located within the Milky Way; and the extragalactic nebulae, other giant systems of stars outside of the Milky Way. The extragalactic nebulae are, in fact, other galaxies. But in 1912 much of this was unknown. Most significantly, distances to these objects were unknown. Until the mid-1920s, astronomers hotly debated whether the nebulae were located within the Milky Way or were instead separate "island universes" beyond.

By 1914, Slipher had measured the velocities of thirteen of the nebulae. More precisely, Slipher had measured the *colors* of the nebulae. What do colors have to do with speeds? When a moving source of light is traveling toward you, its colors shift up in frequency toward the blue end of the spectrum; when it is moving away, its colors shift down to the red. This phenomenon—called the Doppler shift in honor of Christian Johann Doppler, who first discussed it in 1842—is exactly analogous to the change in pitch of the whistle of a moving train. As the train approaches, its pitch rises above the pitch at rest; as the train moves away, its pitch drops. From the amount of the shift in color (for light) or pitch (for sound), one can calculate the speed of the moving object. Using this method, Slipher concluded that the average spiral-shaped nebula was moving *away* from the earth at a speed of about 600 kilometers per second, a hundred times faster than the speed of any other known type of object in the sky.

By the early 1920s, Slipher had measured the recessional velocities of about forty nebulae, with the same general results. His findings were considered important, but no one knew their significance. Were the spiral nebulae relatively small and nearby constellations of stars, like the globular clusters, or instead large stellar systems like the Milky Way, at great distance from our own galaxy? Hubble, as well as other astronomers, puzzled over the meaning of Slipher's results. As the eminent astronomer

Arthur Eddington wrote in his influential *The Mathematical Theory of Relativity* in 1923: "One of the most perplexing problems of cosmogony is the great speed of the spiral nebulae." Perplexing largely because the distances were not known.

Indeed, the major obstacle for much of astronomy was the determination of distance. When we detect light from an astronomical body like a star or a nebula, we measure only its apparent brightness. To know its distance, we must also know its intrinsic luminosity, just as we must know the wattage of a lightbulb in order to infer its distance from how bright it appears to our eye.

As described in detail in Chapter 6, in 1912 Henrietta Leavitt of the Harvard College Observatory discovered a method to measure the distance to a certain kind of star called a Cepheid variable. Cepheids vary in the intensity of their light in a regular and repeating way, with periods (cycle times) between about three and fifty days. In brief, Leavitt found a relationship between a Cepheid's period and its intrinsic luminosity. The distance to a Cepheid could then be computed as follows: Measure its period and apparent brightness. From Leavitt's period-luminosity law, and the measured period, infer its intrinsic luminosity. From its intrinsic luminosity and its measured apparent brightness, infer its distance.

In 1918, the American astronomer Harlow Shapley systematically searched for Cepheids at various locations in the Milky Way to map out the size of the galazy. He concluded that the Milky Way is about 300,000 light-years in diameter. Recall that a light-year is the distance that light can travel in a year, roughly ten thousand billion miles. Another unit of distance sometimes used is the parsec, equal to about 3.3 light-years. For reference, the nearest star to our sun, Alpha Centauri, is about four light-years away.

Having determined the approximate size of the Milky Way, Shapley argued for various reasons that the nebulae were all within our galaxy. According to Shapley there were no island universes, no extragalactic nebulae. It was Edwin Hubble in 1924 who proved Shapley wrong.

After receiving his Ph.D. in astronomy from the University of Chicago in 1917, Hubble was offered a position at the Mount Wilson Observatory of the Carnegie Institution of Washington. But Hubble wanted to serve in World War I. He received a deferment for his position. With his usual bravado, Hubble joined the American Expeditionary Force in France and quickly rose to the rank of major. In

October 1919, he returned to Mount Wilson, just as the hundred-inch telescope was being completed. Hubble was not yet thirty years old, he was smart, he was well trained in astronomy, and he was ambitious. Perhaps most important, he was given access to the largest telescope in the world. Edwin Hubble was in the right place at the right time.

In 1924, Hubble found a Cepheid star in the Andromeda nebula and could thus measure its distance. The result was 900,000 light-years, three times the maximum extent of the Milky Way as determined by Shapley. The Andromeda nebula lay beyond our galaxy. It was another galaxy itself! (As mentioned in Chapter 6, modern measurements give a diameter of 100,000 light-years for the Milky Way and a distance to the Andromeda galaxy of about two million light-years.) Andromeda turned out to be the closest big external galaxy. Hubble went on to measure the distances to many other nebulae, showing that many of them were also entire galaxies far beyond the Milky Way. Instead of calling these objects galaxies, however, Hubble and other astronomers used the term "extragalactic nebulae" for many years.

After his identification of nebulae as extragalactic, Hubble introduced a significant classification system for galaxies, based on their shapes. And, working with fellow astronomer Milton Humason, he began extending the distance scale. Even with the giant hundred-inch telescope, Cepheid stars could not be seen beyond distances of about five million light-years. Farther than that, Hubble used the more luminous O and B supergiant stars, which possess distinctive features and a fairly narrow range of luminosities. Such stars have approximately known intrinsic luminosities and can be seen out to about ten million light-years. However, they are not as reliable for distance measurements as the Cepheids. As one can imagine, the estimates of distance become less and less trustworthy as one goes out to farther objects. By the end of the 1920s, Hubble was ready to embark on his most famous work, the measurement of distances to Slipher's mysterious retreating nebulae.

The story of Edwin Hubble's discovery of the expansion of the universe is filled with ironies, good fortune, ignorance, some tragedies, and missed opportunities. While Slipher, Leavitt, and Hubble were making their discoveries with telescopes, theoretical astronomers were exploring the universe with pencil and paper. Much of this theoretical work was unknown to Hubble.

In 1917, just as Einstein was formulating his cosmological model, a

Dutch theoretical astronomer named Wilhelm de Sitter proposed an alternative model, also based on Einstein's theory of general relativity. De Sitter politely called his model "Solution B" and Einstein's "Solution A." Both solutions solved Einstein's equations of general relativity, modified with the lambda term. Both solutions assumed a static universe, in which the geometry of space did not change in time. De Sitter, however, made the further assumption that the amount of matter in the universe was negligible compared to the lambda term. Indeed, the Dutch astronomer ignored matter altogether. In his idealized model, the lambda term constituted the only force in the universe.

There were two consequences of de Sitter's assumption, neither present in Einstein's model. First, time flowed at different rates in different places. (Do not confuse this effect with Einstein's notion of special relativity, in which time flows at different rates for observers *in motion* relative to each other. In de Sitter's universe, time flows at different rates even for two observers at rest relative to each other, if they are in different locations.) As a result of this temporal variability, the frequency of light emitted at one location would shift upon reaching a second location. Why? Recall that the frequency of light is nothing but the rate of oscillations of the light. Those oscillations, in turn, are like the swings of the pendulum of a clock. If clocks tick at different rates in different locations, then the frequency of the same beam of light must also be different in different places. According to de Sitter's model, the frequency of a light beam *decreases* as the light beam travels through space. Since the highest frequencies visible to the eye correspond to blue colors and the lowest to red, scientists say that a light beam of decreasing frequency undergoes a redshift. The farther the light travels in de Sitter's universe—that is, the farther apart the emission and reception of the light—the greater the redshift. Such a shift would masquerade as a Doppler effect, but it is brought about not by receding velocity but by a slowing down of the passage of time.

The second unusual feature of de Sitter's model was that if a group of particles (or nebulae) were placed anywhere in the universe, they would fly away from each other, repelled by the antigravity-like lambda term. In such a situation, the ordinary Doppler effect would also cause the light emitted by one nebula to be redshifted when it was received by a second nebula.

Together, these two phenomena—the continual redshifting of a traveling beam of light and the mutual repulsion of nebulae—became known as the de Sitter effect. Both phenomena contribute to a redshift of light. As de Sitter commented in his 1917 paper: "The lines in the spectra [colors] of very distant stars or nebulae must therefore be systematically displaced towards the red . . ." Despite this clear statement, many scientists were confused about the meaning of the de Sitter effect and mixed up the two different phenomena leading to a predicted redshift of light from distant nebulae.

De Sitter was much more in touch with observational astronomy than was Einstein. In particular, he knew about Slipher's observations. At the end of his paper, de Sitter listed the observed velocities of three nebulae, Andromeda, NGC 1068, and NGC 4594, and computed an average outward velocity of 600 kilometers per second. Solution A, Einstein's model, could not explain such outward velocities (or redshifts). In Einstein's model, time flowed at the same rate everywhere, and particles remained motionless. De Sitter happily suggested that the data appeared to favor Solution B.

In 1922, Alexander Friedmann, a thirty-four-year-old Russian scientist at the Academy of Sciences in Petrograd (St. Petersburg), decided to mathematically explore cosmological solutions to Einstein's general relativity that *changed in time*. Friedmann pointed out that Einstein's and de Sitter's assumption of staticity was unverified and nonessential. One could start with Einstein's equations for gravity but not require that all the variables remain constant in time. In Friedmann's model, the universe began in a state of extremely high density and then expanded in time, thinning out as it did so. Friedmann's paper went largely unnoticed, except by Einstein, until well after Hubble's discovery of 1929. By 1925, Friedmann was dead of typhoid fever, at age thirty-seven. The Russian physicist did not live to see the ultimate validation of his theory.

Both the de Sitter and Friedmann models offended Einstein's philosophical sensibilities, the de Sitter model because it claimed to have found a solution to Einstein's own equations in the absence of mass, contradicting Einstein's strong belief that the properties of space should be determined by matter; and the Friedmann solution because Einstein believed that the cosmos was static. Einstein published prompt replies to both papers, claiming to have found mathematical inconsistencies in them. However, as Einstein and others realized later, Einstein had reacted too quickly and made errors himself. The great physicist grudgingly acknowledged that the cosmologies of de Sitter and Friedmann were possible solutions to the "cosmological problem," in addition to his own solution. Yet he did not believe either one of them.

In the academic year 1924–1925, a young Belgian abbé named Georges Lemaître, trained in physics as well as theology, came to the Harvard College Observatory as a postdoctoral fellow. At a meeting in Washington later that year, Lemaître heard of Hubble's discovery that the Andromeda nebula resided outside our galaxy. As Slipher's results were also known, Lemaître interpreted Hubble's findings as evidence for a universe in motion. Lemaître hurried back to Louvain and calculated a cosmological model for an expanding universe, essentially the same model produced by Friedmann several years earlier. In his epochal paper of 1927, Lemaître stated, "The receding velocities of extragalactic nebulae are a cosmical effect of the expansion of the universe." Lemaître went on to predict that the retreating velocity of each galaxy should be proportional to its distance from us—a key result not pointed out in Friedmann's earlier and still unknown paper.

It is important to understand Lemaître's prediction, which ultimately became pivotal in deciphering the meaning of Hubble's 1929 paper. Although Lemaître used his solution of Einstein's equations to derive this crucial law, it is easy to grasp without advanced mathematics, as shown in Figure 12.1. Represent the galaxies by points on a ruler, initially separated by one inch (top line). Now begin stretching the ruler, corresponding to the expansion of space. After one minute, suppose that the ruler has doubled its size, so that there are two inches between neighboring points (bottom line). From the viewpoint of *any* galaxy—galaxy B for example—it appears that all the other galaxies are moving *away*. Galaxies to the left move farther to the left, galaxies to the right move farther to the right. Suppose now that we are sitting on galaxy A. Let's consider the view from our vantage. Galaxy B, initially one inch away from us, has moved to two inches away. Thus, it has increased its distance from us by one inch in one minute, that is, it has an outward speed relative to us of 1 inch/minute. Galaxy C, initially two inches from us, has moved to four inches away, for an increase in distance of two inches in one minute, that is, an outward speed relative to us of 2 inches/minute. In sum, galaxies twice as far from us move away at twice the speed. This result is general. In any uniformly expanding space, outward speed is proportional to distance.

Lemaître made the mistake of publishing his seminal paper in the obscure *Annals of the Scientific Society of Brussels*. In 1928, while

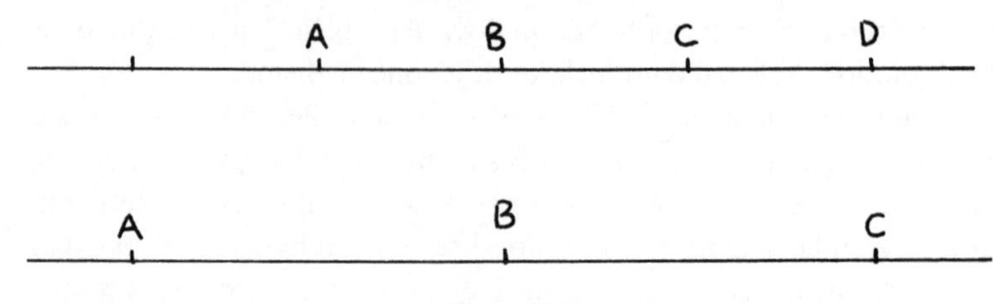

Figure 12.1

Lemaître was at Cambridge University on another fellowship, he placed his paper in the hands of Sir Arthur Eddington. Eddington was one of the best-known and most influential astronomers of the day. Among many accomplishments, in 1919 he had triumphantly measured the bending of starlight by the sun, in an amount predicted by Einstein's theory of general relativity, and in 1923 he had published his widely read book on relativity. For some reason, Eddington paid little attention to Lemaître's paper and apparently misplaced it.

At last, we return to Edwin Hubble. When Hubble began measuring the distances to Slipher's redshifted nebulae in the late 1920s, what he knew was this: these nebulae had enormous redshifts, implying their large outward velocities in all directions. He knew, from his own measurements, that many nebulae were extragalactic and thus important in the big picture of the universe. He knew about the de Sitter effect, popularized in Eddington's 1923 book. He did *not* know about Friedmann's paper or Lemaître's paper. Thus, it seems unlikely that he could have had in his head the concept of an expanding universe. Even with data suggesting a linear relation between outward velocity and distance, it is still a large conceptual and philosophical leap to go from there to the notion of a *dynamic* universe, a universe in which space itself is stretching, an expanding universe. In this regard, it is useful to recall that the de Sitter effect took place within a static cosmos. De Sitter, also, had no notion of an expanding universe.

Hubble begins his 1929 paper by presenting Slipher's results in terms of the "K term." This term was a speed of 600–800 kilometers per second that had to be subtracted from the speeds of all of Slipher's spiral nebulae in order to cancel out their enormous outward velocities and make them appear as an ordinary group of astronomical objects with random velocities. Hubble calls this term a "paradox" because there was no explanation for it. However, there was some evidence that the K term might vary with distance. That is, the outward speeds for the nebulae were not all the same.

Some astronomers, including A. Dose, Knut Lundmark, and Gustaf Strömberg, had previously attempted to see if the redshifts correlated with distance. Indeed, that is the project Hubble has set for himself—and the linear correlation turns out to be his great triumph. Hubble rightfully regards these earlier attempts as "unconvincing." First of all, Lundmark, in his 1925 work, estimated distances to the nebulae by the questionable technique of comparing their apparent diameters and apparent brightnesses to standard galaxies of standard diameter and brightness. This method assumes that all galaxies are the same, according to which a galaxy that appears half as big is twice as far away. Hubble, for the most part, relies on the far more reliable method of Cepheid variables and Leavitt's period-luminosity law. And, of course, he has the enormous advantage of sitting beneath the hundred-inch Hooker telescope. Second, Lundmark, in his attempt to see if the K term varied with distance, fit for both a linear and quadratic dependence on distance and reached the baffling conclusion that the K term should actually start *decreasing* beyond a certain distance.

Note that Hubble refers to the redshifts as "apparent radial velocities." Hubble is very much an observational (experimental) astronomer, not a theorist, and it is the redshift, not the radial velocity, that is directly measured. Indeed, Hubble is skeptical of all theories.

Hubble goes on to mention the various methods that he will use to measure distances. Besides using the reliable Cepheid variables and the O stars, he will assume that the brightest stars in a nebula all have the same luminosity, of about 30,000 times the luminosity of the sun. He expresses this maximum luminosity in terms of the standard astronomical notation, $M = -6.3$, where M, the "absolute magnitude," is related to the luminosity L (expressed in units of the luminosity of the sun) by $M = 4.75 - 2.5 \log(L)$. Hubble later also uses the "apparent magnitude," m, which depends, like apparent brightness, on distance as well as luminosity. In particular, if r is the distance to the object in parsecs, then, $m = M - 5 + 5 \log(r)$. Some of these astronomical notations have been previously discussed in Chapter 6.

Table 1 of the paper gives Hubble's principal results. In the first column are the twenty-four nebulae in the sample, arranged in order of increasing distance. The third column gives the distance to each nebula, the fourth gives the velocity, in kilometers per second. A positive velocity means the nebula is moving away from the earth and its colors are shifted to the red. A negative velocity means the nebula is moving toward the earth and its colors are shifted to the blue. From the distance and apparent magnitude, *m*, the latter being directly measured, Hubble can calculate the absolute magnitude, *M*, of each nebula. (The absolute magnitude is equivalent to its intrinsic luminosity.)

The redshift of a nebula measures its velocity relative to the earth. But the earth is attached to the sun, and the sun is moving through the Milky Way at some to-be-determined speed and direction, called the solar motion. To get the velocities of the nebulae relative to the Milky Way, and not just to the sun, the solar motion must be subtracted out. That is the purpose of the long equation with *X*, *Y*, and *Z*. The *A*, *D*, and V_0 represent the direction and speed of the sun through the Milky Way. Here and elsewhere, Hubble is working and thinking very much within the mind-set of galactic structure. He uses the mathematical notation and concepts of motion within the Milky Way, even though he will ultimately apply them to objects far beyond the Milky Way. We must remember that only a few years earlier, it was not established that the nebulae lay beyond the Milky Way. Extragalactic astronomy, at this time, is an extremely new field, one that Hubble pioneered.

As can be seen from the table, although there are some exceptional cases, the nebular velocities are mostly outward and mostly *increase with increasing distance*. Hubble then makes the claim that "the data in the table indicate a linear correlation between distances and velocities . . ." That is, Hubble is proposing something more than simply that the velocities increase with increasing distance. He proposes that his data suggest a particular law: the velocities are proportional to distance. Double the distance and the velocity doubles. That is the meaning of a "linear correlation."

Hubble has made a great leap of faith in his proposal, even though he is aware of the "scanty material, so poorly distributed." In fact, the data points in Figure 1 of his paper have fairly large deviations about the straight (linear) line that Hubble has drawn through them. But the situation is actually even more precarious than Hubble realized at the time. Lemaître's result that the recessional velocity of the nebulae should be proportional to their distance holds only for a universe whose mass is evenly distributed, so that the cosmos can expand uniformly in all directions. Hubble's data went out to a distance of about two million parsecs, or about six million light-years. Astronomers now know that the distribution of matter in the universe does not begin to look uniform until one reaches distances of at least 100 million light-years. At that distance, the lumpiness caused by individual galaxies begins to average out and disappear, just as the individual grains of sand on a beach disappear when viewed from a height of twenty feet or more. For smaller distances, the linear relationship between outward velocity and distance predicted for a homogeneous (uniform) and expanding universe is not valid. Hubble doesn't know about the expanding universe model of Lemaître or its predictions. He has made a lucky guess in claiming a linear law, and even his own data don't support the guess very well.

By 1931, Hubble and Milton Humason used new methods to extend their observations to 100 million light-years. Figure 12.2a shows Hubble's data of 1929, Figure 12.2b gives the extended data of Hubble and Humason two years later. As can be seen, the linear law in Figure 12.2b is more obvious and now justified.

As an important check on his proposal, Hubble then applies his linear law to nebulae too faint for their distances to be determined directly. From their outward velocities, he can see where they fall on his graph and assign them distances. From the distances, he can compute their luminosities, an intrinsic property of the nebulae. He can then compare those luminosities with the luminosities of nebulae whose distances can be determined directly and finds much similarity.

All of the theoretical discussion in Hubble's paper is relegated to the very last paragraph. Here, he mentions de Sitter's model and only de Sitter's model. De Sitter, apparently, exerted some influence over Hubble. First, the Dutch scientist's work had been published in English and

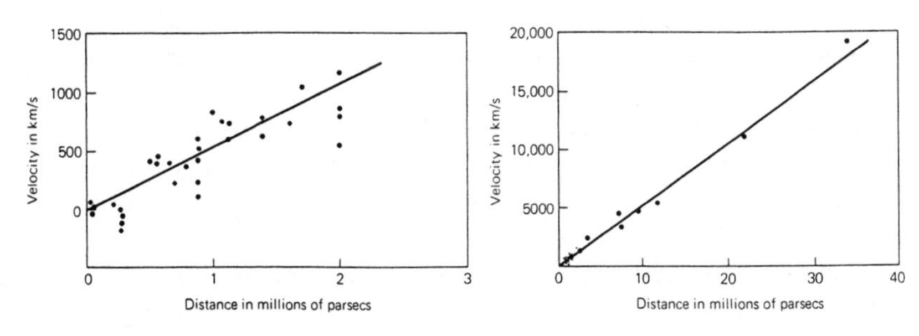

Figure 12.2a Figure 12.2b

popularized by Eddington. Second, de Sitter and later Eddington had applied de Sitter's mathematical and abstruse results to actual data, the redshifts of known nebulae. This gesture must have warmed the heart of the data-oriented Hubble. And finally, Hubble visited Leiden in the summer of 1928, where de Sitter personally encouraged him to extend Slipher's redshifts to higher redshifts and fainter nebulae.

Thus, through the de Sitter effect, Hubble is aware that his "numerical data may be introduced into discussions of the general curvature of space" and thus may be of great importance to cosmology. However, Hubble does not mention the idea of an expanding universe. It seems likely that he did not know of that concept.

There are several ironies here. First, as has been mentioned, in 1929 Hubble did not have nearly good enough data to support his claim of a linear relation between velocity and distance. Second, the sole theory that he used to interpret his findings, de Sitter's model, was later overthrown in favor of Friedmann's and Lemaître's models. Thus, even if his claim turned out to be true, Hubble probably did not understand its significance at the time.

In late 1929, well after Hubble's paper had appeared in print, Lemaître sent Eddington a *second* copy of his 1927 model for an expanding universe. Now, with Hubble's results published, everything clicked in Eddington's mind. Here was a (theoretical) solution of Einstein's equations predicting a linear relation between redshift and distance as a consequence of a homogeneous and expanding universe. On the experimental side, Hubble had recently claimed such a relation in the wispy glimmers of the distant nebulae. Eddington immediately publicized Lemaître's paper and convinced him to publish it in English. Friedmann's paper was recalled by Einstein and others and duly celebrated. By early 1931, Einstein was prepared to make his announcement in the library of the Mount Wilson Observatory, recanting his static model, tossing away for good measure the lambda term—which he had always considered an ugly appendage to his graceful theory of gravity—and honoring Edwin Hubble. ("Your husband's work is beautiful," Einstein said to Hubble's wife in a trip to Pasadena later in 1931.) Within a year or two, the idea of an expanding universe, with a beginning in time, was being anxiously digested by the public. The journalist George Gray, writing in the February 1933 issue of the *Atlantic*, described the discovery as "a radically new picture of the cosmos—a universe in expansion, a vast bubble blowing, distending, scattering, thinning out into gossamer, losing itself."

Hubble, like many scientists, drew a sharp line between what he considered the objective world of science and the subjective world of the humanities. In his major autobiography, *Realm of the Nebulae*, published in 1936, Hubble wrote:

> Science is the one human activity that is truly progressive. The body of positive knowledge is transmitted from generation to generation, and each contributes to the growing structure . . . Agreement is secured by means of observation and experiment. The tests represent external authorities which all men must acknowledge . . . Science, since it deals only with such judgments, is necessarily barred from the world of values. There, no external authority is known. Each man appeals to his private god and recognizes no superior court of appeal.

Modern philosophers and historians of science would partly disagree with Hubble's severe distinction between science and other professions. Although the actual data of science may be objective, the "human activity" involved in the enterprise of science is full of the same prejudices, passions, and personal judgments that mark other human endeavors. Indeed, such personal factors may be essential to empower and propel a man like Edwin Hubble through his scientific career.

A RELATION BETWEEN DISTANCE AND RADIAL VELOCITY AMONG EXTRA-GALACTIC NEBULAE

Edwin Hubble

Mount Wilson Observatory, Carnegie Institution of Washington

Communicated January 17, 1929

Proceedings of the National Academy of Sciences (1929)

DETERMINATIONS OF THE MOTION OF THE SUN with respect to the extra-galactic nebulae have involved a *K* term of several hundred kilometers which appears to be variable. Explanations of this paradox have been sought in a correlation between apparent radial velocities and distances, but so far the results have not been convincing. The present paper is a re-examination of the question, based on only those nebular distances which are believed to be fairly reliable.

Distances of extra-galactic nebulae depend ultimately upon the application of absolute-luminosity criteria to involved stars whose types can be recognized. These include, among others, Cepheid variables, novae, and blue stars involved in emission nebulosity. Numerical values depend upon the zero point of the period-luminosity relation among Cepheids, the other criteria merely check the order of the distances. This method is restricted to the few nebulae which are well resolved by existing instruments. A study of these nebulae, together with those in which any stars at all can be recognized, indicates the probability of an approximately uniform upper limit to the absolute luminosity of stars, in the late-type spirals and irregular nebulae at least, of the order of *M* (photographic) = –6.3.[1] The apparent luminosities of the brightest stars in such nebulae are thus criteria which, although rough and to be applied with caution, furnish reasonable estimates of the distances of all extra-galactic systems in which even a few stars can be detected.

Finally, the nebulae themselves appear to be of a definite order of absolute luminosity, exhibiting a range of four or five magnitudes about an average value *M* (visual) = –15.2.[1] The application of this statistical average to individual cases can rarely be used to advantage, but where considerable numbers are involved, and especially in the various clusters

TABLE 1

NEBULAE WHOSE DISTANCES HAVE BEEN ESTIMATED FROM STARS INVOLVED OR FROM MEAN LUMINOSITIES IN A CLUSTER

OBJECT	m_s	r	v	m_t	M_t
S. Mag.	..	0.032	+ 170	1.5	–16.0
L. Mag.	..	0.034	+ 290	0.5	17.2
N. G. C. 6822	..	0.214	– 130	9.0	12.7
598	..	0.263	– 70	7.0	15.1
221	..	0.275	– 185	8.8	13.4
224	..	0.275	– 220	5.0	17.2
5457	17.0	0.45	+ 200	9.9	13.3
4736	17.3	0.5	+ 290	8.4	15.1
5194	17.3	0.5	+ 270	7.4	16.1
4449	17.8	0.63	+ 200	9.5	14.5
4214	18.3	0.8	+ 300	11.3	13.2
3031	18.5	0.9	– 30	8.3	16.4
3627	18.5	0.9	+ 650	9.1	15.7
4826	18.5	0.9	+ 150	9.0	15.7
5236	18.5	0.9	+ 500	10.4	14.4
1068	18.7	1.0	+ 920	9.1	15.9
5055	19.0	1.1	+ 450	9.6	15.6
7331	19.0	1.1	+ 500	10.4	14.8
4258	19.5	1.4	+ 500	8.7	17.0
4151	20.0	1.7	+ 960	12.0	14.2
4382	..	2.0	+ 500	10.0	16.5
4472	..	2.0	+ 850	8.8	17.7
4486	..	2.0	+ 800	9.7	16.8
4649	..	2.0	+1090	9.5	17.0
Mean					–15.5

m_s = photographic magnitude of brightest stars involved.

r = distance in units of 10^6 parsecs. The first two are Shapley's values.

v = measured velocities in km./sec. N. G. C. 6822, 221, 224 and 5457 are recent determinations by Humason.

m_t = Holetschek's visual magnitude as corrected by Hopmann. The first three objects were not measured by Holetschek, and the values of m_t represent estimates by the author based upon such data as are available.

M_t = total visual absolute magnitude computed from m_t and r.

of nebulae, mean apparent luminosities of the nebulae themselves offer reliable estimates of the mean distances.

Radial velocities of 46 extra-galactic nebulae are now available, but individual distances are estimated for only 24. For one other, N. G. C. 3521, an estimate could probably be made, but no photographs are available at Mount Wilson. The data are given in Table 1. The first seven distances are the most reliable, depending, except for M 32 the companion of M 31, upon extensive investigations of many stars involved. The next thirteen distances, depending upon the criterion of a uniform upper limit of stellar luminosity, are subject to considerable probable errors but are believed to be the most reasonable values at present available. The last four objects appear to be in the Virgo Cluster. The distance assigned to the cluster, 2×10^6 parsecs, is derived from the distribution of nebular luminosities, together with luminosities of stars in some of the later-type spirals, and differs somewhat from the Harvard estimate of ten million light years.[2]

The data in the table indicate a linear correlation between distances and velocities, whether the latter are used directly or corrected for solar motion, according to the older solutions. This suggests a new solution for the solar motion in which the distances are introduced as coefficients of the *K* term, i.e., the velocities are assumed to vary directly with the distances, and hence *K* represents the velocity at unit distance due to this effect. The equations of condition then take the form

$$rK + X \cos \alpha \cos \delta + Y \sin \alpha \cos \delta + Z \sin \delta = \mathrm{v}.$$

Two solutions have been made, one using the 24 nebulae individually, the other combining them into 9 groups according to proximity in direction and in distance. The results are

	24 OBJECTS	9 GROUPS
X	-65 ± 50	$+3 \pm 70$
Y	$+226 \pm 95$	$+230 \pm 120$
Z	-195 ± 40	-133 ± 70
K	$+465 \pm 50$	$+513 \pm 60$ km./sec. per 10^6 parsecs.
A	286°	269°
D	+ 40°	+ 33°
V_0	306 km./sec.	247 km./sec.

For such scanty material, so poorly distributed, the results are fairly definite. Differences between the two solutions are due largely to the four Virgo nebulae, which, being the most distant objects and all sharing the peculiar motion of the cluster, unduly influence the value of K and hence of V_0. New data on more distant objects will be required to reduce the effect of such peculiar motion. Meanwhile round numbers, intermediate between the two solutions, will represent the probable order of the values. For instance, let $A = 277°$, $D = +36°$ (Gal. long. = 32°, lat. = +18°), $V_0 = 280$ km./sec., $K = +500$ km./sec. per million parsecs. Mr. Strömberg has very kindly checked the general order of these values by independent solutions for different groupings of the data.

A constant term, introduced into the equations, was found to be small and negative. This seems to dispose of the necessity for the old constant K term. Solutions of this sort have been published by Lundmark,[3] who replaced the old K by $k + lr + mr^2$. His favored solution gave $k = 513$, as against the former value of the order of 700, and hence offered little advantage.

TABLE 2

NEBULAE WHOSE DISTANCES ARE ESTIMATED FROM RADIAL VELOCITIES

OBJECT	V	V_s	r	m_t	M_t
N.G.C. 278	+ 650	−110	1.52	12.0	−13.9
404	− 25	− 65	. .	11.1	. .
584	+1800	+ 75	3.45	10.9	16.8
936	+1300	+115	2.37	11.1	15.7
1023	+ 300	− 10	0.62	10.2	13.8
1700	+ 800	+220	1.16	12.5	12.8
2681	+ 700	− 10	1.42	10.7	15.0
2683	+ 400	+ 65	0.67	9.9	14.3
2841	+ 600	− 20	1.24	9.4	16.1
3034	+ 290	−105	0.79	9.0	15.5
3115	+ 600	+105	1.00	9.5	15.5
3368	+ 940	+ 70	1.74	10.0	16.2
3379	+ 810	+ 65	1.49	9.4	16.4
3489	+ 600	+ 50	1.10	11.2	14.0
3521	+ 730	+ 95	1.27	10.1	15.4
3623	+ 800	+ 35	1.53	9.9	16.0
4111	+ 800	− 95	1.79	10.1	16.1
4526	+ 580	− 20	1.20	11.1	14.3
4565	+1100	− 75	2.35	11.0	15.9
4594	+1140	+ 25	2.23	9.1	17.6
5005	+ 900	−130	2.06	11.1	15.5
5866	+ 650	−215	1.73	11.7	−14.5
Mean				10.5	−15.3

The residuals for the two solutions given above average 150 and 110 km./sec. and should represent the average peculiar motions of the individual nebulae and of the groups, respectively. In order to exhibit the results in a graphical form, the solar motion has been eliminated from the observed velocities and the remainders, the distance terms plus the residuals, have been plotted against the distances. The run of the residuals is about as smooth as can be expected, and in general the form of the solutions appears to be adequate.

The 22 nebulae for which distances are not available can be treated in two ways. First, the mean distance of the group derived from the mean apparent magnitudes can be compared with the mean of the velocities corrected for solar motion. The result, 745 km./sec. for a distance of 1.4×10^6 parsecs, falls between the two previous solutions and indicates a value for K of 530 as against the proposed value, 500 km./sec.

Secondly, the scatter of the individual nebulae can be examined by assuming the relation between distances and velocities as previously

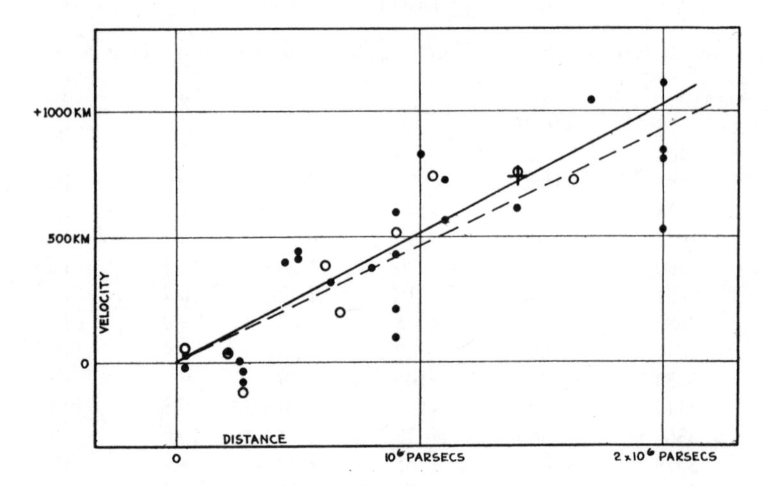

Figure 1
Velocity-Distance Relation among Extra-Galactic Nebulae.

Radial velocities, corrected for solar motion, are plotted against distances estimated from involved stars and mean luminosities of nebulae in a cluster. The black discs and full line represent the solution for solar motion using the nebulae individually; the circles and broken line represent the solution combining the nebulae into groups; the cross represents the mean velocity corresponding to the mean distance of 22 nebulae whose distances could not be estimated individually.

determined. Distances can then be calculated from the velocities corrected for solar motion, and absolute magnitudes can be derived from the apparent magnitudes. The results are given in Table 2 and may be compared with the distribution of absolute magnitudes among the nebulae in table 1, whose distances are derived from other criteria. N. G. C. 404 can be excluded, since the observed velocity is so small that the peculiar motion must be large in comparison with the distance effect. The object is not necessarily an exception, however, since a distance can be assigned for which the peculiar motion and the absolute magnitude are both within the range previously determined. The two mean magnitudes, −15.3 and −15.5, the ranges, 4.9 and 5.0 mag., and the frequency distributions are closely similar for these two entirely independent sets of data; and even the slight difference in mean magnitudes can be attributed to the selected, very bright, nebulae in the Virgo Cluster. This entirely unforced agreement supports the validity of the velocity-distance relation in a very evident matter. Finally, it is worth recording that the frequency distribution of absolute magnitudes in the two tables combined is comparable with those found in the various clusters of nebulae.

The results establish a roughly linear relation between velocities and distances among nebulae for which velocities have been previously published, and the relation appears to dominate the distribution of velocities. In order to investigate the matter on a much larger scale, Mr. Humason at Mount Wilson has initiated a program of determining velocities of the most distant nebulae that can be observed with confidence. These, naturally, are the brightest nebulae in clusters of nebulae. The first definite result,[4] v = +3779 km/sec for N. G. C. 7619, is thoroughly consistent with the present conclusions. Corrected for the solar motion, this velocity is +3910, which, with $K = 500$, corresponds to a distance of 7.8×10^6 parsecs. Since the apparent magnitude is 11.8, the absolute magnitude at such a distance is −17.65, which is of the right order for the brightest nebulae in a cluster. A preliminary distance, derived independently from the cluster of which this nebula appears to be a member, is of the order of 7×10^6 parsecs.

New data to be expected in the near future may modify the significance of the present investigation or, if confirmatory, will lead to a solution having many times the weight. For this reason it is thought premature to discuss in detail the obvious consequences of the present results. For example, if the solar motion with respect to the clusters represents the rotation of the galactic system, this motion could be subtracted from the results for the nebulae and the remainder would represent the motion of the galactic system with respect to the extra-galactic nebulae.

The outstanding feature, however, is the possibility that the velocity-distance relation may represent the de Sitter effect, and hence that numerical data may be introduced into discussions of the general curvature of space. In the de Sitter cosmology, displacements of the spectra arise from two sources, an apparent slowing down of atomic vibrations and a general tendency of material particles to scatter. The latter involves an acceleration and hence introduces the element of time. The relative importance of these two effects should determine the form of the relation between distances and observed velocities; and in this connection it may be emphasized that the linear relation found in the present discussion is a first approximation representing a restricted range in distance.

1. *Mt. Wilson Contr.*, No. 324; *Astroph. J., Chicago, Ill.*, **64,** 1926 (321).
2. *Harvard Coll. Obs. Circ.*, 294, 1926.
3. *Mon. Not. R. Astr. Soc.*, **85,** 1925 (865–894).
4. These PROCEEDINGS, **15,** 1929 (167).

13

ANTIBIOTICS

> Not many days after [the Lacedaemonians'] arrival in Attica the plague first began to show itself among the Athenians. It was said that it had broken out in many places previously in the neighborhood of Lemnos and elsewhere; but a pestilence of such extent and mortality was nowhere remembered. Neither were the physicians at first of any service, ignorant as they were of the proper way to treat it, and they died themselves the most thickly . . . people in good health were all of a sudden attacked by violent heats in the head, and redness and inflammation in the eyes, the inward parts such as the throat or tongue becoming bloody and emitting an unnatural and fetid breath. These symptoms were followed by sneezing and hoarseness, after which the pain soon reached the chest and produced a hard cough. When it fixed in the stomach, it upset it; and discharges of bile of every kind named by physicians ensued, accompained by very great distress. In most cases an ineffectual retching followed, producing violent spasms . . . Externally the body was reddish, livid, and breaking out into small pustules and ulcers . . . [Victims] succumbed, in most cases, on the seventh or eighth day of the internal inflammation . . . the bodies of dying men lay one upon another, and half-dead creatures reeled about the streets.

SO READS A PASSAGE FROM *The History of the Peloponnesian War*, in which Thucydides recounts firsthand a plague that swept through Athens in 430 B.C., the second year of the war between Athens and Sparta. A mystery to science for the next 2,300 years, the plague was caused by the bacterium *Yersinia pestis* and transmitted to humans by rat fleas, then from human to human. In the fourteenth century it was called the Black Death. A quarter of the population of Europe, or about 25 million people, succumbed to its agonies. From 1894 to 1914, a new outbreak of plague spread from southern ports in China to the rest of the world and killed another ten million people.

Plague is one example of an infectious disease. Others include influenza, which killed 20 to 40 million people in the world epidemic of 1918, pneumonia, syphilis, typhoid, diphtheria, meningitis, gonorrhea, hepatitis, polio, smallpox, measles, cholera, tetanus, yellow fever. Another, tuberculosis, begins when the bacterium *Mycobacterium tuberculosis* inhabits and destroys lung tissue, leaving dead mass and holes. Throughout modern history, tuberculosis has claimed more lives than any other affliction. Tuberculosis and all infectious diseases are caused by microorganisms, tiny living creatures invisible to the eye.

In the spring of 1928, a Scottish biologist named Alexander Fleming was at work in his small laboratory at St. Mary's Hospital Medical School in London when he noticed something strange. One of his colonies of staphylococcus bacteria had been contaminated by a white fluffy mold. Staphylococcus is a common genus of bacteria, discovered in the late nineteenth century. The bacteria can be found in the mucous membranes and skin of many warm-blooded animals and get their name from their spherical shape (*coccus*) and tendency to congregate in grapelike clusters (*staphyle* means "bunch of grapes" in Greek). Since the glass plates on which Fleming cultured his bacteria were often exposed to open air, known to be brimming with other microbes and spores, it was not uncommon to find such contaminations. But in this instance, the patch of staphylococci closest to the mold had been miraculously dissolved. Where there should have been opaque yellow clumps, the staphylococci were as clear as dew. Their cell walls had burst open. Some staphylococci infect wounds and cause food poisoning. Others produce an infectious disease. And they had been slaughtered by the white mold.

At this time, Fleming was a seasoned biologist, forty-seven years old. Blue-eyed, fair-haired, and sufficiently short to be called "Little Flem" by his colleagues at St. Mary's, he was immensely respected for his quiet intelligence and keen observation. One of his associates, John Freeman, commented that Little Flem "could be more eloquently silent than any man I have ever known." Another friend, C. A. Pannett, with whom Fleming had shared all the top honors when they were medical students at St. Mary's, wrote that Fleming "never liked talking, but when he did make up his mind to express a judgment in words, you could be perfectly certain that it would be in the highest degree intelligent."

Although quiet and excessively modest, Fleming possessed a quirky, mischievous sense of humor. Pannett recalled that his friend "delighted in making difficulties for himself, just for the fun of overcoming them. He once undertook to play a round of golf using only one club." On another occasion, Fleming, who had elevated the task of making laboratory glassware to an art form, created a lifelike cat out of glass. Then he made a group of tiny creatures running away from the cat.

One contrary quality turned out to be of paramount importance. Fleming dedicated himself to a kind of studied disorder. Often chiding his colleagues for being too neat and tidy, for carefully putting away their test tubes and plates at the end of the day, Fleming left his festering dishes of bacteria lying about his lab for weeks at a time. Then he would look at them carefully to see if anything unexpected or "interesting" had occurred.

The appearance of white fluff in the culture of staphylococci was such an occurrence. Fleming had not been searching for antibacterial agents. Rather he had been investigating the abnormal forms of staphylococci for a routine academic article. This mold was a surprise. Yet Fleming was ready. Since his 1908 medical school thesis on "Acute Bacterial Infections," he had devoted himself to finding the means of fighting bacterial infections, which he considered the most dangerous illnesses threatening the human race.

Fleming immediately transplanted the mysterious mold and began testing it. Growing atop a nourishing broth, the white fluffy substance became a dark green felted mass, then turned the broth beneath a bright yellow. Growth slowed at 37°C and raced at 20°C (room temperature). Soon, Fleming had identified the mold as a *Penicillium* fungus. What were its effects on other bacteria? In a well-practiced procedure, the Scottish biologist cut a trough in a glass dish of agar, which is a gelatinous nutrient, filled the furrow with mold-laden broth, then streaked cultures of various bacteria at right angles. The *Penicillium* fungus killed not only staphylococci but also the bacteria that caused streptococcus infections, pneumonia, syphilis, gonorrhea, gas gangrene, cerebral meningitis, and diphtheria. Most importantly, the mold was not toxic. Fleming injected relatively large quantities of the stuff into rabbits and mice without harmful results. By contrast, almost all previously known antibacterial agents, most of them harsh chemicals, destroyed white blood cells and other parts of an animal's natural immune system. Penicillin, as Fleming named the active ingredient of his mold, seemed to be the ideal antiseptic. But it would be over a decade before the mold extract could be sufficiently concentrated, purified, and stabilized for medicinal use.

When a colleague, Merlin Pryce, visited Fleming shortly after the first appearance of the white fluff, he found the biologist in his small laboratory completely surrounded by plates and dishes of bacterial colonies—growing quilts of reds, greens, and yellows. Other culture dishes were stacked up at random in corners. Test tubes and glass slides littered the counters. As usual, Fleming had left his door open, so that any young researcher could walk in and borrow a smear of staphylococci or pneumococci or some other microbe. "Take a look at that," said Little Flem, pointing at his powerful mold. "Things fall out of the air."

Pryce later recounted: "What struck me was that he didn't confine himself to observing, but took action at once. Lots of people observe a phenomenon, feeling that it may be important, but they don't get beyond being surprised—after which they forget. That was never the case with Fleming."

It is believed that the first microorganisms were discovered about 1670, by the Dutch civil servant and scientist Antoni van Leeuwenhoek. Peering through his new microscope, van Leeuwenhoek must have been astonished to find an entire world of tiny animals in a drop of pond water—gyrating, throbbing, wriggling and squirming, propelling themselves along with waving hairs and rotors and pulsating plasma blobs—just as Galileo, a half century earlier, had been astonished to find craters on the moon with his new telescope.

Van Leeuwenhoek could see two different kinds of microorganisms: bacteria, such as the staphylococcus; and protozoa, such as the amoeba. Protozoa are single-celled organisms with a complete cell and cell nucleus. Bacteria, also single-celled organisms, lack a cell nucleus. Bacteria date back 3.5 billion years and are the most ancient life forms on earth. Bacteria are also the most plentiful organisms on the planet. A handful of rich soil contains billions of them. Both protozoa and bacteria are typically 3×10^{-4} centimeters in diameter, roughly ten times smaller than the smallest object that can be seen by the naked eye. At the end of the nineteenth century a third kind of microorganism was discovered, the virus. Viruses are even smaller than bacteria and can reproduce only by commandeering the metabolic machinery of a host cell.

The idea that microorganisms were the cause of infectious diseases arose in the 1860s and 1870s, championed most strongly by Louis Pasteur (1822–1895). The great French scientist arrived at his theory round about. Pasteur began his career as a chemist. In the early 1850s, he became fascinated by the discovery that organic molecules rotated the plane of polarized light, while inorganic compounds did not. Organic molecules, of course, are associated with life. Following this seemingly esoteric piece of data, Pasteur began studying alcoholic fermentation when he learned that one of its by-products had the peculiar light-rotating property. Upon examining fermentation more closely, he found that it was indeed associated with a living organism, an organism of microscopic proportions, namely yeast. By 1863, Pasteur had turned his attention to the practical problem of the diseases of wine, closely related to fermentation. With his microscope, he was able to identify a particular microorganism with each of the various diseases.

Gradually, Pasteur developed his own "germ theory" of disease: that infectious diseases are caused by microscopic living organisms. Deadly anthrax was one of the first human diseases to be identified with a microorganism, the anthrax bacillus. Over the next two decades, Pasteur's idea was extended to tuberculosis, cholera, diphtheria, typhoid, gonorrhea, pneumonia, and plague. Biologists now realized that all these diseases, and many others, were caused and transmitted by microorganisms. Humanity would need means to defeat them.

Two kinds of weapons presented themselves: vaccination and chemotherapy. In vaccination, the body uses its own natural defenses to attack an invading microbe, defenses triggered by injection of a dead or reduced form of the microorganism. In 1880, for example, Pasteur developed a vaccine against anthrax, as well as vaccines for chicken cholera and rabies. German bacteriologist Emil von Behring, one of the first immunologists, developed vaccines against tetanus and diphtheria in the late 1880s.

In chemotherapy, chemicals such as mercury or compounds of arsenic were applied or ingested to kill dangerous microorganisms. In 1881, the German physician and chemist Paul Ehrlich found that a chemical dye called methylene blue was absorbed by particular bacteria but not others, leading him to propose the concept of a "magic bullet," a drug that would selectively destroy particular invading organisms. In early 1909, Ehrlich himself developed a drug called Salvarsan 606 that was effective against syphilis. A problem with chemotherapy was (and still is) that most chemicals are toxic. Thus mercury, used to treat syphilis, quinine for malaria, compounds of arsenic to treat amoebic dysentery, carbolic acid to kill staph infections and gangrene in wounds, phenol, methylene blue, boric acid—all kill healthy cells and tissues in addition to the invading organisms.

Vaccination and chemotherapy represented two philosophically different approaches to disease. In the former, one trusted the natural defense mechanisms of the body—a living organism was a miraculous thing, for the most part equipped with all that it needed to survive. In the latter, one attacked diseases with every possible means, including inorganic chemicals, synthetic compounds, and even electromagnetic radiation.

In 1877, Pasteur and his collaborator Jules François Joubert found that some bacteria from the air seemed to inhibit the growth of anthrax bacteria. Here was a new idea. Evidently, one microorganism could kill another, in a miniature version of the larger dog-eat-dog world. The idea also suggested a kind of intermediate way between fighting disease with the body's internal natural defenses versus with chemicals from the outside. Pasteur had discovered a third weapon, and wrote that "in the inferior organisms, still more than in the great animal and vegetable species, life hinders life . . . these facts may, perhaps, justify the greatest hopes from the therapeutic point of view." However, Pasteur did not follow up his great hopes.

In 1885, two French bacteriologists, A. V. Cornil and V. Babes, restated Pasteur's idea and further suggested that tuberculosis in particular, the greatest killer of all, might be defeated by a rival bacterium. Soon, a Swiss botanist found that the bacterium *Pseudomonas fluorescens* could inhibit the growth of the deadly *Bacterium typhusum*, which causes typhoid fever. In 1899, two German bacteriologists, Rudolf von Emmerich and O. Loew, found that *P. fluorescens* also helped treat skin wounds, presumably killing or suppressing the microorganisms causing surface infections.

The work of all of these scientists helped extend natural history to the microbial level, creating a kind of ecology of the microscopic world. Here, minuscule organisms lived together in dynamic relationships, sometimes helping each other, sometimes hindering, just as in the macroscopic world. In particular, by 1900 it was well known that some microorganisms could fight other microorganisms. The process was called bacterial antagonism. However, few biologists at this time pursued bacterial antagonism as a treatment for infectious diseases. A major reason for the restraint was that it was hard to imagine intentionally swallowing germs. Killing a germ with another germ on a glass culture plate in the lab was one thing. Injecting germs into a living body to kill other germs was quite another.

Alexander Fleming was born in Ayrshire, Scotland, in 1881, son of a farmer of modest means. Young Fleming walked daily to school at Dorval, four miles away. At thirteen he moved to London to live with his brothers and a sister. In 1900, he joined the London Scottish Regiment. As he was especially good at rifle shooting, he remained with the regiment until 1914.

There were twelve great medical schools in London. Fleming lived near three of them. He later wrote: "I had no knowledge of any of these three, but I had played water polo against St. Mary's, and so to St. Mary's I went" (on scholarship). The blue-eyed young man received his Conjoint Board Diploma in 1906 and in 1908 passed his final medical examinations, receiving a Gold Medal from the University of London. Already he was considered brilliant.

In 1906, at the age of twenty-five, Fleming joined the laboratory of Almroth Edward Wright, one of the pioneers of immunology. Wright had successfully developed a vaccine against typhoid fever that was tested on soldiers in India. Since 1902 he had been a professor of pathology at St. Mary's. Like Rutherford, Bohr, Heisenberg, and Otto Warburg, whom we will encounter in Chapter 14, Wright created a kind of atelier for budding scientists, and his students were devoted to him. A big, bearlike man with rounded shoulders and expressive eyebrows, Wright hulked about his laboratory slowly and almost clumsily. Like Warburg, he believed in hard work and often toiled through the night with his microscope and bacteria. He was also charming and erudite and spouted great chunks from the Bible or *Paradise Lost* or Shakespeare or Dante. Furthermore, he was passionate, outspoken, dogmatic, sometimes overstating his point—the very opposite of his quiet young apprentice Alexander Fleming. Other Wright students in this period, with varying temperatures, included Stuart Douglas, Leonard Noon, Bernard Spilsbury, and John Freeman. At teatime, Wright would install himself in his chair and hold forth on immunology and the world, while his disciples huddled around him on the floor and listened.

Almroth Wright was enormously committed to the idea of vaccine therapy, and he was skeptical of other treatments of disease. At first, young Fleming followed his teacher's doctrine and worked with him to test the blood's specific disease-fighting resistances to specific infections. When World War I broke out, Fleming joined the Royal Army Medical Corps in a wound research lab and demonstrated the antibacterial action of pus. Over the next few years he witnessed firsthand that almost all externally applied chemical germicides and antiseptics killed as many beneficial white blood cells as harmful microorganisms. However, Fleming would eventually part ways with his teacher. With his quiet determination, he embarked upon a mission to find an external antiseptic that was harmless to the body.

In 1921, Fleming became assistant director of the Inoculation Department at St. Mary's, still working under "the Old Man," as Wright was called. Shortly thereafter, the Scot made an important discovery, a prelude to his later work with penicillin. One day when he was suffering from a cold, he sneezed on a bacterial culture of *Micrococcus lysodeikticus*. Following his usual routine of leaving his bacterial colonies lying about, ten days later he observed that the bacteria near his own nasal mucus had been dissolved.

Over the next few weeks, Fleming and a colleague, Dr. V. D. Allison, found that the same antibacterial agent in mucus was also present in other bodily secretions, including human saliva and tears. (Allison later reported that he and Fleming produced many of their own tears for their experiments by buying and cutting up large quantities of lemons.) Evidently, the two scientists had found an "internal" antiseptic, made in the body. The Old Man was delighted. Using his knowledge of Latin, Wright named the antibacterial substance lysozyme, since it was some kind of enzyme and it "lysed," or dissolved, certain bacteria. It was no surprise that lysozyme was nontoxic, since it emerged from the body itself. Unfortunately, the microbes on which lysozyme acted most strongly, such as *Micrococcus lysodeikticus*, were not particularly dangerous. Fleming's work on lysozyme received little attention, although he published five papers on the substance between 1922 and 1927.

Great discoveries in science are sometimes accidental and sometimes intentional. Bayliss and Starling, and Rutherford, for example, stumbled on their dramatic findings by accident. Loewi, Planck, and Bohr pretty much knew what they were looking for. Even the accidents, however, generally require prepared minds.

In the very first paragraph of his landmark paper of 1929, Fleming acknowledges the accidental nature of his discovery. Almost immediately, we become acquainted with his modesty of style. There is a reserve in his voice, a removal, a disembodied character, almost as if he were not involved with his own experiments. "It was noticed that . . . ," "It was found that . . . ," and so on. And the all-important point 8 in the summary: "It is suggested that . . ." Compare this tone to the self-confident and active voice of Ernest Rutherford in his paper: "We shall first examine . . . ," or to Einstein: "We will raise this conjecture . . ." Evidently, there is a wide variation in personal style among great scientists.

Despite his humility, Fleming is a keen observer and not afraid of using vivid language. The first paragraphs, especially, describe with much sensuousness the colors and textures of his organisms: the "white fluffy mass" of the fungus, the "bright yellow color" which "diffuses into the medium," the change to a "dark green felted mass." These descriptions, which followed a narrative convention in bacteriology, were so different from the abstract passages in the papers in twentieth-century physics, where the objects of study were becoming further and further removed from human sensory perception.

Fleming carefully distinguishes between "bacteriolytic" and "bactericidal," technically meaning able to dissolve bacteria and kill bacteria, respectively, but the two come down to the same thing. Occasionally, he refers to his substance's pH, which is a measure of the degree of acidity (low pH) or alkalinity (high pH), with a pH of 7 being neutral.

At the bottom of the first page, Fleming says that his antibacterial species of *Penicillium* most closely resembles *P. rubrum* (a suggested identification for which he thanks the mycologist Mr. la Touche at the end of the paper). Later, the species was found to be *P. notatum*, first described in 1911 by Richard Westling in his Ph.D. thesis from the University of Stockholm. In good scientific practice, Fleming tests other molds and finds that only *Penicillium* inhibits bacteria. He then names the active agent penicillin.

In the next section of the paper, Fleming uses some of the same methods of testing the antibacterial action of penicillin (e.g., the vertical trough of penicillin, the horizontal smears of bacteria) that he did eight years earlier for lysozyme, although he doesn't mention lysozyme anywhere in this paper. However, many years later in his Nobel Prize lecture, he credits the discovery of lysozyme as "of great use to me" in his work with penicillin. Lysozyme was also an antibacterial agent, also discovered by accident as a "contamination" of a bacterial culture, also harmless to healthy cells. Indeed, Fleming's earlier work with lysozyme served as a kind of mental template, just as Krebs's work on the ornithine cycle would serve as a template for his later discovery of the citric acid cycle (see Chapter 14).

Figure 2 in the paper gives the first indication of which bacteria penicillin kills and which not. As can be seen, staphylococcus, streptococcus, pneumococcus, gonococcus, and diphtheriae all succumb to penicillin. They have stopped growing near the vertical trough filled with penicillin. Here, we instantly grasp the critical difference between penicillin and lysozyme: the inhibited microbes in Figure 2 are among the most harmful to humans.

As can also be seen in Figure 2, however, penicillin doesn't kill all harmful bacteria. *B. influenzae* and *B. coli* are unaffected and grow right up to the trough of penicillin. (The *B.* stands for *Bacterium*, the genus; *influenzae* and *coli* are different species of the same genus.) Later, it was found that typhus, plague, and tuberculosis, among other bacteria, are also insensitive to penicillin. In addition, penicillin, like streptomycin and other antibiotics to be discovered, is not effective at all against viruses.

Fleming next measures the antibacterial action of penicillin by the opacity, or clarity, of a mixture of penicillin and various bacteria. When penicillin is effective, it dissolves the bacteria, making the solution clear. He goes on to measure various properties of penicillin, such as its effectiveness when heated and its solubility in various liquids. Note that the experiments are described in enough detail so that they can be quantitatively reproduced by other scientists.

Fleming measures the power of penicillin at various strengths, or dilutions, and at various time periods. Evidently, the antibacterial power initially increases and then decreases with time. This fading behavior will turn out to be a serious obstacle to Fleming and other scientists in their attempts to make medical applications of penicillin.

Table III again shows which bacteria are affected by penicillin, this time of varying dilutions. Most affected are bacteria of the pyogenic cocci type, where "pyogenic" refers to those bacteria that cause pus infections and "cocci" are those that are physically round. Least affected are the Gram-negative bacteria, which are those bacteria that have an additional outer membrane covering their cell walls and thus resist the "Gram stain." Years later, biologists learned that penicillin works by preventing certain bacteria from manufacturing a protein needed to construct their cell wall. Without a cell wall, they dissolve.

The short section titled "Toxicity of Penicillin" is enormously important. As previously remarked, almost all previously known antibacterial agents killed leucocytes (white blood cells) and thus lowered an animal's natural resistance.

In the next two sections, Fleming makes much use of the result that some bacteria are sensitive to penicillin and some not. Penicillin can be used to isolate bacteria in a mix of bacteria, killing those that are sensitive and leaving the rest to grow, much like a weed killer. *B. influenzae*, in particular, can be isolated and easily identified in this way. (Ironically, it was later found that a virus, not *B. influenzae*, causes the flu. Thus Fleming had isolated a harmless bacterium.) As the title of his paper suggests, Fleming considers this "isolating" application of penicillin to be a major result of his work. In hindsight, of course, the treatment and cure of disease is by far the more important application.

In the discussion section, Fleming compares penicillin to other antibacterial agents, both bacterial and chemical, and argues that it is both more powerful and less toxic.

The most important point in the summary is number 8, worded in Fleming's typically tentative style: "It is suggested that it [penicillin] may be an efficient antiseptic for application to, or injection into, areas infected with penicillin-sensitive microbes." Here Fleming clearly shows that he understands the potential importance of his fungus to fight disease. Furthermore, he is willing for us to consider that penicillin may be not only superficially applied, as carbolic acid is applied to skin wounds, but also "injected." When Fleming showed Wright his paper before publication, the Old Man asked his former student to delete point 8. Point 8 was a defiance, a heresy. The master opposed any suggestion that the body's internal natural defenses were insufficient to fight disease. Wright also knew that all external antiseptics to date had proven toxic. Fleming quietly argued with his teacher. In the end, Little Flem stood his ground. Point 8 remained.

But the story was far from over. Although Fleming had discovered a powerful antibiotic, the substance had to be isolated, concentrated, and chemically purified before it could be of any medicinal use. And it had to be cured of its annoying property of fading after a week. Fleming was not a chemist. For these next chemical procedures, he enlisted the help of a young colleague in Wright's laboratory named Frederick Ridley. Ridley had not been trained as a chemist, but he knew more chemistry than anyone else in the lab. Ridley did not succeed.

On February 13, 1929, Fleming read his paper on penicillin to the Medical Research Club in London. The audience yawned. Not a single question was asked. Most biologists were still not convinced that it might be beneficial to put one germ into the body to kill another germ. And the charismatic Almroth Wright, who could have shaken up anyone to pay attention to a new idea, did not support this challenge from a former student. Finally, Fleming himself lacked the personality to forcefully champion his discovery. According to Sir Henry Dale, then chairman of the Medical Research Club, Fleming "was very shy, and excessively modest in his presentation, he gave it in a half-hearted way, shrugging his shoulders."

For some time, Fleming's penicillin went largely unnoticed. Three other British chemists, Harold Raistrick, R. Lovell, and P. W. Clutterbuck, attempted to isolate and purify the active agent in penicillin. They also failed, because of the instability and delicacy of the substance. In particular, as Lovell later recounted, the three chemists did not realize that switching the pH over to the alkaline side at a crucial point would have allowed them to counter the deadening effects of ether in the extraction process. In 1934, still convinced of the potential therapeutic value of his mold, Fleming hired yet another biochemist, again without success. Languidly, Fleming continued to mention penicillin in his publications.

The difficult problem of chemically purifying and stabilizing penicillin needed more attention from the scientific community. And that attention, in turn, required a change in thinking about the viability of external antiseptics. Over the next five years such a transformation occurred—brought about in part by the limited success of the synthetic sulfonamide drugs and also by René Dubos's discovery of the antibacterial agent gramicidin. These new developments, as Fleming noted in his Nobel Prize lecture of 1945, "completely changed the medical mind in regard to chemotherapy of bacterial infections."

In 1938, two British scientists at Oxford, Howard Florey and Ernst Boris Chain, both leaders of the changed "medical mind," again took up the challenge of penicillin and succeeded. Eventually, they produced penicillin in its pure crystalline form, about 40,000 times more concentrated than Fleming's original substance. By this time, World War II had begun, and antibiotics were badly needed on the battlefield. In 1941, the new drug was successfully tested on sick people. After that triumph, Florey enlisted the help of researchers and industries in the United States to produce penicillin in huge quantities.

In 1943, prompted by his study of penicillin, the American microbiologist Selman Abraham Waksman discovered streptomycin. Streptomycin defeated tuberculosis and plague. Now, the age of antibiotics had truly arrived. Penicillin, the first effective and nontoxic antibiotic, and the antibiotics that followed have saved millions of lives.

Beyond its immediate and vital applications, the discovery of penicillin was part of an important new notion that external chemicals and biological agents could be injected into the body to combat illness. Ever more advanced pieces of medical technology—organic and inorganic, natural and synthetic, chemical and biological—could be developed outside the body and then used within.

Fleming's work also helped change the conception of bacteriology and of the ecology of the microbial world in general. Prior to penicillin, bacteriologists assumed that bacteria stood apart from the rest of the natural world, that bacteria "contaminated," and contamination was to be avoided at all costs. Fleming invited contamination. He left the lids off his petri dishes, just as he left his office door open. By doing so, he helped create a larger notion in which bacteria are part of the complete ecological system. Bacteria live together, they grow together, they compete. One bacteria killing another is just a part of this larger performance.

Finally, penicillin contributed to a new sense of empowerment. For the first time in history, people felt that science could conquer the deadly scourge of infectious disease. Human beings had advanced one big step in their endless struggle with mortality.

Such an empowerment and hope can be heard in the words of Professor G. Liljestrand of the Royal Caroline Institute in Sweden, in his presentation of the 1945 Nobel Prize to Fleming, Chain, and Florey: "In a time when annihilation and destruction through the inventions of man have been greater than ever before in history, the introduction of penicillin is a brilliant demonstration that human genius is just as well able to save life and combat disease."

ON THE ANTIBACTERIAL ACTION OF CULTURES OF A PENICILLIUM, WITH SPECIAL REFERENCE TO THEIR USE IN THE ISOLATION OF *B. INFLUENZÆ*

Alexander Fleming, F.R.C.S.

From the Laboratories of the Inoculation Department, St Mary's Hospital, London

Received for publication May 10th, 1929

British Journal of Experimental Pathology (1929)

WHILE WORKING WITH STAPHYLOCOCCUS VARIANTS a number of culture-plates were set aside on the laboratory bench and examined from time to time. In the examinations these plates were necessarily exposed to the air and they became contaminated with various micro-organisms. It was noticed that around a large colony of a contaminating mould the staphylococcus colonies became transparent and were obviously undergoing lysis (see Fig. 1).

Subcultures of this mould were made and experiments conducted with a view to ascertaining something of the properties of the bacteriolytic substance which had evidently been formed in the mould culture and which had diffused into the surrounding medium. It was found that broth in which the mould had been grown at room temperature for one or two weeks had acquired marked inhibitory, bactericidal

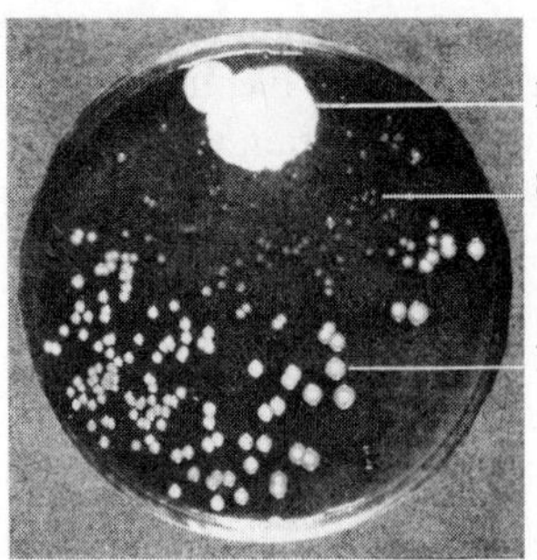

Figure 1. Photograph of a culture-plate showing the dissolution of staphylococcal colonies in the neighbourhood of a penicillium colony.

and bacteriolytic properties to many of the more common pathogenic bacteria.

CHARACTERS OF THE MOULD

The colony appears as a white fluffy mass which rapidly increases in size and after a few days sporulates, the centre becoming dark green and later in old cultures darkens to almost black. In four or five days a bright yellow colour is produced which diffuses into the medium. In certain conditions a reddish colour can be observed in the growth.

In broth the mould grows on the surface as a white fluffy growth changing in a few days to a dark green felted mass. The broth becomes bright yellow and this yellow pigment is not extracted by $CHCl_3$. The reaction of the broth becomes markedly alkaline, the pH varying from 8.5 to 9. Acid is produced in three or four days in glucose and saccharose broth. There is no acid production in 7 days in lactose, mannite or dulcite broth.

Growth is slow at 37°C. and is most rapid about 20°C. No growth is observed under anaerobic conditions.

In its morphology this organism is a penicillium and in all its characters it most closely resembles *P. rubrum*. Biourge (1923) states that he has never found *P. rubrum* in nature and that it is an "animal de laboratoire." This penicillium is not uncommon in the air of the laboratory.

IS THE ANTIBACTERIAL BODY ELABORATED IN CULTURE BY ALL MOULDS?

A number of other moulds were grown in broth at room temperature and the culture fluids were tested for antibacterial substances at various intervals up to one month. The species examined were: *Eidamia viridiscens*, *Botrytis cineria*, *Aspergillus fumigatus*, *Sporotrichum*, *Cladosporium*, *Penicillium*, 8 strains. Of these it was found that only one strain of penicillium produced any inhibitory substance, and that one had exactly the same cultural characters as the original one from the contaminated plate.

It is clear, therefore, that the production of this antibacterial substance is not common to all moulds or to all types of penicillium.

In the rest of this article allusion will constantly be made to experiments with filtrates of a broth culture of this mould, so for convenience and to avoid the repetition of the rather cumbersome phrase "Mould broth filtrate," the name "penicillin" will be used. This will denote the filtrate of a broth culture of the particular penicillium with which we are concerned.

METHODS OF EXAMINING CULTURES FOR ANTIBACTERIAL SUBSTANCE

The simplest method of examining for inhibitory power is to cut a furrow in an agar plate (or a plate of other suitable culture material), and fill this in with a mixture of equal parts of agar and the broth in which the mould has grown. When this has solidified, cultures of various microbes can be streaked at right angles from the furrow to the edge of the plate. The inhibitory substance diffuses very rapidly in the agar, so that in the few hours before the microbes show visible growth it has spread out for a centimetre or more in sufficient concentration to inhibit growth of a sensitive microbe. On further incubation it will be seen that the proximal portion of the culture for perhaps one centimetre becomes transparent, and on examination of this portion of the culture it is found that practically all the microbes are dissolved, indicating that the anti-bacterial substance has continued to diffuse into the agar in sufficient concentration to induce dissolution of the bacteria. This simple method therefore suffices to demonstrate the bacterio-inhibitory and bacteriolytic properties of the mould culture, and also by the extent of the area of inhibition gives some measure of the sensitiveness of the particular microbe tested. Fig. 2 shows the degree of inhibition obtained with various microbes tested in this way.

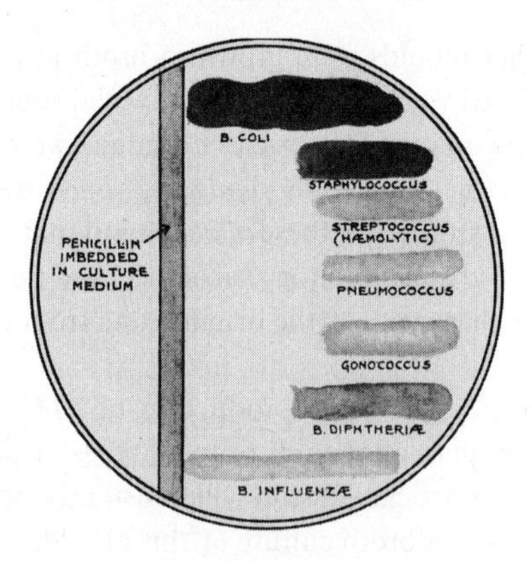

Figure 2.

The inhibitory power can be accurately titrated by making serial dilutions of penicillin in fresh nutrient broth, and then implanting all the tubes with the same volume of a bacterial suspension and incubating them. The inhibition can then readily be seen by noting the opacity of the broth.

For the estimation of the antibacterial power of a mould culture it is unnecessary to filter as the mould grows only slowly at 37°C., and in 24 hours, when the results are read, no growth of mould is perceptible. Staphylococcus is a very suitable microbe on which to test the broth as it is hardy, lives well in culture, grows rapidly, and is very sensitive to penicillin.

The bactericidal power can be tested in the same way except that at intervals measured quantities are explanted so that the number of surviving microbes can be estimated.

PROPERTIES OF THE ANTIBACTERIAL SUBSTANCE

EFFECT OF HEAT. Heating for 1 hour at 56° or 80° C. has no effect on the antibacterial power of penicillin. Boiling for a few minutes hardly affects it. Boiling for 1 hour reduces it to less than one quarter its previous strength if the fluid is alkaline, but if it is neutral or very slightly acid then the reduction is much less. Autoclaving for 20 minutes at 115° C. practically destroys it.

EFFECT OF FILTRATION. Passage through a Seitz filter does not diminish the antibacterial power. This is the best method of obtaining sterile active mould broth.

SOLUBILITY. It is freely soluble in water and weak saline solutions. My colleague, Mr. Ridley, has found that if penicillin is evaporated at a low temperature to a sticky mass the active principle can be completely extracted by absolute alcohol. It is insoluble in ether or chloroform.

RATE OF DEVELOPMENT OF INHIBITORY SUBSTANCE IN CULTURE. A 500 c.c. Erlenmeyer flask containing 200 c.c. of broth was planted with mould spores and incubated at room temperature (10° to 20° C.). The inhibitory power of the broth to staphylococcus was tested at intervals.

After 5 days complete inhibition in 1 in 20 dilution.
„ 6 „ „ „ „ 1 in 40 „
„ 7 „ „ „ „ 1 in 200 „
„ 8 „ „ „ „ 1 in 500 „

Grown at 20° C. the development of the active principle is more rapid and a good sample will completely inhibit staphylococci in a 1 in 500 or

TABLE I.

EFFECT OF KEEPING AT ROOM TEMPERATURE ON THE ANTI-STAPHYLOCOCCAL POWER OF PENICILLIN.

	Growth of staphylococcus in dilutions of penicillin as under.											
	1/20.	1/40.	1/60.	1/80.	1/100.	1/200.	1/300.	1/400.	1/600.	1/800.	1/1000.	Control.
At time of filtration	–	–	–	–	–	–	–	–	–	±	++	++
After 4 days	–	–	–	–	–	–	–	–	–	±	++	++
„ 7 „	–	–	–	–	–	–	–	±	+	+	++	++
„ 9 „	–	–	–	–	–	–	–	±	+	+	++	++
„ 13 „	–	–	–	–	–	+	+	+	+	+	++	++
„ 15 „	–	±	+	+	+	+	+	+	+	+	++	++

1 in 800 dilution in 6 or 7 days. As the culture ages the antibacterial power falls and may in 14 days at 20° C. have almost disappeared.

The antibacterial power of penicillin falls when it is kept at room temperature. The rate of this fall can be seen from Table I.

If the reaction of penicillin is altered from its original pH of 9 to a pH of 6.8 it is much more stable.

The small drops of bright yellow fluid which collect on the surface of the mould may have a high antibacterial titre. One specimen of such fluid completely inhibited the growth of staphylococci in a dilution of 1 in 20,000 while the broth in which the mould was growing, tested at the same time, inhibited staphylococcal growth in 1 in 800.

If the mould is grown on solid medium and the felted mass picked off and extracted in normal salt solution for 24 hours it is found that the extract has bacteriolytic properties.

If this extract is mixed with a thick suspension of staphylococcus suspension and incubated for 2 hours at 45° C. it will be found that the opacity of the suspension has markedly diminished and after 24 hours the previously opaque suspension will have become almost clear.

INFLUENCE OF THE MEDIUM ON THE ANTIBACTERIAL TITRE OF THE MOULD CULTURE. So far as has been ascertained nutrient broth is the most suitable medium for the production of penicillin. The addition of glucose or saccharose, which are fermented by the mould with the production of acid, delays or prevents the appearance of the antibacterial substance. Dilution of the broth with water delays the formation of the antibacterial substance and diminishes the concentration which is ultimately reached.

INHIBITORY POWER OF PENICILLIN ON THE GROWTH OF BACTERIA

Tables II [omitted here] and III show the extent to which various microbes, pathogenic and non-pathogenic, are inhibited by penicillin. The first table shows the inhibition by the agar plate method and the second shows the inhibitory power when diluted in nutrient broth.

Certain interesting facts emerge from these Tables. It is clear that penicillin contains a bacterio-inhibitory substance which is very active towards some microbes while not affecting others. The members of the coli-typhoid group are unaffected as are other intestinal bacilli such as *B. pyocyaneus*, *B. proteus* and *V. cholerae*. Other bacteria which are insensitive to penicillin are the enterococcus, some of the Gram-negative cocci of the mouth, Friedländer's pneumobacillus, and *B. influenzae* (Pfeiffer), while the action on *B. dysenteriae* (Flexner), and *B. pseudotuberculosis rodentium* is almost negligible. The anthrax bacillus is completely inhibited in a 1 in 10 dilution but in this case the inhibitory influence is trifling when compared with the effect on the pyogenic cocci.

It is on the pyogenic cocci and on bacilli of the diphtheria group that the action is most manifest.

Staphylococci are very sensitive, and the inhibitory effect is practically the same on all strains, whatever the colour or type of the staphylococcus.

Streptococcus pyogenes is also very sensitive. There were small differences in the titre with different strains, but it may be said generally that it is slightly more sensitive than staphylococcus.

Pneumococci are equally sensitive with *Streptococcus pyogenes*.

The green streptococci vary very considerably, a few strains being almost unaffected while others are as sensitive as *S. pyogenes*. Gonococci, meningococci, and some of the Gram-negative cocci found in nasal catarrhal conditions are about as sensitive as are staphylococci. Many of the Gram-negative cocci found in the mouth and throat are, however, quite insensitive.

B. diphtheriæ is less affected than staphylococcus but is yet completely inhibited by a 1% dilution of a fair sample of penicillin.

It may be noted here that penicillin, which is strongly inhibitory to many bacteria, does not inhibit the growth of the original penicillium which was used in its preparation.

TABLE III.

INHIBITORY POWER OF PENICILLIN ON DIFFERENT BACTERIA

	Dilution of penicillin in broth.											
	1/5.	1/10.	1/20.	1/40.	1/80.	1/100.	1/200.	1/400.	1/800.	1/1600.	1/3200.	Control.
Staphylococcus aureus	0	0	0	0	0	0	0	0	±	++	++	++
,, *epidermidis*	0	0	0	0	0	0	0	0	±	++	++	++
Pneumococcus	0	0	0	0	0	0	0	0	0	++	++	++
Streptococcus (hæmolytic)	0	0	0	0	0	0	0	0	0	±	++	++
,, *viridans* (mouth)	0	0	0	0	0	0	±	++	++	++	++	++
,, *fæcalis*	++	++	++	++	++	++	++	++	++	++	++	++
B. anthracis	0	0	+	+	++	++	++	++	++	++	++	++
B. pseudo-tuberculosis rodentium	+	+	++	++	++	++	++	++	++	++	++	++
B. pullorum	+	+	++	++	++	++	++	++	++	++	++	++
B. dysenteriæ	+	++	++	++	++	++	++	++	++	++	++	++
B. coli	++	++	++	...	...	...	...	...	...	...	...	++
B. typhosus	++	++	++	...	...	...	...	...	...	...	...	++
B. pyocyaneus	++	++	++	...	...	...	...	...	...	...	...	++
B. proteus	++	++	++	...	...	...	...	...	...	...	...	++
V. choleræ	++	++	++	...	...	...	...	...	...	...	...	++

		1/60.	1/120.	1/300.	1/600.	Control.
B. diphtheriæ (3 strains)		0	±	++	++	++
Streptococcus pyogenes (13 strains)		0	0	0	++	++
,, ,, (1 ,,)		0	0	±	++	++
,, *fæcalis* (11 ,,)		++	++	++	++	++
,, *viridans* at random from fæces	(1 strain)	0	0	0	++	++
,, ,, ,, ,,	(2 strains)	0	0	±	++	++
,, ,, ,, ,,	(1 strain)	0	±	++	++	++
,, ,, ,, ,,	(1 ,,)	+	++	++	++	++
,, ,, ,, ,,	(1 ,,)	++	++	++	++	++
,, ,, at random from mouth	(1 ,,)	0	±	++	++	++
,, ,, ,, ,,	(2 strains)	0	0	++	++	++
,, ,, ,, ,,	(1 strain)	0	0	0	++	++

0 = no growth; ± = trace of growth; + = poor growth; ++ = normal growth.

The Rate of Killing of Staphylococci by Penicillin

Some bactericidal agents like the hypochlorites are extremely rapid in their action, others like flavine or novarsenobillon are slow. Experiments were made to find into which category penicillin fell.

To 1 c.c. volumes of dilutions in broth of penicillin were added 10 c.mm. volumes of a 1 in 1000 dilution of a staphylococcus broth culture. The tubes were then incubated at 37° C. and at intervals 10 c.mm. volumes were removed and plated with the following result:

	Number of colonies developing after sojourn in penicillin in concentrations as under:				
	Control.	1/80.	1/40.	1/20.	1/10.
Before	27	27	27	27	27
After 2 hours	116	73	51	48	23
,, 4½ ,,	∝	13	1	2	5
,, 8 ,,	∝	0	0	0	0
,, 12 ,,	∝	0	0	0	0

It appears, therefore, that penicillin belongs to the group of slow acting antiseptics, and the staphylococci are only completely killed after an interval of over 4½ hours even in a concentration 30 or 40 times stronger than is necessary to inhibit completely the culture in broth. In the weaker concentrations it will be seen that at first there is growth of the staphylococci and only after some hours are the cocci killed off. The same thing can be seen if a series of dilutions of penicillin in broth are heavily infected with staphylococcus and incubated. If the cultures are examined after four hours it may be seen that growth has taken place apparently equally in all the tubes but when examined after being incubated overnight, the tubes containing penicillin in concentrations greater than 1 in 300 or 1 in 400 are perfectly clear while the control tube shows a heavy growth. This is a clear illustration of the bacteriolytic action of penicillin.

TOXICITY OF PENICILLIN

The toxicity to animals of powerfully antibacterial mould broth filtrates appears to be very low. Twenty c.c. injected intravenously into a rabbit were not more toxic than the same quantity of broth. Half a c.c. injected intraperitoneally into a mouse weighing about 20 gm. induced no toxic symptoms. Constant irrigation of large infected surfaces in man was not accompanied by any toxic symptoms, while irrigation of the human conjunctiva every hour for a day had no irritant effect.

In vitro penicillin which completely inhibits the growth of staphylococci in a dilution of 1 in 600 does not interfere with leucocytic function to a greater extent than does ordinary broth.

USE OF PENICILLIN TO DEMONSTRATE OTHER BACTERIAL INHIBITIONS

When materials like saliva or sputum are plated it is not uncommon to see, where the implant is thick, an almost pure culture of streptococci and pneumococci, and where the implant is thinner and the streptococcal colonies are more widely separated, other colonies appear, especially those of Gram-negative cocci. These Gram-negative cocci are inhibited by the streptococci (probably by the peroxide they produce in their growth) and it is only when the mass effect of the streptococci is reduced that they appear in the culture.

Penicillin may be used to give a striking demonstration of this inhibition of bacteria by streptococci and pneumococci. Sputum is spread thickly on a culture plate, and then 5 or 6 drops of penicillin is spread over one half of it. After incubation it may be seen that on the half untreated with penicillin there is a confluent growth of streptococci and pneumococci and nothing else, while on the penicillin-treated half many Gram-negative cocci appear which were inhibited by the streptococci and pneumococci, and can only flourish when these are themselves inhibited by the penicillin.

If some active penicillin is embedded in a streak across an agar plate planted with saliva an interesting growth sometimes results. On the portion most distal from the penicillin there are many streptococci, but these are obscured by coarsely growing cocci, so that the resultant growth is a copious confluent rough mass. These coarse growing cocci are extremely penicillin sensitive and stop growing about 25 mm. from the embedded penicillin. Then there is a zone of about 1 cm. wide of pure streptococci, then they are inhibited by the penicillin, and as soon as that happens Gram-negative cocci appear and grow right up to the embedded penicillin. The three zones of growth produced in this way are very striking.

USE OF PENICILLIN IN THE ISOLATION OF B. INFLUENZÆ (PFEIFFER) AND OTHER ORGANISMS

It sometimes happens that in the human body a pathogenic microbe may be difficult to isolate because it occurs in association with others which grow more profusely and which mask it. If in such a case the first microbe is insensitive to penicillin and the obscuring microbes are sensitive, then by the use of this substance these latter can be inhibited while the former are allowed to develop normally. Such an example occurs in the body, certainly with *B. influenzæ* (Pfeiffer) and probably with Bordet's whooping-cough bacillus and other organisms. Pfeiffer's bacillus, occurring as it does in the respiratory tract, is usually associated with streptococci, pneumococci, staphylococci and Gram-negative cocci. All of these, with the exception of some of the Gram-negative cocci, are highly sensitive to penicillin and by the addition of some of this to the medium they can be completely inhibited while *B. influenzæ* is unaffected. A definite quantity of the penicillin may be incorporated with the molten culture medium before the plates are made, but an easier and very satisfactory method is to spread the infected material, sputum, nasal mucus, etc., on the plate in the usual way and then over one half of the plate spread 2 to 6 drops (according to potency) of the penicillin. This small amount of fluid soaks into the agar and after cultivation for 24 hours it will be found that the half of the plate without the penicillin will show the normal growth while on the penicillin treated half there will be nothing but *B. influenzæ* with Gram-negative cocci and occasionally some other microbe. This makes it infinitely easier to isolate these penicillin-insensitive organisms, and repeatedly *B. influenzæ* has been isolated in this way when they have not been seen in films of sputum and when it has not been possible to detect them in plates not treated with penicillin. Of course if this method is adopted then a medium favourable for the growth of *B. influenzæ* must be used, *e. g.* boiled blood agar, as by the repression of the pneumococci and the staphylococci the symbiotic effect of these, so familiar in cultures of sputum on blood agar, is lost and if blood agar alone is used the colonies of *B. influenzæ* may be so minute as to be easily missed. . . .

DISCUSSION

It has been demonstrated that a species of penicillium produces in culture a very powerful antibacterial substance which affects different bacteria in different degrees. Speaking generally it may be said that the least sensitive bacteria are the Gram-negative bacilli, and the most susceptible are the pyogenic cocci. Inhibitory substances have been described in old cultures of many organisms; generally the inhibition is more or less specific to the microbe which has been used for the culture, and the inhibitory substances are seldom strong enough to withstand even slight dilution with fresh nutrient material. Penicillin is not inhibitory to the original penicillium used in its preparation.

Emmerich and other workers have shown that old cultures of *B. pyocyaneus* acquire a marked bacteriolytic power. The bacteriolytic agent, pyocyanase, possesses properties similar to penicillin in that its heat resistance is the same and it exists in the filtrate of a fluid culture. It resembles penicillin also in that it acts only on certain microbes. It differs however in being relatively extremely weak in its action and in acting on quite different types of bacteria. The bacilli of anthrax, diphtheria, cholera and typhoid are those most sensitive to pyocyanase, while the pyogenic cocci are unaffected, but the percentages of pyocyaneus filtrate necessary for the inhibition of these organisms was 40, 33, 40 and 60 respectively (Bocchia, 1909). This degree of inhibition is hardly comparable with 0.2% or less of penicillin which is necessary to completely inhibit the pyogenic cocci or the 1% necessary for *B. diphtheriæ*.

Penicillin, in regard to infections with sensitive microbes, appears to have some advantages over the well-known chemical antiseptics. A good sample will completely inhibit staphylococci, *Streptococcus pyogenes* and pneumococcus in a dilution of 1 in 800. It is therefore a more powerful inhibitory agent than is carbolic acid and it can be applied to an infected surface undiluted as it is non-irritant and non-toxic. If applied, therefore, on a dressing, it will still be effective even when diluted 800 times which is more than can be said of the chemical antiseptics in use. Experiments in connection with its value in the treatment of pyogenic infections are in progress.

In addition to its possible use in the treatment of bacterial infections penicillin is certainly useful to the bacteriologist for its power of inhibiting unwanted microbes in bacterial cultures so that penicillin-insensitive bacteria can readily be isolated. A notable instance of this is the very easy isolation of Pfeiffers bacillus of influenza when penicillin is used.

In conclusion my thanks are due to my colleagues, Mr. Ridley and Mr. Craddock, for their help in carrying out some of the experiments described in this paper, and to our mycologist, Mr. la Touche, for his suggestions as to the identity of the penicillium.

SUMMARY

1. A certain type of penicillium produces in culture a powerful antibacterial substance. The antibacterial power of the culture reaches its

maximum in about 7 days at 20° C. and after 10 days diminishes until it has almost disappeared in 4 weeks.

2. The best medium found for the production of the antibacterial substance has been ordinary nutrient broth.

3. The active agent is readily filterable and the name "penicillin" has been given to filtrates of broth cultures of the mould.

4. Penicillin loses most of its power after 10 to 14 days at room temperature but can be preserved longer by neutralization.

5. The active agent is not destroyed by boiling for a few minutes but in alkaline solution boiling for 1 hour markedly reduces the power. Autoclaving for 20 minutes at 115° C. practically destroys it. It is soluble in alcohol but insoluble in ether or chloroform.

6. The action is very marked on the pyogenic cocci and the diphtheria group of bacilli. Many bacteria are quite insensitive, *e.g.* the coli-typhoid group, the influenza-bacillus group, and the enterococcus.

7. Penicillin is non-toxic to animals in enormous doses and is non-irritant. It doses not interfere with leucocytic function to a greater degree than does ordinary broth.

8. It is suggested that it may be an efficient antiseptic for application to, or injection into, areas infected with penicillin-sensitive microbes.

9. The use of penicillin on culture plates renders obvious many bacterial inhibitions which are not very evident in ordinary cultures.

10. Its value as an aid to the isolation of *B. influenzæ* has been demonstrated.

REFERENCES

BIOURGE.—(1923) 'Des moissures du group *Penicillium* Link.' Louvain, p. 172.
EMMERICH, LOEUW AND KORSCHUN.—(1902) *Zbl. Bakt.*, **30,** 1.
BOCCHIA.—(1909) *Ibid.*, **50,** 220.
FILDES, P.—(1920) *Brit. J. Exp. Path.*, **1,** 129.

14

THE MEANS OF PRODUCTION OF ENERGY IN LIVING ORGANISMS

ENERGY IS THE CURRENCY OF NATURE. Nothing takes place without energy. The swing of a bat, the cooking of a soufflé, the faint chirp of a sparrow—all demand energy.

As I sit at my keyboard this moment, the flutter of my fingers requires about half a joule of energy each minute. (I'm a slow typist.) The joule, named after the British physicist James Prescott Joule, is defined as the motional energy in a one-kilogram brick moving at a speed of one meter per second. At a lecture at St. Anne's Church Reading Room in Manchester in 1847, Joule proposed that total energy is constant. The total energy in any self-contained system never increases or decreases, although that energy may appear in many different forms, and it can change from one form to another. For example, some of the heat energy in a burning coal furnace may be transformed into the mechanical energy of a rotating shaft, which can change into the electrical energy surging through a municipal power line.

As I sit typing, where does the energy reside before it flows to my fingers? That energy, relayed by the contractions of muscles, comes from the chemical energy stored in molecules of adenosine triphosphate (ATP) in my muscle cells. To animate my fingers at this moment, seven million million million molecules of ATP are severing their atomic bonds every minute. And where did that energy come from? From the breakdown of carbohydrates in the bagel I ate early this morning. The energy in those carbohydrates, in turn, came from the sunlight that glistened on certain wheat fields last spring. And that light energy originated in the nuclear reactions within the bowels of the sun. To be exact, the energy for one minute of my typing was provided by the fusions of one hundred billion atoms of hydrogen in the sun. In a sense, my fingers are being powered by nuclear energy.

The concept of energy has always been fundamental in physics. Empedocles and the ancient Greeks had the notion of various kinds of energy and even a rudimentary idea of the conservation of total energy. Leonardo da Vinci measured muscle power, cocked springs, and gunpowder in terms of their equivalent gravitational energies and ability to lift weights. Gottfried Wilhelm Leibniz, a contemporary and bitter rival of Isaac Newton, proposed a quantitative measure for the energy of moving masses, which he called *vis viva*, or living force. In the mid-nineteenth century, the British physiologist and physicist Julius Robert Mayer reiterated the equivalence of all kinds of energy, including heat.

By contrast, the importance of energy in biology has a much shorter history. As discussed in Chapter 2, the application of physics and chemistry to biology struggled against the philosophical belief that living matter obeys different laws than does nonliving matter. Thus, the importance of energy for biology gained recognition only as the living organism came to be viewed as a kind of mechanistic machine.

Much of that transformation in thinking occurred in the nineteenth century, led by scientists in Germany. In particular, as the modern law of the conservation of energy was being articulated in the 1840s, the chemist Justus von Liebig and Julius Mayer independently proposed that the energy needs of animals were supplied solely by the chemical breakdown of food. A gallop, a grinding of teeth, a warm breath on a cold winter's night could not possibly occur without ingestion of food. No more than a ball on level ground could begin rolling without a push, energy in a living thing could not be created from nothing. The physicist Hermann von Helmholtz, a strong supporter of the mechanistic view of life and an admirer of Liebig's ideas, showed that heat energy is released in working muscles. That energy as well as the mechanical energy of the moving muscle would have to be previously stored in food.

In the late nineteenth century, two German physiologists, Adolf Eugen Fick and Max Rubner, began testing Mayer's and Liebig's plausible hypothesis in more quantitative detail. The energies required for body heat, muscle contractions, and other physical activities were tabulated and compared against the chemical energy stored in food. Each gram of fat, carbohydrate, and protein had its energy equivalent. By the end of the century, Rubner concluded that the energy used by a living creature exactly equaled the energy consumed in food. In other words, the physicists' law of conservation of energy was also true for biology. On the ledger sheets of energy, a living being could be considered a container of so many coiled springs, balls in motion, weights on cantilevers, and electrical repulsions.

But Rubner's decree didn't mark the end of investigations. Scientists are driven to know not only how things work in general, but also how they work in detail. Sometimes insights leap from the details. How, exactly, is a glucose molecule from a candy bar manipulated in the body to yield energy? What are the chemical steps in detail? That was a problem for biochemistry.

In 1937, a thirty-seven-year-old German biochemist named Hans Adolf Krebs discovered the specific process by which most of the energy is liberated from food. Most profoundly, he and other scientists showed that this process, now called the Krebs cycle, operates in every kind of animal and plant on earth, from human beings down to single-celled bacteria. The precise molecules, the precise chemical steps, are the same. Even plants, whose initial input of energy is light rather than food, produce organic molecules and later use the Krebs cycle to break down those molecules for their stored energy. Evidently, the Krebs cycle is the principal energy-releasing mechanism in all living things, and its universality offers a strong argument that all life on the planet had a common beginning. The Krebs cycle, like DNA, is an ancient hieroglyph of life.

Details aside, the broad-brush picture of how food is converted into energy was first painted in the late eighteenth century by the great Antoine-Laurent Lavoisier (1743–1794), widely considered the father of modern chemistry. Lavoisier showed that organic substances undergo combustion with oxygen, releasing energy and leaving carbon dioxide and water as waste. (Organic molecules, made chiefly by living organisms, contain carbon, hydrogen, and often oxygen, plus additional atoms.) For example, suppose we begin with glucose, a high-energy carbohydrate in food. Lavoisier's chemical reaction would be represented by the equation

$$C_6H_{12}O_6 + 6O_2 \rightarrow 6CO_2 + 6H_2O + \mathit{Energy}$$

This equation, like all equations, is a kind of shorthand. A molecule of glucose, denoted by $C_6H_{12}O_6$, has six atoms of carbon (C), twelve atoms of hydrogen (H), and six atoms of oxygen (O). And so on. Lavoisier's equation says that one molecule of glucose combines with six molecules of oxygen to produce six molecules of carbon dioxide, six molecules of water, and energy. That energy could be in the form of heat, as in Lavoisier's initial conception, but it could also reside in any other form.

The process of combining glucose with oxygen is called oxidation. One might say also that glucose "burns" in the body to yield energy, since Lavoisier's reaction is quite similar to what happens when a log burns in air. (Wood is made largely of cellulose, which is a form of glucose.) In both cases, glucose combines with oxygen to yield carbon dioxide and water. The difference is that in open fires the energy is released erratically and at high temperature, where in the body the burning is more controlled and much cooler.

In all kinds of burning, the ultimate source of energy lies in the electrical repulsions between electrons in the various atoms involved. Those repulsive forces can be compared to compressed springs. They release their coiled energy when the electrons rearrange themselves to form new molecules. Because of the particular atomic geometries and forces, the "compressed springs" in a molecule of glucose are under far more repulsive tension than those in several molecules of water. Thus, net energy is liberated when glucose changes to water. Oxygen is critical in this process because it receives the hydrogen atoms from the high-energy glucose to form low-energy water. Without oxygen, Lavoisier's reaction cannot occur. We die when we stop breathing because we have stopped energy production in our bodies.

By 1930, scientists had a fair understanding of the energies in chemical bonds. Biochemists were also well aware that the Lavoisier reaction could not proceed in one step. Like skaters on an ice rink, molecules tend to bump into each other only two at a time. Thus, it is extremely unlikely that six molecules of oxygen would simultaneously meet with a molecule of glucose. Far more probable would be a sequence of intermediate steps and molecules, moving hydrogen atoms to oxygen one or two atoms at a time. That choreography is what Krebs set out to find.

In his autobiography *Reminiscences and Reflections*, published the year of his death in 1981, Hans Krebs describes himself in his early years as "self conscious, timid, and solitary . . . never aggressive or rebellious—on the contrary, I was anxious to conform." Apparently, a young scientist does not have to be a self-confident revolutionary, like Einstein, to accomplish great work. Krebs remembers his parents as "stern" people who frowned upon any show of emotion. Although young Krebs studied hard and typically finished in the top quarter of his class, his father expressed skepticism about his son's intellectual potential and often quoted to his children, with a sigh of resignation, "You cannot make a silk purse out of a sow's ear." Despite painful memories like these, Krebs credits his father with arousing his first interest in living things by taking him on long walks through the countryside near Hildesheim.

At Göttingen University, Krebs learned the importance of chemistry to biology. In particular, he studied with Franz Knoop, who was investigating how fat is metabolized in intermediate steps and who would later do crucial work pointing to the Krebs cycle. Intent upon entering medicine like his father, an ear, nose, and throat surgeon, Krebs obtained his M.D. degree from the University of Hamburg in 1925. Then, from 1926 to 1930, he worked as one of the many assistants to Otto Warburg at the Kaiser Wilhelm Institute for Biology at Berlin-Dahlem. At the time, Warburg was showing how catalysts aid in the combustion of food—work that would garner him the Nobel in 1931.

Krebs considered Warburg the most influential teacher of his life. Already inclined by nature to accept authority, Krebs practically worshiped Warburg, who ruled his laboratory like a king and demanded absolute obedience and respect from his students. (Warburg had himself been an apprentice of the great Nobel chemist Emil Fischer, who also ruled his minions with an iron hand.) Krebs's characterization of Warburg can be read as a statement of what he most admired in a scientist:

> He set an example of high standards in research and in general conduct . . . His dedication manifested itself in his long and regular working hours and his contempt for those who tried to further their careers by jockeying for position, by hobnobbing with and courting the influential, or by publishing trivia for publishing's sake. He was prepared to take infinite pains with every aspect of his work . . . He also took pride in the fact that when he found a mistake (which did not happen often) he would admit it and publish a correction without delay.

At the end of Krebs's apprenticeship, Warburg did not help his thirty-year-old student find a job. As Krebs recalls, so poignantly reminiscent of his father's attitude, "[Warburg] did not think I had sufficient ability for a successful research career . . . I came to the conclusion that my talents were rather mediocre. It was only my keen interest that drove me to keep trying for a position which would give me scope for research."

Finally, the "mediocre" student obtained a position back in hospital work, first at the Municipal Hospital at Altona and later at the Medical Clinic of the University of Freiburg. It was there, in 1932, that he discovered one of the first metabolic cycles in biology, called the ornithine cycle. In this process, an organic molecule named ornithine is changed to citrulline, which is changed to arginine, which is changed back to ornithine. Along the path of this loop, the intermediate molecules absorb ammonia and give off urea. Ammonia, a by-product of other biochemical reactions, is a toxin. Thus the ornithine cycle is a way for a living organism to rid itself of internal poisons.

In June 1933, Krebs, who was Jewish, lost his job under the National Socialist government and was forced to leave Germany. Thanks to an invitation from Sir Frederick Gowland Hopkins, the leading biochemist in England, Krebs secured a post in Cambridge. Upon entering a private English home, he was overwhelmed by "British friendliness and human warmth. I had never met anything like it before."

In 1935, Krebs was appointed lecturer in pharmacology at the University of Sheffield, a small institution with only eight hundred students. And there, two years later, he discovered the famous biochemical cycle that carries his name.

In the early 1930s the production of energy from the oxidation of food, called respiration, was a major area of study in biochemistry. Indeed, a number of the intermediate steps in the Krebs cycle had been previously discovered by other scientists. Until Krebs's work, however, nobody knew how these isolated reactions were related. Krebs recognized that a cyclical process was probably involved and discovered a crucial missing step of the cycle, thus combining all previous work in a unified picture.

Perhaps most critical of all earlier work was the research of the Hungarian-American biochemist Albert Szent-Györgyi. Szent-Györgyi had found that for the study of respiration, the flight muscle of pigeons was ideal because of its intense metabolic activity. Pigeon flight muscle burns food at a very high rate. (Ounce for ounce, the flight muscle of a hummingbird also burns food at a very high rate, but hummingbirds are tiny and not easy to come by.) Since respiration requires oxygen, the rate of respiration can be measured by the rate of oxygen consumption. In 1935, Szent-Györgyi discovered that four particular organic molecules—succinic acid, fumaric acid, malic acid, and oxaloacetic acid—when added to pigeon muscle greatly increase the rate of respiration. Furthermore, the increased oxygen consumption is far larger than needed to extract the energy from these added molecules. Evidently, they themselves are not the source of the energy. Rather, these molecules are helping, or "catalyzing," the energy-producing reactions of other molecules. To do so, they must be used over and over again. These four molecules are similar in structure, all having four carbon atoms, and they can be converted into each other by the removal of hydrogen atoms and the addition of water molecules in the following way:

$$\underset{\text{Succinic}}{C_4H_6O_4} \xrightarrow{-2H} \underset{\text{Fumaric}}{C_4H_4O_4} \xrightarrow{+H_2O} \underset{\text{Malic}}{C_4H_6O_5} \xrightarrow{-2H} \underset{\text{Oxaloacetic}}{C_4H_4O_5}$$

The next important piece of work was provided in early 1937 by Franz Knoop, with whom Krebs had studied in Göttingen, and C. Martius. Knoop and Martius discovered some of the chemical steps in the oxidation of citric acid, not a foodstuff itself but found in small quantities in many foods. In particular, Knoop and Martius found that citric acid is converted to aconitic acid, which is converted into isocitric acid (the same atoms as citric but with different bonds between atoms), which is converted to α-oxoglutaric acid in the following reactions:

$$\underset{\text{Citric}}{C_6H_8O_7} \xrightarrow{-H_2O} \underset{\text{Aconitic}}{C_6H_6O_6} \xrightarrow{+H_2O} \underset{\text{Isocitric}}{C_6H_8O_7} \xrightarrow{-2H} \underset{\alpha\text{-Oxoglutaric}}{C_5H_6O_5} + CO_2$$

The above reaction represents an "oxidation" of citric acid because hydrogen atoms are pulled off the isocitric acid and later combined with oxygen atoms (not shown) to form water. The same is true for the Szent-Györgyi chain of reactions. It was already known that α-oxoglutaric acid, with the addition of an atom of oxygen, could be transformed into succinic acid plus carbon dioxide. Thus, putting together the two above sets of reactions, Krebs knew that *there was a continuous metabolic pathway from citric to oxaloacetic acid.*

At Sheffield, Krebs's most valuable assistants were Leonard Eggleston and William Arthur Johnson. Eggleston began working for Krebs in 1936, at age seventeen, and remained his faithful helper and collaborator until 1974. Johnson, fresh from his undergraduate degree in chemistry at the University of Sheffield, found himself at the right place at the right time. His joint work with Krebs on the Krebs cycle became part of his doctoral dissertation.

After a discussion of biochemical methods, Krebs and Johnson begin their paper by showing that citrate (citric acid), like the four molecules of Szent-Györgyi, is a catalyst in respiration. That is, small quantities of citrate greatly increase the oxygen absorption in active pigeon muscle, far more than would be needed to oxidize the citrate. Following Table I, which measures oxygen consumption with and without added citrate, Krebs points out that after 150 minutes, citrate has increased oxygen consumption in 460 milligrams of pigeon muscle from 1,187 to 2,080 microliters (μl), an increase of 893 microliters—whereas only 302 microliters of oxygen would be needed to completely oxidize, or use up, the citrate. Thus, like other catalysts, the citrate must be acting over and over again, without being permanently consumed. But since Knoop and Martius had already shown that citrate is definitely used up to form α-oxoglutaric acid in the presence of oxygen, there must be some other process that continually *replenishes* the citrate. This idea is essential to Krebs's thinking. In section VI he will show himself how citrate is regenerated.

A short digression on experimental methods. To measure the amount of oxygen consumed in their biochemical reactions, Krebs and Johnson use a U-shaped tube, called a manometer, which contains a known quantity of liquid and gas. As the gas (e.g., oxygen) is absorbed, the pressure in the tube changes by a corresponding amount, causing the height of the liquid to vary. The change in height tells the oxygen consumed. The amount of citric acid and other chemicals produced is measured by a colorimeter, which uses light and colored filters to accurately measure the color of a liquid. The chemical product to be quantified is reacted with other chemicals to produce a colored substance. The amount of that substance (in turn a measure of the amount of initial product) determines the color of a liquid solution, which is measured by the colorimeter.

In section IV, Krebs refers to the work of Martius and Knoop, already mentioned. Now, however, he knows that citric acid, and hence possibly the entire Martius-Knoop chain of reactions, is a key player in respiration.

Section V draws on the earlier work of the Swedish biochemist Thorsten Thunberg, who showed that the chemical named malonate blocks the oxidation of succinate to fumurate (part of the Szent-Györgyi sequence of reactions). Krebs repeats Thunberg's experiment and now proves that malonate also inhibits the oxidation of citrate (turning to aconitic, etc.) in respiration. This result is further evidence that citrate is part of the same chain of reactions involving succinate and fumurate. If these two sets of reactions were not part of one long chain of reactions, then there would be no reason that the blocking of fumurate would also stop the production of aconitic, just as a traffic jam on a highway in Canada should have no effect on the flow of cars in Cuba. *Thus, Krebs has now shown that the reactions discovered by Szent-Györgyi and by Martius and Knoop are probably linked in one chain.*

Krebs begins section VI by laying out what he knows of the chemical steps in respiration (leaving out the aconitic and isocitric acid intermediaries and calling α-oxoglutaric acid by its close cousin α-ketoglutaric acid): citric acid → α-ketoglutaric acid → succinic acid → fumaric acid → *l*-malic acid → oxaloacetic acid. From his earlier work, Krebs knows that citric acid must somehow be regenerated. And, in fact, he proves that with his next experiment, summarized with the statement that "muscle is capable of forming large quantities of citric acid if oxaloacetic acid is present." This result is a new contribution of Krebs and key to his conclusions. If citric acid were not regenerated, then the entire chain of reactions would come to a halt. Citric acid leads to oxaloacetic acid along the Knoop-Martius and Szent-Györgyi chains. Oxaloacetic acid, in turn, leads to citric acid.

Thus, Krebs has now found a metabolic cycle, or repeating loop. Some unknown two-carbon molecule present in muscle tissue, which Krebs tentatively calls "triose," combines with the four-carbon molecule oxaloacetic acid to form the six-carbon molecule citric acid. The citric acid then follows the various biochemical steps of Martius and Knoop and of Szent-Györgyi to be oxidized to oxaloacetic acid, which then combines with fresh triose to repeat the cycle. In section VII, Krebs presents the cycle in its skeletal and full form. (A modern version of the cycle, completed in 1951, appears in Figure 14.1.) Note that Krebs calls his newly discovered cycle the citric acid cycle. Only later did it become known as the Krebs cycle.

In item 6 of section VII, Krebs finds that the citric acid cycle occurs in many other animal tissues. In fact, we now know that the cycle occurs *in every single cell* of a living organism. Each living cell is like a city of its own, exchanging raw materials and products with the outer world through the cell wall, and the Krebs cycle is the central power plant of that city. Although Krebs failed to find his cycle in yeast or in the

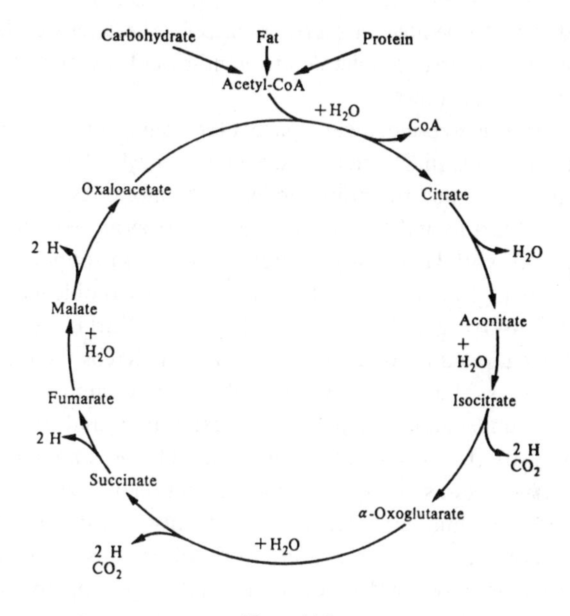

Figure 14.1

bacterium *B. coli*, that was only because his experiments at the time did not include the carrier protein that accompanies citrate through the cell wall for these cells. Later experiments proved that the Krebs cycle indeed occurs in these organisms as well.

In item 7, Krebs performs the important follow-up experiment to show that the rate of citric acid synthesis from "triose" and oxaloacetic acid is sufficiently fast to power the entire process of respiration. (The cycle must go around twice to oxidize each incoming molecule of glucose. Therefore, as can be seen from Lavoisier's equation, each molecule of citrate produced should be accompanied by the consumption of three molecules of oxygen.) This result supports Krebs's proposal that the citric acid cycle is the dominant process in respiration.

The last sentence of the paper clearly shows that Krebs understands the significance of his work: "The quantitative data suggest that the 'citric acid cycle' is the preferential pathway through which carbohydrate is oxidised in animal tissues." Krebs had found a universal process in biology. Ten years earlier, the Dutch scientist Albert Jan Kluyver had argued that the large range of metabolic processes in microorganisms could all be reduced to the simple process of oxidation of organic molecules. Now, Krebs had drawn the fine lines of that process and further proposed that it operated in every living cell. The Krebs cycle represented one of the first complete understandings of a biochemical process of major importance.

It is fascinating to ponder why a particular scientist makes a discovery at a particular time. In Chapter 8, I speculated on why it was Niels Bohr who proposed the first quantum model of the atom. As we have seen, the work of Martius and Knoop, in early 1937, was an essential precursor of Krebs's work later that year. But Krebs was surrounded by other talented scientists interested in the same or similar problems. Martius, Knoop, Szent-Györgyi, Warburg, Otto Meyerhof, Karl Lohmann, Karl Meyer, Fritz Lipmann, to name a few. Why was Krebs the one to discover the process? Besides his excellent laboratory techniques, his determination to understand the process of respiration, and his complete familiarity with the work of other biochemists before him, a large part of Krebs's success was undoubtedly his recognition of a cyclical process. Hans Krebs was thinking in terms of cycles. Just as von Laue pictured the overlap of waves in the miniature world of crystals, Krebs was picturing cycles, processes that go round in a circle, repeating over and over again. And here, his previous experience with the ornithine cycle was essential. As he wrote later, "In visualizing the cycle mechanism it was of major relevance that five years earlier I had been concerned with the first metabolic cycle to be discovered, the ornithine cycle of urea synthesis." Since Krebs, the idea of cyclical processes in biology has been of immense importance.

For some years after the Krebs-Johnson paper of 1937, there was doubt about the meaning and sequence of Krebs's cycle. In particular, some biochemists thought that citrate was only a minor side step of the process and that it was aconitate rather than citrate that formed from triose and oxaloacetate. If true, that belief would falsify many of Krebs's suppositions. By the late 1940s, however, theoretical studies by the British biochemist Alexander Ogston showed that Krebs had been right all along.

There remained the puzzling identity of Krebs's "triose," some two-carbon derivative of food. It had long been known that a molecule of glucose, $C_6H_{12}O_6$, must be broken down into two molecules of pyruvic acid, $C_3H_4O_3$, before it can be oxidized in respiration. In 1951, the German-American biochemist Fritz Albert Lipmann showed that pyruvic acid interacts with an enzyme he called coenzyme A to form the two-carbon molecule acetyl, C_2H_3O. Coenzyme A shepherds acetyl into the Krebs cycle to combine with oxaloacetic acid and water to form citric acid. Thus acetyl is Krebs's "triose." Lipmann and Krebs shared the 1953 Nobel Prize in physiology or medicine.

Finally, it was shown that not only carbohydrates, like glucose, but also proteins and fats could be broken down to acetyl attached to coenzyme A (denoted in Figure 14.1 by Acetyl-CoA) and processed through the Krebs cycle. Thus, essentially all of the food that we eat passes through this cycle. Foodstuffs come in at the top, water and carbon dioxide flow from the bottom, and each of the intermediate organic molecules is assembled and disassembled over and over again. Roughly one-third of the energy of food is released in preparation for the Krebs cycle and the other two-thirds via the cycle itself.

As already mentioned, the energy in the cycle is obtained by combining hydrogen atoms with oxygen atoms to form low-energy water. Note in Figure 14.1 the output of hydrogen atoms at four points along the loop. Functionally, one can think of the Krebs cycle as a means of plucking hydrogen atoms off organic molecules in order to fuel oxygen. Ultimately, the energy liberated by this process is stored in molecules of ATP, discovered by Karl Lohmann in 1929. This molecule, like glucose, has a great deal of energy coiled in its electrical repulsions, but the energy in ATP can be released quickly and easily, without need of other molecules and a long chain of reactions. Approximately ten molecules of ATP are formed for each turn of the Krebs cycle. Thus, every minute of my typing requires about a million million million loops of the cycle. Those loops are happening at each moment, invisibly, and in every cell of my body. If each loop of the Krebs cycle faintly tinkled like the drop of a pin, I would turn deaf from the roar.

With the discoveries of ATP, the details of muscular metabolism, and the Krebs cycle, the period from 1920 to 1950 in biochemistry might be called the era of energy. In the 1950s, the discovery of the structure of DNA ushered in the era of information.

In 1967, Sir Hans Krebs was obliged to give up his Oxford professorship, for he had reached the mandatory retirement age of sixty-seven. But he felt fit and resourceful, and he had no intention of retiring from science. And he had ingested the work ethic of his mentor Otto Warburg. From 1968 until his death in 1981, Krebs published another 116 papers in biochemistry. Near the end of that period he was asked by a journalist who knew he loved music why he continued to do scientific research when he could enjoy his retirement by listening to more music. Wasn't listening to good music the greatest pleasure for a music lover? Krebs replied that "there can be one pleasure still greater—the creating of music."

THE ROLE OF CITRIC ACID IN INTERMEDIATE METABOLISM IN ANIMAL TISSUES

Hans Krebs and W. A. Johnson

From the Department of Pharmacology, University of Sheffield

29 June 1937

Enzymologia (1937)

DURING THE LAST DECADE MUCH PROGRESS has been made in the analysis of the anaerobic fermentation of carbohydrate, but very little is so far known about the intermediate stages of the oxidative breakdown of carbohydrate. A number of reactions are known in which derivatives of carbohydrate take part and which are probably steps in the breakdown of carbohydrate; we know furthermore, from the work of SZENT-GYÖRGYI[20] that succinic acid, fumaric acid and oxaloacetic acid play some role in the oxidation of carbohydrate, but the details of this role are still obscure.

In the present paper experiments are reported which throw new light on the problem of the intermediate stages of oxidation of carbohydrate; in conjunction with the work of SZENT-GYÖRGYI,[20] STARE and BAUMANN[19] and MARTIUS and KNOOP,[13, 14] the new experiments allow us to outline the principal steps of the oxidation of sugar in animal tissues.

I. METHODS.

1. Tissue. Pigeon breast muscle was used for the majority of the experiments described in this paper. The tissue was minced in a LATAPIE mincer immediately after killing and usually suspended in 3–7 volumes of 0.1 M sodium phosphate buffer (ph = 7.4). On further dilution of the suspension the rate of metabolism decreased.

2. Quantitative determination of citric acid. The method of PUCHER, SHERMAN and VICKERY[17] was used; citric acid is oxidised to pentabromoacetone which is subsequently converted by means of sodium sulphide to a coloured material suitable for colorimetric determination. As pointed out by PUCHER c.s. the method is suitable for the determination

of quantities between 0.1 and 1 mg. It is fairly specific; there are only a few other substances which yield also a yellow coloured material, and the only substance which interfered in our experiments was oxaloacetic acid. In pure solution 0.5 millimol oxaloacetate yielded 1.98×10^{-3} millimol "citric acid." The yield of "citric acid" is increased by 50% if pyruvic acid is present at the same time and this suggests that bromine, like hydrogen peroxide (MARTIUS and KNOOP[13]) brings about a synthesis of citric acid from oxaloacetic and pyruvic acid. This interfering reaction can be removed if oxaloacetic acid is decomposed by heating the solution for one hour in neutral or weakly acid medium after deproteinisation. Although oxaloacetate is not completely decomposed if heated in pure solution the destruction is practically complete in the deproteinised tissue extract. This effect may be explained by the observations of POLLAK[16] and LJUNGGREN[10] which demonstrate a catalytic influence of amino compounds on the decompositions of β-ketonic acids.

3. Quantitative determination of succinic acid. The method used was a modification of SZENT-GYÖRGYI's[20] manometric method. The details will be described in full elsewhere.

4. Quantitative determination of α-ketoglutaric acid. The α-ketoglutaric acid was quantitatively determined by estimation of the amount of succinic acid formed on oxidation. In an aliquot sample the preformed succinic acid is determined and the figure obtained is substracted from the sample treated with the oxidising agent. Suitable oxidising agents are cold permanganate or ceric sulphate in acid medium. Under our conditions both reagents gave identical results. The ceric sulphate method is better if significant amounts of α-hydroxyglutaric acid or glutamic acid are present, since these two substances react more readily with permanganate than with ceric sulphate to yield succinic acid. Citric acid, iso-citric acid and cis-aconitic acid do not yield succinic acid if treated with permanganate or ceric sulphate.

5. Quantitative determination of succinic acid in the presence of malonic and α-ketoglutaric acid. Malonic acid interferes with the manometric determination of succinic acid and has therefore to be removed first. The removal is brought about by a principle suggested by WEIL-MALHERBE.[22] To the neutral solution containing the three substances are added 1 ccm 2 M sodium sulphite and 2 ccm 3 M tartaric acid and the solution is then extracted continuously with ether in a KUTSCHER-STEUDEL[9] extractor. Under our conditions extraction of succinic acid was complete in 30 minutes, whilst α-ketoglutaric acid remains quantitatively in the aqueous phase. The ethereal extract which also contains

malonic acid is freed from ether, dissolved in water and treated with permanganate in acid solution in order to destroy the malonic acid and is then extracted with ether again. This second ethereal extract contains the succinic acid ready for the manometric determination.

6. Metabolic quotients. According to the usual convention the quantities of metabolites are expressed in μ*l* even if they are not gases; 1 millimol citric acid, for instance, is considered equivalent to 22400 μ*l*. The rate of metabolic reaction is expressed by the quotient

$$\frac{\text{substrate metabolised}}{\text{mg dry tissue} \times \text{hour}}$$

In the case of muscle the dry weight was considered to be 20% of the wet weight.

II. CATALYTIC EFFECT OF CITRATE ON RESPIRATION.

If muscle tissue is minced and suspended in 6 volumes of phosphate buffer a high rate of respiration is observed initially, but after 20–40 minutes the rate begins to fall off. If citrate is added the rate of respiration is often increased and the falling off of respiration is always much retarded. This effect is brought about by small quantities of citrate, and comparing the extra respiration with the citrate added we find that the extra oxygen uptake is by far greater than can be accounted for by the complete oxidation of citrate. An example is the following experiment:

TABLE I.

EFFECT OF CITRATE ON RESPIRATION OF MINCED PIGEON BREAST MUSCLE. (MANOMETRIC EXPERIMENT)

Time (min.)	μ*l* O_2 absorbed by 460 mg muscle (wet weight) suspended in 3 ccm phosphate saline	
	No substrate added	0.15 ccm 0.02 M sodium citrate added
30	645	682
60	1055	1520
90	1132	1938
150	1187	2080

In this experiment the citrate caused an increased respiration of 893 μ*l*, whilst 302μ*l* O_2 are calculated for the complete oxidation of the citrate added.

The magnitude of the effect of citrate shows considerable variations from experiment to experiment; the effect appears to be dependent on the amounts of citrate and other substrates preformed in the tissue. The effect is more pronounced if glycogen, or hexosediphosphate, or α-glycerophosphate are added to the muscle, and we presume therefore that the substrate the oxidation of which is catalysed by citrate, is a carbohydrate or a related substance.

An example in which citrate promotes catalytically the oxidation of α-glycerophosphate is given in Table II. There is only a small effect of citrate in this experiment if added alone to the suspension, but a very considerable effect is observed if α-glycerophosphate is present.

SZENT-GYÖRGYI[20] and STARE and BAUMANN[19] have shown that fumarate, oxaloacetate, and succinate have a similar catalytic effect under the same experimental conditions, a fact of great importance to which we shall refer later.

The problem of the mechanism of this citrate catalysis can be approached in various ways.

TABLE II.

EFFECT OF CITRATE ON THE RESPIRATION OF PIGEON BREAST MUSCLE IN THE PRESENCE OF α-GLYCEROPHOSPHATE.

(40°; 140 min.; 460 mg muscle (wet weight) in 3 ccm phosphate saline per flask; manometric experiment)

Substrates added	μl O_2 absorbed
—	342
0.15 ccm 0.02 M citrate	431
0.3 ccm 0.2 M glycerophosphate	757
0.3 ccm 0.2 M glycerophosphate + 0.15 ccm 0.02 M citrate	1385

We have chosen the investigation of the intermediate stages of the breakdown of citrate in the tissues. If all the stages are known the mechanism of the catalytic effect will be clear.

III. RATE OF DISAPPEARANCE OF CITRIC ACID IN MUSCLE.

Since citric acid reacts catalytically in the tissue it is probable that it is removed by a primary reaction but regenerated by a subsequent reaction. In the balance sheet no citrate disappears and no intermediate products accumulate. The first object of the study of intermediates is therefore to find conditions under which citrate disappears in the balance sheet. We find that some poisons bring about this effect, for instance arsenite (Table III) or malonate. If one of these two substances is present, very large amounts of citric acid disappear provided that oxygen is available. Obviously the poisons leave the breakdown of citric acid unaffected whilst they check the synthesis of citric acid.

TABLE III.

DISAPPEARANCE OF CITRIC ACID IN PIGEON-BREAST MUSCLE IN THE PRESENCE OF ARSENITE (3×10^{-3} MOL.).

(3 ccm muscle suspension containing 750 mg wet muscle were shaken for 40 min. at 40°)

μ*l* citrate added	μ*l* citrate found after 40 min.	μ*l* citrate used	$Q_{citrate}$
1120	30	1090	−10.9
2240	972	1268	−12.7
4480	2790	1690	−16.9

IV. CONVERSION OF CITRIC ACID INTO α-KETOGLUTARIC ACID.

The oxidation of citric acid in the presence of arsenite or malonate is not complete. Only one or two molecules of oxygen are absorbed for each molecule of citric acid removed and the solution must therefore contain intermediate products of oxidation of citric acid.

Although it has long been known that citric acid is readily metabolised (see Östberg,[15] Sherman c.s.[18]), the pathway of the breakdown remained obscure until early 1937, when Martius and Knoop[13, 14] working with citrico dehydrogenase from liver discovered that the oxidation of citric acid by methylene blue yields α-ketoglutaric acid. We are able to confirm Martius and Knoop's results with other tissues and with molecular oxygen as the oxidising agent.

In previous work it had been shown that arsenite is a specific inhibitor for the oxidation of α-ketonic acid in animal tissues. It had been possible, for example, to stop the oxidation of glutamic acid[6] and of proline[23] at the stage of α-ketoglutaric acid. We find now that large quantities of α-ketoglutaric acid are present in those suspensions in which citric acid was oxidised in the presence of arsenite.

For example: 46 grammes (wet weight) of minced muscle was suspended in 145 ccm phosphate buffer and shaken in an atmosphere of oxygen for one hour in three large flasks of the shape described previously ([5]) with 11.5 ccm 1 M citrate and 6 ccm 0.1 M arsenite. After one hour 40 ccm of trichloroacetic acid (30%) was added and 190 ccm filtrate was treated with 1 gramme 2.4 dinitrophenylhydrazine dissolved in 100 ccm 2 N HCl. A precipitate was formed immediately. It was collected after two hours on a sintered glass filter, and thoroughly washed with 0.1 N hydrochloric acid and water. The precipitate weighed 1.199 grammes and proved to be practically pure 2.4-dinitrophenylhydrazone of α-ketoglutaric acid (M.P. 217° C). On recrystallisation from aqueous alcohol a substance was obtained which gave the correct melting and mixed melting points (222° C). The calculated total yield of dinitrophenylhydrazone is $1.199 \times (\frac{249}{190}) = 1.57$ grammes, or 4.82 millimol.

The quantitative analysis of the deproteinised filtrate with the manometric method gave no succinic acid, but 0.0612 millimol α-ketoglutaric acid per 3 ccm filtrate or 5.07 millimol in the total volume. Both methods, isolation and manometric method, thus agree very well, the former as expected giving a slightly lower figure.

The determination of citric acid in another aliquot of the filtrate showed that 4.64 mg of citric acid were left per ccm liquid, or 6.02 millimol in the total volume. The amount of citric acid added was 11.5 millimol so that the amount metabolised was 5.48 millimol. The yield of α-ketoglutaric acid as will be seen from Table IV is thus as complete as could be expected in the circumstances.

TABLE IV.

Citric acid metabolised	5.48 millimol	
α-Ketoglutaric acid formed (manometrically)	5.07 ,,	(yield 93%)
α-Ketoglutaric acid formed (as hydrazone)	4.82 ,,	(yield 88%)

V. CONVERSION OF CITRIC ACID INTO SUCCINIC ACID.

In the presence of malonate the oxidation of citrate is checked at the stage of succinic acid as shown by the following experiment: 7.5 grammes (wet weight) minced pigeon muscle were suspended in 22.5 ccm phosphate buffer (0.1 M; ph = 7.4) and 3 ccm 0.2 M sodium citrate and

1 ccm 1 M malonate were added. The suspension was shaken for 40 min. in an atmosphere of oxygen, and then deproteinised by adding 34 ccm water, 2 ccm 50% sulphuric acid and 2 ccm 15% sodium tungstate. In the filtrate succinic and α-ketoglutaric acids were determined manometrically. 3 ccm contained 472 μ*l* succinic acid and 80 μ*l* α-ketoglutaric acid; α-ketoglutaric acid was also identified by the isolation of the 2.4-dinitrophenylhydrazone.

VI. SYNTHESIS OF CITRIC ACID IN THE PRESENCE OF OXALOACETIC ACID.

The new results of the citric acid breakdown, in conjunction with previous work on the oxidation of succinic acid in tissues may be summarised by the following series:

citric acid → α-ketoglutaric acid → succinic acid → fumaric acid → *l*-malic acid → oxaloacetic acid → pyruvic acid.

If it is true that the oxidation of citric acid is a stage in the catalytic action of citric acid then it follows that citric acid must be regenerated eventually from one of the products of oxidation. We are thus led to examine whether citric acid can be resynthesised from any of the intermediates of the citric acid breakdown.

Systematic experiments show that indeed large quantities of citric acid are formed if oxaloacetic acid is added to muscle anaerobically, whilst all the other intermediates, including pyruvic acid yield no citric acid under the same conditions. It is because the synthesis of citric acid from oxaloacetic acid does not require molecular oxygen and because citric acid is stable in the tissue anaerobically that it is possible to demonstrate the synthesis of citric acid in a simple experiment.

Minced pigeon breast muscle was suspended as usual in 3 volumes phosphate buffer and 3 ccm suspension were measured into a conical manometric flask the sidearm of which contained 0.3 ccm 1 M oxaloacetate. In the centre chamber a stick of yellow phosphorus was placed and the gas space was filled with nitrogen. After the removal of oxygen the oxaloacetate was added to the tissue and the flask was shaken in the water bath for 20 mins. During this period about 1000 μ*l* CO_2 were evolved. After the incubation, the suspension was quantitatively transferred into 25 ccm 6% trichloracetic acid and the volume was made up to 50 ccm. Citric acid was determined in the filtrate and 0.0131 millimol (293 μ*l*) citric acid were found. $Q_{citrate}$ is thus $\frac{293 \times 3}{150} = 5.86$. No citrate was present in the controls.

This experiment shows that muscle is capable of forming large quantities of citric acid if oxaloacetic acid is present and the question arises from which substance the two additional carbon atoms of the citric acid molecule are derived. Addition of various possible precursors such as acetate, or pyruvate, or of α-glycerophosphate had no effect on the rate of citric acid synthesis, but this negative result is no proof against the participation in the synthesis of one of these substances. Pyruvic acid and acetic acid arise rapidly from oxaloacetic acid and it may be that the tissue is already saturated with these substances if oxaloacetic acid alone has been added.

The fact that the catalytic effect of citrate is more pronounced if glycogen, or hexosemonophosphate, or α-glycerophosphate are present suggests that the substance condensing with oxaloacetate is derived from carbohydrate. We may term it provisionally as "triose," leaving it open whether triose reacts as such or as a derivative for example as a phosphate ester, or pyruvic acid or acetic acid.

A synthesis of citric acid from a C_4-dicarboxylic acid and a second substance has often been discussed, especially with reference to the citric acid fermentation of moulds,[1, 24] though it has not been shown before to occur in animal tissues.

MARTIUS and KNOOP[12] showed recently that citric acid is formed in vitro if oxaloacetate and pyruvate are treated with hydrogen peroxide in alkaline medium. This model reaction is an interesting analogy and it suggests that the synthesis of citric acid may be a comparatively simple reaction[3]).

VII. ROLE OF CITRIC ACID IN THE INTERMEDIATE METABOLISM.

I. CITRIC ACID CYCLE. The relevent facts concerning the intermediate metabolism of citric acid may now be summarised as follows:

1. Citrate promotes catalytically the oxidations in muscle tissue, especially if carbohydrates have been added to the tissue.

2. Similar catalytic effects are shown by succinate, fumarate, malate, oxaloacetate (SZENT-GYÖRGYI,[20] STARE and BAUMANN[19]).

3. The oxidation of citrate in muscle passes through the following stages: citric acid → α-ketoglutaric acid → succinic acid → fumaric acid → *l*-malic acid → oxaloacetic acid.

4. Oxaloacetic acid reacts with an unknown substance to form citric acid.

These facts suggest that citric acid acts as a catalyst in the oxidation of carbohydrate in the following manner:

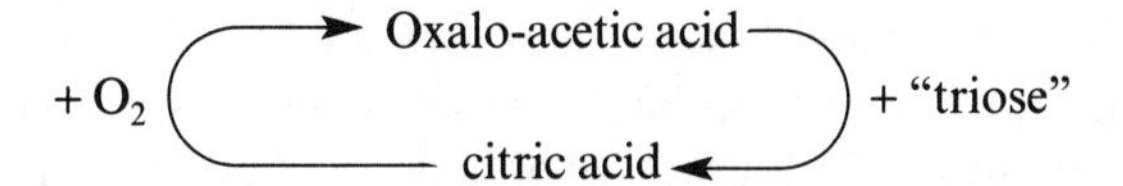

According to this scheme oxaloacetic acid condenses with "triose" to form citric acid, and by oxidation of citric acid oxaloacetic acid is regenerated. The net effect of the "citric acid cycle" is the complete oxidation of "triose."

The synthesis of citric acid from oxaloacetic acid as well as the oxidation of citric acid to oxaloacetic has been experimentally verified. The only hypothetical point in the scheme is the term "triose," though we may consider it as certain that the substance condensing with oxaloacetic acid is related to carbohydrate.

The proposed scheme outlines a pathway for the oxidation of carbohydrate. Many details must necessarily be left open at the present time, but a few points will be discussed in the following sections.

2. ORIGIN OF THE C_4-DICARBOXYLIC ACID. According to the scheme succinic acid or a related compound is necessary as "carrier" for the oxidation of carbohydrate and the question of the origin of succinic acid arises. We have shown previously ([8]) that succinic acid can be synthesised by animal tissues in small amounts if pyruvic acid is available. The physiological significance of the synthesis is now clear: it provides the carrier required for the oxidation of carbohydrate.

3. FURTHER INTERMEDIATE STAGES. (a) iso-Citric acid. WAGNER-JAUREGG and RAUEN[22] and MARTIUS and KNOOP[13, 14] have suggested that iso-citric acid is an intermediate in the oxidation of citric acid. We find that iso-citric acid is indeed readily oxidised in muscle, the rates of oxidation of citric acid and iso-citric acids being about the same.

(b) cis-Aconitic acid. cis-aconitic acid, discovered by MALACHOWSKI and MASLOWSKI,[11] was first discussed as an intermediate by MARTIUS and KNOOP[13] and MARTIUS[14] showed that it yields readily citric acid with liver. We have examined the behaviour of cis-aconitic acid in muscle and other tissues and find that it is oxidised as readily as citric acid. The conversion of cis-aconitic acid into citric acid is also brought about by tissue extracts. One milligramme muscle tissue (dry weight) converts up to 0.1 mg cis-aconitic acid into citric acid per hour (40°; ph = 7.4).

MARTIUS and KNOOP[13, 14] assume that the reaction cis-aconitic ⇆ citric acid is reversible and believe that it plays a role in the breakdown of citric acid. It cannot yet be said, however, whether the reaction is an intermediate step in the breakdown or in the synthesis of citric acid.

(c) Oxalo-succinic acid. The oxidation of iso-citric acid would be expected to yield in the first stage oxalo-succinic acid (MARTIUS and KNOOP). This β-ketonic acid is only known in the form of its esters, since the free acid is unstable in a pure state. In acid solution it is readily decarboxylated and yields α-ketoglutaric acid (BLAISE and GAULT[2]).

(d) Detailed citric acid cycle. The information available at present about the intermediate steps of the cycle may be summarised thus:

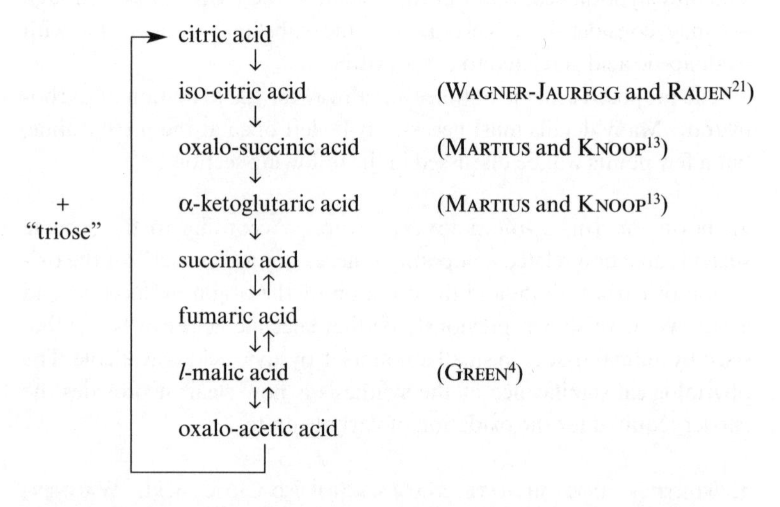

4. REVERSIBLE STEPS. Succinic acid arises according to our scheme by oxidative reactions from oxaloacetic acid, via citric and α-ketoglutaric acids. Anaerobic experiments, however, show succinic acid can also be formed by reduction from oxaloacetic acid (see also SZENT-GYÖRGYI[20]). The reactions succinic acid → fumaric acid → *l*-malic acid → oxaloacetic acid are thus reversible under suitable conditions.

The outstanding problem in this connection is the question of the oxidative equivalent of the reduction. At least a partial answer may be given. The synthesis of citric acid as shown in section VI takes place anaerobically, although it is an oxidative process. A reductive process equivalent to the oxidation must therefore occur at the same time. The reduction of oxaloacetic acid to succinic acid is the only reduction of sufficient magnitude (see the next section) known so far to occur simultaneously with the citric acid synthesis and we assume therefore it is the equivalent for the synthesis of citric acid.

5. EFFECT OF MALONATE. It follows from the preceding paragraph that succinic acid can arise from oxaloacetic acid in two different ways (a) oxidatively via citric and α-ketoglutaric acids (b) reductively via *l*-malic and fumaric acids. That two different ways and therefore two different enzymic systems bring about the conversion of oxaloacetic into succinic acid can be demonstrated with the aid of malonate. Malonate inhibits specifically the reaction succinic acid ⇆ fumaric acid. Aerobically it will therefore increase the yield of succinic acid from oxaloacetic acid since it prevents its secondary breakdown. Anaerobically, on the other hand, it will inhibit the formation of succinic acid, since in this case the succinic dehydrogenase is concerned with the formation of the succinic acid. The following experiment shows that the results are as expected.

TABLE V.

EFFECT OF MALONATE ON THE AEROBIC AND ANAEROBIC CONVERSION OF OXALOACETIC INTO SUCCINIC ACID.

(0.75 grammes wet muscle in 3 ccm phosphate buffer; 40° C; ph = 7.4)

Experimental conditions (final concentration of the substrates)	μ*l* succinic acid formed in 40 min.
1. O_2; 0.1 M oxaloacetate;	1086
2. O_2; 0.1 M oxaloacetate; 0.06 M malonate	1410
3. N_2; 0.1 M oxaloacetate;	1270
4. N_2; 0.1 M oxaloacetate; 0.06 M malonate	834

6. CITRIC ACID CYCLE IN OTHER TISSUES. We have tested the principal points of the citric acid cycle in various other animal tissues and find that brain, testis, liver and kidney of the rat are capable of oxidising citric acid as well as synthesising it from oxaloacetic acid. Of these four tissues testis shows the highest rate of synthesis and this is of interest in view of the work of THUNBERG's school on the occurrence of citric acid in spermatic fluid. 1 mg (dry weight) rat testis forms anaerobically up to 0.02 mg citric acid per hour if oxaloacetic acid is present.

Whilst the citric acid cycle thus seems to occur generally in animal tissues, it does not exist in yeast or in *B. coli*, for yeast and *B. coli* do not oxidise citric acid at an appreciable rate.

7. QUANTITATIVE SIGNIFICANCE OF THE CITRIC ACID CYCLE. Though the citric acid cycle may not be the only pathway through which carbohydrate is oxidised in animal tissues the quantitative data of the oxidation and resynthesis of citric acid indicate that it is the preferential pathway. The quantitative significance of the cycle depends on the rate of the slowest partial step, that is for our experimental conditions the synthesis of citric acid from oxaloacetic acid. According to the scheme one molecule of citric acid is synthesised in the course of the oxidation of one molecule of "triose," and since the oxidation of triose requires 3 molecules O_2, the rate of citric acid synthesis should be one third of the rate of O_2 consumption if carbohydrate is oxidised through the citric acid cycle. We find for our conditions:

Rate of respiration (Q_{O_2}) = –20
Rate of citric acid synthesis ($Q_{citrate}$) = + 5.8

The observed rate of the citric acid synthesis is thus a little under the expected figure (–6.6), but it is very probable that the conditions suitable for the demonstration of the synthesis (absence of oxygen) are not the optimal conditions for the intermediate formation of citric acid, and that the rate of citric acid synthesis is higher under more physiological conditions. This is suggested by the experiments on the aerobic formation of succinic acid from oxaloacetic acid (Table IV). $Q_{succinate}$, in the presence of malonate and oxaloacetate is +14.1, and if citrate is an intermediate stage the rate of citrate formation must be at least the same. But even the observed minimum figures of the rate of the synthesis justify the assumption that the citric acid cycle is the chief pathway of the oxidation of carbohydrate in pigeon muscle.

8. THE WORK OF SZENT-GYÖRGYI. SZENT-GYÖRGYI[20] who first pointed out the importance of the C_4-dicarboxylic acids in cellular respiration, came to the conclusion that respiration, in muscle, is oxidation of triose by oxaloacetic acid. In the light of our new experiments it becomes clear that SZENT-GYÖRGYI's view contained a correct conception, though the manner in which oxaloacetic acid reacts is somewhat different from what SZENT-GYÖRGYI visualised. The experimental results of SZENT-GYÖRGYI can be well explained by the citric acid cycle; we do not intend, however, to discuss this in full in this paper.

SUMMARY.

1. Citric acid catalytically promotes oxidations in muscle, especially in the presence of carbohydrate.

2. The rate of the oxidative removal of citric acid from muscle was measured. The maximum figure for $Q_{citrate}$ observed was –16.9.

3. α-Ketoglutaric acid and succinic acid were found as products of the oxidation of citric acid. These experiments confirm MARTIUS and KNOOP's results obtained with liver citric dehydrogenase.

4. Oxaloacetic acid, if added to muscle, condenses with an unknown substance to form citric acid. The unknown substance is in all probability a derivative of carbohydrate.

5. The catalytic effect of citrate as well as the similar effects of succinate, fumarate, malate and oxaloacetate described by SZENT-GYÖRGYI and by STARE and BAUMANN are explained by the series of reactions summarized in section VII 3 d.

6. The quantitative data suggest that the "citric acid cycle" is the preferential pathway through which carbohydrate is oxidised in animal tissues.

We wish to thank the Medical Research Council and the ROCKEFELLER Foundation for grants and Professor E. J. WAYNE for his help and advice.

1. Bernhauer, Ergebn. Enzymf. **3,** 185 (1934).
2. Blaise, Gault, C. R. **147,** 198 (1908).
3. Claisen, Hori, Ber. Chem. Ges. **24,** 120 (1891)
4. Green, Biochem. Jl. **30,** 2095 (1936).
5. Krebs, Zs. phys. Chem. **217,** 191 (1933).
6. Krebs, Zs. phys. Chem. **218,** 151 (1933).
7.. Krebs, Biochem. Jl. **29,** 1620 (1935).
8. Krebs, Johnson, Biochem. Jl. **31,** 645 (1937).
9. Kutscher, Steudel, Zs. phys. Chem. **39,** 474 (1903).
10. Ljunggren, Katalytisk Kolsyreavspjälkning ur Ketokarbonsyror, Lund (1925).
11. Malachowski, Maslowski, Ber. Chem. Ges. **61,** 2524 (1928).
12.. Martius, Knoop, Zs. phys. Chem. **242,** I (1936).
13. Martius, Knoop, Zs. phys. Chem. **246,** 1 (1937).
14. Martius, Zs. phys. Chem. **247,** 104 (1937).
15. Östberg, Skand. Arch. Phys. **62,** 81 (1931).
16. Pollak, Hofmeisters Beitr. **10,** 232 (1907).

17. Pucher, Sherman, Vickery, Jl. of Biol. Chem. **113,** 235 (1936).

18. Sherman, Mendel, Vickery, Jl. of Biol. Chem. **113,** 247, 265 (1936).

19. Stare, Baumann, Proc. Roy. Soc. **B 121,** 338 (1936).

20. Szent-Györgyi c.s., Bioch. Zs. **162,** 399 (1925); Z. phys. Chem. **224,** 1 (1934), **236,** 1 (1935), **244,** 105 (1936) **247,** I (1937).

21.. Wagner-Jauregg, Rauen, Zs. phys. Chem. **237,** 227 (1935).

22. Weil-Malherbe, Biochem. Jl. **31,** 299 (1937).

23. Weil-Malherbe, Krebs, Biochem. Jl. **29,** 2077 (1935).

24. Wieland, Sonderhoff, Ann. Chem. Pharm. Liebig **503,** 61 (1933).

15

NUCLEAR FISSION

On Christmas day of 1938, Otto Frisch went to visit his aunt, Lise Meitner, at a small hotel in the town of Kungälv on the west coast of Sweden. Both were nuclear physicists. Frisch, a good-looking young man, thirty-four years old, of Jewish stock like his aunt, had been dismissed from his post in Hamburg and now worked in Copenhagen. Meitner, who had forged a thirty-year scientific career at the Kaiser Wilhelm Institute in Berlin, who had been director of her own laboratory and compared to Madame Curie by Einstein, had also been forced to flee Germany and now occupied a humble position in Stockholm. That Christmas morning, Frisch found his aunt at breakfast brooding over a letter. It was a letter from Otto Hahn, her longtime collaborator in Berlin.

For some years, Hahn and Meitner had been bombarding uranium nuclei with neutrons, producing new families of radioactive elements. Everyone had always assumed that such artificially created nuclei would be close to uranium in mass. Thorium, for example, with 97 percent of the mass of uranium. Or actinium. By all logical reasoning, a diminutive neutron, a single subatomic particle traveling at low speed, did not have the energy to chip off more than a tiny piece of the huge uranium nucleus. Hahn's letter claimed that in a recent experiment, in the radioactive debris after neutron bombardment, he had found barium. Barium is an atom with half the mass of uranium. It was as if the uranium nucleus had been broken in two! It was as if a stone from a slingshot had cracked open a mountain. "Perhaps you can come up with some sort of fantastic explanation," Hahn wrote in his note. "We know ourselves that it can't actually burst apart into barium."

Meitner and her nephew took a walk in the snow, he on skis, she on foot. Gradually, the idea took shape in their heads that the uranium nucleus could indeed be split in two by a tiny neutron, not by being chipped or cracked but by being slightly deformed from its normal spherical shape. Once deformed, the nucleus would become even more elongated from the imbalance of its internal forces, eventually "fissioning" in two. As the two scientists decided, the uranium nucleus was in fact a cocked mousetrap, waiting for a slight prod to explode. The energy was not in the prod but in the coiled repulsive forces already there.

Meitner calculated the expected energy release from a single fission: 200 million electron volts. In more telling terms, fissioning a gram of uranium would produce about ten million times as much energy as burning a gram of coal, or detonating a gram of TNT. Ernest Rutherford had found the atomic nucleus in 1911. Now Meitner and Hahn had discovered how to release the colossal energy locked within. Like Prometheus, Meitner and Hahn had brought fire to the human race. The consequences would be both evil and good.

Lise Meitner was born in Vienna in 1878. Her parents, Hedwig and Philipp, were of Jewish ancestry, but Judaism played no part in their lives, and they had all of their eight children baptized and brought up as Protestants. Philipp was a lawyer and an avid chess player. He and his wife moved in social circles with writers, lawyers, and other intellectuals. As a testament to that environment, the five Meitner daughters all received advanced education, in an era when it was extremely difficult for girls to be trained beyond secondary school. As Lise recalled many years later, "Thinking back to the time of my youth, one realizes with some astonishment how many problems then existed in the lives of ordinary young girls . . . among the most difficult of these problems was the possibility of normal intellectual training."

Even as a child of eight and nine years old, Lise kept a math book under her pillow and demonstrated the kind of skepticism and independence of mind that characterized the scientist. According to the recollections of a family friend, when Lise's grandmother cautioned that she should not sew on the Sabbath or else the heavens would come tumbling down, Lise decided to do an experiment. The little girl lightly touched her knitting needle to some embroidery and looked up. Nothing happened. Then she took a single stitch, waited, looked up. Again, nothing. Finally, she was satisfied that her grandmother had been mistaken and went happily about her sewing.

Meitner burned to become a scientist. However, her official schooling in Vienna ended when she was fourteen. By extreme persistence, she managed to get her Matura (high school) certification in the summer of 1901 and entered the University of Vienna a few months later. In 1905 she received her Ph.D., only the second woman in Vienna to achieve a doctorate in physics.

In September 1907, Meitner traveled to Berlin for further study with the great Max Planck. At that time, women were still excluded from German universities. To attend Planck's lectures Meitner had to ask special permission. Already, she was fascinated with the new field of radioactivity, the mysterious process by which some atoms spontaneously emit high-energy particles and rays. At the University of Vienna, Meitner had designed and performed one of the first experiments showing that the subatomic "alpha particles" could be deflected in passing though matter. (See Chapter 5.)

In the fall of 1907, Meitner was introduced to Otto Hahn. Hahn, a chemist who researched radioactivity at the Chemical Institute of the University of Berlin, was charming, informal, and not opposed to working with females. And so began one of the most dramatic collaborations in scientific history.

To begin with, women were forbidden at the Chemical Institute. What to do? The only thing possible was for Meitner to join Hahn in his basement laboratory, converted from a carpentry shop. Hahn was free to go to other laboratories upstairs, but Meitner was required to remain in the basement of the institute. After a year, a new law permitted women in Prussian universities. A year after that, Meitner was allowed to travel to the higher regions of the building. Yet even with her new freedom, Meitner was clearly a woman in a man's world. Most men could barely conceive of women as scientists. When Ernest Rutherford first met Meitner at the end of 1908, after she had published a number of important papers, he said, "Oh, I thought you were a man!"

In 1911, Hahn was offered the position of heading the radiochemistry department of the new Kaiser Wilhelm Institute for Chemistry in Dahlem. Soon after, Meitner was invited there.

Hahn was four months younger than Meitner. The son of a well-off Prussian tradesman, he had received a degree in chemistry from the University of Marburg and a doctorate in 1901. In 1904, Hahn went to London and was introduced to the new field of radioactivity by Sir William Ramsay. There, Hahn discovered radiothorium, a radioactive form of thorium. The following year, he worked with Rutherford in Canada. In 1906, Hahn returned to Germany with the position of privatdozent (lecturer). He was the only chemist in Berlin engaged with radioactivity.

Like the medical team of Bayliss and Starling, Meitner and Hahn complemented each other well. Meitner's background was in physics, Hahn's in chemistry. Although she designed and performed experiments, Meitner had strength in mathematical and graphical skills, in conceptual thinking, in making generalizations. Hahn's great ability was in detailed and meticulous chemical laboratory work, particularly in separating and identifying different substances by their chemical properties.

A photograph of Meitner and Hahn in their laboratory in 1910 suggests further differences. Hahn, with his Wilhelminian handlebar mustache, looks full face at the camera with an easy and confident air, hands in his vest pocket, a gold chain draped across his vest. Meitner, a tiny and slender woman, turns sideways, almost hiding, a demure and timid expression on her face as if she were a child who has just been scolded by a parent, her eyes sunken in deep hollows that would deepen and darken with age.

For many years, Meitner was painfully shy. Although she shared equally in the collaboration and became the intellectual leader of the team, she initially assumed a subservient role to Hahn. A glimmer of that relationship, and Meitner's self-effacing manner, can be seen in a letter she wrote to Hahn in early 1917, when he was off in World War I and she was left doing their experiments alone:

> Dear Herr Hahn! The pitchblende experiment is of course important and interesting but I cannot do it right now—don't be angry please . . . Yesterday I gave a colloquium. I thought of you and spoke loudly and looked at the people and not at the blackboard . . . Be well and *please don't be angry* about the pitchblende delay.

Over the course of their thirty years of work together, Meitner and Hahn were among the world leaders in the science of radioactivity. They discovered dozens of new radioactive substances, charted the rates of disintegration of atoms, measured the penetrating powers of subatomic alpha and beta particles. Gradually, Meitner was accorded the stature she deserved. In 1917, she was appointed head of her own laboratory at the Kaiser Wilhelm Institute (KWI). Now there was a Laboratorium Meitner in physics, as well as a Laboratorium Hahn in chemistry. In 1919, at the age of forty-one, Meitner was given the title of Professor at

KWI, the first woman in Germany to hold that title. (Hahn had been made Professor nine years earlier.) Meitner was also gaining independence from Hahn. Although she contined to work with him on and off until 1938, from 1921 to 1934 Meitner authored fifty-six papers on her own, without Hahn. Such achievement and recognition made her forced exile in 1938 even more bitter. But, to her mind, her inadequate credit for the discovery of nuclear fission would be the most bitter experience of all.

Radioactivity was first found, accidentally, in 1896, when the French physicist Antoine-Henri Becquerel realized that a covered photographic plate became fogged up by uranium salts lying on a table nearby. In 1898 Polish scientist Marie Sklodovska Curie and her French husband Pierre Curie discovered several new elements, including polonium and radium, that emitted high-energy particles and rays. By 1900 it was known that such "radioactive" elements emitted two kinds of electrically charged particles: alpha particles, with a positive charge and about four times the mass of a hydrogen atom; and beta particles, with a negative charge and an extremely small mass. Indeed, beta particles eventually proved identical to subatomic electrons, discovered in 1897 by J. J. Thomson.

When Rutherford and his coworkers discovered the atomic nucleus, a hundred thousand times smaller than the atom as a whole, the physicist from New Zealand made a number of prescient speculations. First, the high-energy particles emitted in radioactivity originated within the atomic nucleus. Thus, Rutherford predicted, radioactivity was strictly a nuclear process. Second, some attractive force must counteract the huge repulsive force that would exist among the positively charged particles (protons) in the nucleus. Otherwise, the protons would fly apart like tomcats thrown together in a barrel. Rutherford also suspected that a population of electrically neutral particles might share the cramped living quarters with the protons. Such particles, called neutrons, were discovered by the British physicist James Chadwick in 1932.

With the discovery of the neutron, nuclear physics took a leap forward. Now, scientists had a more refined geography of the powerful center of the atom. There were two kinds of subatomic particles in the atomic nucleus, protons and neutrons. Protons were positively charged, neutrons neutral. Neutrons were slightly heavier than protons. It was well known that the chemical properties of atoms—the way in which they react with other atoms—were determined completely by the number of negatively charged electrons in the outer portions of atoms (see Chapter 11 on Pauling). Since atoms normally were electrically neutral, the number of electrons must be balanced by an equal number of protons within the nucleus. Thus, in effect, the number of protons in the nucleus fixed the chemical identity of the atom, that is, the particular element it was. Hydrogen, the lightest element, had one proton in its nucleus. Carbon had six. Uranium had ninety-two. Much of the confusion in the early days of radioactivity arose from the fact that each atomic element, with a specific number of protons in its nucleus, could have a variable number of neutrons. For example, one form of uranium had 143 neutrons in its nucleus. Another form had 146 neutrons. Forms of the same element with different numbers of neutrons were called isotopes of that element. The number of protons in the nucleus was called the atomic number and denoted by the letter Z. The total number of protons plus neutrons was called the atomic mass and denoted by the letter A.

With these concepts, radioactive emissions and transformations now became partly a bookkeeping problem. An alpha particle, weighed and measured by the curve of its trajectory in a magnetic field, consisted of two protons and two neutrons. A beta particle was simply an electron, created as a by-product when a neutron changed into a proton. The nuclear arithmetic would then be as follows: when a radioactive atom emitted an alpha particle, its atomic number decreased by two and its atomic mass by four. When an atom emitted a beta particle, its atomic number *increased* by one, while its atomic mass remained the same. These processes are illustrated in the sequence of reactions in which an atom of uranium gradually transforms into lead:

$$
{}^{238}_{92}\mathrm{U} \rightarrow \alpha + {}^{234}_{90}\mathrm{Th} \rightarrow \beta + {}^{234}_{91}\mathrm{Pa} \rightarrow \beta + {}^{234}_{92}\mathrm{U} \rightarrow \alpha + {}^{230}_{90}\mathrm{Th}
\rightarrow \alpha + {}^{226}_{88}\mathrm{Ra} \rightarrow \ldots {}^{206}_{82}\mathrm{Pb}
$$

Here, alpha and beta particles are denoted by α and β, respectively. The Latin letters are symbols of elements. Thus U stands for uranium, Th for thorium, Pa for proactinium (discovered by Hahn and Meitner in 1918), Ra for radium, and Pb for lead. The subscript preceding each element denotes its atomic number; the superscript is its atomic mass.

Beta particles emitted in radioactive disintegrations were measured with a Geiger counter, first developed in the early 1900s by Hans Geiger and perfected in the next couple of decades. A Geiger counter consisted of an electrically charged wire placed lengthwise in a gas-filled tube. Flying through a window into the tube, a beta particle would strip off the electrons in the gas atoms, and the released electrons would flow to the positively charged wire and alter the current passing through it. From the strength and rapidity of the variation in current, one could gauge the fluxes of the beta particles entering the detector. Alpha particles were normally measured by an instrument with a thinner window called an ionization chamber. In this device, a voltage was applied between two parallel plates. As in the Geiger counter, an incoming alpha particle would dislodge electrons from gas atoms between the plates; the resulting electrically charged atoms, or ions, would then move to the plates of opposite electrical charge. (Remember that like charges repel; opposite charges attract.)

The counting rate of the Geiger counter or ionization chamber directly measured the rate of disintegration of radioactive atoms. (By disintegration, we do not mean that the atomic nucleus completely decomposes but only that it emits an alpha or beta particle.) A key feature of these disintegrations is the half-life, also called the period. The period is the amount of time it takes for half of the atoms to disintegrate. For example, suppose a certain isotope of uranium has a period of twenty-four minutes and we begin with 1,000 atoms of this substance. After twenty-four minutes, about 500 atoms will have disintegrated, leaving 500 undisintegrated atoms. After another twenty-four minutes, about half of those, or about 250, will disintegrate. And so on. No two radioactive substances have the same period. Thus Meitner and Hahn and other scientists could measure periods to discover and identify new radioactive species, like fingerprinting a population.

The question remained: why did some atomic nuclei disintegrate and not others? A carbon nucleus with six protons and six neutrons could sit peacefully intact forever, while a uranium nucleus with 92 protons and 147 neutrons disintegrated after twenty-four minutes. The explanation lay in the complex competition of forces within the nucleus—a repulsive electrical force between protons and an attractive nuclear force between all nuclear particles, both protons and neutrons. All systems of nature attempt to achieve the lowest energy possible, just as a ball rolling around on an uneven floor tends to come to rest at the lowest point of the floor. Some nuclei, with a particular number of neutrons and protons, find themselves already at a very low energy, like a ball in a deep hollow in the floor. These nuclei are "stable" and will remain unchanged for a very long time. Others, with a different number of neutrons and protons and thus a different arrangement of forces and energies, can lower their energy by emitting an alpha or beta particle. These are the radioactive nuclei. Indeed, a study of radioactivity helped Hahn and Meitner and other scientists understand the nature of the forces struggling against each other at the core of the atom.

In 1934, the Curies' daughter Irène and her husband Frédéric Joliot discovered "artificial radioactivity." In this surprising phenomenon, an otherwise stable, nonradioactive atom could be rendered radioactive when bombarded by alpha particles. Evidently, an alpha particle, once absorbed by an atomic nucleus, disturbs the balance of forces and energies there. Such a disturbed nucleus relieves itself by spitting out subatomic particles.

Shortly thereafter, the Italian physicist Enrico Fermi reasoned that neutrons would make even better projectiles than alphas. Neutrons, being electrically neutral, would not be repelled by the repulsive force of the nuclear protons. When Fermi fired neutrons at uranium, the nucleus with the largest known atomic number, he believed that he had created new elements, which he named with "eka-" prefixes. (In Greek, *eka* means "beyond.") For example, an artificially created element with atomic number 94, produced when a disturbed uranium nucleus emitted two beta particles, Fermi named eka-osmium—the osmium because such an element would be chemically similar to osmium (which has atomic number 76). As a group, elements with atomic numbers larger than 92 were called transuranic elements. Fermi, like all other scientists, took as a given that the new nuclei produced by bombarding uranium with neutrons would be close to uranium in mass.

In 1935, Hahn and Meitner took up Fermi's lead with the uranium experiments, collaborating again after a number of years of pursuing their individual research projects. Here, Hahn's superb skills as a chemist were essential, as the scientists wished to determine which new elements they had produced in their neutron bombardments. Ordinary chemical methods were useless because the amount of any new radioactive elements was tiny. To identify the new radioactive substances, Hahn and Meitner took advantage of the fact that when radioactive atoms of an element are dissolved in a solution with the same element in nonradioactive form (a different isotope), they will behave like the nonradioactive atoms in a chemical separation. For example, if iron is bombarded with neutrons and a new radioactive substance is produced, the likeliest candidates for the new substance might be chromium, manganese, and cobalt, all near iron in atomic number. Small amounts of these candidates could be added to a solution of nitric acid, along with the new radioactive substance, and individually precipitated out by the usual chemical means of combining them with other substances. If the unknown radioactive substance precipitated out with the manganese (as determined by testing the precipate with a Geiger counter, for example), then one could assume that it was an isotope of manganese.

Soon, Meitner and Hahn were joined by Fritz Strassmann, one of Hahn's former assistants, who had risked his career and even his life by refusing to join Nazi organizations. Strassmann had been blacklisted from most professional employment, but Meitner persuaded Hahn to hire him at half pay.

To their delight and consternation, the three scientists found a crowded zoo of new radioactive species, with many different periods and sequences of beta particle emissions. In one sequence, $^{239}_{92}\mathrm{U}$ ($^{238}_{92}\mathrm{U} + n$) disintegrated with a period of 10 seconds, in another with a period of 40 seconds, in still another 23 minutes. How could a single radioactive nucleus lead to so many different results? Also confusing was the fact that the reactions seemed to be more probable the *slower* (less energetic) the neutron projectiles, exactly opposite to expectations.

In March 1938, right at the height of these perplexing and provocative experiments, German troops occupied Austria. At this point, Meitner, unwilling to hide her Jewish ancestry and no longer protected by her Austrian citizenship, became an obvious target of the anti-Semitic laws of the Nazis. She would soon lose her job; furthermore, Heinrich Himmler, head of the secret police, had issued an order that no university teachers would be allowed to leave Germany. Meitner seemed to be trapped. But with the help of several scientists, including Hahn, Meitner managed to escape across the Dutch border on July 13. She had no passport, no job, and very little money. After a short stay in Holland, Meitner moved to Denmark, under the hospitality of her friend Niels Bohr and his wife Margrethe. Soon, at the invitation of the Swedish physicist Manne Siegbahn, she moved on to the Nobel Institute for Physics in Sweden.

Meanwhile, in late 1938, Hahn and Strassmann had obtained even more puzzling results. The two men believed that they had witnessed $^{239}_{92}\mathrm{U}$ decay into radium, $^{231}_{88}\mathrm{Ra}$, with the emission of two alpha particles:

$$^{238}_{92}\mathrm{U} + n \rightarrow {}^{239}_{92}\mathrm{U} \rightarrow \alpha + {}^{235}_{90}\mathrm{Th} \rightarrow \alpha + {}^{231}_{88}\mathrm{Ra}$$

At least the scientists had determined that a product of neutron bombardment of uranium apparently had the *chemical properties of radium*. Thus, they proposed the chain of reactions above. In fact, none of the intermediaries between uranium and radium had been observed. Thorium had not been detected. The alpha particles had not been detected. Instead, these intermediates were *inferred*, in a necessary sequence of steps to get from uranium to radium. And, from their chemical tests, the scientists thought they had produced radium.

To be absorbed by an atomic nucleus, an alpha particle must approach with such a high speed and energy that it can overcome the repulsive electrical force, called the Coulomb barrier, and become trapped in the nucleus by the attractive nuclear force. Like glue, the attractive nuclear force is powerful but can be felt only at very close range, so the alpha must actually touch and enter the nucleus in order to become "glued" in. Conversely, it would seem that a great deal of energy must be *added* to a nucleus to allow an alpha particle to pull free from the strong nuclear glue and escape the nucleus. And here was Hahn proposing that a single neutron, traveling at low speed, could impart enough energy to the uranium nucleus to liberate *two* alpha particles. It seemed impossible!

On November 13, Hahn met Meitner in Copenhagen to discuss the proposed double α results. Hahn had been invited to give a lecture at Bohr's institute, and Meitner took the train from Stockholm. It had been four months since the two scientists had last seen each other. According to Meitner's biographer Ruth Lewin Sime, the meeting between Meitner and Hahn was kept secret outside Copenhagen, because of the fear of jeopardizing Hahn's precarious situation in Germany. (In various ways, he had already challenged the Nazis.) Meitner, the physicist, told Hahn that his proposed reaction was extremely implausible on physical grounds and that he should return to Berlin and repeat his chemical tests for radium. As Strassmann later recounted: "[Meitner] urgently requested that these experiments be scrutinized very carefully and intensively one more time. Fortunately, L. Meitner's opinion and judgment carried so much weight with us in Berlin that the necessary control experiments were immediately undertaken."

What Meitner had asked Hahn to do was to make sure that it was indeed radium that had been produced by bombarding uranium with neutrons. To extract "radium" from the radioactive debris, Hahn and Strassmann had used the element barium. Because barium (Z = 56) and radium (Z = 88) have similar chemical properties, lying in the same column of the Periodic Table (see Chapter 11), radium would act like barium in a chemical solution and be precipitated in a barium compound (like barium sulfate or barium carbonate), forming compounds like radium sulfate or radium carbonate. In this way, the radium would be separated from other elements of different chemical properties.

Now, following Meitner's urgent request, Hahn and Strassmann attempted to further separate the "radium" from barium. To do this, they made use of the fact that radium bromide is less soluble than barium bromide. That is, in a liquid solution containing barium, radium, and bromide atoms, radium bromide will precipitate out more readily and in higher concentration than barium bromide. Thus, the solid, saltlike precipitate at the bottom of the container will have a higher ratio of radium to barium than in the original solution. In a second step, the precipitate at the bottom is redissolved in a second solution, and more bromide is added. Now, a second precipitate sinks to the bottom of the container. This second precipitate should have an even higher ratio of radium to barium than the first, since the second solution had a higher ratio of radium to barium than the first solution. The process is repeated. Each successive precipitation is called a new "fraction," with an increasing proportion of radium to barium in each successive fraction. The greatest proportion of radium to barium should be in the last precipitate, the least in the first.

To their surprise, Hahn and Strassmann found the same proportion of "radium" in all of the fractions, saying in their paper that "the activity [radioactive measure of the amount of 'radium'] was distributed evenly among all the barium fractions." Evidently, the "radium" bromide and barium bromide were identically soluble. Hahn had reasonably assumed that a remnant of the neutron bombardment with the chemical properties of barium must be radium, an element close to uranium. But now, he had evidence that the remnant, the so-called "radium," might be barium itself. A barium nucleus is roughly half the mass of a uranium nucleus. How could that be?

Hahn was a superb chemist, probably at that time the best radiochemist in the world, and he was confident in his chemical experiments. Yet he had now discovered something impossible for him to understand on physical grounds. His trepidation is clear in the language he uses to introduce his new experiments, "which we publish rather hesitantly due to their peculiar results." Some scientists might not have published such revolutionary and startling results, afraid of making fools of themselves. Hahn is well aware that if he has indeed found barium in the remnants of the neutron bombardment—a result he cannot quite acknowledge despite the evidence of his own careful experiments—then he has split the uranium nucleus in two. That awareness is underscored in the next-to-last paragraph of the paper, when he suggests a new experiment to find technetium. Technetium would be the other fission fragment partnered with barium, since the mass number of the likeliest isotope of technetium, 101, combined with the mass number of the most common isotope of barium, 138, adds up to the mass number of the original uranium, 239. (Hahn's "mass number" is what we have called atomic mass before.)

Hahn ends his paper with a statement that is both anguished and modest: "we cannot bring ourselves yet to take such a drastic step [the definite identification of barium in the remnants] which goes against all previous experience in nuclear physics. There could perhaps be a series of unusual coincidences which has given us false indications."

As Meitner and her nephew Otto Frisch walked in the snow on Christmas Day 1938, pondering Hahn's "fantastic" results, they recalled a beautiful metaphor proposed by Niels Bohr. The legendary Danish physicist had compared an atomic nucleus to a drop of liquid. Like the molecules within a drop of water, the protons and neutrons inside an atomic nucleus would share energy so rapidly that they would move together as a fluid rather than as individual particles. The mobile outer surface of an atomic nucleus would be held and shaped by the cohesive nuclear force, much as the molecular surface tension force holds together an actual drop of liquid. However, as Meitner and Frisch realized, the enormous repulsive force that existed between 92 protons in a uranium nucleus would greatly weaken the "surface tension" and allow the nucleus to change shape easily. As they say in their paper, "the uranium nucleus has only a small stability of form."

The basic physical idea is this: recall that there are two competing forces in the atomic nucleus. Because the attractive nuclear force is *short range*, only nuclear particles very close to one other are held together. By contrast, the repulsive electrical force is *long range* and can be felt between protons far apart. When the nucleus is spherical, these two opposing forces are balanced. But when a uranium nucleus is slightly deformed, the nuclear particles, on average, become farther apart. That increase in separation significantly weakens the total pull of the short-range attractive force but barely alters the total effect of the long-range repulsive force, thus giving an upper hand to the latter. Once dominant, the repulsive force pushes the particles of the nucleus even farther apart, the nucleus becomes even more elongated, making the repulsive force even more dominant over the attractive force, and so on, until the nucleus divides into two pieces. The evolution is illustrated in Figure 15.1, where the sequence of steps goes from A to E. After the last step, E, the two fission halves go flying apart, propelled by the repulsive electrical force between them.

The incident neutron does not need to supply much energy—the energy for fission is already available in the huge repulsive electrical forces temporarily held at bay. The initial neutron needs only to agitate the nucleus slightly, to deform it slightly. (In a fascinating footnote to history, fission had actually been proposed several years earlier, in 1934, by the German chemist Ida Noddack. In an article published obscurely and dismissed as theoretically implausible, Noddack suggested that the likeliest result of bombarding heavy nuclei with neutrons was their splitting into several large fragments rather than slightly changing their masses, as Fermi and others believed.)

The calculation of the energy release, from the known size of the uranium nucleus and the amount of electrical charge it contains, involves only pre-quantum science and can be done by any college freshman in physics. In their paper Meitner and Frisch discuss the available energy also in terms of the "packing fraction," which is related to the average energy with which each proton or neutron is bound together in the nucleus. The packing fraction varies with the numbers of protons and neutrons, as the relative contributions of the energies due to the attractive and repulsive forces vary.

Another technical term that Meitner and Frisch use is "isomer." An isomer is an atomic nucleus that has been excited to a higher internal

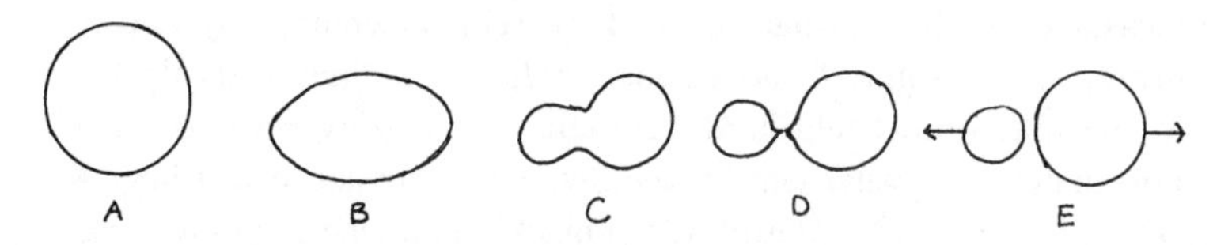

Figure 15.1

energy. Such an excited nucleus usually rids itself of the excess energy by the emission of high-energy photons. Two atomic nuclei with the same numbers of neutrons and protons may have different internal energy levels and thus be different isomers.

In the seventh paragraph of their brief paper, Meitner and Frisch use for the first time in history the word "fission" to describe the splitting of the atom. After speaking to an American biologist, Frisch had coined the usage by analogy to the process in which a living cell divides in two.

Meitner and Frisch go on to discuss the likely disintegrations that would follow the fission of uranium into barium and krypton. They also bravely suggest that many of the earlier claims of Meitner, Hahn, Strassmann, and Fermi may have been erroneous. In particular, Fermi's "eka" elements, with Z larger than 92, may in fact have atomic numbers much smaller than 92. Perhaps all of their experiments of the last few years, bombarding uranium with neutrons, have in fact fissioned uranium, producing much lighter and more common elements than the many exotic new nuclear species they had proposed. Since different numbers of neutrons might be expected to cleave to the two fission fragments after each bombardment, the many different isotopes of these lighter elements could masquerade as many new nuclear species and be misidentified as eka elements. (Meitner and Frisch's suspicions were correct. Although Enrico Fermi made a number of discoveries worthy of the Nobel Prize, he did not "discover new radioactive elements beyond uranium," as the citation read for his Nobel Prize of 1938. That Nobel was awarded in error.)

After Meitner and Frisch's revelatory walk in the woods, Frisch returned to Copenhagen to tell Bohr about the fission idea, just as the latter was boarding a ship to America. According to Frisch's recollections, the great atomic physicist immediately slapped his forehead and said, "Oh, what fools we have been! We ought to have seen that before."

Frisch went to his own laboratory and repeated Hahn's experiment, measured the fission fragments as they went careening away, and confirmed that they indeed had the energies he and Meitner had calculated. Meanwhile, Bohr could hardly contain the amazing news to himself. Fortunately, his stateroom on the Swedish-American ocean liner MS *Drottningholm* was equipped with a blackboard. One can only imagine Bohr's excitement as he outlined the discovery to Leon Rosenfeld, a young physicist on board with him. When the ship docked in New York on January 16, 1939, Bohr was met by Enrico Fermi and John Wheeler and told them the news. That night, Rosenfeld electrified a small seminar at Princeton with the discoveries. Ten days later, Bohr and Fermi announced the results at the Fifth Washington (D.C.) Conference on Theoretical Physics, taking care to give credit to the four original scientists. Everyone rushed back to their laboratories to repeat Hahn's experiment. During February 1939, at least fifteen new papers on fission were submitted to *Nature* in Great Britain, to *Physical Review* in the United States, and to *Comptes Rendus* in France. By the end of 1939, more than a hundred papers on fission had been published, from scientists all over the world. The atomic nucleus had been split, with the promise of staggering amounts of energy. The nuclear age had begun.

Almost like characters in an ancient Greek tragedy, Meitner and Hahn seemed destined for different fates. For many years, they were good friends as well as close colleagues. At the same time, world events and their differences in character eventually drove them apart. Meitner was politically liberal. Hahn fervently supported German nationalism and power, although he opposed the Nazis and defied them by supporting disenfranchised scientists, including Meitner. (In this regard, there were some similarities with von Laue and Heisenberg, whom we have encountered in earlier chapters.) Hahn remained in Berlin and became a leading figure in postwar Germany. In the words of colleague Rod Spence, who wrote an appreciation of the German chemist for the *Biographical Memoirs* of the British Royal Society, Hahn was "trusted for his human qualities, simplicity of manner, transparent honesty, common sense, and loyalty." Meitner was forced to flee Germany, with little more than the clothes on her back, banished to a country whose language she didn't speak, and never able to fully establish herself in research again.

The discovery of fission created another fork in the road. Hahn's successful experiments with Strassmann to find barium were a direct result of Meitner's collaboration both before and after her exile to Sweden. Yet it was politically impossible for Hahn to place Meitner's name on his discovery paper with Strassmann. (He did refer to her in the acknowledgments in a follow-up paper four weeks later.) Meitner was devastated. In February 1939, she wrote to her brother Walter: "I have no self confidence . . . Hahn has just published absolutely wonderful things based on our work together . . . much as these results make me happy for Hahn, both personally and scientifically, many people here must think I contributed absolutely nothing to it—and now I am so discouraged."

In 1945, in a decision that was much contested, Hahn alone was awarded the 1944 Nobel Prize for the discovery of fission. His prize was in the chemistry division. The Nobel Prize in physics that year went to Wolfgang Pauli, for his discovery of the exclusion principle in quantum mechanics (see Chapter 11). In fact, a number of scientists felt that the prize in physics should have gone to Meitner, for her role in the discovery of fission. Meitner herself felt slighted. In a letter to Birgit Broomé Aminoff, a scientist and wife of a board member of the Nobel Foundation, she wrote:

> Surely Hahn fully deserved the Nobel Prize in chemistry. There is really no doubt about it. But I believe that Frisch and I contributed something not insignificant to the clarification of the process of uranium fission—how it originates and that it produces so much energy, and that was something very remote from Hahn. For this reason I found it a bit unjust that in the newspapers I was called a *Mitarbeiterin* [junior associate] of Hahn's in the same sense that Strassmann was.

The issues here are complex. If Meitner had been able to remain in Germany in good standing, she certainly would have been a coauthor with Hahn and Strassmann on the new fission work. Even from a distance, she had played a crucial role in Hahn's careful reanalysis of his experiment and conclusion that he had indeed split uranium into barium. However, the conclusive determination was a *chemical* accomplishment, and Hahn achieved that accomplishment alone. As it turned out, Meitner's most striking contribution to the scientific drama was the *interpretation* of Hahn's results, the understanding in terms of physics. Thus, one might say that aside from their differences as a chemist and a physicist, Hahn *discovered* nuclear fission while Meitner *interpreted* that discovery. And, rightly or wrongly, the Nobel committee in Stockholm has a long history of preferring experimental discovery over theoretical interpretation. (A similar story involves Dicke versus Penzias and Wilson in the discovery of the cosmic background radiation. See Chapter 19.) For his part, Hahn wished to reserve the discovery of fission as a triumph of chemistry. The physicists, Hahn repeated over and over for years, had declared fission "impossible."

After winning the Nobel, Otto Hahn joined Werner Heisenberg as a leading statesman of science in Germany and was appointed president of the Kaiser Wilhelm Gesellschaft in Göttingen. Meitner also received honors, although none equal to the Nobel. In 1946 she was proclaimed Woman of the Year by the American Women's National Press Club. In 1947 she received the Vienna Prize for Science and Art, in 1949 the Max Planck Medal (shared with Hahn), in 1962 a medal from Göttingen, and numerous honorary degrees. In 1947 she was given her own lab at the Royal Institute of Technology in Sweden but never her once-promised professorship there. Sweden would be her uneasy home until her retirement to England in 1960. In 1966, at the age of eighty-eight, she shared the Enrico Fermi Prize with Hahn and Strassmann. For years, Hahn and Meitner wrote congratulatory articles on each other's major birthdays.

The creation of the atomic bomb will, of course, haunt all discussion of nuclear fission forever. On August 6, 1945, Otto Hahn was informed that an atomic bomb had been built by the Allies and just dropped on Hiroshima. Hahn and other German nuclear scientists were then in captivity at Farm Hall, an English country house near Cambridge. In the words of British Major T. H. Rittner, "Hahn was completely shattered by the news and said he felt personally responsible for the deaths of hundreds of thousands of people, as it was his original discovery which had made the bomb possible. He told me that he had originally contemplated suicide when he realized the terrible potentialities of his discovery." About fifteen years later, in a personal letter to a colleague, Lise Meitner gave her own assessment of her work in nuclear physics, in the years before the war: "[at that time] one could love one's work and not always be tormented by the fear of the ghastly and malevolent things that people might do with beautiful scientific findings."

CONCERNING THE EXISTENCE OF ALKALINE EARTH METALS RESULTING FROM NEUTRON IRRADIATION OF URANIUM

Otto Hahn and Fritz Strassmann

Die Naturwissenschaften (1939)

IN A RECENT PRELIMINARY ARTICLE in this journal[1] it was reported that when uranium is irradiated by neutrons, there are several new radioisotopes produced, other than the transuranic elements—from 93 to 96—previously described by Meitner, Hahn, and Strassmann. These new radioactive products are apparently due to the decay of ^{239}U by the successive emission of two alpha particles. By this process the element with a nuclear charge of 92 must decay to a nuclear charge of 88; that is, to radium. In the previously mentioned article a tentative decay scheme was proposed. The three radium isotopes, with their approximate half-lives given, decayed to actinium, which in turn decayed to thorium isotopes.

A rather unexpected observation was pointed out, namely that these radium isotopes, which are produced by alpha emission and which in turn decay to thorium, are obtained not only with fast but also with slow neutrons.

The evidence that these three new parent isomers are actually radium was that they can be separated together with barium salts, and that they have all the chemical reactions which are characteristic of the element barium. All the other known elements, from the transuranic ones down through uranium, protactinium, thorium, and actinium have different chemical properties than barium and are easily separated from it. The same thing holds true for the elements below radium, that is, bismuth, lead, polonium, and ekacesium (now called francium). Therefore, radium is the only possibility, if one eliminates barium itself. . . .

When uranium is bombarded with slow neutrons, it is not easy to understand from energy considerations how radium isotopes can be produced. Therefore, a very careful determination of the chemical properties of the new artificially made radioelements was necessary. Various analytic groups of elements were separated from a solution containing the irradiated uranium. Besides the large group of transuranic elements, some radioactivity was always found in the alkaline-earth group (barium carrier), the rare-earth group (lanthanum carrier), and also with elements in group IV of the periodic table (zirconium carrier). The barium precipitate was the first to be investigated more thoroughly, since it apparently contains the parent isotopes of the observed isomeric series. The goal was to show that the transuranic elements, and also U, Pa, Th, and Ac, could always be separated easily and completely from the activity which precipitates with barium. . . .

To summarize our results, we have identified three alkaline earth metals which are designated as Ra II, Ra III, and Ra IV. Their half-lives are 14 ± 2 min, 86 ± 6 min, and 250–300 h. It should be noted that the 14-min activity was not designated as Ra I nor the other isomers as Ra II and Ra III. The reason is that we believe there is an even more unstable "Ra" isotope, although it has not been possible to observe it so far. . . .

The decay scheme which was given in our previous article must now be corrected. The following scheme takes into account the needed changes, and also gives the more accurately determined half-lives for the parent of each series:

$$\text{“Ra I”?} \xrightarrow[<1\text{ min}]{\beta} \text{Ac I} \xrightarrow[<30\text{ min}]{\beta} \text{Th?}$$

$$\text{“Ra II”} \xrightarrow[14 \pm 2\text{ min}]{\beta} \text{Ac II} \xrightarrow[2.5\text{ h}]{\beta} \text{Th?}$$

$$\text{“Ra III”} \xrightarrow[86 \pm 6\text{ min}]{\beta} \text{Ac III} \xrightarrow[\text{several days?}]{\beta} \text{Th?}$$

$$\text{“Ra IV”} \xrightarrow[250\text{–}300\text{ h}]{\beta} \text{Ac IV} \xrightarrow[<40\text{ h}]{\beta} \text{Th?}$$

The large group of transuranic elements so far bears no known relation to these isomeric series.

The four decay series listed above can be regarded as doubtlessly correct in their genetic relationship. We have already been able to verify some of the "thorium" end products of the isomeric series. However, since the half-lives have not been determined with any accuracy yet, we have decided to refrain altogether from reporting them at the present time.

Now we still have to discuss some newer experiments, which we publish rather hesitantly due to their peculiar results. We wanted to identify beyond any doubt the chemical properties of the parent members of the radioactive series which were separated with the barium and which have been designated as "radium isotopes." We have carried out fractional crystallizations and fractional precipitations, a method which is well-known for concentrating (or diluting) radium in barium salt solutions.

Barium bromide increases the radium concentration greatly in a fractional crystallization process and barium chromate even more so when the crystals are allowed to form slowly. Barium chloride increases the concentration less than the bromide, and barium carbonate decreases it slightly. When we made appropriate tests with radioactive barium samples which were free of any later decay products, the results were always negative. The activity was distributed evenly among all the barium fractions, at least to the extent that we could determine it within an appreciable experimental error. . . .

Next the "indicator (i.e., tracer) method" was applied to a mixture of purified long-lived "Ra IV" and pure $MsTh_1$; this mixture with barium bromide as a carrier was subjected to fractional crystallization. The concentration of $MsTh_1$ was increased, and the concentration of "Ra IV" was not, but rather its activity remained the same for fractions having an equivalent barium content. We come to the conclusion that our "radium isotopes" have the properties of barium. As chemists we should actually state that the new products are not radium, but rather barium itself. Other elements besides radium or barium are out of the question.

Finally we have made a tracer experiment with our pure separated "Ac II" (half-life about 2.5 h) and the pure actinium isotope $MsTh_2$. If our "Ra isotopes" are not radium, then the "Ac isotopes" are not actinium either, but rather should be lanthanum. Using the technique of Curie,[2] we carried out a fractionation of lanthanum oxalate, which contained both of the active substances, in a nitric acid solution. Just as Mme. Curie reported, the $MsTh_2$ became greatly concentrated in the end fractions. With our "Ac II" there was no observable increase in concentration at the end. We agree with the findings of Curie and Savitch[3] for their 3.5-h activity (which was however not just a single species) that the product resulting from the beta decay of our radioactive alkaline earth metal is not actinium. . . .

The "transuranic group" of elements are chemically related but not identical to their lower homologs, rhenium, osmium, iridium, and platinum. Experiments have not been made yet to see if they might be chemically identical with the even lower homologs, technetium, ruthenium, rhodium, and palladium. After all one could not even consider this as a possibility earlier. The sum of the mass numbers of barium + technetium, 138 + 101, gives 239!

As chemists we really ought to revise the decay scheme given above and insert the symbols Ba, La, Ce, in place of Ra, Ac, Th. However, as "nuclear chemists," working very close to the field of physics, we cannot bring ourselves yet to take such a drastic step which goes against all previous experience in nuclear physics. There could perhaps be a series of unusual coincidences which has given us false indications.

It is intended to carry out further tracer experiments with the new radioactive decay products.

1. O. Hahn and F. Strassmann, *Naturwiss.* **26,** 756 (1938).
2. Mme. Pierre Curie, *J. Chim. Phys.* **27,** 1 (1930).
3. I. Curie and P. Savitch, *Compt. Rend.* **206,** 1643 (1938).

DISINTEGRATION OF URANIUM BY NEUTRONS: A NEW TYPE OF NUCLEAR REACTION

Lise Meitner and Otto Frisch

Nature (1939)

ON BOMBARDING URANIUM with neutrons, Fermi and collaborators[1] found that at least four radioactive substances were produced, to two of which atomic numbers larger than 92 were ascribed. Further investigations[2] demonstrated the existence of at least nine radioactive periods, six of which were assigned to elements beyond uranium, and nuclear isomerism had to be assumed in order to account for their chemical behaviour together with their genetic relations.

In making chemical assignments, it was always assumed that these radioactive bodies had atomic numbers near that of the element bombarded, since only particles with one or two charges were known to be emitted from nuclei. A body, for example, with similar properties to those of osmium was assumed to be eka-osmium ($Z = 94$) rather than osmium ($Z = 76$) or ruthenium ($Z = 44$).

Following up an observation of Curie and Savitch,[3] Hahn and Strassmann[4] found that a group of at least three radioactive bodies, formed from uranium under neutron bombardment, were chemically similar to barium and, therefore, presumably isotopic with radium. Further investigation,[5] however, showed that it was impossible to separate these bodies from barium (although mesothorium, an isotope of radium, was readily separated in the same experiment), so that Hahn and Strassmann were forced to conclude that *isotopes of barium* ($Z = 56$) *are formed as a consequence of the bombardment of uranium* ($Z = 92$) *with neutrons.*

At first sight, this result seems very hard to understand. The formation of elements much below uranium has been considered before, but was always rejected for physical reasons, so long as the chemical evidence was not entirely clear cut. The emission, within a short time, of a large number of charged particles may be regarded as excluded by the small penetrability of the "Coulomb barrier," indicated by Gamov's theory of alpha decay.

On the basis, however, of present ideas about the behaviour of heavy nuclei,[6] an entirely different and essentially classical picture of these new disintegration processes suggests itself. On account of their close packing and strong energy exchange, the particles in a heavy nucleus would be expected to move in a collective way which has some resemblance to the movement of a liquid drop. If the movement is made sufficiently violent by adding energy, such a drop may divide itself into two smaller drops.

In the discussion of the energies involved in the deformation of nuclei, the concept of surface tension of nuclear matter has been used[7] and its value has been estimated from simple considerations regarding nuclear forces. It must be remembered, however, that the surface tension of a charged droplet is diminished by its charge, and a rough estimate shows that the surface tension of nuclei, decreasing with increasing nuclear charge, may become zero for atomic numbers of the order of 100.

It seems therefore possible that the uranium nucleus has only small stability of form, and may, after neutron capture, divide itself into two nuclei of roughly equal size (the precise ratio of sizes depending on finer structural features and perhaps partly on chance). These two nuclei will repel each other and should gain a total kinetic energy of *c.* 200 Mev., as calculated from nuclear radius and charge. This amount of energy may actually be expected to be available from the difference in packing fraction between uranium and the elements in the middle of the periodic system. The whole 'fission' process can thus be described in an essentially classical way, without having to consider quantum-mechanical "tunnel effects," which would actually be extremely small, on account of the large masses involved.

After division, the high neutron/proton ratio of uranium will tend to readjust itself by beta decay to the lower value suitable for lighter elements. Probably each part will thus give rise to a chain of disintegrations. If one of the parts is an isotope of barium,[5] the other will be krypton ($Z = 92 - 56$), which might decay through rubidium, strontium and yttrium to zirconium. Perhaps one or two of the supposed barium-lanthanum-cerium chains are then actually strontium-yttrium-zirconium chains.

It is possible,[5] and seems to us rather probable, that the periods which have been ascribed to elements beyond uranium are also due to light elements. From the chemical evidence, the two short periods (10 sec. and 40 sec.) so far ascribed to ^{239}U might be masurium isotopes ($Z = 43$) decaying through ruthenium, rhodium, palladium and silver into cadmium.

In all these cases it might not be necessary to assume nuclear isomerism; but the different radioactive periods belonging to the same chemical element may then be attributed to different isotopes of this element, since varying proportions of neutrons may be given to the two parts of the uranium nucleus.

By bombarding thorium with neutrons, activities are obtained which have been ascribed to radium and actinium isotopes.[8] Some of these periods are approximately equal to periods of barium and lanthanum isotopes[5] resulting from the bombardment of uranium. We should therefore like to suggest that these periods are due to a 'fission' of thorium which is like that of uranium and results partly in the same products. Of course, it would be especially interesting if one could obtain one of these products from a light element, for example, by means of neutron capture.

It might be mentioned that the body with half-life 24 min.[2] which was chemically identified with uranium is probably really ^{239}U, and goes over into an eka-rhenium which appears inactive but may decay slowly, probably with emission of alpha particles. (From inspection of the natural radioactive elements, ^{239}U cannot be expected to give more than one or two beta decays; the long chain of observed decays has always puzzled us.) The formation of this body is a typical resonance process[9]; the compound state must have a life-time a million times longer than the time it would take the nucleus to divide itself. Perhaps this state corresponds to some highly symmetrical type of motion of nuclear matter which does not favour 'fission' of the nucleus.

Lise Meitner

Physical Institute, Academy of Sciences, Stockholm.

O. R. Frisch

Institute of Theoretical Physics, University, Copenhagen.

January 16.

1. Fermi, E., Amaldi, F., d'Agostino, O., Rasetti, F., and Segrè, E. *Proc. Roy. Soc.*, A, **146,** 483 (1934).

2. See Meitner, L., Hahn, O., and Strassmann, F., *Z. Phys.*, **106,** 249 (1937).

3. Curie, I., and Savitch, P., *C.R.*, **206,** 906, 1643 (1938).

4. Hahn, O., and Strassmann, F., *Naturwiss.*, **26,** 756 (1938).

5. Hahn, O., and Strassmann, F., *Naturwiss.*, **27,** 11 (1939).

6. Bohr, N., NATURE, **137,** 344, 351 (1936).

7. Bohr, N., and Kalckar, F., *Kgl. Danske Vid. Selskab, Math. Phys. Medd.*, **14,** Nr. 10 (1937).

8. See Meitner, L., Strassmann, F., and Hahn, O., *Z. Phys.*, **109,** 538 (1938).

9. Bethe, A. H., and Placzek, G., *Phys. Rev.*, **51,** 450 (1937).

16

THE MOVABILITY OF GENES

On May 13, 1947, Barbara McClintock rose early to finish the spring planting of her corn, an annual ritual for the past twenty years. "I was so interested in what I was doing I could hardly wait to get up in the morning and get at it," she once said. Now, from her lumpy acre of land, she could look in one direction and see her laboratory window at the Carnegie Institution's Department of Genetics. In the other direction, she could gaze out to the water and beyond to the little village of Cold Spring Harbor on Long Island. Most likely, she would have been wearing baggy pants and a short-sleeved white shirt. Forty-four years old, with short tousled hair, wire-rimmed glasses, and an impish smile, she cut a slight figure, less than a hundred pounds altogether. Who could have guessed from her looks and demeanor that she was one of the greatest biologists in the world, the foremost authority on the genetics of maize, only the third woman to have been elected to the National Academy of Sciences, the first female president of the Genetics Society of America? A hint of such power might be glimpsed in her eyes, which radiated a piercing and fearless intelligence. But her most important discovery was in progress.

Alone, she walked the rows of her seedings. Each of her several hundred plants she knew personally. She knew the parents she had mated to produce the plant, she knew the genetic makeup of the parents, she knew their chromosomes under the microscope. A wooden paddle stuck in the ground by each seed identified its heritage. Over the course of the growing season, from May to October, she would come every day to this field to water her plants, to nourish them, and to peer at their patterns of colors and stripes and waxiness. Indeed, she was famous among her colleagues for her extraordinary powers of observation. When the time came to mate the new generation, she would take all the precautions of the most meticulous matchmaker, to ensure that the right pollen fertilized the right eggs.

Corn is peculiar in its two-step life cycle. In the first step, a seed in the ground creates a plant that has both male and female parts, the male pollen within tassels at the top of the plant, the female eggs at the base of the sticky "silks" that emerge from special leaves on the stalk. In the second part of the cycle, pollen from the original plant or from any nearby plant fertilizes the eggs. A shift in the wind might impregnate neighboring eggs with different pollens. And each egg becomes a single kernel of corn. Thus different kernels on the same cob may have different parents.

To get the crossings, or matings, you want, you cover the shoots carrying the eggs with pleated glassine bags, protecting them from wayward pollen. Within the plastic, the shoots continue to grow. You put brown sacks around the tassels at the tops of the plants, to keep the pollen from spreading in the wind. When you're ready to make a match, you carefully collect the powdery, saffron-colored pollen from the plant you want and sprinkle it on the appropriate silk. It is a tedious, exacting procedure.

McClintock did all the fieldwork by herself, even the routine tasks. She didn't trust anyone with her plants, and she didn't tolerate fools. As she once wrote to a colleague, the Department of Genetics "had a greenhouse man but he is not too bright."

Indeed, since a young age, McClintock had been both prickly and proud, fiercely independent, and later brilliant. She was born Eleanor McClintock in June 1902, in Hartford, Connecticut, the daughter of Henry McClintock, a physician, and Sara Handy, a Boston Brahmin. According to McClintock's accounts, her parents decided that Eleanor was too "delicate" and "feminine" a name for her temperament and changed it to Barbara. As a child, she recalled, she "used to love to be alone . . . just thinking about things." During McClintock's freshman year at Cornell, she refused to join a sorority when she realized that some people were invited and some not. As she said to Evelyn Fox Keller, who conducted extensive interviews with her in the late 1970s, "Here was a dividing line that put you in one category or another. And I couldn't take it." A year later, at the age of eighteen, she became a standout maverick on campus overnight by having her hair cut short.

In her junior year, McClintock took a genetics course and loved it. She stayed on at Cornell as a graduate student, in a program created by Rollins A. Emerson, the most famous maize (corn) geneticist of the day.

After receiving her Ph.D. in botany in 1927, at the age of twenty-four, McClintock continued her research at Cornell. By this time, she knew that genetics was her passion, her self-fulfillment, her life. She had friends, but she kept them at a distance. Years later, she told Keller, "There was not that strong necessity for a personal attachment to anybody. I just didn't feel it. And I could never understand marriage."

By 1934, the string of research positions at Cornell had dried up. McClintock, now a world-renowned geneticist, had no financial support and no job offers. She grew bitter, blaming much of her trouble on being a woman in a man's world. "At this stage," she recalled, "in the mid thirties, a career for women did not receive very much approbation. You were stigmatizing yourself by being a spinster and a career woman, especially in science." In fact, the professional opportunities for women in science in America did not much improve until after World War II, and then only slowly. Eventually, McClintock received employment at the University of Missouri, and then, in 1942, a research position at Cold Spring Harbor.

As was common for cell geneticists but uncommon for most other biologists, McClintock performed both the breeding work in the field and also the laboratory work, studying the microscopic chromosomes that contained the genes. Before her work at Cornell in the late 1920s and early 1930s, the principal organism for such studies was *Drosophila*, the fruit fly, useful because of its rapid life cycle. Maize, on the other hand, had the advantages of highly visible genetic traits, such as kernel colors and leaf markings, and its chromosomes were larger than *Drosophila*'s and thus more easily studied under the microscope. (Eventually, large chromosomes in the salivary gland cell of *Drosophila* were found.) Refining new cell-staining methods, McClintock was the first person to identify and characterize the ten different chromosomes of maize. By the early 1930s, she had made maize, or *Zea mays*, of equal importance to geneticists as *Drosophila*.

McClintock was especially curious about the way in which maize chromosomes could break in particular places and then rejoin, in the process introducing mutations and changes in genetic traits. By careful breeding, she found that she could produce plants whose chromosomes regularly went through these "breakage-fusion" mutations. (See Figure 16.4 for an illustration.)

One particular plant named B-87, harvested in 1944 in her fields at Cold Spring Harbor, tipped McClintock off that something unusual was happening in these particular mutations. For example, she found that in some of these plants, single kernels of corn were speckled in color. The phenomenon, found also in markings on leafs, was called variegation. A nascent kernel begins with a single cell, which then divides into two cells, each of which divides into two cells, and so on until the full kernel is created, each cell supposedly passing down an exact copy of the original cell's genes. Exact copies of genes from one cell to the next should produce a kernel with a uniform color. Speckled kernels, with patches of purple, yellow, and red, meant that *a pigment-governing gene was turning on and off at various points in the series of cell divisions.* On, and a patch of purple would begin growing; off, and the purple would give way to the background yellow. Such an on-off gene was called a "mutable gene," or an "unstable gene." Mutations could occur in the standard theory of inheritance, of course, but they were thought to be permanent rather than transient, and they were also thought to be random.

Astonishingly, these mutations were not random. With her keen eye, McClintock noticed that the size of the interloping spots of color and the relative numbers of spots of each size were *constant* throughout the life cycle of the plant. The size of the spots indicated how early in the cell division process the mutation had occurred, with larger spots having come from earlier (and thus longer-acting) mutations. The relative number of spots indicated the rate of the mutation. Randomly occurring mutations would not have produced such features with any constancy and uniformity.

Evidently, these mutations were being controlled. Something was altering the genes on the maize chromosome in a regular and systematic way. That idea, already, was a revolution. Until then, biologists had viewed genes as fixed links on the chain of the chromosome, unchanging except for random mutations, sending information and instructions in a one-way path from the chromosome to the rest of the organism.

McClintock spent the next several years trying to find out what was controlling the orderly variegations in corn. During this period, she became obsessed with the problem. She worked day and night on it, almost always alone, sometimes sleeping the fitful nights on a cot in her lab. She had always loved puzzles. As she once said of her high school science classes in Brooklyn, "I would solve some of the problems in ways that weren't the answers the instructor expected . . . It was a tremendous joy, the whole process of finding that answer, just pure joy."

Now, on the late spring morning of May 13, 1947, as she stood among her new plantings, McClintock felt that she was close to the answer of what controlled the regular mutations in corn. She collected

Linus Pauling.
California Institute of Technology, courtesy American Institute of Physics Emilio Segrè Visual Archives, Segrè Collection

Edwin Hubble.
Hale Observatories, courtesy American Institute of Physics Emilio Segrè Visual Archives

Alexander Fleming, 1932.
With permission of the Alexander Fleming Laboratory Museum (St. Mary's NHS Trust)

Hans Krebs.
Courtesy University of Sheffield

Lise Meitner (right) and Otto Hahn (left) in Hahn's laboratory in the basement of the Chemical Institute of the University of Berlin, 1909.
Courtesy Archiv zur Geschichte der Max Planck-Gesellschaft, Berlin-Dahlem

Rosalind Franklin serving coffee in crucibles in Jacques Mering's laboratory in Paris, ca. 1948. *Photograph by Vittorio Luzzati, reprinted with permission from* Physics Today*, 56(3), March 2003, p. 44. Copyright © 2003 American Institute of Physics. Courtesy Lynne Osman Elkin*

Barbara McClintock (right) with L. C. Dunn, ca. 1940.
Courtesy American Philosophical Society

Francis Crick (left) and James Watson (right)
during a walk along the college backs in Cambridge, ca. 1952.
Courtesy Cold Spring Harbor Laboratory Archives

Max Perutz with his model of hemoglobin.
Courtesy Medical Research Council, London, England

Robert Dicke in the early 1950s.
Courtesy American Institute of Physics Emilio Segrè Visual Archives

Steven Weinberg at Cambridge University, 1975.
Courtesy Steven Weinberg

Arno Penzias (right) and Robert Wilson (left) standing in front of the Crawford radio telescope in Holmdel, New Jersey.
Lucent Technologies' Bell Laboratories, courtesy American Institute of Physics Emilio Segrè Visual Archives, Physics Today *Collection*

Jerome Friedman at a control panel of the Stanford Linear Accelerator, late 1960s.
Courtesy Jerome Friedman

Paul Berg in his laboratory at Stanford University.
Courtesy Stanford University

her four-by-six-inch index cards, on which she would write the pedigrees of each plant, and returned to her lab. In a few months she would look at her new offsprings' chromosomes under her microscope. And there she would make the greatest discovery of her life.

Evelyn Witkin, then a young researcher at the Carnegie Department of Genetics, remembers that one day in March or early April 1948 McClintock telephoned her. "[McClintock] was just beside herself with excitement, and almost incoherent, she was talking so fast. She had drawn the conclusion that this thing was moving around." "This thing" was a genetic element on the chromosome. McClintock had found that *genetic elements actually changed positions on the chromosomes*, rearranging themselves in a controlled way. Those changes in position, in turn, were causing the variegations.

No longer could one think of genes as fixed links on a chain, or of the chromosome as a static warehouse of instructions. *The chromosome, and the genes on it, were a dynamic system, changing during a single lifetime, both controlling and being controlled by the rest of the organism*. McClintock fathomed some of these ideas, but not all, at the time. Even today, biologists don't understand the details of how information from the developing organism is relayed back to the chromosomes.

The modern science of genetics began in the late 1850s, with the work of the Austrian monk and botanist Gregor Mendel. By examining the visible traits of garden peas through many generations, Mendel arrived at the idea that each trait was controlled by a *pair* of individual factors, later called genes, one inherited from each parent. Each factor could come in several varieties, for example blue eyes or brown eyes. When the inherited pair of factors for a trait were of two different varieties, one factor dominated and was called "dominant," while the other was called "recessive."

Figure 16.1 illustrates a typical Mendelian experiment and its analysis. Here we consider only a single trait, the color of the flower petals, and only the genes that govern that one trait. Suppose the flower color of a plant may be red or white. Begin with two groups of "purebred" plants, one group that has shown only red flowers through many generations and the other that has shown only white. When a purebred red is crossed with a purebred white, the offspring are not a blended shade of pink, but all red. Evidently, the red factor completely dominates the white factor, as if it acted alone. The situation is shown in diagram A,

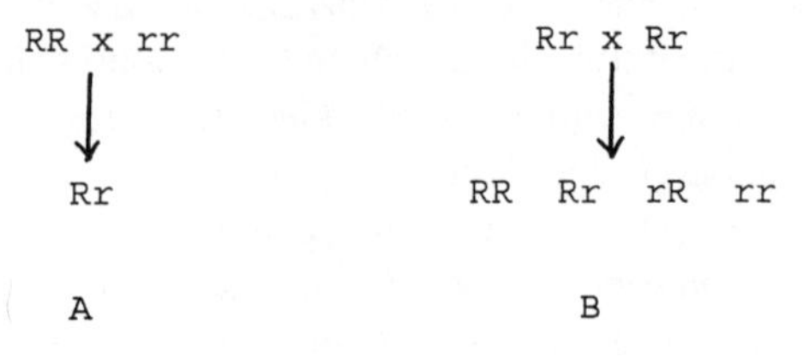

Figure 16.1

where *R* denotes the dominant (red) form of the gene and *r* the recessive (white) form of the gene. Each purebred red plant evidently has two red genes (*RR*) while each purebred white has two whites (*rr*). Each offspring, receiving one gene from each parent, must be *Rr*. As red dominates white, these plants are all red. Diagram B shows the results of incestuously mating the *Rr* offspring with each other. There are four possibilities for the gene combinations of the children, each equally likely. Thus, in many such matings, three-quarters of these children are observed to have red flowers (with at least one red gene) and one-quarter to have white flowers (with two white genes). From such experiments Mendel arrived at his ideas.

By 1890, the German zoologist Theodor Boveri, using a microscope to study the changes that cells undergo upon division and reproduction, hypothesized that Mendel's factors of inheritance were located on chromosomes, the sausagelike bodies seen in the nucleus of each cell of each living organism. A typical chromosome is about a thousandth of a centimeter long, or a few times smaller than what can be seen with the naked eye.

In the period 1910–1915, the American biologist Thomas Hunt Morgan confirmed Boveri's suggestion and showed that the genes are laid out in a linear way on the chromosome. Morgan's student, A. H. Sturtevant, only an undergraduate, first mapped out a half dozen genes on a single chromosome of *Drosophila*.

McClintock spent a great deal of time studying the behavior of chromosomes in her microscope. To understand her work, it is necessary to understand how chromosomes, and the genetic information they contain, are passed down from one cell to the next. This transfer occurs in two different ways. The first, during normal cell division within the life of a single organism, is called mitosis. The second, called meiosis, occurs in the formation of a sperm or egg cell for mating with another organism.

Mitosis is illustrated in Figure 16.2. The full set of chromosomes (ten in maize, twenty-three in humans) comes in Mendelian pairs, one chromosome of each pair inherited from the father and one from the mother. Mendelian pairs, also called homologous pairs, have genes for the same traits in the same positions on the chromosomes but may have different varieties (such as red versus white flowers) for each gene. In part A of Figure 16.2, only a single homologous pair is shown, the white chromosome inherited from one parent and the black from the other. (The colors black and white are used only to distinguish the two chromosomes.) Each chromosome has a little knot along its length, called the centromere, indicated by a circle. The centromere functions like a handle that pulls the chromosome about in the cell during the cell division process. In part B, each member of the homologous pair has doubled, producing an exact copy of itself called its sister chromatid. In part C, as cell division begins, sister chromatids are separated in a process called anaphase. Finally, in D, cell division is complete. The original cell has divided into two new cells. Each of the new cells has exactly the same chromosomes as the original. Organisms use mitosis to increase in size, to manufacture more tissue, to grow.

Meiosis is illustrated in Figure 16.3. Here parts A and B are the same as in mitosis, Figure 16.2. But in part C, a new feature appears. The legs of two chromatids of a homologous pair "cross over" and actually transfer genes between them. In steps E and F, the cell divides, with each doubled member of a homologous pair going to *separate* new cells. In G, the sister chromatids of each chromosome separate, and in H, we finally have *four* new cells, each with a *single chromosome*. Each of these four cells is prepared to mate and join with a cell of the opposite sex. Note that meiosis produces twice as many new cells as mitosis (four versus two), and each final cell has half as many chromosomes. Each mem-

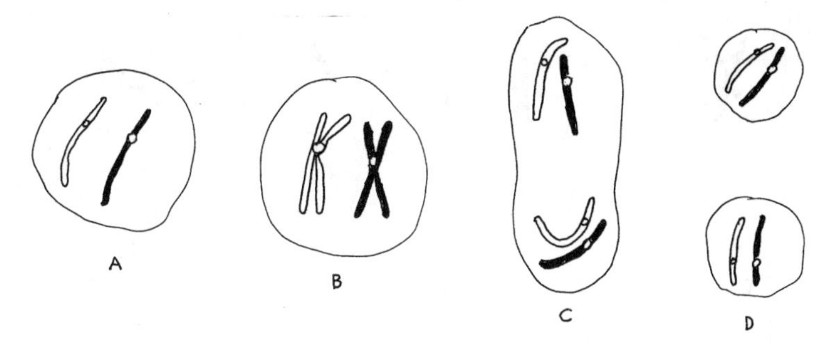

Figure 16.2

ber of a homologous, or Mendelian, pair has been isolated in a separate cell. Furthermore, some of these chromosomes have become hybrids, with new genes, from the crossover process.

It is important to emphasize a critical distinction between meiosis and mitosis as sketched above: genes can change positions in meiosis but not in mitosis. According to the traditional understanding of cell biology, with the exception of random mutations, genes passed down from one cell to the next within the life cycle of a single organism, in mitosis, stay in fixed positions on the chromosome. McClintock was to overthrow that view.

In a crucial technique, McClintock and other biologists used the

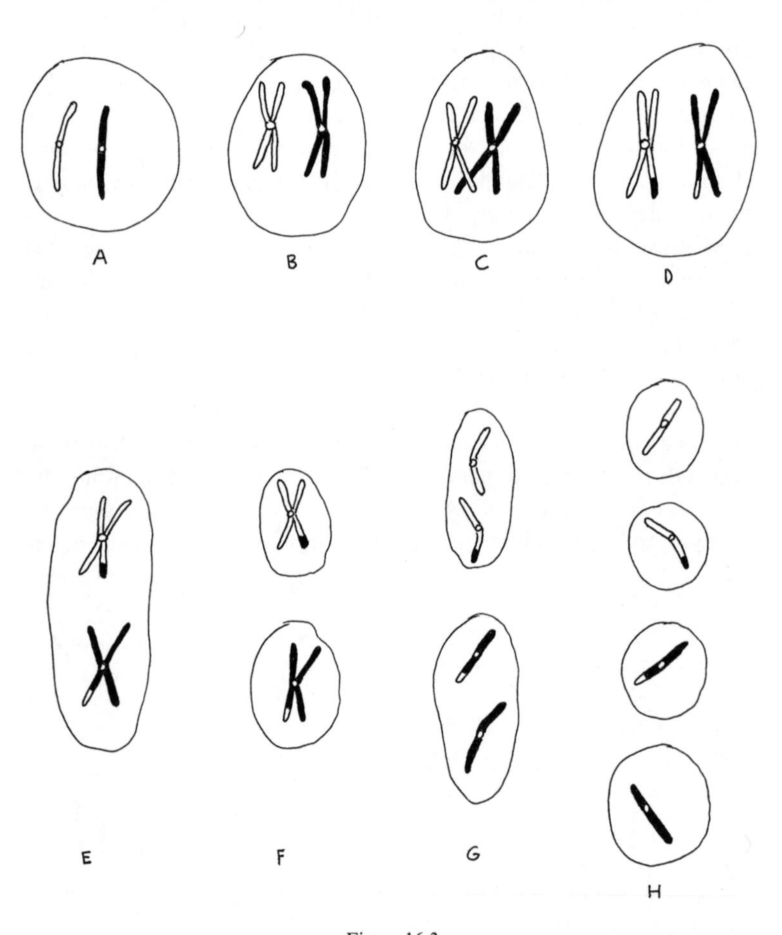

Figure 16.3

crossover process in meiosis to map the location of one gene relative to another on the length of the chromosome. As can be seen in parts C and D of Figure 16.3, a crossover acts like a scissors that cuts a chromosome in two, removing the section below the cut and replacing it with something else. Suppose two genes, say *Q* and *T*, are initially on the same chromosome. The probability of their being separated by the crossover cut is just proportional to their distance apart. If *Q* and *T* are very close together, the chances of the cut landing between them and thus separating them is small. If *Q* and *T* are far apart, the chance of the cut landing between them is large. By analyzing the relative numbers of chromosomes containing both *Q* and *T* after a crossover—as evidenced by inherited traits of the offspring—one can calculate how far apart *Q* and *T* were on the original chromosome.

The first hint that genes could move on chromosomes was something McClintock called the twin-sector phenomenon. In 1946 she noticed that some sections of maize plants showed a rate of mutations greatly different from that of the plant as a whole. For example, if the average number of yellow streaks on a variegated leaf of corn was five per square centimeter, some sections had two, or eight. Most importantly—and a minute pattern that would be sighted only by a most acute observer—these odd sections came in adjacent *pairs*, with one member of the pair having a higher rate of mutations than the average and the other neighboring member, side by side, lower than the average. Here was a big clue! Since each of the sections of the leaf originated from a different progenitor cell, it appeared as if one of the progenitor cells had given something to the other during mitosis. Years later, McClintock recalled that the twin-sector phenomenon "was so striking that I dropped everything, without knowing—but I felt sure that I would be able to find out what it was that one cell gained and the other cell lost, because that was what it looked like."

The twin-sector phenomenon suggested to McClintock that genetic elements might change locations during the life cycle of a single organism. As she would discover, the phenomenon was caused by genetic material moving from a chromosome to its sister chromatid during cell division. She would also eventually show that *genetic elements could change positions on a single chromosome*, altering functions as they did so.

In addition to analyzing patterns of variegation, McClintock was studying cell nuclei under the microscope. There, she found that the chromosomes of her variegated plants were showing regularly recurring breakages at specific places on the chromosomes. This finding again suggested to her some external control mechanism. Something was controlling the breaks in the chromosomes, which in turn might be associated with the variegations.

Just as McClintock knew her plants like her own clothes, she knew her plants' chromosomes. It was always chromosome 9 that was breaking, and chromosome 9 was always breaking one-third of the distance from the centromere to the end of the short arm. (The centromere, seldom in the middle, divides the chromosome into a short arm and a long arm.) McClintock called this position of breakage the *Ds* locus, "*Ds*" being shorthand for dissociation. In a confusing terminology, McClintock and other geneticists used the word "locus" to mean physical genetic material as well as position on the chromosome. Thus, "*Ds* locus" could be used to mean either the position of the breakage or the genetic element that caused the break.

After many crossings and observations, McClintock discovered that the *Ds* locus came in both a dominant form (called *Ds*) and a recessive form (called *ds*). Furthermore, even *Ds* did not always act to produce breaks in chromosome 9. McClintock concluded that a second controlling element was controlling the action of *Ds*. This second element she called the *Ac* locus, "*Ac*" being shorthand for activator. When *Ac* was present, *Ds* acted to produce breakage; when *Ac* was absent, *Ds* did not act.

Of perhaps the most significance of all, McClintock's work suggested that variegation appeared when *Ds* moved, or "transposed," to the location of the pigment gene, blocking its action.

McClintock's landmark 1948 paper, her first announcement that genetic elements could change positions on a chromosome within the life cycle of a single organism, was published in the annual report of her institution, as was Henrietta Leavitt's paper of 1912. Such annual reports, although not as widely circulated as national journals, were read by other experts in the field. Note that McClintock makes no references to any other scientists or their work in this paper, nor did Leavitt in her paper. These institutional publications very much represented work in progress, almost as if they were lab notes.

In general, McClintock's papers are dense and hard to read, and this one is no exception. Often, it seems as if she is burying the reader in a hodgepodge of results, with little guiding organization or unifying analysis. Another complicating factor is that in this paper McClintock was most interested in the process of control rather than the transposition of genes.

She begins the paper by summarizing the manner in which mutable genes, which she also calls mutable loci and unstable loci, can be turned on and off in their expression of traits. Observable genetic traits, such as pigment color, stripe marks, and degree of waxiness, are called phenotypes. She lists a number of genes showing "instability." Here, as with the flower color before, a lowercase letter means a recessive form of the gene, while a capital letter means a dominant form. For example, *c*, the recessive form of the pigment color, results in a colorless outer layer of the kernel, called the aleurone, while *C* results in a colored aleurone. McClintock then reviews her findings that some mutable genes do not mutate without the presence of a second controlling factor, *Ac*.

In the section titled "Nature of the *Ac* Action: The *Ds* locus," McClintock discusses the location of the *Ds* on chromosome 9, with distances measured in "cross-over units." She then describes the particular way that *Ds* causes a mutation, illustrated in Figure 16.4, where the steps proceed from A to C. Some other terminology in this section: "Sporophytic tissues" comprise the material of the plant proper, such as leaves and stalk. "Endosperm tissue" is the substance within the kernel used as food by the growing embryo.

Later in this section, McClintock mentions for the first time that "the *Ds* locus may change its position in the chromosome," a process she later calls "transposition." (In a few pages, she suggests that *Ac* can also change location.) She elaborates that *Ds* sometimes shifts from "a posi-

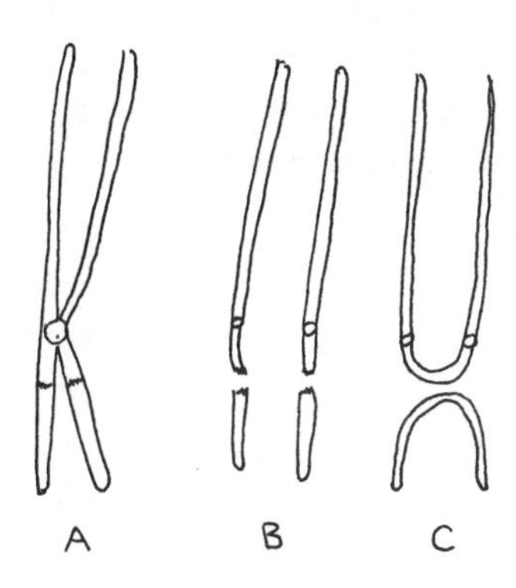

Figure 16.4

tion a few units to the right of *Wx* [a gene associated with waxiness] to a position between *I* [a gene blocking color] and *Sh* [a gene causing shrunken endosperm in the kernel]." The situation is illustrated in Figure 16.5, where only the short arm of the chromosome, to the left of the centromere, is shown. The top and bottom illustrations show the initial and final positions of *Ds*, respectively.

In the subsection "The effects of dosage of *Ac*," McClintock presents her finding that the effects of the activating element *Ac* vary with how much *Ac* is present. Unlike the case in most cells, where chromosomes come in homologous pairs, the chromosomes in the endosperm come in homologous triplets (called 3*n*), two maternal and one paternal. Thus each trait may have zero, one, two, or three dominant Ac genes controlling it, represented by *ac ac ac*, *Ac ac ac*, *Ac Ac ac*, and *Ac Ac Ac*, respectively.

Near the end of this section, McClintock comments that some of the effects of variable doses of *Ac* are obtained even when only a single *Ac* gene is present in the initial cell. How could that be? Picking up an idea previously proposed by other biologists, she conjectures that even a single gene of *Ac* must come in forms of variable strength. Evidently, she pictures *Ac* as consisting of "a number of identical and probably linearly arranged units," like a row of pennies. In this picture, an *Ac* gene can transfer some of its pennies to a sister chromatid during the mitosis process, so that "one chromatid gains units that the sister chromatid loses." The total number of pennies remains the same. The physical transfer of genetic material, with one chromatid gaining and another losing, might at last explain McClintock's earlier, revelatory observation of the twin-sector phenomenon. Note that McClintock has not proven that *Ac* comes in variable strengths. She "assumes" that it does so, in order to "fit" her observations. She is relying on some recent observations of the variable effects of Ac, her knowledge of the twin-sector phenomenon, and a great deal of intuition.

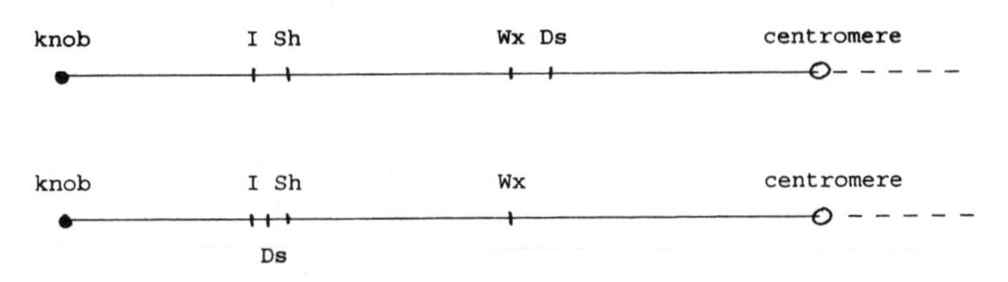

Figure 16.5

In the section titled "The Mutable *c* Loci," McClintock discusses other *Ac*-controlled genes related to pigment color, c^{m-1} and c^{m-2}. She notes that the *Ds* and *c* loci seem to be controlled by *Ac* in an "amazingly alike" manner. Again, she makes a hypothesis. She is led to "suspect" that the *c* locus may act through chromosome breakage just as *Ds* does. McClintock's suspicion here is important, and it is also an inspired leap of faith. Since she knows that *Ds* causes breakages and that *Ds* moves on the chromosome, one additional step (which she does not yet make) would conclude that *the transposition of Ds causes all the mutations that she has considered.*

A few more notes on notation: a slash between combinations of genes signifies that they come from different chromosomes of a homologous pair or triple. Thus, for example, "*C ds/ c ds/ c ds*" means that the first chromosome in the triple has genes *C* and *ds*, the second has *c* and *ds*, and the third has *c* and *ds*. For indicating different forms of the same gene from a homologous set of chromosomes, a simpler notation is used without slashes, such as "*ac Ac Ac*."

In the "Conclusions" section, McClintock repeats her evidence that some of the variegations are caused by "graded" increases or decreases in a gene like *C* or *Wx*. She again hypothesizes that some gains and losses of "identical units" in sister chromatids may explain the twin-sector phenomenon.

In this last section, McClintock is cautious about reaching beyond her data, labeling her conjectures as "interpretations" and "working hypotheses." At the same time, just like a theoretical physicist or chemist, she is clearly searching for a simplifying principle that will unify and explain all of the disparate phenomena she has observed. As she writes near the end, "With so many mutable loci behaving in very much the same manner, it is unlikely that many different, unrelated mechanisms are involved."

Barbara McClintock's 1948 paper is far more tentative and unresolved than any of the landmark papers we have previously considered. This paper reads very much like an internal report. And it is difficult to read. The American geneticist has some hard results, some hypotheses, some unanswered questions.

Furthermore, with hindsight, the paper seems unfocused. Although this is the paper in which McClintock first reports her discovery of transposition—that genes and genetic elements can change position on the chromosome—transposition is by no means a central issue of the paper. Rather, McClintock is mostly concerned with the process of "control," a word she uses throughout. The biologist wants to know what controls the orderly mutations that she's observed in maize. The "Loci" in her title, in fact, refers to the mutable genes themselves rather than to the positions of the genes. And her sharpest tentative conclusion is that genetic elements are graded and can be passed piecemeal from one sister chromatid to another. That phenomenon is related to but distinct from transposition, where a gene moves from one location of a chromosome to another location on the same chromosome, with an accompanying change in function. In this 1948 paper, the idea of transposition, which would ultimately become the most important discovery of her career, is almost hidden in the thicket of her many other results. Certainly, its significance was not well understood.

A year later, in her 1949 report for the Carnegie Institution, McClintock has already come to realize that transposition may be responsible for many of the orderly mutations in corn. There, she writes that "continued study . . . has revealed a type of event involving the *Ds* locus that appears to be responsible for the origin and subsequent behavior of all *Ac*-controlled mutable loci. This event brings about a transposition of the *Ds* locus from one location in the chromosome complement to another." Thus we see McClintock's thought evolving and progressing. By 1951, in a highly influential paper titled "Chromosome Organization and Gene Expression," published in *Cold Spring Harbor Symposia on Quantitative Biology*, she has identified transposition as a major phenomenon relating to the organization and dynamic nature of the chromosome. In a letter to her colleague George Beadle earlier that year, she wrote, "It seems to me that we might as well stop dealing with genes in the old sense and attempt to view the nucleus as an organized system with various types of controls of the action of the determiners [genes]."

The full recognition of McClintock's discovery of transposable genes was delayed for at least two reasons. First, the phenomenon was initially thought to apply only to maize. Second, just as biologists were beginning to embrace the idea, in 1953 James Watson, Francis Crick, and Rosalind Franklin discovered the structure of DNA, the molecule that contains the genetic information. At that point, many geneticists turned their attention to the molecular details of inheritance and information transfer rather than to the organization and systemic behavior of the chromosome as a whole, McClintock's interest. Transposition was overshadowed by DNA.

Over the next twenty-five years, however, more and more evidence pointed to the importance and universality of transposition. In the late 1960s, biologists found that some bacteria develop immunity to antibiotics by changing the position of their genes on the chromosome. The position of a gene, in fact, partly governs its function, the function of neighboring genes, and the gene's interaction with the rest of the organism.

Gradually, transposition was recognized as a universal mechanism, important in many forms of life. In the 1970s it was found that a process similar to transposition allows an organism to develop the huge required number of antibodies from a relatively small number of genes. Rearrangement of genes allows a vastly larger amount of information to be encoded and brought into action, just as the rearrangement of the letters *t*, *e*, *a*, and *m* allows the spelling of the words "tame," "meat," and "mate," in addition to the original "team." In 1983, McClintock was awarded the Nobel Prize for her discovery of genetic transposition.

Two aspects of McClintock's genius, especially, helped her solve scientific puzzles: her ability to see patterns in the genetic features of corn, just as the physicists of the late 1800s had noticed patterns in the spectral emissions of atoms, and her ability to study an organism both at the microscopic level of its chromosomes and at the global level of the entire plant. In a rare and remarkable comment to Keller in the late 1970s, McClintock described her creative moments of discovery: "When you suddenly see the problem, something happens—you have the answer before you are able to put it into words. It is all done subconsciously. This has happened too many times to me, and I know when to take it seriously. I'm so absolutely sure. I don't talk about it, I don't have to tell anybody about it, I'm just sure this is it."

MUTABLE LOCI IN MAIZE

Barbara McClintock

Carnegie Institution of Washington Yearbook (1948)

PREVIOUS REPORTS HAVE STATED that a number of unstable loci have recently arisen in the maize cultures. In a particular cell of a plant, a normal "wild-type" locus becomes altered; the normal, dominant expression of this locus changes and gives rise to a recessive expression (or, in several cases, a recessive locus becomes unstable and mutates toward a dominant expression). This expression of the locus need not be permanent. In some descendent cells, a second change may occur within the locus that results in the restoring of the capacity of the locus to express the dominant phenotype or brings about an intermediate expression between full recessive and full dominant. In the latter case, a third alteration may occur in some descendent cells that steps up the phenotypic expression toward the full dominant or reduces it toward the full recessive.

When a locus changes from a stable to an unstable state, recognition of the occurrence depends upon several factors. A clear-cut phenotypic expression of the changed locus is essential. This limits ease of recognition to those gene loci associated with the production of some obvious plant constituent, such as the chlorophyll pigments or other plant pigments, or to contrasting morphological characters striking enough to allow the contrasts to be detected in single small sectors of some part of the plant. The presence of a changed locus that gives a recessive expression often may remain hidden in a heterozygous state until a self-pollination or some special cross is made that allows its presence to be uncovered. No attempt has been made to detect the presence of these hidden changed loci, and thus the frequency of occurrence of such changes is not known. The unstable loci that have been discovered have all appeared in the progeny of crosses designed for other purposes. It is suspected that the number of mutable loci still undetected may be quite great, with numerous independent occurrences of unstabilization of the same locus. The detected unstable loci include many previously unknown to maize geneticists. Some well known loci in maize have also become unstable in these cultures. These include two independently arising unstable *yg* loci (yellow-green chlorophyll), four independently arising unstable *c* loci (*c*, colorless aleurone; *C*, colored aleurone), three independently arising unstable *wx* (waxy) loci (*wx*, starch staining red with iodine; *Wx*, starch staining blue with iodine), and one unstable a_2 locus (anthocyanin in aleurone and plant). With the exception of the two *yg* loci, all these unstable loci have originated by a change in a normal, previously quite stable locus showing dominant expression. In the case of unstable *yg*, the recessive *yg* locus mutates to form chlorophyll of a much darker color than the normal *yg* locus expresses.

These mutable loci fall into two distinct classes: (1) those that require the presence of a second locus, the activator locus (*Ac*), for instability to be expressed and maintained, and (2) those that do not require such a second locus for instability to be expressed. The *Ac* locus itself is unstable and resembles in this respect the second class of unstable gene loci.

Two recognizable subdivisions of the mutation process are shown by all the unstable loci: (1) control of the time and the apparent frequency of mutations in a tissue by a factor capable of changing during a mitosis so that altered rates of mutation will be expressed in descendent cells following such a change; and (2) the subsequent change at the locus that occurs in a particular cell and gives rise immediately to recognizable altered phenotypic expression in this cell and its descendent cells. The term "state of a locus" has been used to distinguish the differing potentialities of a mutable locus for expressing visible mutations in descendent cells. The mutations giving rise to changes in state of a locus are readily recognized by the altered rates of the second type of mutation that follow from such an event. In the second class of mutable loci—those controlled by *Ac*—the time and apparent frequency of mutations of the affected locus are controlled by the states of both loci: the locus controlled by *Ac* and the *Ac* locus itself.

During the past year, attention has been directed mainly to those loci that require the *Ac* locus for continued expression of instability. All these loci that are known are in the short arm of chromosome 9, possibly because the genetic methods being used allow certain mutable loci in this arm to be readily detected. Those receiving particular study include the *Ds* (Dissociation) locus, mentioned in previous reports, two independently arising mutable *c* loci, and two independently arising *wx* loci. In each case, the mutable recessive state of the previously stable dominant locus arose in a particular cell of a plant that also possessed an *Ac* locus. The origin of this potent *Ac* locus is quite unknown. It has been traced to a culture grown in the summer of 1942, but its presence in still earlier cultures cannot be traced. It cannot be stated that *Ac* induces the initial state of instability of a locus, although a causal connection may be suspected.

None of these *Ac*-controlled mutable loci will show any type of instability if the *Ac* locus is absent from the nucleus. As soon as *Ac* is introduced into a nucleus having such a susceptible locus, instability is resumed. Mutations at the susceptible locus sometimes may occur in the initial nucleus following the introduction of *Ac*. To illustrate the control of mutability of a locus by *Ac*, the mutable c^{m-1} locus may be used. This locus has been given the symbol c^{m-1} because it was the first of the several mutable *c* loci isolated; the normal *c* locus, used for many years in genetic experiments, is designated c^{8} because it is a stable locus not mutating under the influence of *Ac*. The behavior of this unstable recessive *c* locus in the designated constitutions is indicated below:

c^{m-1} *ac*: no mutations; aleurone layer colorless
c^{8} *ac*: no mutations; aleurone layer colorless
c^{8} *Ac*: no mutations; aleurone layer colorless
c^{m-1} *Ac*: mutations to *C* occur; aleurone layer variegated for color

It should be emphasized that, with similar constitutions, the same type of responses will be registered by any of the other loci that are activated by *Ac*.

NATURE OF THE *Ac* ACTION

THE *Ds* LOCUS. The *Ds* locus was discovered before any of the other *Ac*-controlled unstable loci. *Ds* is located at the position demarking the proximal third of the short arm of chromosome 9, between one and two crossover units to the right (toward the centromere) of the *Wx* locus. The normal, nonmutating locus has been designated *ds*. An observable *Ds* mutation arises as the consequence of breakage of two sister chromatids within the *Ds* locus in each case. This breakage is followed by fusions of broken ends to give rise to a U-shaped acentric fragment composed of the two sister segments of the distal two-thirds of the short arm, and a U-shaped dicentric chromosome composed of the proximal third of the short arm, the centromere, and all of the long arm of each of the two sister chromatids. Nearly all the known loci of this chromosome are carried in the distal two-thirds of the short arm of chromosome 9 and are thus included in the acentric fragment after a *Ds* mutation. The genetic methods of observing the time and frequency of the *Ds* mutations in various parts of the plant have been explained in previous reports (Year Books Nos. 45 [1945–1946] and 46 [1946–1947]).

The *Ac* locus acts upon *Ds*, or any other locus susceptible to it, usually quite late in the development of any of the sporophytic tissues. In contrast, action of *Ac* on susceptible loci may be apparent at all stages in the development of the endosperm tissues. In this respect, the endosperm behaves like an extension of the sporophytic tissues.

As stated above, the detectable *Ds* mutations are associated with breaks at this locus that are followed by 2-by-2 fusions of those broken ends of sister chromatids that lie adjacent to one another. This type of fusion following breakage is not the only one that could occur. Restitutions could take place, re-establishing the previous organization of the chromatids; or crisscross fusions of broken ends could occur, simulating crossing over but involving sister chromatids in somatic nuclei. It is not known whether or not the latter types of fusion do occur with expected frequencies at the *Ds* locus. The evidence suggests that they may occur, at least occasionally. It is known that breaks at the *Ds* locus may be followed directly or indirectly by fusions of broken ends other than those giving rise to the U-shaped dicentric chromatids. This is shown by the fusions that occur after an occasional coincidental break in another part of the chromosome. It is possible then for a broken end associated with a *Ds* mutation to fuse with one of these other newly broken ends. This gives rise to a gross chromosomal aberration that can be analyzed. Such chromosomal aberrations have been observed in individual cells or clusters of cells in the examined sporocytes of the *Ds Ac* plants. Several of these translocations have been found in individual plants, and have received further study. These aberrations have been useful in interpreting occasional inconsistencies in the behavior and the location of the *Ds* and the *Ac* loci. It is now known that the *Ds* locus may change its position in the chromosome after such coincidental breaks have occurred. One very clear case has been analyzed, and, through appropriate selection of crossover chromatids, strains having morphologically normal chromosomes 9 have been obtained. As a consequence of this aberration, the *Ds* locus in these strains has been shifted from a position a few units to the right of *Wx* to a position between *I* and *Sh*. This is a very favorable position for showing the nature of the *Ds* mutation process. When the usual type of *Ds* mutation occurs at the locus in this new position, a U-shaped dicentric chromosome is formed. At the succeeding anaphase a chromatin bridge is produced, which subsequently breaks. The breakage-fusion-bridge cycle is thus initiated and can be genetically expressed in succeeding nuclear divisions. This is because the loci of *Sh*, *Bz*, and *Wx* now lie between *Ds* and the centromere. With appropriate genic constitutions of chromosome 9, detection of the breakage-fusion-bridge cycle is certain after a *Ds* mutation has occurred in this new location. These observations have been made, and the genetic analysis completely confirms the cytological analysis of the nature of the *Ds* mutation process.

THE INHERITANCE OF Ac AS A SEPARATE LOCUS. *Ac* behaves in inheritance as a single independent locus. Tests for the presence of *Ac* in F_2 progenies of *ds ds Ac ac* plants have given ratios of 1 *Ac Ac* to 2 *Ac ac* to 1 *ac ac*; backcrosses of *Ac ac* plants by *ac ac* plants have given ratios of 1 *Ac ac* to 1 *ac ac* plant. Crosses of *Ac Ac* plants by *ac ac* plants have given progenies all of which were found to be *Ac ac*. In the many crosses that have been made with plants that were heterozygous for both *Ds* and *Ac*, independent inheritance of the *Ac* and *Ds* loci was clearly established, except in three independently arising cases. In these three cases, *Ac* was obviously linked with *Ds* and could be located 6 to 20 crossover units to the right of it. Such linkages were maintained in later tests of the progeny of one of these plants, when both the crossovers and the noncrossover chromatids were tested. Similar tests are now under way for the other two cases. Cytological examination of heterozygous plants in each of the three cases showed no observable chromosomal abnormalities. Chromomere matching between synapsed homologues was perfect for all of chromosome 9.

Either of two possibilities may explain this change in genetic location of *Ac*. First, *Ac* may be located toward the end of the long arm of chromosome 9 and normally show no genetic linkage with *Ds* because the crossover distance is too great. In the case of *Ds*-*Ac* linkage, a newly arising crossover modifier, not associated with a gross chromosomal abnormality, would need to be invoked to explain the observations. This modifier would need to be closely linked with *Ac*, as it has not been removed in the crossover tests so far made. Second, the *Ac* locus may have been removed from its former position and inserted into a new position in chromosome 9 in a manner similar to that observed for the transposition of the *Ds* locus, described above. Because *Ac* induces breaks at specific loci and gives evidence of undergoing a specific breakage process itself, this latter explanation is not improbable. Tests are now under way to determine whether or not either of these alternatives can apply.

THE EFFECTS OF DOSAGE OF *Ac*. Studies of the effects of various dosages of *Ac* have shown that the time and apparent frequency of *Ac*-controlled mutations is in large measure a function of dosage of *Ac*. This applies both to the sporophytic tissues and to the endosperm tissues. The endosperm tissue is 3*n*. Here, one, two, or three doses of *Ac* can be obtained and observations made of the effects of each of these dosages on any of the *Ac*-controlled mutable loci. All *Ac*-controlled mutable loci respond alike to the various *Ac* dosages. The description given below of responses of the *Ds* locus in endosperm tissues can be applied equally well to the responses of any of the other *Ac*-controlled mutable loci.

In *Ds ds ds* kernels, the mutation rates of the single *Ds* locus have been compared in *Ac ac ac*, *Ac Ac ac*, and *Ac Ac Ac* endosperm constitutions. One dose of *Ac* allows mutations to occur relatively early; considerable irregularity in both time and frequency of mutations is apparent, for the endosperm is often divided into various larger or smaller sectors each showing its own special rate of mutation. This irregularity is only apparent, as subsequent evidence will show. Two doses of *Ac* delay the timing of *Ds* mutations so that they occur relatively late in the development of the endosperm. The frequency may be very high, however. When three doses of *Ac* are present, only relatively few, very late-occurring *Ds* mutations are recognized in these kernels. Three doses of *Ac* appear to decrease the frequency of *Ds* mutations, but this appearance is probably deceptive. It is more likely that the increased dosage of *Ac* so delays the timing of *Ds* mutations that only the very earliest-occurring of these can possibly show in the endosperm, and that if the endosperm cells had continued to divide, very large numbers of *Ds* mutations would have appeared. In other words, the tissues mature before these potential *Ds* mutations can be expressed. The validity of this interpretation will be apparent when the various states of the *Ac* locus are defined and their dosage responses compared. It may be concluded, then, that the absolute dosage of *Ac* is one main factor controlling the time and apparent frequency of mutations: the higher the *Ac* dosage, the later the occurrence of *Ds* mutations. . . .

It is obvious that something occurs in the early divisions of the endosperm that results in nuclei with differing potentialities for expressing the time and frequency of *Ds* mutations. Immediately upon examination of these kernels, one is impressed with the resemblance of the mutation patterns in the various sectors to the patterns that have been obtained by combining various dosages of *Ac*. Here, however, only one

Ac locus was introduced into the primary endosperm nucleus. Obvious questions present themselves: Do these sectors arise as the consequence of some abnormality that results in segregations of various dosages of *Ac* in these early nuclei? If so, what abnormalities occur and what is the segregation mechanism? Neither nondisjunction of chromatids containing *Ac*, nor transposition of the *Ac* locus to another position in the chromosome set followed by mitotic segregation, satisfactorily explains the observed ratios of the various types of kernels. A satisfactory fit is obtained if it is assumed that (1) the *Ac* locus is composed of a number of identical and probably linearly arranged units, and (2) changes in the number of units can take place at the locus during or after chromosome reduplication such that one chromatid gains units that the sister chromatid loses. . . .

THE MUTABLE *c* LOCI

MUTABLE c^{m-I}. One of the two mutable *c* loci being studied originated from the cross of a female plant homozygous for *c sh ds ac* by a male plant homozygous for *C Sh Ds* and heterozygous for *Ac* (*Ac ac*). The same male plant was used in making a number of similar crosses. On one of the resulting ears, a single aberrant kernel was observed. Instead of showing either a complete *C* color (*C Ds/c ds/c ds*, *ac ac ac* constitution) or a variegation pattern composed of colored aleurone with colorless sectors, owing to *Ds* mutations (*C Ds/c ds/c ds*, *Ac ac ac* constitution), this kernel was obviously composed of colorless aleurone in which colored areas were present. In pattern of variegation, it was the reverse of expectation: a *c* locus appeared to be mutating to *C*. This kernel was removed, a plant was grown from it, and various crosses were made to determine the nature of the altered expression of aleurone color. Appropriate genetic analysis indicated that a mutable *c* locus was present which had arisen from the normal *C* locus present in the male parent. This *C* locus had changed to a mutable *c* locus capable of mutating back to the original *C*. This mutable *c* locus is *Ac*-controlled and responds in precisely the same manner as the *Ds* locus to various doses of *Ac* and to various states and observable changes in state of the *Ac* locus. With respect to *Ac*, the responses of the two loci are amazingly alike. *Ds* mutations, however, are known to be the result of some mechanism leading to chromosomal breakage and fusion, whereas the mutations at the *c* locus appear to involve quite a different mechanism, leading to changes in expression of the locus from recessive to dominant.

Similarities in response of *Ds* and c^{m-1} to *Ac*, together with the known breakage mechanism at the *Ds* locus and also the obvious changes in state of the *Ac* locus that can be explained on the basis of a chromosomal breakage mechanism, lead one to suspect that some kind of chromosomal breakage and fusion mechanism may likewise be responsible for these reverse mutations. . . .

MUTABLE c^{m-2}. The second mutable *c* locus, c^{m-2}, arose in a somatic cell of a plant of the constitution *I Sh ds*/*C Sh ds*, *Ac Ac*. The plant was sectorial for this mutable *c* locus. With respect to dosage of *Ac*, the production of sections, etc., the c^{m-2} locus responds exactly as do the *Ds* and c^{m-1} loci. The phenotypic expression of mutations at the c^{m-1} and at the c^{m-2} locus is distinctly different. The mutations of c^{m-2} give rise to sectors showing great variation in color intensity, from a very faint to an intensely deep color—often much deeper than that produced by a single, double, or even triple dose of *C*. Any one intensity may appear as a mutant sector in any part of the aleurone layer on any one kernel; also, any one kernel may show a number of different mutations each having its own particular color intensity.

The very great differences in phenotypic expression of the two independently arising mutable *c* loci suggest that the normal *C* locus may be composed of at least two blocks, each having its own particular organization and function but both necessary for producing the *C* phenotype. They are not strictly complementary, for combinations of c^{m-1} and c^{m-2} in the same endosperm do not result in the production of *C* color. The mutable c^{m-1} locus is assumed to have arisen from a normal *C* locus after alterations had occurred in only one block. On the other hand, the mutable c^{m-2} arose from a normal *C* locus after alterations that involved only the second block. The c^{m-1} locus undergoes mutational changes only in the one affected block, whereas the mutations of c^{m-2} involve changes only in the second block. A possible alternative to this interpretation will be considered in the concluding paragraphs of this report. . . .

CONCLUSIONS

It may be premature to consider in detail the question asked in the concluding sentence above. Because so many mutable loci have recently appeared in the maize cultures, because in many respects they all behave in very much the same way, and because this behavior is similar also to that described for other mutable loci both in maize and in other organisms, it may be profitable to review briefly the pertinent facts about the cases described in this report, in order to ascertain the similarities and dissimilarities among these cases. They involve the mutable locus *Ac* and the mutable loci it controls—*Ds*, c^{m-1}, c^{m-2}, and wx^{m-1}.

Ds stands alone in that the chromosomal consequences of a mutation at this locus are known. The detectable *Ds* mutations unquestionably arise as a consequence of some mechanism that either brings about breakage and fusion between sister chromatids at the *Ds* locus or simulates this mechanism in its consequences; for dicentric chromatids are produced after a *Ds* mutation. In the pattern of mutations, c^{m-1} and *Ds* are strictly comparable. Variegation patterns produced by *Ds* mutations in *C ds*/*C ds*/*I Ds*, *Ac ac ac* constitutions can be so similar to variegation patterns produced by *c* to *C* mutations in c^8 c^8 c^{m-1}, *Ac ac ac* constitutions that they may be indistinguishable by mere observation of the kernels. Yet the formation of dicentric chromatids alone cannot explain most of the observed c^{m-1} mutations. This likewise applies to the c^{m-2} mutations and to many of the mutations of wx^{m-1}. Some form of chromosome breakage and fusion, however, may be involved.

As stated earlier, mutations of c^{m-1} and c^{m-2} differ greatly in phenotypic expression. Visible mutations of c^{m-1} give rise each time to the full dominant expression expected from a single dose of *C*, whereas those of c^{m-2} are quantitatively expressed, with color intensities varying from faint to extremely deep. The normal *C* locus is known to give dosage effects: the more whole *C* loci present, the greater the depth of color. By means of duplications of the short arm of chromosome 9, it has been possible to observe effects of doses up to and including six *C* loci. With the highest dose, the color of the aleurone is unusually deep. Some of the mutations of c^{m-2} result in intensities greater than that shown with three doses of the normal *C* locus, whereas others are so faint that they obviously have produced much less pigment than is produced by a single normal *C* locus. The various consequences of mutations of the c^{m-2} locus are in complete agreement with the hypothesis that the mutations result from graded increases in the number of units within a depleted locus and that they express themselves phenotypically by graded increases in the substance or substances responsible for the phenotypic character. The observed mutations of c^{m-1} do not lead to this hypothesis, for they show no quantitative subdivisions. Similarity of the c^{m-1} mutations to *Ds* mutations in basic response to *Ac*, and their conformity to the general pattern of the *Ac*-controlled mutations that do give quantitative effects, has led to formulation of a subsidiary hypothesis rather than abandonment of the general hypothesis. The validity of this subsidiary hypothesis is subject to tests that are now being conducted. Assuming the fundamental mutation process to be similar for all *Ac*-controlled mutable loci, the genes in the mutating c^{m-1} block of the *C* locus could be related to some chemical process that requires a threshold number of units for expression to be fulfilled; or it may be that the mechanism bringing about a change in units at this mutable *c* locus assures a specific increase in number of units.

The mutations occurring at the c^{m-1} and wx^{m-1} loci are amazingly similar in every respect. The mutations fall into a graded quantitative series in both cases. Also, both give twin or adjacent sectors showing the same grades of contrasting intensities of expression of the dominant character. There can be little doubt that the same mechanism is responsible for the mutations occurring at these two different loci. With regard to the wx^{m-1} locus, there is evidence for believing that the unit number within the locus may be responsible for the expression of unit increases of the dominant phenotypic expression. Considering the obvious similarities between the two cases, it would be difficult to avoid concluding that the same conditions apply to mutations at the c^{m-2} locus. In this connection it may be recalled that mutations at the *Ac* locus likewise suggest a mutation mechanism involving changes in the numbers of units at the locus.

Because many of the mutations occurring at the two mutable *c* loci, the wx^{m-1} locus, and the mutable *Ac* locus do not result in dicentric chromatids, as do *Ds* mutations, and do not lead to detectable gross chromosomal aberrations, the mechanism, if it is a breakage phenomenon, must restore the normal chromosome morphology. Unequal breakage within the locus, followed by the crisscross type of fusion mechanism, could accomplish this end.

Any mechanism that gives rise to an increase in numbers of units at a locus might also give rise to the reverse condition; that is, to chromatids with decreased numbers of units. If so, chromosomes with loci having various unit numbers should appear as isolates in these cultures. Some of these isolates should show more visible mutations than others under given conditions. The frequency of visible mutations would depend upon the initial number of units present in the locus before mutation occurred. The more units were present, the greater would be the chance that the increase in the units during any one mutation would be sufficient to produce a visible effect; and the converse would also be true. Isolates from the different mutable loci, showing just these expected variations in the rates of visible mutations, have been obtained. The term "state of the locus" applies as well to the mutable loci controlled by *Ac* as to the *Ac* locus itself. What has been termed a high-state locus gives high rates of mutation, and a low-state locus gives low rates of mutation. It is possible, therefore, that the number of units present at a mutable locus is correlated with the state of the locus as well as with the expression of visible effects. It is a matter of degree. If the initial number is high, but not high enough to produce a visible effect, the state of the locus may be considered high. Conversely, if only a few units remain in the locus, the state of the locus may be considered low.

The above conclusions are supported by the many observations that have been made especially of the chlorophyll-producing types of mutable loci. As mentioned in previous reports, the motivation for this study of mutable loci was the observation of twin sectors, apparently arising from sister nuclei, that showed inverse rates of visible mutations. The most extreme of these twin sectors showed a mutation to dominant in one sector, and in the sister sector either a complete recessive or a very much reduced rate of mutation as compared with that in the surrounding tissue. In these observed cases of twin sectoring, it is obvious that the factor or factors controlling the rate of mutation and the visible mutations themselves are of the same general nature, if not actually different resultants of the same mechanism. The factor responsible for twin sectoring acts at a mitosis, and the apparent result is that one chromatid gains something that the sister chromatid loses. In the interpretation given, this gain and loss are considered to be an increase and a decrease, respectively, of identical units at the locus, which, in turn, controls not only the appearance of visible mutations but also the state of the locus as reflected in the time and frequency of occurrence of future visible mutations.

The above interpretations are being used as a working hypothesis in continuing studies of mutable loci. The evidence at present points toward the presence of reduplicated units within a locus, these units often expressing themselves in a quantitative manner through unitary action on substrates responsible for phenotypic characters. The hypothesis that the phenotypic expression of a locus depends upon the number of such units present in any one chromosome, and that some mechanism or mechanisms can alter this number and give rise to visible mutations, is both sufficiently simple and sufficiently integrative to afford precise direction in some types of experimentation. It is believed that this approach to one phase of the over-all mutation problem may be productive, even though it is realized that the details will need clarification and may be subject to degrees of modification. With so many mutable loci behaving in very much the same manner, it is unlikely that many different, unrelated mechanisms are involved. It is more likely that one general type of condition exists in all these mutable loci and that this condition may be altered in any of these loci by one kind of mechanism.

In this report, it has not been possible to include a discussion of the many other observations and conclusions that are relevant to the subject, such as the restabilization of a mutable locus, the behavior of *Ac*-controlled mutable loci when two or more are present in the same nucleus, or the changes at the *Ac* locus that often appear to accompany an *Ac*-induced mutation. Nor has it been possible to describe the accumulated evidence on some of the non-*Ac*-controlled mutable loci, or to mention the many new mutations that are constantly arising in these cultures. These studies are being continued and will be reported later.

17

THE STRUCTURE OF DNA

IN ONE OF THE EPISODES OF *Star Trek, the Next Generation,* Commander Data fractures a part of himself, a hand as I remember, and stares at the bare tangle of wires and computer chips protruding from his wrist. Although Data is a machine, we have come to regard him as human. He looks human. He acts toward other characters with compassion and sweetness. He appears to know right from wrong. And thus something disturbs us about this scene, not so much because Data is hurt but because he sees inside his own mechanism. The secret of his being hangs open in the air. All the complexities of his bodily actions and thoughts, the subtle depths of his feelings, the seemingly infinite mysteries of a living being, have been graphically reduced to so many amperes of current flowing through these protruding wires, to particular patterns of zeros and ones within these computer components. We feel some kind of violation of the natural order of things. How can a creature know itself in this way, gaze into its own inner workings? Conceivably, such a creature could build itself, in a dizzy circular loop without beginning, a maker that makes itself, a universe that imagines itself into existence.

In the last fifty years, we human beings have discovered much of our own mechanism—in the chemical structure of a molecule called DNA. Most molecules have a fixed chemical structure. By contrast, the DNA molecule has pieces that vary, and the particular variations and arrangements of those pieces, like the order of letters to form words, chemically spell out the instructions for how to make a particular human being, or any living organism. Other molecules carry out the directions encoded in the DNA molecule, creating bones and muscles, blood, livers, brains, lungs or gills, skin or shells, hair or feathers, flowers and stalks.

The structure of DNA was discovered in 1953 by James Watson, Francis Crick, and Rosalind Franklin. Watson and Crick's landmark paper in *Nature*, arguably the most important paper in biology in the twentieth century, is so famous that often one cannot find it in a library copy of the journal. Pages 737 and 738 of volume 171 have been ripped out like stones stolen from the Church of the Holy Sepulchre in Jerusalem. The paper is little more than one page long, scarcely a thousand words altogether. Franklin's article follows two pages later.

A molecule of DNA is shaped like a twisted ladder. Each of the two legs of the ladder forms a helix. Each rung of the ladder consists of two small molecules fastened together and may be one of only four types: C-G, G-C, A-T, or T-A. The particular sequence of rungs up the ladder contains all the genetic information to create an organism. For example, A-T followed by G-C followed by A-T is the code for serine, one of the twenty amino acids that form proteins; G-C followed by C-G followed by A-T codes for arginine, another amino acid.

Molecule by molecule, a human or a mouse is manufactured from a specific recipe of amino acids and other biochemicals ordered up from the sequence of rungs on the DNA ladder. The complete recipe to make a human being, for example, requires about five billion rungs on the ladder. A typical bacterium requires only about five million.

And how are the directions read? By touch. Different molecules differ in shape. The building-block molecules and their contractors rove up and down the DNA ladder, binding momentarily when their shape fits snugly with the shape of a particular rung or group of rungs in the ladder. In this way, pieces of life are partly assembled, then leave the DNA and assemble more pieces.

The DNA ladder is extremely long relative to its width. For humans, the total DNA is about two meters long but only 0.2 millionths of a centimeter wide. DNA molecules reside within each living cell, in the chromosomes (see Chapter 16 on McClintock). To fit within a cell, thousands of times smaller in length, the ladder must be wound up and folded upon itself many times. A DNA ladder goes nowhere, yet it leads to everything.

As in other scientific collaborations we have seen, Watson, Crick, and Franklin brought different skills and personal temperaments to their work on DNA. Crick, born in 1916 in Northampton, Britain, was trained at University College in London in mathematics and physics. After World War II, during which he worked for the British Admiralty designing circuits for magnetic mines, Crick lost interest in physics and decided to switch to biology. In 1951, at age thirty-five, he went to Cambridge. There he hoped to bring his considerable mathematical abilities to the study of the structure of protein molecules for his Ph.D. At that time and for the previous decade, the most successful method for probing the structure of proteins and other complex molecules was X-ray diffraction. In this method, the pattern of X-rays scattered from a molecule reveals the arrangement of its parts (see Chapters 7 and 18).

Rosalind Franklin, an accomplished experimentalist in X-ray diffraction, arrived at nearby King's College in London also in 1951. Born in 1920 to a prosperous merchant banker in London, Franklin received her Ph.D. from Cambridge in 1945, then worked in Paris for three years, where she refined her techniques in X-ray diffraction and became a world expert on the structure of coal.

James Watson, an American born in 1928 in Chicago, was trained in biology and the genetics of bacteria. After graduate school at the University of Indiana, studying with Salvador Luria, Watson went to Copenhagen, where he hoped to learn some chemistry.

Since the beginning of the twentieth century and earlier, genetics, inheritance, and embryonic development had been major themes of biology. Watson's influential mentor Luria, although not a chemist himself, suspected that an understanding of the working of genes probably required a detailed knowledge of their chemical structure. An essential clue was provided in 1944, when the American biologist Oswald Avery and colleagues found strong evidence that the active molecule in genes was deoxyribonucleic acid, or DNA for short. The chemical components of DNA had been known since the 1920s: five-carbon sugars called deoxyribose (Figure 17.1); phosphate groups, consisting of a central phosphate atom bound to four oxygen atoms (Figure 17.2); two double-ring nitrogen-containing compounds called guanine and adenine (Figure 17.3); and two single-ring nitrogen-containing compounds called thymine and cytosine (Figure 17.4). (Recall that in chemical notation, C, H, O, and N stand for carbon, hydrogen, oxygen, and nitrogen, the first three being the principal atoms of all organic compounds.) The four nitrogen compounds are collectively called bases.

The chemical constituents of DNA had been relatively easy to determine. But the spatial arrangement of those constituents—how they fit together in three-dimensional space—was another matter entirely. Since the work of the Dutch physical chemist Jacobus Hendricus van't Hoff in the 1870s, it had been known that the same group of atoms could bond together in different ways, leading to different structures and shapes for the resulting molecule. Most importantly, these different forms had different behaviors and properties. Therefore, it was crucial to determine the particular spatial arrangement of atoms of DNA. Uncovering that structure, in turn, would almost certainly require X-ray diffraction.

In Copenhagen, Watson became obsessed with understanding the structure of DNA. In the fall of 1951, at age twenty-three, he went to Cambridge to learn about X-ray diffraction from Max Perutz, who was then using that technique to study the structure of proteins (see Chapter 18).

Thus, within less than a year of each other, Watson and Crick arrived in Cambridge and Franklin in London, an hour's train ride away. As it turned out, Franklin would make the crucial X-ray pictures of DNA, while Watson and Crick would use their scientific intuition and Franklin's photographs to build three-dimensional models out of cardboard and metal. Although in close proximity, Watson and Crick never worked directly with Franklin. Indeed, Watson describes his relationship with Franklin as "sticky." Their personalities and styles couldn't have been more different.

In his memoir *What Mad Pursuit*, Francis Crick writes that "Jim and I hit it off immediately, partly because our interests were astonishingly similar and partly, I suspect, because a certain youthful arrogance, a ruthlessness and an impatience with sloppy thinking came naturally to both of us." One might argue that Jim Watson and Francis Crick were impatient not only with sloppy thinking. They were simply impatient. They spoke their minds quickly and without trepidation. Watson, at this time sporting a great mass of frizzy and unruly hair, recalls in his own memoir that Crick "talked louder and faster than anyone else and, when he laughed, his location within the Cavendish was obvious."

While Watson and Crick both possessed a "youthful arrogance" and loved to joke with friends at the Cambridge pubs, Franklin was more serious and less sociable. Frederick Dainton, who taught her as an undergraduate at Cambridge University, wrote to her biographer Anne Sayre that Franklin was a "very private person with very high personal and scientific standards, and uncompromisingly honest." She was also, in the words of her Ph.D. supervisor Ronald Norrish (who won the Nobel in chemistry in 1967), "stubborn and difficult to supervise." In all respects, Franklin was an independent person. She could be a stoic. During the war she accidentally thrust a sewing needle deep into her knee and then proceeded to walk a long distance, alone, to a hospital to have the needle removed.

Franklin went to King's College in London to work on X-ray diffraction of DNA. At that time the physicist Maurice Wilkins, also at King's, had been studying DNA with X-ray diffraction for several years. According to Watson, molecular studies of DNA in England in 1950 were "for all practical purposes the property of Maurice Wilkins." So far, Wilkins had made only limited progress, in part because, unbeknownst to him, his samples of DNA contained a mixture of two different forms of the substance, confusing the X-ray pictures.

Franklin and Wilkins clashed almost immediately. Subsequently, they shared the same lab quarters but worked separately. From time to time Watson would take the train from Cambridge to London to talk to Wilkins or hear a lecture by Franklin. While Watson and Crick constantly erupted with new ideas and models for DNA, relying a great deal on their instincts and common sense, Franklin was both closer to the data and more cautious in her scientific style. As she once told Dainton, "the facts will speak for themselves."

Unraveling the structure of DNA would require a knowledge of chemistry more than anything else. And the chemist supreme in 1950 was Linus Pauling in America (see Chapter 11). In the spring of 1951, Pauling and Robert Corey published a paper showing that many protein molecules are arranged like a helix, what Pauling called an alpha-helix. The alpha-helix was the first helical structure known in biology. It was beautiful, and it excited the imaginations of other biochemists. DNA was not a protein, but it was another complex organic molecule, and some biologists thought that perhaps it, too, was shaped like a helix. Pauling began working to crack the structure of DNA, the most prized molecule of biology.

From the moment that Watson and Crick encountered each other, in the fall of 1951, they decided to race Pauling to unveil the secrets of DNA. As Watson writes in his celebrated book *The Double Helix*, "Our lunch conversations quickly centered on how genes were put together. Within a few days after my arrival, we knew what to do: imitate Linus Pauling and beat him at his own game."

A principal technique would be to try out different three-dimensional models built out of pieces of paper, cardboard, and metal, cut to the shapes of the various constituents of DNA. Watson describes their approach:

> I soon was taught [by Crick] that Pauling's accomplishment was a product of common sense, not the result of complicated mathematical rea-

> soning. The alpha-helix had not been found by only staring at X-ray pictures; the essential trick, instead, was to ask which atoms like to sit next to each other. In place of pencil and paper, the main working tools were a set of molecular models superficially resembling the toys of pre-school children. We could thus see no reason why we should not solve DNA in the same way. All we had to do was to construct a set of molecular models and begin to play.

The toys that Watson and Crick proceeded to play with, the known constituents of DNA, are shown in Figures 17.1–17.4.

In 1950, at least a century of biological discovery had laid the groundwork for unraveling DNA. First was the essential idea that complete living organisms are not preformed inside eggs, as had been believed by many prominent biologists of the seventeenth and eighteenth centuries,

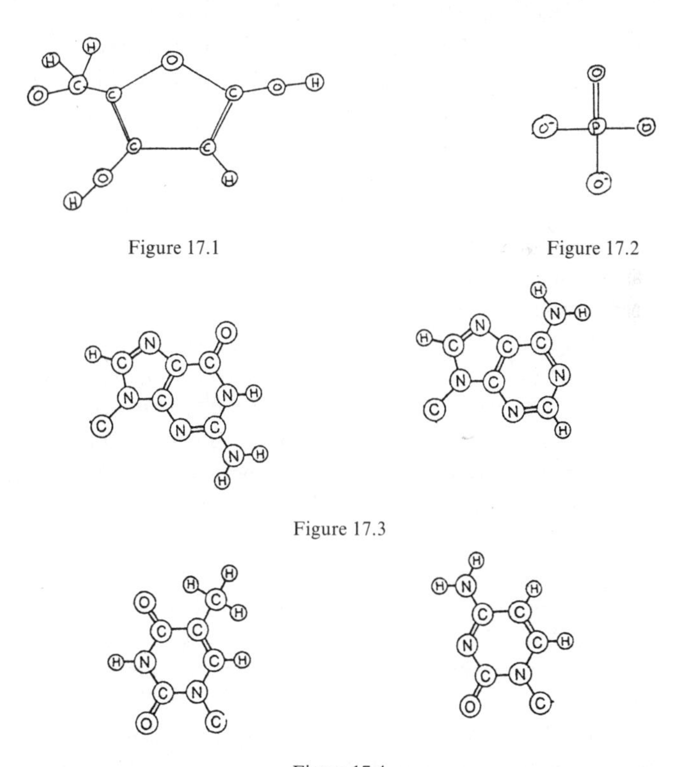

Figure 17.1

Figure 17.2

Figure 17.3

Figure 17.4

but are created piece by piece in the developing embryo. Such a revised picture required that *instructions* for making the new pieces be passed along from one cell to the next, and from one generation to the next. Second was the emerging study of cell biology and the gradual recognition that the cell nucleus contained creational instructions. (Cells in some primitive organisms, such as bacteria, do not contain nuclei but carry creational instructions nonetheless.) And third, the finding that the particular molecule that harbored the instructions was deoxyribonucleic acid, DNA.

The cell nucleus was discovered in 1831 by the British scientist Robert Brown. The all-important cell division process, in which one cell becomes two cells in the growth of an organism, was first observed by the Swiss botanist Karl Wilhelm von Nägeli in 1842. During cell division, enormous change took place in the cell nucleus, leading to the suspicion that the nucleus played a critical role in the process of growth and reproduction. (Recall from Chapter 16 that there are two different processes in which genetic information is passed along from one cell to the next: ordinary cell division in the *growth* of a developing organism, and the combination of egg and sperm to make a new cell in the *reproduction* of a new organism.) A critical question was whether there existed some conserved substance, a genetic blueprint, physically conveyed from one cell to the next in growth or reproduction.

Early observations suggested that the cell nucleus dissolved during cell division and fertilization. If so, then it would be hard to support the idea of a conserved genetic substance being passed along. However, in the 1860s, with more careful experiments and observations through the microscope, German biologist Edward Strassburger concluded that the cell nucleus does not dissolve but instead divides in two, with each "daughter cell" receiving some material from the "parent cell."

Simultaneously with Strassburger's work, Gregor Mendel showed that the traits of inheritance are embodied in a pair of discrete factors, later called genes, with one factor coming from the father and one from the mother. In a familiar tragedy of scientific research, Mendel's work was unknown by cell biologists until 1900 (he published only obscurely), when it was rediscovered by Hugo de Vries of Holland, Karl Correns of Germany, and Erich Tschermak von Seysenegg of Austria. Thus Mendel's work, ironically, played no role in the development of genetics until after 1900.

In 1879, Walther Flemming of the University of Kiel, in very close microscopic observations, found that certain rodlike bodies in the cell nucleus, called chromosomes, split longitudinally during cell division. A dramatic clue! At this point, there was good reason to hypothesize that the chromosomes contained the critical genetic information. In 1890, the German biologist Theodor Boveri showed that chromosomes maintained their identity during the entire life cycle of the cell. Boveri argued that here was the home of the genes.

As discussed in Chapter 16, Boveri's speculation was proven around 1910 by Thomas Hunt Morgan and his students, who found that certain inherited traits, such as sex and eye color, were passed along in groups, as they would be if physically located on chromosomes. Furthermore, Morgan's student A. H. Sturtevant was able to map the physical location of the Mendelian factors, that is, the genes, on particular chromosomes. Now, there was little doubt that the genes resided within the chromosomes. The genes were physical substances, with known addresses.

In 1928, in experiments with mice, the British biologist Fred Griffith found that virulent bacteria, even after being killed with heat, could transform nonvirulent bacteria into virulent bacteria. He concluded that genetic material was being passed along from the first kind of bacteria to the second. Then, in 1944, Oswald Avery and colleagues grew tons of virulent bacteria and separated them into their various biochemical constituents: proteins, fats, carbohydrates, DNA, and RNA (a molecule closely related to DNA). After careful experiments, Avery concluded that the constituent responsible for converting nonvirulent bacteria to virulent in Griffith's experiments was DNA.

The DNA molecule, if indeed the carrier of genetic information, would need at least two features: a chemical means of encoding information, and a means of copying itself for the cell division process.

In mid-November of 1951, Watson took the train from Cambridge to London to hear the first of "Rosy's" occasional talks on her X-ray pictures of DNA. She spoke to an audience of about fifteen in an ancient and creaky lecture hall. Little of what Franklin said seems to have made much of an impression on Watson at this time. However, he did pay attention to her style and appearance, recalling that her delivery was "quick, nervous . . . without a trace of warmth or frivolity . . . The years of careful, unemotional crystallographic training had left their mark." Watson would always claim that Franklin didn't know how to interpret her own data, didn't have the needed intuition and insight.

Following the lecture, Watson and Wilkins walked through the Strand to Choy's restaurant in Soho. According to Watson's recollections, Wilkins seemed pleased in his belief that Franklin had made little progress since arriving at King's. He stated to Watson that her X-ray photographs, although sharper than his, did not reveal much about the structure of DNA. What Watson was most interested in was whether DNA was shaped like a helix, the mesmerizing shape found by Pauling. Crick had performed detailed mathematical calculations indicating what the X-ray diffraction picture of a helical molecule should look like, but these properties were not apparent in Franklin's pictures at this time.

In fact, Franklin had already made the breakthrough discovery that DNA could exist in two different configurations, which she called A and B. The A form was crystalline. The B form contained more water and was looser and more stretched out. In most samples of DNA, the A and B forms were mixed, leading to a complex and undecipherable X-ray diffraction picture.

By the summer of 1952, with painstaking laboratory procedures that involved bubbling hydrogen gas through salt solutions and then carefully controlling the humidity of the DNA sample, Franklin was able to obtain very pure samples of both forms of DNA. She then had to pull out thin single fibers of the material, design a "tilting microfocus camera," and precisely orient the fibers in the X-ray camera's collimated beam to make the X-ray pictures.

A particular X-ray photograph of the B form, labeled #51, was especially telling and clearly indicated a helical structure to the cognoscenti. That photo is shown in her paper. The large X-shape formed by the dark spots is evidence of a helical structure. The spacing between successive dark spots in each arm of the X gives the distance of each complete turn of the helix, 34 Å, where "Å" stands for angstrom, 10^{-8} centimeters. (The angstrom, a unit of distance on the atomic scale, is named after the Swedish physicist Anders Ångström.) The larger distance from the center of the X to the top of the figure gives the distance between successive nitrogen bases, a tenth of a complete turn, or 3.4 Å. The angle between the X's upper arms relates to the diameter of the molecule, 20 Å.

Franklin quietly analyzed her new photos of DNA. In her section of the King's 1952 Medical Research Council Report, she correctly proposed a double-stranded helical structure for DNA, with the two twisted legs of the ladder made of phosphates and sugars and the rungs made of the nitrogen bases. Furthermore, she was able to deduce all of the quantitative information mentioned above, including the pitch angle, or tilt, of the helix. What she didn't know was how the bases fit together to make the rungs—a critical detail not revealed by her X-ray diffraction pictures.

In mid-January of 1953, Pauling proposed a three-stranded helical model of DNA. Scientists soon realized that Pauling's model could not be right, not because of the number of strands but because the molecule did not properly act as an acid, releasing positive hydrogen atoms when dissolved in water. Watson and Crick were elated. The great chemist had blundered, and blundered on chemical rather than structural grounds. With fanfare, the two Cambridge friends retired to the Eagle pub "to drink a toast," as Watson recounts, "to the Pauling failure."

Within a couple of weeks, Watson saw Franklin's photograph #51 for the first time. (It was Wilkins who showed the photo to Watson, without Franklin's permission or knowledge.) As Watson recalls in his memoir, "the instant I saw the picture my mouth fell open and my pulse began to race . . . The pattern was unbelievably simpler than those obtained previously."

Most obvious in the pattern was the helical structure, revealed by the dark cross. If Watson had analyzed the photo more carefully, with Crick's mathematical knowledge of X-ray diffraction patterns, he might also have inferred from the *relative intensities* of the dark spots that the molecule was double-stranded, a double helix. In particular, each leg of the X has a missing black spot in fourth position, counting out from the center. That absence suggests that the molecule is double-stranded. Instead, Watson simply assumed that DNA was double-stranded, because of a vague intuition that "important biological objects come in pairs."

Over the next weeks, Watson and Crick struggled to find out how the nitrogen bases fit together to make the rungs of the twisted ladder. From stiff cardboard, Watson cut out figures in the shapes of the four bases. Two essential problems confronted the scientists: (1) Since the four bases were of different shapes and sizes, how could they possibly produce successive rungs of *the same width*? And if the rungs were not of the same width, the ladder would buckle in and out, producing a horribly complex shape that could not lead to the simple pattern of the X-ray diffraction picture. (2) How did the bases fasten to each other and to the phosphate-sugar legs of the ladder? Known experimental results regarding the degree of acidity of DNA and other evidence suggested that each rung of the ladder was made of two or more bases joined to each other by hydrogen atoms. But these facts were only a small part of the solution.

At first, Watson tried to make each rung of the ladder out of two identical bases fastened together—in other words, rungs of C-C, A-A, T-T, and G-G, where C, A, T, and G stand for cytosine, adenine, thymine, and guanine, respectively. Such rungs were of unequal width and therefore unacceptable. Furthermore, Watson had disregarded an important hint discovered by the American chemist Erwin Chargaff in 1950: in DNA, the amount of A is equal to the amount of T, and the amount of C equals the amount of G. We will return to that fact shortly.

A few days later, Watson's and Crick's office mate, Jerry Donahue, provided another critical clue that Watson did not ignore. The textbooks were slightly wrong on the structure of the bases. Watson would need to place the hydrogen atoms in different locations, changing the molecules from the "enol" form to the "keto." (Note the dangling hydrogen atoms in Figures 17.3 and 17.4.) The placement of these hydrogen atoms governed the way in which the bases could bond to each other.

According to Watson's memoir, "When I got to our still empty office the following morning [mid-February 1953], I quickly cleared away the papers from my desk so that I would have a large, flat surface on which to form pairs of bases held together by hydrogen bonds." Watson couldn't wait for the metal figures from the machine shop and had cut out cardboard bases himself. "Suddenly I became aware that an adenine-thymine pair held together by two hydrogen bonds was identical in shape to a guanine-cytosine pair." (See Figure 17.5.) The new pair-

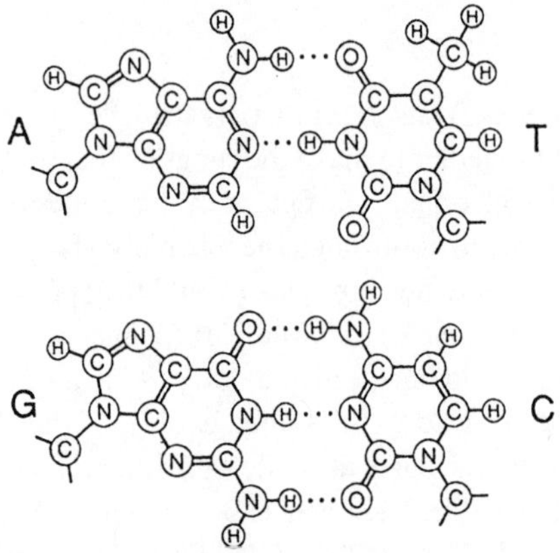

Figure 17.5

ing would produce rungs of the same width and would also automatically satisfy Chargaff's result, since each A would always partner with a T, and each G with a C. Whatever the number of A molecules in a section of DNA, the number of T molecules would be equal, etc.

The overall shape of the resulting double helix model is shown in the single figure in Watson and Crick's paper. This difficult figure was, in fact, drawn by Crick's wife Odile, already known for her paintings of nudes. The double helix was to be her most famous artistic creation by far.

A close-up illustration of the DNA structure is shown in Figure 17.6. Here, the two twisted strands of the helix are made of phosphates (each indicated by P inside a circle) and sugars (each indicated by a pentagon). The bases are the squares. The dashed line between paired bases in each rung represents a weak "hydrogen bond" that holds the two bases together.

From Watson and Crick's model, which has been confirmed in detail, we immediately understand the two essential features of DNA: how it codes information and how it copies itself. The coding is done by the particular sequence of nitrogen bases in the rungs, as discussed earlier. And the copying is done by breaking the ladder down the middle, through the dashed hydrogen bonds. After such a breakage, each remnant has one leg and one half of each rung. Since C always pairs with

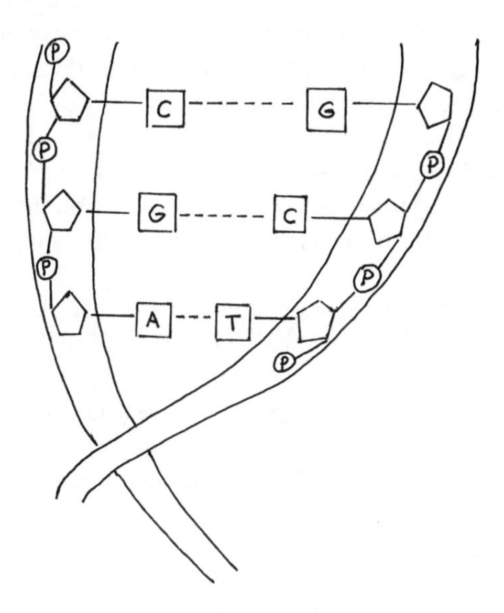

Figure 17.6

G, and A with T, each half of the ladder contains all the information of the full ladder. From the sea of biochemical ingredients surrounding the chromosomes, the missing half of each remnant can be manufactured, producing two full ladders from the original ladder. In this way, DNA copies itself.

Watson and Crick begin their landmark paper by criticizing the recent three-stranded model of Pauling and Corey. Then they propose their own model, which they call "radically different." Their style here and for the rest of the paper is brisk and concise.

To understand some of the terminology: an acid is a substance that releases positive hydrogen atoms (ions) when dissolved in water, and a salt is an acid where the H^+ ion has been replaced by another positively charged ion. The most common salt is sodium chloride, NaCl, also known simply as table salt. A hydrogen bond is a weak bond between the positively charged H^+ atom of one molecule and a negatively charged atom of another molecule. Such weak bonds are easy to break in the copying process.

The manner in which the sugar and phosphate groups bonded together in the ladder legs, or "chains," was already well known. Counting clockwise from the oxygen atom at the peak of each sugar molecule (Figure 17.1), the fifth carbon on a sugar bonds to the oxygen of a phosphate group (Figure 17.2), which in turns bonds to the third carbon of the next sugar. Such a sequence is called a "5′3′ linkage." Because of the positions of the various bonds linking the bases to the legs on both sides, the two chains of sugars and phosphates must run in opposite directions, 5′3′ linkages on one leg and 3′5′ linkages on the other. (Note in Figure 17.6 that the pentagon sugars point in opposite directions on the two legs.)

Watson and Crick next give some quantitative data about the dimensions of the molecule, all previously discovered by Franklin. Here, each "residue" is one sugar-phosphate group, comprising one unit of the chain. Each strand makes a complete turn every ten residues, in a distance of 34 Å.

Next, Watson and Crick discuss the "novel feature of the structure," namely the nature of the rungs that hold the two ladder legs together. *The rungs, in fact, contain all the genetic information and also the secret of how the molecule copies itself.* The four nitrogen bases are divided into two groups: the purines, which are the two double-ring bases, guanine and adenine; and the pyrimidines, the two single-ring bases, thymine and cytosine. Using Donahue's commandment that nature uses only the "keto" forms of the bases, not the "enol," the bases can pair together by hydrogen bonds only in certain ways.

In hindsight, Watson and Crick now refer to Chargaff's results about the proportion of A, T, C, and G. It is fascinating that they ignored this hint until after they had found a model in accordance with it. As an aside, such neglect is not uncommon in the way science is actually practiced. In an analogous story, the striking similarity in shape of the opposite coasts of South America and Africa, as if two fitting pieces of a jigsaw puzzle, was ignored for centuries as long as geologists believed that landmasses could move only vertically over time. However, once Alfred Wegener proposed the idea of (horizontal) continental drift in 1912, allowing the possibility that South America and Africa could once have been joined, the complementarity in their coastal shapes became a powerful fact in support of the new theory. There are many other such examples in which striking but unexplained observations are largely ignored until a new theory comes long that explains them.

Toward the end of their paper, the scientists refer to previously published X-ray diffraction data, calling it "insufficient for a rigorous test" of their proposed model. This statement is a little misleading. Franklin's X-ray picture of DNA B, which not only partly confirmed Watson and Crick's model but helped motivate its creation in the first place, had already been seen by the two scientists and was published two pages later in the same issue of *Nature*. It is true that none of the X-ray diffraction pictures revealed much about the organization of the bases.

Now for Franklin's paper, some of which has already been discussed. She collaborated with Raymond Gosling, a research assistant six years her junior. The "sodium thymonucleate" in the title refers to a salt of DNA taken from the calf thymus, a common source of DNA.

Franklin begins by discussing the distinction between DNA A and DNA B, the two forms that she discovered a year and a half earlier. She presents her famous X-ray picture of DNA B, which shouted "helical structure!" to all those who had seen it.

Franklin interprets the X-ray picture in terms of a "structure function," denoted by F_n. The structure function is a theoretical formula—worked out previously by Stokes, Crick, Cochran, and Vand—that relates the structure of the molecule to the amplitude (strength) and phase (crest or trough) of the X-ray wave emerging from the molecule. "Reflexions" are simply the X-ray spots on her photographic film. By comparing the locations of those spots to the structure function formula, she can determine the dimensions of the DNA molecule, as discussed previously.

She next gives evidence that the phosphate groups lie on the outside legs of the double helix and that the helix is double-stranded.

Note the cautiousness of the language here, especially in comparison to that of Crick and Watson in their paper. For example, while Watson and Crick propose a "radically different" structure for DNA, with no qualifications, Franklin says that the evidence "cannot, at present, be taken as direct proof" of the helical structure and that the molecule is "probably" double-stranded.

In summarizing the various contributions, one could say that Franklin's work provided the overall shape and dimensions of the molecule, particularly the legs of the double-stranded ladder, while Watson and Crick figured out the rungs of the ladder. In the opinion of many historians of science, Watson and Crick did not give enough credit to Franklin, either in this paper or in subsequent writings and accounts.

Only five years after the discovery of the structure of DNA, following distinguished work on RNA viruses at Birbeck College, Franklin died from cancer. She was thirty-seven years old. In 1962, Watson, Crick, and Wilkins shared the Nobel Prize for discovering the structure of DNA. The prize cannot be given posthumously, and it cannot be given to more than three people in any category in any given year. It is interesting to speculate on what the Swedish Nobel committee would have done had Franklin still been alive.

After continued significant work on DNA and RNA in England, in 1976 Crick went to the Salk Institute for Biological Sciences in southern California. In later years, Crick wrote about the origin of life on earth, the brain, and the nature of consciousness. He died in the summer of 2004. Watson served on the faculty of Harvard, then went to the Cold Spring Harbor Laboratory in New York, where he is still president. In 1988, Watson became the first director of the Human Genome Project.

In 1953 the time was ripe for uncovering the structure of DNA. If Watson and Crick and Franklin had not done their work when they did, it is likely that other scientists would have made the same discovery within a year. Sometimes intuition, ambition, and luck are more important than brilliance. Watson and Crick had a clear idea of what they wanted to achieve, they were highly focused on that achievement, and they were eager. They also had the advantage of the superb X-ray photographs of Franklin, and they had Donahue's important tip about the proper configuration of the bases. Like many, but not all, of the other scientists we have encountered, Watson and Crick immediately recognized the significance of what they had done. It required a couple of years for the rest of the world to do so.

The full impact of the discovery of the structure of DNA has not yet been measured. In 1961, Crick and his colleague Sydney Brenner deciphered the triplet code of bases in DNA that governs the production of amino acids. In the mid-1970s, Frederick Sanger at Cambridge figured out how to determine the order of bases in a stretch of DNA. In 1972 the American Paul Berg and colleagues did the first "recombinant DNA" experiments, in which a new piece of DNA is manufactured in the lab (see Chapter 22). The first "gene therapy," a process of curing disease by modifying a person's DNA, occurred in 1990 for severe combined immune deficiency (SCID). In 1995, Craig Venter announced that he had determined the complete sequence of rungs (bases) for the DNA of a single-celled bacterium called *Haemophilus influenzae*. In 2002, the complete DNA of a human was similarly decoded. Among other results, scientists discovered that our human genetic blueprint contains only about 30,000 genes, far fewer than expected.

It is impossible to say what lies in the future. The knowledge of DNA and how to modify it could help cure disease, alter personality and behavior, create new life forms, produce new kinds of computers, produce creatures half animal and half machine.

Beyond the almost limitless range of possible applications of DNA research, two fundamental questions remain. First, given its universality in all living forms we know, how did DNA first arise on our planet? This question is closely related to the origin of life. And second, given that the DNA is identical in each living cell of a single organism, how do cells in a developing embryo know how to differentiate themselves, with some cells becoming liver cells, some heart cells, some brain and muscle? These questions should keep biologists busy for decades to come.

MOLECULAR STRUCTURE OF NUCLEIC ACIDS

A STRUCTURE FOR DEOXYRIBOSE NUCLEIC ACID

James D. Watson and Francis H. C. Crick

Nature (1953)

WE WISH TO SUGGEST A STRUCTURE for the salt of deoxyribose nucleic acid (D.N.A.). This structure has novel features which are of considerable biological interest.

A structure for nucleic acid has already been proposed by Pauling and Corey.[1] They kindly made their manuscript available to us in advance of publication. Their model consists of three intertwined chains, with the phosphates near the fibre axis, and the bases on the outside. In our opinion, this structure is unsatisfactory for two reasons: (1) We believe that the material which gives the X-ray diagrams is the salt, not the free acid. Without the acidic hydrogen atoms it is not clear what forces would hold the structure together, especially as the negatively charged phosphates near the axis will repel each other. (2) Some of the van der Waals distances appear to be too small.

Another three-chain structure has also been suggested by Fraser (in the press). In his model the phosphates are on the outside and the bases on the inside, linked together by hydrogen bonds. This structure as described is rather ill-defined, and for this reason we shall not comment on it.

We wish to put forward a radically different structure for the salt of deoxyribose nucleic acid. This structure has two helical chains each coiled round the same axis (see diagram). We have made the usual chemical assumptions, namely, that each chain consists of phosphate diester groups joining β-D-deoxyribofuranose residues with 3′,5′ linkages. The two chains (but not their bases) are related by a dyad perpendicular to the fibre axis. Both chains follow righthanded helices, but owing to the dyad the sequences of the atoms in the two chains run in opposite directions. Each chain loosely resembles Furberg's[2] model No. 1; that is, the bases are on the inside of the helix and the phosphates on the outside. The configuration of the sugar and the atoms near it is close to Furberg's "standard configuration," the sugar being roughly perpendicular to the attached base. There is a residue on each chain every 3·4 A. in the z-direction. We have assumed an angle of 36° between adjacent residues in the same chain, so that the structure repeats after 10 residues on each chain, that is, after 34 A. The distance of a phosphorus atom from the fibre axis is 10 A. As the phosphates are on the outside, cations have easy access to them.

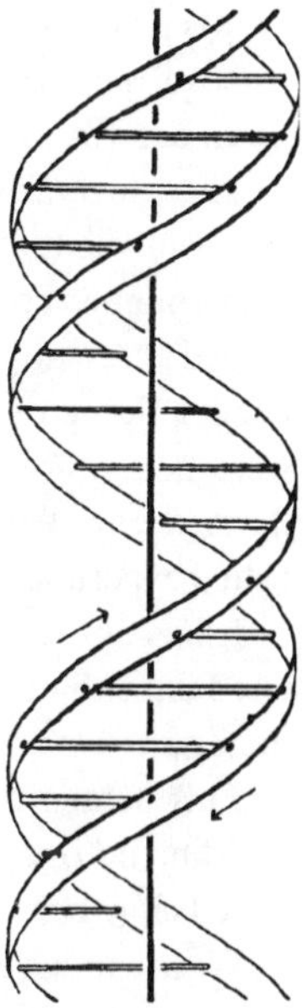

The structure is an open one, and its water content is rather high. At lower water contents we would expect the bases to tilt so that the structure could become more compact.

The novel feature of the structure is the manner in which the two chains are held together by the purine and pyrimidine bases. The planes of the bases are perpendicular to the fibre axis. They are joined together in pairs, a single base from one chain being hydrogen-bonded to a single base from the other chain, so that the two lie side by side with identical z-co-ordinates. One of the pair must be a purine and the other a pyrimidine for bonding to occur. The hydrogen bonds are made as follows: purine position 1 to pyrimidine position 1; purine position 6 to pyrimidine position 6.

Figure 1 This figure is purely diagrammatic. The two ribbons symbolize the two phosphate–sugar chains, and the horizontal rods the pairs of bases holding the chains together. The vertical line marks the fibre axis.

If it is assumed that the bases only occur in the structure in the most plausible tautomeric forms (that is, with the keto rather than the enol configurations) it is found that only specific pairs of bases can bond together. These pairs are: adenine (purine) with thymine (pyrimidine), and guanine (purine) with cytosine (pyrimidine).

In other words, if an adenine forms one member of a pair, on either chain, then on these assumptions the other member must be thymine; similarly for guanine and cytosine. The sequence of bases on a single chain does not appear to be restricted in any way. However, if only specific pairs of bases can be formed, it follows that if the sequence of bases on one chain is given, then the sequence on the other chain is automatically determined.

It has been found experimentally[3,4] that the ratio of the amounts of adenine to thymine, and the ratio of guanine to cytosine, are always very close to unity for deoxyribose nucleic acid.

It is probably impossible to build this structure with a ribose sugar in place of the deoxyribose, as the extra oxygen atom would make too close a van der Waals contact.

The previously published X-ray data[5,6] on deoxyribose nucleic acid are insufficient for a rigorous test of our structure. So far as we can tell, it is roughly compatible with the experimental data, but it must be regarded as unproved until it has been checked against more exact results. Some of these are given in the following communications. We were not aware of the details of the results presented there when we devised our structure, which rests mainly though not entirely on published experimental data and stereochemical arguments.

It has not escaped our notice that the specific pairing we have postulated immediately suggests a possible copying mechanism for the genetic material.

Full details of the structure, including the conditions assumed in building it, together with a set of co-ordinates for the atoms, will be published elsewhere.

We are much indebted to Dr. Jerry Donohue for constant advice and criticism, especially on interatomic distances. We have also been stimulated by a knowledge of the general nature of the unpublished experimental results and ideas of Dr. M. H. F. Wilkins, Dr. R. E. Franklin and their co-workers at King's College, London. One of us (J. D. W.) has been aided by a fellowship from the National Foundation for Infantile Paralysis.

J. D. Watson

F. H. C. Crick

Medical Research Council Unit for the Study of the Molecular Structure of Biological Systems, Cavendish Laboratory, Cambridge.

April 2.

1. Pauling, L., and Corey, R. B., *Nature*, 171, 345 (1953); *Proc. U.S. Nat. Acad. Sci.*, 39, 84 (1953).

2. Furberg, S., *Acta Chem. Scand.*, 6, 634 (1952).

3. Chargaff, E., for references see Zamenhof, S., Brawerman, G., and Chargaff, E., *Biochim. et Biophys. Acta*, 9, 402 (1952).

4. Wyatt, G. B., *J. Gen. Physiol.*, 36, 201 (1952).

5. Astbury, W. T., Symp. Soc. Exp. Biol. 1, *Nucleic Acid*, 65 (Camb. Univ. Press, 1947).

6. Wilkins, M. H. F., and Randall, J. T., *Biochim. et Biophys. Acta*, 10, 192 (1953).

MOLECULAR CONFIGURATION IN SODIUM THYMONUCLEATE

Rosalind E. Franklin and R. G. Gosling

Nature (1953)

SODIUM THYMONUCLEATE FIBRES GIVE two distinct types of X-ray diagram. The first corresponds to a crystalline form, structure *A*, obtained at about 75 per cent relative humidity; a study of this is described in detail elsewhere.[1] At higher humidities a different structure, structure *B*, showing a lower degree of order, appears and persists over a wide range of ambient humidity. The change from *A* to *B* is reversible. The water content of structure *B* fibres which undergo this reversible change may vary from 40–50 per cent to several hundred per cent of the dry weight. Moreover, some fibres never show structure *A*, and in these structure *B* can be obtained with an even lower water content.

The X-ray diagram of structure *B* (see photograph) shows in striking manner the features characteristic of helical structures, first worked out in this laboratory by Stokes (unpublished) and by Crick, Cochran and Vand.[2] Stokes and Wilkins were the first to propose such structures for nucleic acid as a result of direct studies of nucleic acid fibres, although a helical structure had been previously suggested by Furberg (thesis, London, 1949) on the basis of X-ray studies of nucleosides and nucleotides.

While the X-ray evidence cannot, at present, be taken as direct proof that the structure is helical, other considerations discussed below make the existence of a helical structure highly probable.

Structure *B* is derived from the crystalline structure *A* when the sodium thymonucleate fibres take up quantities of water in excess of about 40 per cent of their weight. The change is accompanied by an increase of about 30 per cent in the length of the fibre, and by a substantial re-arrangement of the molecule. It therefore seems reasonable to suppose that in structure *B* the structural units of sodium thymonucleate (molecules on groups of molecules) are relatively free from the influence of neighbouring molecules, each unit being shielded by a sheath of water. Each unit is then free to take up its least-energy configuration independently of its neighbours and, in view of the nature of the long-chain molecules involved, it is highly likely that the general form

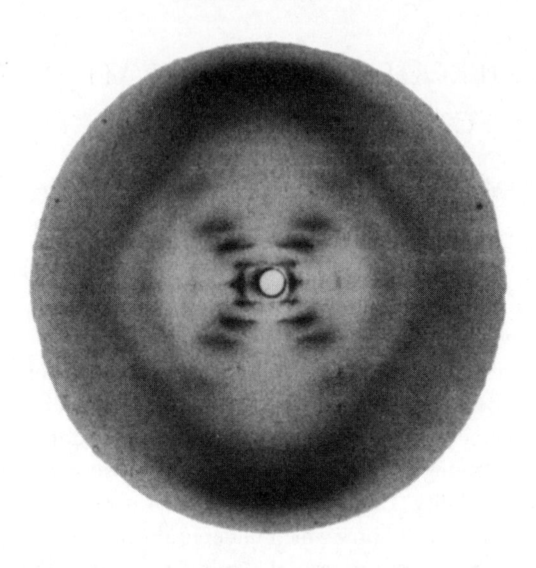

Sodium deoxyribose nucleate from calf thymus. Structure *B*

will be helical.[3] If we adopt the hypothesis of a helical structure, it is immediately possible, from the X-ray diagram of structure *B*, to make certain deductions as to the nature and dimensions of the helix.

The innermost maxima on the first, second, third and fifth layer lines lie approximately on straight lines radiating from the origin. For a smooth single-strand helix the structure factor on the *n*th layer line is given by:

$$F_n = J_n(2\pi r R) \exp i\, n(\psi + \tfrac{1}{2}\pi),$$

where $J_n(u)$ is the *n*th-order Bessel function of u, r is the radius of the helix, and R and ψ are the radial and azimuthal co-ordinates in reciprocal space[2]; this expression leads to an approximately linear array of intensity maxima of the type observed, corresponding to the first maxima in the functions J_1, J_2, J_3, etc.

If, instead of a smooth helix, we consider a series of residues equally spaced along the helix, the transform in the general case treated by Crick, Cochran and Vand is more complicated. But if there is a whole number, m, of residues per turn, the form of the transform is as for a smooth helix with the addition, only, of the same pattern repeated with its origin at heights mc^*, $2mc^*$ . . . etc. (c is the fibre-axis period).

In the present case the fibre-axis period is 34 A. and the very strong reflexion at 3.4 A. lies on the tenth layer line. Moreover, lines of maxima radiating from the 3.4-A. reflexion as from the origin are visible on the fifth and lower layer lines, having a J_5 maximum coincident with that of the origin series on the fifth layer line. (The strong outer streaks which apparently radiate from the 3.4-A. maximum are not, however, so easily explained.) This suggests strongly that there are exactly 10 residues per turn of the helix. If this is so, then from a measurement of R_n the position of the first maximum on the nth layer line (for $n \leq 5$), the radius of the helix, can be obtained. In the present instance, measurements of R_1, R_2, R_3 and R_5 all lead to values of r of about 10 A.

Since this linear array of maxima is one of the strongest features of the X-ray diagram, we must conclude that a crystallographically important part of the molecule lies on a helix of this diameter. This can only be the phosphate groups or phosphorus atoms.

If ten phosphorus atoms lie on one turn of a helix of radius 10 A., the distance between neighbouring phosphorus atoms in a molecule is 7.1 A. This corresponds to the P . . . P distance in a fully extended molecule, and therefore provides a further indication that the phosphates lie on the outside of the structural unit.

Thus, our conclusions differ from those of Pauling and Corey,[4] who proposed for the nucleic acids a helical structure in which the phosphate groups form a dense core.

We must now consider briefly the equatorial reflexions. For a single helix the series of equatorial maxima should correspond to the maxima in $J_0(2\pi rR)$. The maxima on our photograph do not, however, fit this function for the value of r deduced above. There is a very strong reflexion at about 24 A. and then only a faint sharp reflexion at 9.0 A. and two diffuse bands around 5.5 A. and 4.0 A. This lack of agreement is, however, to be expected, for we know that the helix so far considered can only be the most important member of a series of coaxial helices of different radii; the non-phosphate parts of the molecule will lie on inner co-axial helices, and it can be shown that, whereas these will not appreciably influence the innermost maxima on the layer lines, they may have the effect of destroying or shifting both the equatorial maxima and the outer maxima on other layer lines.

Thus, if the structure is helical, we find that the phosphate groups or phosphorus atoms lie on a helix of diameter about 20 A., and the sugar and base groups must accordingly be turned inwards towards the helical axis.

Considerations of density show, however, that a cylindrical repeat unit of height 34 A. and diameter 20 A. must contain many more than ten nucleotides.

Since structure *B* often exists in fibres with low water content, it seems that the density of the helical unit cannot differ greatly from that of dry sodium thymonucleate, 1.63 gm./cm.^{3},[1,5] the water in fibres of high water-content being situated outside the structural unit. On this basis we find that a cylinder of radius 10 A. and height 34 A. would contain thirty-two nucleotides. However, there might possibly be some slight inter-penetration of the cylindrical units in the dry state making their effective radius rather less. It is therefore difficult to decide, on the basis of density measurements alone, whether one repeating unit contains ten nucleotides on each of two or on each of three co-axial molecules. (If the effective radius were 8 A. the cylinder would contain twenty nucleotides.) Two other arguments, however, make it highly probable that there are only two co-axial molecules.

First, a study of the Patterson function of structure *A*, using superposition methods, has indicated[6] that there are only two chains passing through a primitive unit cell in this structure. Since the $A \rightleftarrows B$ transformation is readily reversible, it seems very unlikely that the molecules would be grouped in threes in structure *B*. Secondly, from measurements on the X-ray diagram of structure *B* it can readily be shown that, whether the number of chains per unit is two or three, the chains are not equally spaced along the fibre axis. For example, three equally spaced chains would mean that the *n*th layer line depended on J_{3n}, and would lead to a helix of diameter about 60 A. This is many times larger than the primitive unit cell in structure *A*, and absurdly large in relation to the dimensions of nucleotides. Three unequally spaced chains, on the other hand, would be crystallographically non-equivalent, and this, again, seems unlikely. It therefore seems probable that there are only two co-axial molecules and that these are unequally spaced along the fibre axis.

Thus, while we do not attempt to offer a complete interpretation of the fibre-diagram of structure *B*, we may state the following conclusions. The structure is probably helical. The phosphate groups lie on the outside of the structural unit, on a helix of diameter about 20 A. The structural unit probably consists of two co-axial molecules which are not equally spaced along the fibre axis, their mutual displacement being such as to account for the variation of observed intensities of the innermost maxima on the layer lines; if one molecule is displaced from the other by about three-eighths of the fibre-axis period, this would account for the absence of the fourth layer line maxima and the weakness of the sixth. Thus our general ideas are not inconsistent with the model proposed by Watson and Crick in the preceding communication.

The conclusion that the phosphate groups lie on the outside of the structural unit has been reached previously by quite other reasoning.[1] Two principal lines of argument were invoked. The first derives from the work of Gulland and his collaborators,[7] who showed that even in aqueous solution the —CO and —NH_2 groups of the bases are inaccessible and cannot be titrated, whereas the phosphate groups are fully accessible. The second is based on our own observations[1] on the way in which the structural units in structures *A* and *B* are progressively separated by an excess of water, the process being a continuous one which leads to the formation first of a gel and ultimately to a solution. The hygroscopic part of the molecule may be presumed to lie in the phosphate groups ($(C_2H_5O)_2PO_2Na$ and $(C_3H_7O)_2PO_2Na$ are highly hygroscopic[8]), and the simplest explanation of the above process is that these groups lie on the outside of the structural units. Moreover, the ready availability of the phosphate groups for interaction with proteins can most easily be explained in this way.

We are grateful to Prof. J. T. Randall for his interest and to Drs. F. H. C. Crick, A. R. Stokes and M. H. F. Wilkins for discussion. One of us (R. E. F.) acknowledges the award of a Turner and Newall Fellowship.

Rosalind E. Franklin
R. G. Gosling
Wheatstone Physics Laboratory, King's College, London.
April 2.

1. Franklin, R. E., and Gosling, R. G. (in the press).
2. Cochran, W., Crick, F. H. C., and Vand, V., *Acta Cryst.*, 5, 501 (1952).
3. Pauling, L., Corey, R. B., and Bransom, H. R., *Proc. U.S. Nat. Acad. Sci.*, 37, 205 (1951).
4. Pauling, L., and Corey, R. B., *Proc. U.S. Nat. Acad. Sci.*, 39, 84 (1953).
5. Astbury, W. T., Cold Spring Harbor Symp. on Quant. Biol., 12, 56 (1947).
6. Franklin, R. E., and Gosling, R. G. (to be published).
7. Gulland, J. M., and Jordan, D. O., Cold Spring Harbor Symp. on Quant. Biol., 12, 5 (1947).
8. Drushel, W. A., and Felty, A. R., *Chem. Zent.*, 89, 1016 (1918).

18

THE STRUCTURE OF PROTEINS

A COUPLE OF YEARS AGO, on a tour through Germany, I visited the great cathedral in Cologne. Inside, the massive structure seemed to float on its delicate ribbings and shafts, the impossibly high ceiling hung like a mountaintop on the peaked arches, the stained glass windows sang with color and light. I was an ant, in the presence of something far larger than myself. A similar sensation comes over me in the towering cathedrals of Chartres, Notre Dame, Amiens, and Salisbury. These are buildings built to inspire, and they do. These are forms that serve function.

Form often serves function in the animal world as well. The great blue heron, for example, wades in shallow water for its food and has long slender legs for the purpose. So as not to break those legs upon landing, it must be able to fly slowly. Large wings accomplish that feat. The long necks of giraffes allow these animals to browse on foliage from the tall trees of their environment. And so on.

As scientists have found in the last fifty years, form also serves function in the invisible world of the atom. The organic molecules that make life are not flat connections of carbon and oxygen and hydrogen atoms, as might be suggested by chemical diagrams, but instead complex structures that twist and turn through three-dimensional space. In a single molecule, one might find arches and gates and stairways and spirals—all ultrasmall but there nonetheless. When biologists began unraveling the structures of organic molecules in the 1930s, they suspected that these exquisite architectural features were not simply accidents.

In 1959, working at the famous Cavendish Laboratory of Cambridge University, the biochemist Max Perutz and his colleagues succeeded in discovering the structure of hemoglobin. Hemoglobin is the molecule that carries oxygen to living tissues and cells. As we saw in Chapter 14 on Hans Krebs, oxygen helps produce the energy required for life. The hemoglobin molecule must be able to absorb oxygen in the high-oxygen climate of the lungs and then release it in the low-oxygen environment of individual cells. (Hemoglobin also facilitates the transport of carbon dioxide in the opposite direction.) To carry out this task, the molecule consists of about ten thousand atoms, grouped in four chains that bend and twist about like four tangled snakes. In its own way, a molecule of hemoglobin is a tiny cathedral. By 1970, Perutz was able to show how the fabulous structure of this vital molecule indeed serves its function.

Hemoglobin is a protein. In many respects, proteins are the workhorses of the body. Enzymes, which promote biochemical reactions, are proteins. Hormones are proteins. Gamma globulins, soldiers of the immune system, are proteins. Some proteins cause muscle contractions. Some store nutrients in milk. Some store iron in the spleen. Before Perutz's work, no one knew what a protein looked like or how it worked. Hemoglobin, and its smaller cousin, myoglobin, were the first proteins whose structures became known.

It took Max Perutz twenty-two years to unravel the structure of hemoglobin and another ten to figure out how it worked. During much of that period, from the mid-1940s to the end of the 1950s, Perutz helped lead two revolutions in science. First, the tools and thinking of physics were applied to biology. Here were the beginnings of molecular biology, the study of biological systems at the ultrasmall level of atoms and molecules. Other actors in this new drama included Linus Pauling, James Watson, and Francis Crick, scientists we have previously met. In 1947, Perutz became founding director of the new Medical Research Council for Molecular Biology at the Cavendish, an extraordinary synthesis of biology and physics.

The second revolution was the transition from the "small science" of the nineteenth century and earlier to the "big science" of the mid-twentieth century and beyond. Big science is driven largely by the complex instruments and equipment needed to pursue science at the cutting edge of today—such instruments as the X-ray diffraction machines and analyzers employed by Perutz and his colleagues or the subatomic particle accelerators used by physicists or the earth-orbiting telescopes launched by astronomers. Unlike the more modest experimental demands of earlier science, the equipment and analyses of big science require large teams of scientists and huge financial support. Thus, in the 1960s and later, we begin to see scientific papers coauthored by half a dozen people or more.

Max Perutz oversaw all of these developments in his own career. He was a fiercely committed scientist who spent over thirty years on the hemoglobin problem alone, a man who fought passionately and sometimes bitterly for his social convictions as well as his science, a man described by a reporter for *The Guardian* as possessing "a mind like a razor and an elegant command of language," described by colleague Alexander Rich as having an "outward manner that was reserved and quiet, but underneath a finely honed sense of humor," a man whose sense of fairness and generosity led him to include in his Nobel Prize lecture of 1962 a detailed description of the contributions of twenty-one scientists who had helped him along the way.

Perutz revealed something of the personal qualities he admired in his eulogy of John Kendrew, who became his research student in 1945 and later discovered the structure of myoglobin: "I found in Kendrew an outstandingly able, resourceful, meticulous, brilliantly organized, knowledgeable, hard worker and a stimulating companion with wide interests in science, literature, music and the arts." Like Otto Loewi, Perutz himself was uncommonly cultured and well read. He was also an excellent writer and a regular contributor to the literary *New York Review of Books*.

Perutz worked right up to his death in early 2002, at the age of eighty-seven. Even after his official retirement in 1980, he published over one hundred scientific papers. When once asked why he hadn't stopped working when he retired, Perutz replied that he was "tied up in some very interesting research at the time."

Hemoglobin was discovered in 1864 by the German physiologist and chemist Felix Hoppe-Seyler. Hoppe-Seyler understood the function of hemoglobin and, using the chemical methods of the time, was also able to determine its chemical composition.

As with all organic molecules, hemoglobin is made mostly of carbon, hydrogen, and oxygen atoms. Carbon is the primary atom of life. Because of its ability to share a relatively large number of electrons with other atoms (see Chapter 11), carbon can form lots of chemical bonds, in a variety of ways, and thus create the complex molecules needed for life. Proteins, fats, and carbohydrates all have rings and chains of atoms held together by carbon.

The proteins, like hemoglobin, are further distinguished by being constructed of twenty building blocks called amino acids. All amino acids have the composition shown in Figure 18.1: a central carbon atom

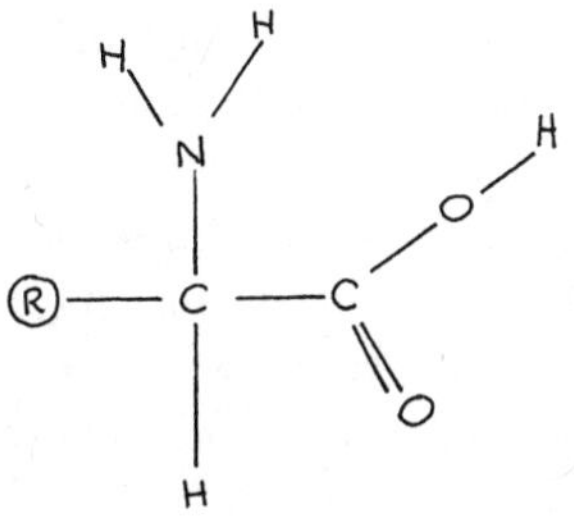

Figure 18.1

bonded to a nitrogen atom on top, another carbon atom to the right, and a hydrogen atom below. (The nitrogen and second carbon atoms are further bonded to hydrogen and oxygen atoms as shown.) The twenty amino acids differ from one another in the complex of atoms denoted by R, which contains from one to eighteen atoms. Particular amino acids have names like serine, asparagine, histidine, cysteine. A single molecule of hemoglobin harbors some 574 amino acids, linked together in long chains.

What gives hemoglobin its name are its additional "heme" (spelled "haem" in Perutz's paper) groups of atoms. A heme group is shown in Figure 18.2. Here we see a network of ringlike structures, mostly made out of carbon, surrounding a central black dot. That dot is an atom of iron. The iron atom is the single atom that captures oxygen, and the four nitrogen atoms garrisoned around it serve as gates that usher in or

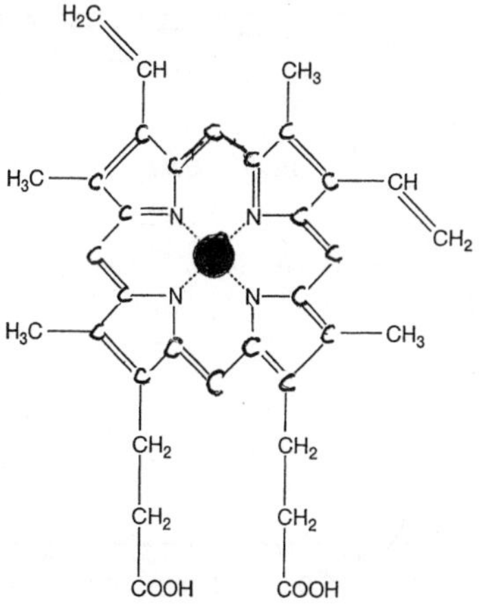

Figure 18.2

exclude the oxygen. Each molecule of hemoglobin contains four heme groups. Thus, out of more than ten thousand atoms in all, only four of those atoms are iron. The four iron atoms are the power brokers of hemoglobin. They form a Roman quadrumvirate of sorts.

Finally, the "globin" derives from the shape of the molecule as a whole. Each heme group connects to a long chain of amino acids, with four chains in all. Although these chains weave and twist all around, together they form roughly a sphere, or a "globular." About 300 million of these tiny spheres inhabit each red blood cell.

We can better understand the function of hemoglobin and some of its mysteries by measuring how it absorbs oxygen at different pressures. We are all familiar with taking our blood pressure. As blood moves through the arteries and veins, pushed by the contractions of the heart, it experiences a varying pressure in different parts of the body. Such pressures are usually measured in units of millimeters of mercury. A pressure of 100 millimeters of mercury, for example, is the pressure exerted by the weight of a vertical column of mercury 100 millimeters tall. The pressure of the air around us, at sea level, is 760 millimeters of mercury.

Figure 18.3 shows hemoglobin's percentage of saturation with oxygen at a range of pressures. For example, at a pressure of 100 millimeters of mercury, typical of the pressures in the arteries where hemoglobin first receives oxygen from the lungs, the molecules absorb 95 percent of the maximum amount of oxygen they can hold. At the lower pressure of 30 millimeters of mercury, typical of the pressure in the veins where hemo-

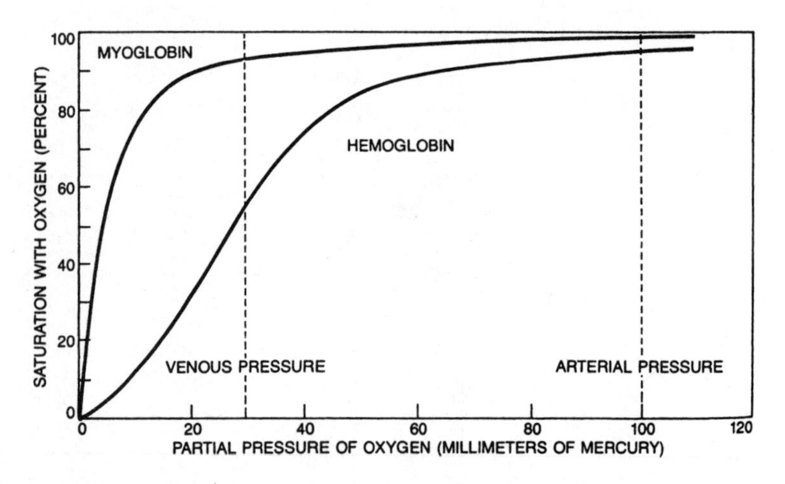

Figure 18.3

globin must release its oxygen to the needy tissues and cells, it has let go of almost half of its oxygen, with an absorption down to about 50 percent.

The saturation properties of another oxygen-carrying molecule, myoglobin, are shown for comparison. Myoglobin, found in red muscle, combines with oxygen released by red cells and transports it to the place where energy is produced. Myoglobin contains a single heme group. It is very similar in composition to any one of the four chains of hemoglobin. And here we come to a revealing difference between these two molecules: myoglobin clings to its oxygen much more jealously than does hemoglobin at low pressures, as can be seen in the figure. For this reason, myoglobin cannot be the principal oxygen carrier in animals. It does not release its oxygen easily enough. A person with only myoglobin in his blood would quickly die from asphyxiation.

Thus we arrive at a major puzzle of hemoglobin. If hemoglobin were simply four myoglobin molecules tied together, its oxygen-binding properties would be identical to myoglobin and therefore unworkable. Yet each of its four chains is similar to myoglobin. Somehow, the four chains together work in a way that none can alone. How do they perform such a trick? And what is the architectural arrangement required? These were questions that haunted Max Perutz.

Perutz was born in Vienna in May 1914. Both his father and mother descended from families of wealthy textile producers, who had introduced mechanical weaving into the Austrian monarchy. Max's parents urged him to study law, so that he could promote the family business, but young Perutz decided to try his mind at chemistry instead. At the University of Vienna, Perutz, in his words, "wasted five semesters in an exacting course of inorganic analysis" before turning to organic biochemistry. One of Perutz's early interests was the mechanics and structure of ice. In 1936, with the financial support of his father, the twenty-two-year-old young man left Vienna for Cambridge and the Cavendish Laboratory.

At the Cavendish, Perutz hoped to pursue his Ph.D. in biochemistry, working for John D. Bernal, an expert in X-ray diffraction. Since the 1920s, X-ray diffraction had been the principal means for exploring molecular structure. (For a review, see Chapter 7.) And the Cavendish was a leading center of this powerful technique. Lawrence Bragg, one of its pioneers, worked at the Cavendish.

In 1937, a friend in Prague suggested to Perutz that he study the structure of hemoglobin. Hemoglobin was a good molecule for analysis because, even at 10,000 atoms, it was one of the smallest proteins. Furthermore, good crystals of it could be made, a prerequisite for X-ray diffraction. Another colleague grew crystals of hemoglobin for Perutz, and Bernal taught the young man how to make X-ray pictures and interpret them. By 1938, Perutz was publishing his first X-ray diffraction pictures of hemoglobin. However, the pictures were inconclusive, and it soon became clear that the road to understanding would be difficult and long. Fortunately, Perutz began receiving encouragement and support from Bragg, the new director of the Cavendish, a Nobel laureate and a father figure for Perutz.

For a budding scientist, the Cavendish Laboratory was a dream. With its cobbled courtyard, stone archways, and massive oak gates that were locked and unlocked twice a day with the clanking of keys, the Cavendish was probably the most famous laboratory for experimental physics in the world. Founded in 1871, the Cavendish had seen four directors before Bragg: the great electromagnetic theorist, James Maxwell (1871–1879); John William Strutt, Lord Rayleigh (1879–1884); the discoverer of the electron, Joseph John Thomson (1884–1919); and the booming-voiced nuclear physicist, Ernest Rutherford (1919–1937), whom we have already encountered. The Cavendish was producing Nobel Prizes as if they were grammar school spelling awards. Rayleigh, Thomson, and Rutherford had all won Nobels. So had Thomson's student at the Cavendish, Francis Aston, as had Rutherford's student James Chadwick. Two other Rutherford students at the Cavendish, John Cockcroft and Patrick Blackett, were destined to win the great prize. At the associated Physiology Lab of Cambridge, Alan Hodgkin and Andrew Huxley were to win Nobels. Watson and Crick, with the help of X-ray diffraction, were to win their Nobels at the Cavendish. And Perutz would live to see his own laboratories at the Cavendish produce nine Nobel Prizes, including his own.

The year was 1938. World War II had drastic consequences for Perutz, as it did for so many other European scientists. He had barely begun his work on hemoglobin when Hitler invaded Austria. Perutz's family business in Vienna was appropriated by the Germans, his parents became refugees, and he was left with dwindling funds, saved only by an assistantship with Bragg. However, the worst was not over. In the spring of 1940, Perutz, along with other Germans and Austrians living in England, was interned. A few months later he was banished to Newfoundland aboard the transatlantic liner *Arandora Star*. At the beginning of July the ship was torpedoed by a German submarine. Perutz, clinging to floating debris in water aflame with burning diesel oil, nearly drowned. Out of 1,800 people on board, most lost their lives. When the young biochemist was finally rescued and nursed back to health, his "enemy alien" status was reversed with the help of the British Broadcasting Corporation, for whom he had worked as a journalist. He returned to his research at the Cavendish.

The process of producing and analyzing the X-ray photographs of a molecule like hemoglobin, without the sophisticated computers of today, was a Herculean task. Photographs had to be taken from different angles, and the number of diffraction "spots" ran into the tens of thousands. In the preface to his book *I Wish I'd Made You Angry Earlier* (1998), Perutz gives some sense of the labors involved:

> I took several hundred X-ray diffraction pictures of hemoglobin crystals, each taking two hours' exposure. I took some of the pictures during World War II, when I had to spend nights in the laboratory in order to extinguish incendiary bombs in the event of a German air raid. I used these nights to get up every two hours, turn my crystal by a few degrees, develop the exposed films, and insert a new pack of films into the cassette. When all the photographs had been taken, the real labour only began. Each of them contained several hundred little black spots whose degree of blackness I had to measure by eye, one by one. These numbers outlined not a picture of the structure I was trying to solve, but a mathematical abstraction of it: the directions and lengths of all the 25 million lines between the 10,000 atoms in the hemoglobin molecule radiating from a common origin.

To understand Perutz's "mathematical abstraction" a bit better, let us look at a small sample of an X-ray photograph, Figure 18.4. Here we see a number of spots, where the size of each spot represents the intensity of the X-rays landing there. Such spots are formed by the overlapping waves of X-rays deflected from the electrons of a molecule and then landing on the photographic film. (See Chapter 7 on von Laue.) To work back from the spots to a picture of the electrons and atoms that created them is like deducing the positions of upright sticks in a stream from the pattern of water waves downstream. It is not a straightforward proce-

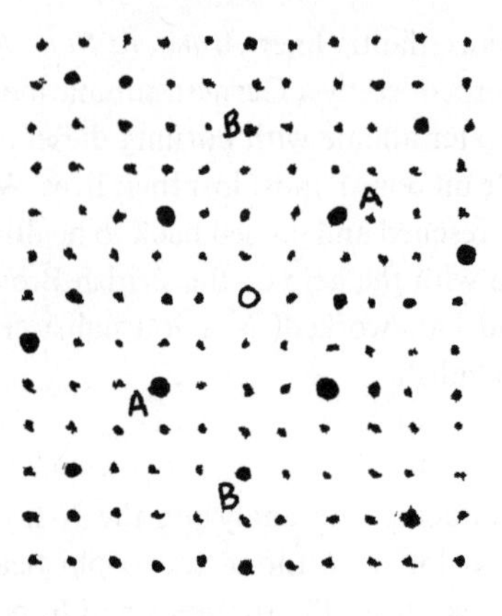

Figure 18.4

dure. As it turns out, the *positions* of the spots are determined by the repeating patterns of many molecules, or "unit cells," while the *intensities* reveal the arrangement of atoms within a single molecule. The latter was what Perutz wanted to know.

The procedure is roughly as follows. Each pair of symmetrical spots in Figure 18.4 is produced by electrons in the planar slices through the molecule shown in Figure 18.5. For example, the spots labeled A are produced by the slices shown in Figure 18.5a, while the spots labeled B are produced by the slices in Figure 18.5b. The intensity of the spots corresponds to the density of electrons in each slice. In turn, the electron density tells us about the placement of atoms and groups of atoms, since each atom contains a known number of electrons. Thus, slice by slice, we create a map of the original molecule.

Perutz's landmark paper of 1960 is one of the most technical treatises we have considered so far. In this paper, he analyzes horse hemoglobin—

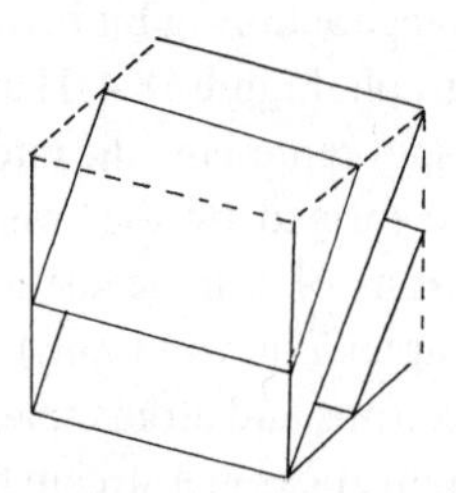

Figure 18.5a

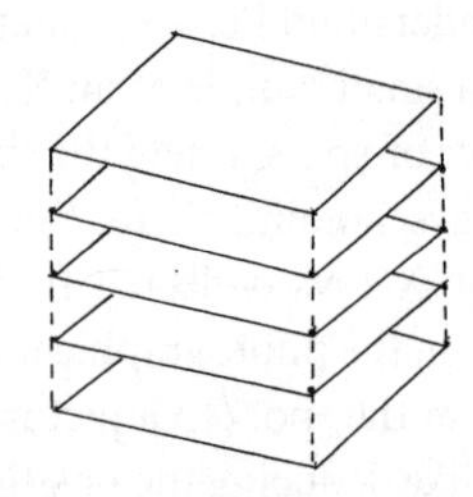

Figure 18.5b

quite similar to the hemoglobin in all vertebrate animals—and in the particular form, called oxyhemoglobin, in which it is saturated with oxygen. The form of hemoglobin without oxygen is called deoxyhemoglobin, or reduced hemoglobin. In the first paragraph, Perutz acknowledges that he will not be able to determine the placement of individual amino acids. That is because the resolution of his X-ray photographs is only 5.5 Å, the smallest size for which he can make out details. (Recall that Å stands for 10^{-8} centimeters, or one hundred millionth of a centimeter, the yardstick of distance in the atomic domain.) On average, a cube of size 5.5 Å might contain 25–50 atoms, equal to several amino acids. Thus, Perutz cannot see individual amino acids. An entire hemoglobin molecule, on the other hand, is about 50 Å in diameter. By analogy, if the hemoglobin molecule were a terrestrial globe, Perutz can pinpoint countries the size of Venezuela and larger. In particular, Perutz's X-ray diffractions have sufficient resolution to discover that hemoglobin consists of four pieces, or subunits. The Fourier synthesis method is the procedure, described earlier, of going backward from the diffraction spots to the positions of groups of atoms in the original molecule.

In the "Method of Analysis" section, Perutz refers to a determination of the "phase angle" of the "reflexions." The reflexions are simply the spots, as in Figure 18.4 above, formed by the deflection and overlap of the incident X-rays. The "phase angle" of a spot refers to whether the X-ray wave was at a crest or a trough or somewhere in between when it struck the photographic film at that point. All waves have phase angles. To work back to a map of the molecule from its X-ray diffraction photograph requires that the phase angles of all the spots be known. This determination is the most difficult and arcane aspect of the project, as the intensity of a spot does not reveal its phase angle. Here, Perutz refers to a method of "isomorphic replacement" that he himself refined in 1953. In this method, particular atoms in hemoglobin are replaced by heavy atoms, which don't change the shape of the molecule but do alter the diffraction photograph in a way that depends on the phase angles. By comparing X-ray photographs with and without heavy atom replacements, one can determine the phase angles.

Figure 1 of Perutz's paper shows a reconstructed section of the molecule, with contours of electron densities like contours of elevation in a mountain range. (Electron densities are quoted in units of electrons/$Å^3$.) Such a section is created by adding together many groups of slices of the form shown in Figure 18.5 above. The "unit cell" is the smallest region that repeats itself throughout a crystal of hemoglobin. Perutz and his colleagues find that a unit cell of hemoglobin has two molecules. The "dyad axis" is an axis of symmetry, like the line through the center of a playing card: a rotation of 180 degrees about a dyad axis brings the structure back to its original form.

In the next section, Perutz describes how he has obtained the precise three-dimensional positions of the four heme groups. These are shown in Figure 4 of his paper.

In reading this paper, one senses that Perutz and his colleagues are molecular geographers. Furthermore, they are toiling in three dimensions, not two. To map the structure of hemoglobin, they want to visualize the molecule in space. Indeed, understanding complex structures in science often requires pictures as well as equations. What computers give the scientists are contours of electron density, rich in information but not sufficient in visual terms. In a fascinating technique outlined in "Configuration of the Polypeptide Chains," Perutz describes a process of rolling out sheets of plastic cut to the shapes of his computer-generated electron density maps and then building up a 3-D model from the sheets of plastic. (In a similar manner, Watson and Crick built toy models of DNA.)

The results are shown in Figures 5–10 of the paper. Hemoglobin is found to consist of four convoluted chains, grouped in two identical pairs called the two white chains and the two black chains. (Alternatively, these pairs are sometimes called the alpha chains and the beta chains.) The white chains and black chains differ by only a few amino acids.

In the section titled "Arrangement of the Four Sub-units," note that the contours of the black chains exactly fit those of the whites, like neighboring pieces of a jigsaw puzzle. As Perutz writes: "The structural complementarity is one of the most striking features of the molecule." In short, the chains fit together. Indeed, in many places of this paper, Perutz and his colleagues must stretch for language to describe the strange and beautiful structures they have discovered. In one section we read about "four tortuous clouds" of high electron density, in another about the "S-shaped bend at the top."

Some details of terminology: The "N" and "C" terminals refer to whether the last amino acid in the long chain of amino acids has a free nitrogen atom or a free carbon atom (see Figure 18.1 above). The "His" in Figure 8 of Perutz's paper refers to the amino acid histidine, which attaches to each heme group and connects it to the rest of the amino acid chain.

Figures 8–10 give different views of the full molecule. In Figure 8 each heme group, represented by a disk, is folded inside a coil of amino acids like a bud within the petals of a flower.

The clear separation of the heme groups poses a problem, for it leaves unanswered the major question of how the heme groups act in concert. As Perutz says in the "Discussion" section, "The haem groups are much too far apart for the combination with oxygen of any one of them to affect the oxygen affinity of its neighbors directly. Whatever interaction between the haem groups exists must be of a subtle and indirect kind that we cannot yet guess." One can almost hear the tone of frustration when Perutz writes, "Little can be said as yet about the relation between structure and function."

Indeed, to answer that question, Perutz and his colleagues need to solve for the structure of the other form of hemoglobin without oxygen, deoxyhemoglobin, and then compare the two. At the end of his paper, Perutz correctly speculates that the oxy and deoxy structures differ not in the makeup of the individual four chains but in the arrangement of the chains relative to one another.

By 1962, Perutz had analyzed both oxy and deoxy forms and proven this conjecture. The beta (black) chains shift by over 7 Å relative to each other in going from the oxy to the deoxy structure.

Yet even after solving for the two structures, Perutz could not fully understand how hemoglobin functioned. In his Nobel Prize lecture of 1962, he explained the continuing challenge in this way: "The wide distances separating the haem groups was perhaps the greatest surprise the structure presented to us, for one would have expected the chemical interaction between them to be due to their close proximity. As it stands, the structure of oxyhemoglobin leaves its physiological properties unexplained."

It was only over the period from 1962 to 1970 that Perutz and his colleagues discovered how the four heme groups work *cooperatively* to bind and unbind with oxygen, the critical function of hemoglobin. The answer involves oxygen "gates" made out of nitrogen atoms, movements in three dimensions, and mechanical linkages between heme groups using amino acids hooked together like Tinkertoys. The four chains are weakly linked together by electrical forces, so that a slight shift in one can cause a shift in the others. Perutz found that when an oxygen molecule is absorbed by one heme group, the iron atom of that group moves down into the two-dimensional plane formed by the four nitrogen atoms attached to it. This movement of the iron atom, through the histidine amino acid attached to it and the linkages to other amino acids and chains, causes the other three iron atoms also to move down to the planes of their associated nitrogen gates. Those gates then widen and allow oxygen to enter more freely. In this way, the four heme groups act together. A partial illustration for a single heme group is suggested in Figure 18.6—although much of the movement occurs perpendicularly to the plane of the figure. Here, the position of oxyhemoglobin is shown by the dashed lines, the position of deoxyhemoglobin by solid lines.

The quest to understand the structure and mechanism of hemoglobin required many more years than did the other scientific discoveries we have considered. In this respect, Perutz's discovery was quite different from Rutherford's discovery of the nucleus of the atom, Loewi's of the mechanism of transmission between nerves, or Fleming's of penicillin. As early as 1938, Perutz had a clear idea of what he wanted to do and its significance. It was the sheer complexity of the project that demanded so many years and collaborative efforts. And, as we have seen, even after Perutz had found the structure of hemoglobin, it still took nearly a decade to relate the structure of this vital molecule to its function.

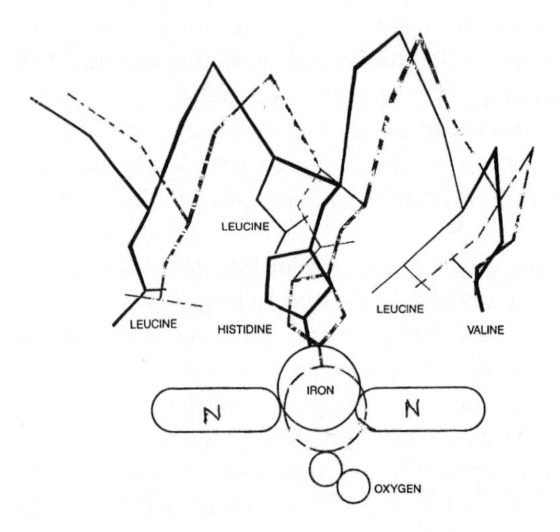

Figure 18.6

In his work on hemoglobin, Perutz improved and refined the techniques of X-ray diffraction, leading to the discovery of the structure of many other proteins. For example, we now know the architecture of amylase, which converts starch to glucose; potassium channel proteins, which regulate the electrical properties of nerve cells; and vasopressin, which increases the retention of water in the kidneys. Today, thousands of protein structures are solved each year. Knowledge of these structures has increased basic understanding of biology as well as facilitated treatment of disease. Perutz himself showed how mutations in hemoglobin could cause sickle-cell anemia. (Linus Pauling had already discovered that sickle-cell amenia was a "molecular disease.") In a broad sense, Perutz helped demonstrate the exquisite intricacy of form in biological molecules. It is amazing that the effective transport of oxygen in animals requires a molecule of more than ten thousand atoms, grouped in a particular and complex way. Perhaps a cathedral without its stained glass windows, without its intimate stone carvings and towering ceiling, could not inspire.

During much of his adult life Max Perutz kept a journal in which he wrote down any wise sayings he found. He called this journal his Commonplace Book, a name that goes back to the ancient Greek and Roman orators, who collected metaphors for their speeches. Perutz's Commonplace Book provides a glimpse into his heart. One of the entries is "Science without conscience is the ruin of the soul," from Rabelais's *Pantagruel.* Another is "It is not true that the scientist goes after the truth. It goes after him," from Robert Musil's *Der Mann ohne Eigenschaften*. Yet another is one of Perutz's own: "What is known for certain is dull."

This last phrase echoes a sentiment Perutz expressed at the end of his Nobel address—a moment during which, although having just received the highest honor in science for solving the structure of hemoglobin, he was still struggling to understand how that structure served function: "Please forgive me for presenting, on such a great occasion, results which are still in the making. But the glaring sunlight of certain knowledge is dull and one feels most exhilarated by the twilight and expectancy of the dawn."

For Perutz, the unknown was more invigorating than the known. Similarly, the novelist is propelled by what he doesn't understand about his characters; the painter constantly searches for some inexplicable quality of atmosphere and being that cannot be captured by the camera. It is the misty and uncertain forms in the twilight, perhaps, that nourish all creative activity. Einstein phrased the thought well when he wrote: “The most beautiful experience we can have is the mysterious. It is the fundamental emotion which stands at the cradle of true art and true science.”

STRUCTURE OF HÆMOGLOBIN

A THREE-DIMENSIONAL FOURIER SYNTHESIS AT 5·5-Å. RESOLUTION, OBTAINED BY X-RAY ANALYSIS

Dr. Max F. Perutz, F.R.S., Dr. M. G. Rossmann, Ann F. Cullis, Hilary Muirhead, and Dr. Georg Will

Medical Research Council Unit for Molecular Biology, Cavendish Laboratory, University of Cambridge

Dr. A.C.T. North

Medical Research Council External Staff, Davy Faraday Research Laboratory, Royal Institution, London, W.1

Nature (1960)

VERTEBRATE HÆMOGLOBIN IS A PROTEIN of molecular weight 67,000. Four of its 10,000 atoms are iron atoms which are combined with protoporphyrin to form four hæm groups. The remaining atoms are in four polypeptide chains of roughly equal size, which are identical in pairs.[1–3] Their amino-acid sequence is still largely unknown.

We have used horse oxy- or met-hæmoglobin because it crystallizes in a form especially suited for X-ray analysis, and employed the method of isomorphous replacement with heavy atoms to determine the phase angles of the diffracted rays.[4–7] The Fourier synthesis which we have calculated shows that hæmoglobin consists of four sub-units in a tetrahedral array and that each sub-unit closely resembles Kendrew's model of sperm whale myoglobin.[6] The four hæm groups lie in separate pockets on the surface of the molecule.

METHOD OF ANALYSIS

Horse oxyhæmoglobin, crystallized from 1.9 *M* ammonium sulphate solution at *p*H 7, has the space group *C*2 with two molecules in the unit cell which lie on dyad axes.[8] In order to determine the phase angles of the 1,200 reflexions contained in the limiting sphere of 5.5 Å.$^{-1}$, six different isomorphous heavy-atom compounds were used (ref. 9 and unpublished work). Intensities were measured photographically and by counter spectrometer (Arndt, U. W., and Phillips, D. C., unpublished work). The relative positions and shapes of the heavy-atom replacement groups were found by correlation functions based on Patterson methods[10] and refined by least squares procedures. For each reflexion the structure amplitudes of all seven compounds were combined in an Argand diagram,[11] and the probability of the phase angle having a value α was calculated for $\alpha = 0, 5, 10, \ldots 355°$. The centroid of the probability distribution, plotted around a circle, was then chosen as the best vector F in the Fourier synthesis.[12] The results were finally plotted on 32 contour maps showing the distribution of electron density in sections spaced 2 Å. apart normal to b (Fig. 1). The absolute configuration of the molecule was determined from anomalous dispersion.[4]

EXTERNAL SHAPE OF THE MOLECULE

More than half the volume of the crystals is taken up by liquid of crystallization, which mainly fills the spaces between the molecules and shows up on the contour maps in the form of flat, featureless regions (Fig. 1). The outlines of the two molecules in the unit cell can be traced by following the boundaries between these regions and the continuous

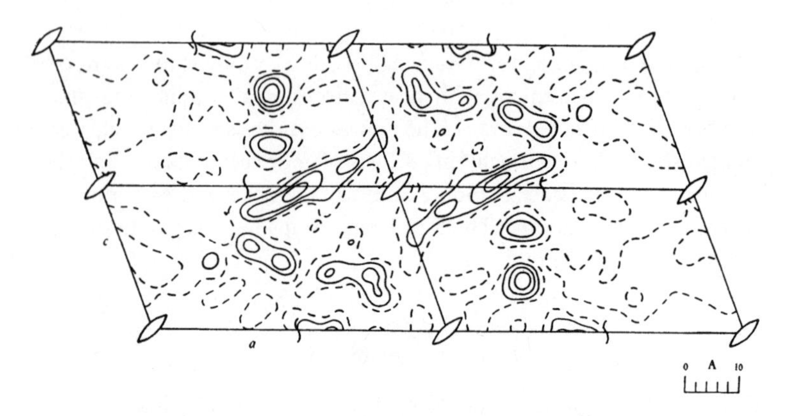

Figure 1. Section at $y = 1/32b$. This cuts through the middle of the molecule on which the diagram is centred. "Flat" areas indicating liquid appear on the left and right. Contours are drawn at intervals of 0.14 electron/Å.[3]. The broken line marks 0.4 electron/Å.[3]. Contours at lower levels are omitted.

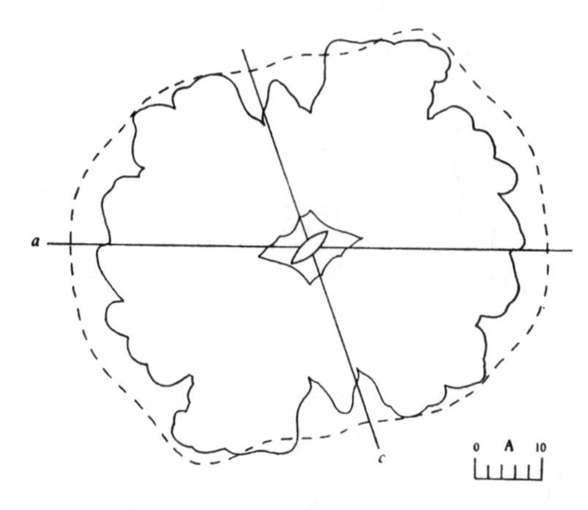

Figure 2. External shape of the molecule. Full lines indicate the boundary derived from the contour at 0·54 electron/Å.^{3}. The broken line shows the boundary derived by Bragg and Perutz from two-dimensional data (ref. 13). Note the hole in the middle.

electron-dense regions described below. In Fig. 2 the outline of one molecule, traced from the periphery of the 0.54 electron/Å.^{3} contour, is seen in projection on the *b*-plane. To a first approximation it can be regarded as a spheroid with a length of 64 Å., a width of 55 Å. and a height of 50 Å. normal to the plane of the paper. Except for a slight shortening along *a*, this shape agrees with the earlier picture obtained from two-dimensional data, even including the dimple in the centre of the molecule (see Fig. 2 of ref. 12).

POSITIONS OF THE HÆM AND SULPHYDRYL GROUPS

Four peaks stand out from the rest, clearly representing the iron atoms with their surrounding porphyrin rings. A flattening of the peaks indicates the approximate orientation of the rings (Fig. 3). Fortunately, their exact orientation was already known from electron spin resonance,[14] and it only remained to assign the correct one of the four alternative orientations of the hæms to each of the electron density peaks. The results are shown in Fig. 4. The iron atoms lie at the corners of an irregular tetrahedron with distances of 33.4 and 36.0 Å. between symmetrically related pairs. The closest approach between symmetrically unrelated iron atoms is 25.2 Å. (Table 1).

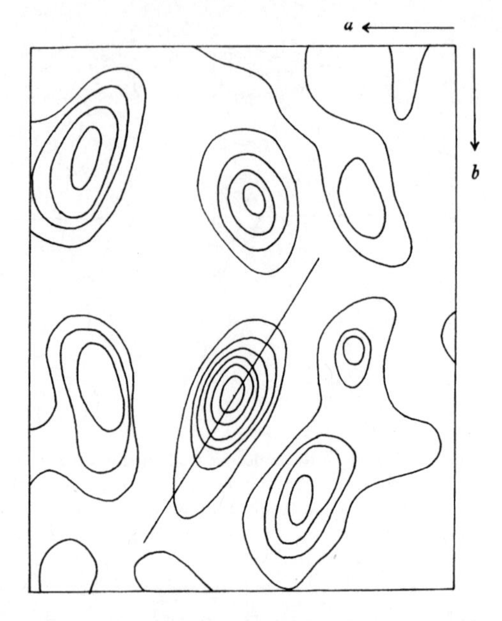

Figure 3. Hæm 1 and 2. Section at $z = \frac{1}{4} c$ showing one of the two hæm groups in the asymmetric unit. The straight line indicates the hæm orientation derived from electron spin resonance (ref. 14). The lowest contour shown is at 0.4 electron/Å.[3].

TABLE 1

	Fe_1	Fe_3	S_1
x	−6.6 Å.	12.3 Å.	5 Å.
y	7.3 Å.	−10.7 Å.	10 Å.
z	13.1 Å.	18.2 Å.	16 Å.
$Fe_1 - Fe_2$ =	33.4 Å.		
$Fe_3 - Fe_4$ =	36.0 Å.		
$Fe_1 - Fe_3$ =	25.2 Å.		
$Fe_1 - S_1$	13 Å.		
$Fe_3 - S_1$	21 Å.		

Horse hæmoglobin contains four cysteine residues, but only two sulphydryls combine with mercury in the native protein.[5] From the positions of the mercury atoms we inferred that each of the two sulphydryl groups is about 13 Å. away from one iron atom and 21 Å. away from another (Fig. 4 and Table 1). The significance of this situation is discussed below.

CONFIGURATION OF THE POLYPEPTIDE CHAINS

The most prominent feature of the Fourier synthesis consists of more or less cylindrical clouds of high density, like the vapour trails of an aeroplane; they are curved to form intricate three-dimensional figures. Sections through various parts of these appear in Fig. 1. To build a model of the figures, we rolled out sheets of a thermo-setting plastic to the thickness of our sections on a scale of 2 Å. = 1 cm., and cut out the shape of each region on the contour maps where the density exceeds 0.54 electron/Å.^{3} (this corresponds to the first full contour line in Fig. 1. From this section, for example, 14 shapes would be cut). The shapes were then assembled in accordance with their positions and heights in the different sections, and the hæm groups were attached in the appropriate orientation. The model was then baked to set it permanently. For comparison, a Fourier synthesis of sperm whale myoglobin was calculated at a resolution of 5.5 Å., using the new X-ray data of Kendrew *et al.*,[15] and a model of the electron density distribution was constructed by the method just described.

From the hæmoglobin Fourier synthesis there emerged four separate

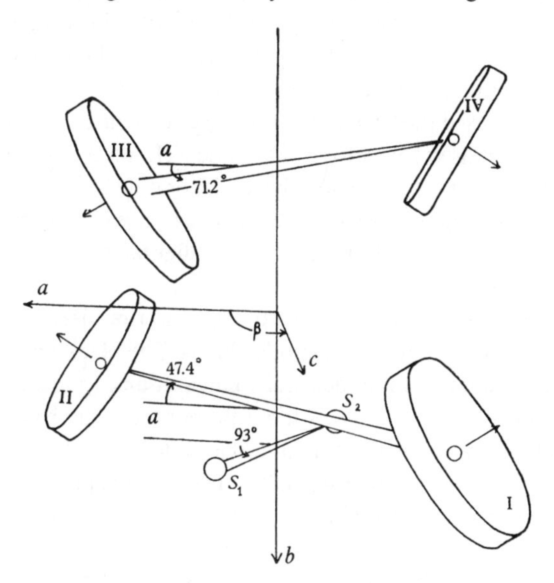

Figure 4. Arrangement of hæm groups in hæmoglobin. Arrows indicate the reactive side of each group. The *c*-axis comes out of the paper towards the observer. This picture and all subsequent ones show the absolute configuration.

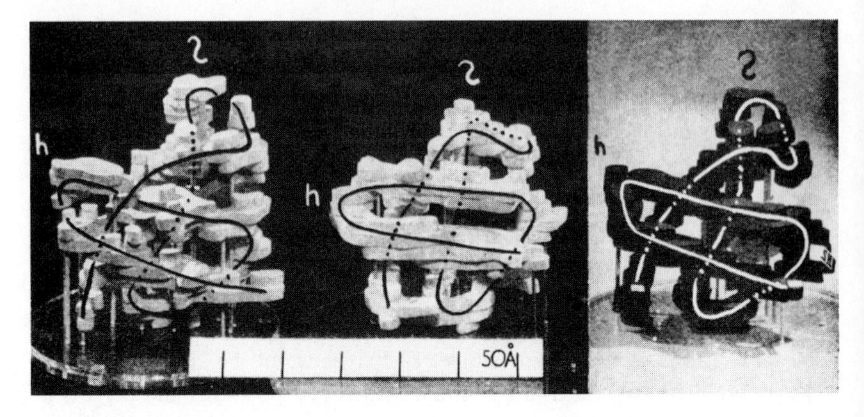

Figure 5. Two different polypeptide chains in the asymmetric unit of hæmoglobin compared with myoglobin (left). The hæm groups are at the back of the chains.

units which are identical in pairs. Fig. 5 shows one member of each pair together with the model of myoglobin on the left. In each unit the cloud of high density describes a complicated figure which, in the white unit (middle), can be traced from end to end by following the superimposed line. Except for two small gaps where the density sinks slightly below 0.54 electron/$Å^3$, the black unit (right) closely resembles the white one. There are several gaps interrupting the myoglobin model where, probably due to increased thermal motion, the electron density falls below the contour-level chosen for cutting the model. However, we know from the recent work of Kendrew and his collaborators[15] that the gaps are bridged by a continuous polypeptide chain, and it is evident from Fig. 5 that apart from the gaps, the model has a configuration closely similar to the hæmoglobin units.

Clearly, the four tortuous clouds of high electron density in hæmoglobin represent the four polypeptide chains. The black and the white chains have similar, but not identical configurations. In the black chain the *S*-shaped bend at the top is more pronounced, the hæm group is lower, and the bend *h* sharper than in the white chain. These, however, are details. The most important result is their resemblance to each other and to sperm whale myoglobin.

ARRANGEMENT OF THE FOUR SUB-UNITS

The first step in the assembly of the molecule is the matching of each chain by its symmetrically related partner (Fig. 6). It will be noted that

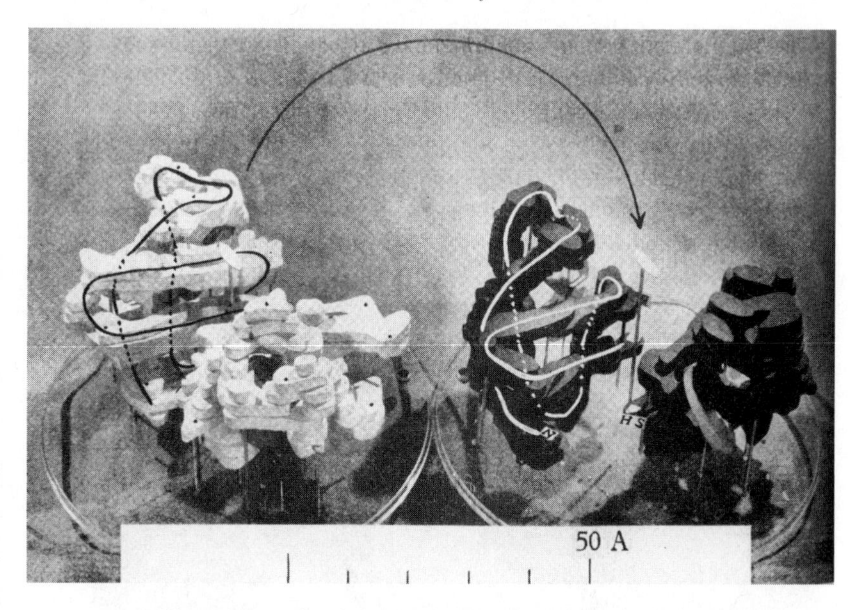

Figure 6. Two pairs of chains symmetrically related by the dyad axis. The arrow shows how one pair is placed over the other to assemble the complete molecule.

there is comparatively little contact between the members of each pair, suggesting rather tenuous linkages. In the next step the white pair is inverted and placed over the black pair as indicated by the arrow. Fig. 7 shows one white chain placed over the pair of black ones and Fig. 8 shows the molecule completely assembled. The resulting arrangement is tetrahedral and has almost, but not quite, the orthorhombic point group symmetry 222. It contains two "pseudo dyads" which lie approximately at right angles to each other and to the true dyad, one emerging from the centre of Fig. 8 and the other from the centre of Fig. 9 [omitted here]. This means that, to a first approximation, each sub-unit can be generated by a rotation of 180° from any of its neighbours. Figs. 7 and 8 also show that the surface contours of the white chains exactly fit those of the black, so that there is a large area of contact between them. This structural complementarity is one of the most striking features of the molecule. Fig. 10 is a view down the true dyad axis and reveals a hole going right through the centre of the molecule, as was to be expected from the Fourier projection.[13] However, the van der Waals radii of the chains are much bigger than appears in the model, and little room may, in fact, be left for water or electrolytes to pass through. Fig. 10 also reveals a dimple at the top where the white chains meet. This is matched by a similar, but larger, hollow at the bottom where the black chains meet.

The hæm groups are seen to lie in separate pockets on the surface of the molecule (Fig. 8). Each pocket is formed by the folds in one of the polypeptide chains, which appears to make contact with the hæm group at four different points at least. The iron atoms in the neighbouring pockets formed by the black and the white chains are 25 Å. apart.

INFORMATION FROM THE FOURIER SYNTHESIS OF MYOGLOBIN AT 2-Å. RESOLUTION (REF. 15)

Thanks to the similarity with myoglobin, the interpretation can be carried further than would have been possible on the basis of our results

Figure 7. Partially assembled molecule showing two black chains and one white.

alone. Kendrew *et al.* have found that the straight stretches of rod indicated in Fig. 5 are α-helices and that the N-terminal end of the chain is at the bottom left. The ends of the two hæmoglobin chains have been labelled *N* and *C* accordingly, as seen in Figs. 7 and 10. The myoglobin Fourier synthesis also reveals the hæm-linked amino-acid side-chain, probably histidine, on one side of the (ferric) iron atom and a small peak, probably representing a water molecule, on the other side. If this information is transferred to hæmoglobin, the reactive side of the hæm group is as indicated by the arrows in Fig. 4 and by the labels O_2 in Figs. 7 and 8*a*.

Figs. 7 and 8*b* show the reactive sulphydryl group (which is absent in myoglobin) to be attached to the portion of the black chain carrying the hæm-linked histidine. It is seen that the histidine and cysteine side-chains point in roughly opposite directions, one towards and the other away from the hæm group. The sulphydryl group may possibly be in contact with the loop of the white chain which lies below the hæm group on the left. This is the situation of one pair of sulphydryl groups in the molecule. The second pair is probably attached to the white chains, but it is unreactive in the native protein and its position is still unknown.

RELIABILITY OF RESULTS

The method of isomorphous replacement makes no assumptions about the structure of the protein, and its results suffer from none of the ambiguities which bedevil the interpretation of vector maps. Ideally, the parent protein, in combination with two different isomorphous heavy-atom compounds, should give accurate phase angles and electron density maps which are free from all except series termination errors. In practice, errors and uncertainties arise from several sources and must be minimized by using more than two heavy-atom compounds. In our case it was possible to estimate the accuracy of the phase angle for each reflexion from the measure of agreement between the angles indicated by the six different compounds. From the standard error in the vector *F*, averaged over all reflexions, the standard error in the final electron density was calculated as 0.12 electron/$Å.^3$, which amounts to 0.85 of the interval between successive contours in Figs. 1 and 3 (0.14 electron/$Å.^3$), or 0.15 of the difference in density between ‘peaks’ and ‘valleys’ (0.7 electron/$Å.^3$) (ref. 12). This calculated error is borne out by the observed fluctuations in the liquid regions between the molecules and by the difference in height between the two iron peaks (0.13 electron/$Å.^3$). An

(*a*)

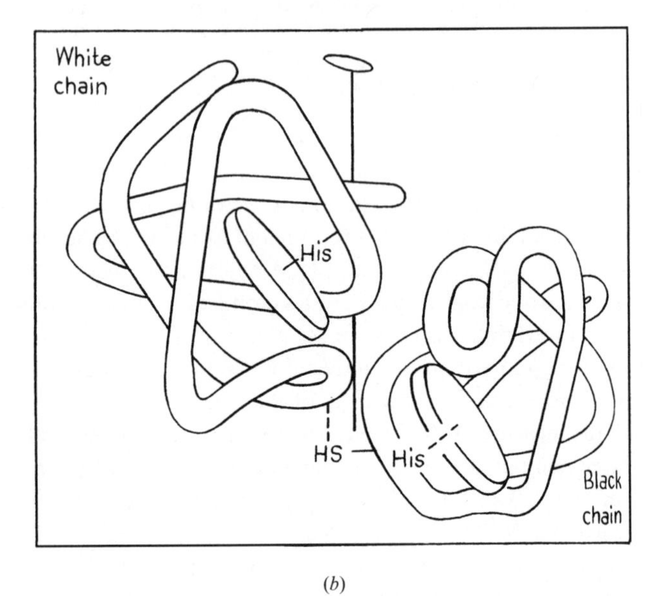

(*b*)

Figure 8. (*a*) Hæmoglobin model viewed normal to *a*. The hæm groups are indicated by grey disks. (*b*) Chain configuration in the two sub-units facing the observer. The other two chains are produced by the operation of the dyad axis.

error of at least five times this magnitude would be needed to turn a peak into a valley and thus to simulate the backbone of a polypeptide chain in a region actually occupied by side-chains.

Normally, the accuracy of a structure is checked by comparison of calculated and observed intensities. In hæmoglobin this is impossible so long as the light atoms are not resolved, but the positions of the iron atoms at least can be checked by their anomalous dispersion effect. Depending on the phase relationship between the structure factor of the four iron atoms by themselves and that of the entire molecule, the total intensity of any given reflexion $I(hkl)$ may be larger or smaller than $I(\bar{h}\bar{k}\bar{l})$. We selected 50 reflexions for which the effect of anomalous dispersion was expected to be largest and measured the intensity of each reflexion in the four symmetrically related quadrants on the counter spectrometer. Statistically significant differences between $I(hkl) + I(\bar{h}k\bar{l})$ and $I(\bar{h}\bar{k}\bar{l}) + I(h\bar{k}l)$ were found in 36 reflexions. In 34 of them the signs of the differences agreed with prediction, indicating that the phase angles of the reflexions as well as the positions of the iron atoms are correct.

As a further check on the positions of the iron atoms, an attempt was made to label them with heavy atoms. This was done by allowing crystals of methæmoglobin, which are isomorphous with oxyhæmoglobin, to react with *p*-iodophenyl-hydroxylamine, which attaches itself to the iron atoms. A different Fourier projection on the *b*-plane showed four prominent peaks which lie well within the calculated distances between the iodine and the iron atoms.

Finally, to avoid any possible bias in the interpretation of the Fourier synthesis, we have tried to construct objective models containing only those features actually found in the electron density maps. Due to series termination errors and inaccuracies in many of the phase angles, these maps must certainly contain errors of detail. Large errors, on the other hand, are unlikely because, quite apart from the checks just described, the results really prove themselves. No combination of errors could have led to the appearance of four distinct chains of roughly equal length, in agreement with chemical evidence, to the similarity between the two pairs of chains which are not related by crystal symmetry, and to the resemblance between hæmoglobin and myoglobin.

DISCUSSION

The polypeptide chain-fold which Kendrew and his collaborators first discovered in sperm whale myoglobin has since been found also in seal myoglobin.[16] Its appearance in horse hæmoglobin suggests that all hæmoglobins and myoglobins of vertebrates follow the same pattern. How does this arise? It is scarcely conceivable that a three-dimensional template forces the chain to take up this fold. More probably the chain, once it is synthesized and provided with a hæm group around which it can coil, takes up this configuration spontaneously, as the only one which satisfies the stereochemical requirements of its amino-acid sequence. This suggests the occurrence of similar sequences throughout this group of proteins, despite their marked differences in amino-acid content. This seems all the more likely, since their structural similarity suggests that they have developed from a common genetic precursor.

Little can be said as yet about the relation between structure and function. The hæm groups are much too far apart for the combination with oxygen of any one of them to affect the oxygen affinity of its neighbours directly. Whatever interaction between the hæm groups exists must be of a subtle and indirect kind that we cannot yet guess. A few observations of possible significance might be mentioned. Kendrew found that the combination of reduced myoglobin with oxygen involves no structural changes detectable by X-ray analysis,[17] both forms being isomorphous with the metmyoglobin normally studied. In the hæmoglobin of horse and of man, on the other hand, the oxygenated and reduced forms are crystallographically different.[18–20] The structure of reduced hæmoglobin is still unknown, but it would not be surprising if loss of oxygen caused the four sub-units to rearrange themselves relative to each other, rather than to change their individual structure to a marked degree.

Riggs has shown that blocking the sulphydryl groups reduces hæm-hæm interaction.[21] As Fig. 8 shows, these groups occupy key positions close to the hæm-linked histidines and to points of contact between two different sub-units. They may well play an important part in the transition between the oxygenated and reduced forms. Incidentally, the cysteine residue should provide a convenient marker for the peptide containing the hæm-linked histidine, and so help to determine the sequence in this important part of the chain.

A full account of this work will be published elsewhere.

We thank the Director and staff of the University of Cambridge Mathematics Laboratory for making their electronic computer *Edsac* II available to us for the many calculations involved in this work. We are also deeply grateful to the Rockefeller Foundation for its long-continued financial support and to Sir Lawrence Bragg for his unfailing enthusi-

Figure 10. A view down the *b*-axis. Note the proximity of the *C* and *N* terminal ends which could serve to form links between the two white chains.

asm and encouragement. Finally, we wish to thank Dr. B. R. Baker, of the Stanford Research Institute, for a gift of organic mercurials, Dr. J. Chatt of Imperial Chemical Industries, Ltd., and Mr. A. R. Powell, of Johnson Matthey, Ltd., for gifts of heavy-atom compounds, and Mrs. Margaret Allen, Miss Ann Jury and Miss Brenda Davies for assistance.

1. Rhinesmith, S. H., Schroeder, W. A., and Pauling, L., *J. Amer. Chem. Soc.*, **79,** 4682 (1957).

2. Braunitzer, G., *Hoppe-Seyl. Z.*, **312,** 72 (1958).

3. Wilson, S., and Smith, D. B., *Can. J. Biochem. Physiol.*, **37,** 405 (1959).

4. Bokhoven, C., Schoone, J. C., and Bijvoet, J. B., *Acta Cryst.*, **4,** 275 (1951).

5. Green, D. W., Ingram, V. M., and Perutz, M. F., *Proc. Roy. Soc.*, A, **225,** 287 (1954).

6. Kendrew, J. C., Bodo, G., Dintzis, H. M., Parrish, R. G., Wyckoff, H. W., and Phillips, D. C., *Nature*, **181,** 662 (1958).

7. Blow, D. M., *Proc. Roy. Soc.*, A, **247,** 302 (1958).

8. Perutz, M. F., *Proc. Roy. Soc.*, A, **225,** 264 (1954).

9. Cullis, A. F., Dintzis, H. M., and Perutz, M. F., Conference on Hæmoglobin, National Academy of Sciences, NAS-NRC Publication 557, p. 50 (Washington, 1958).

10. Rossmann, M. G., *Acta Cryst.* (in the press).

11. Bodo, G., Dintzis, H. M., Kendrew, J. C., and Wyckoff, H. W., *Proc. Roy. Soc.*, A, **253,** 70 (1959).

12. Blow, D. M., and Crick, F. H. C., *Acta Cryst.*, **12,** 794 (1959).

13. Bragg, W. L., and Perutz, M. F., *Proc. Roy. Soc.*, A, **225,** 315 (1954).

14. Ingram, D. J. E., Gibson, J. F., and Perutz, M. F., *Nature*, **178,** 905 (1956).

15. Kendrew, J. C., Dickerson, R. E., Strandberg, B., Hart, R. G., Davies, D. R., Phillips, D. C., and Shore, V. C. (see following article).

16. Scouloudi, H., *Nature*, **183,** 374 (1959).

17. Kendrew, J. C. (private communication).

18. Haurowitz, F., *Hoppe-Seyl. Z.*, **254,** 266 (1938).

19. Jope, H. M., and O'Brien, J. R. P., "Hæmoglobin," 269 (Butterworths, London, 1949).

20. Perutz, M. F., Trotter, I. F., Howells, E. R., and Green, D. W., *Acta Cryst.*, **8,** 241 (1955).

21. Riggs, A. F., *J. Gen. Physiol.*, **36,** 1 (1952).

19

RADIO WAVES FROM THE BIG BANG

ARNO PENZIAS AND BOB WILSON were having difficulties with their radio. For several months, since midsummer 1964, they had been trying to determine what caused the faint hiss in the background.

This was no ordinary radio. Built in 1960 by a man named A. B. Crawford, it was originally designed to detect radio signals from *Echo*, the first communications satellite. The antenna was twenty feet wide and shaped like a horn, to minimize unwanted reception from the sides and the ground. The amplifier consisted not of vacuum tubes or transistors but of an advanced maser, which used quantum effects to magnify weak signals. This particular quantum amplifier was exquisitely tuned to accept only frequencies near 4,080 megacycles per second, falling within the microwave region of the radio band. To reduce internal static, the maser was cooled with liquid helium to 4 degrees above absolute zero, or –269 degrees Celsius. An off-axis parabolic reflector joined the antenna at one end. For rotating the horn, a giant spoked wheel jutted through the middle like an automobile wreck. The whole odd-looking contraption perched uncomfortably atop Crawford Hill in Holmdel, New Jersey.

By 1963, the Crawford radio was no longer employed for communications. At that point, Bell Laboratories, which owned the instrument, turned it over to Penzias and Wilson for pure science. Penzias and Wilson were radio astronomers. They studied celestial bodies by detecting and analyzing radio waves rather than visible light. Now, the Crawford radio would be their telescope. Arno Penzias had been working at Bell Labs for two years and was thirty years old. Robert Wilson, twenty-seven but already balding, son of an oil-well scientist in Houston, had just arrived.

The two young astronomers were thrilled to command such a sensitive instrument and determined to begin measuring radio emissions from gas in the Milky Way. But for such delicate work, they first had to identify and subtract out all sources of static, called "noise." And that's where the problems began. After attempting to account for every crackle and hum in their receiver, a bit of unexplained static remained. Several years earlier, a scientist named E. A. Ohm had also found unexplained noise in the Crawford detector. However, no one paid much attention to Ohm's noise because it was smaller than the experimental errors and other uncertainties. Improved equipment and measurements would be needed. Penzias and Wilson had such equipment.

Radio noise could arise in any number of places. It could come from the ground, warmed by the sun. It could come from the emission of molecules in the earth's atmosphere, or from the gas between stars. Noise could be internally generated by the amplifier or the antenna or the wires connecting the two. One by one, Penzias and Wilson investigated each possible origin of noise. They tilted the telescope in different directions. They monitored it over the seasons as the earth orbited the sun. Yet the static remained constant, ruling out the atmosphere or the Milky Way, both of which would produce variations with direction. The astronomers pointed the antenna toward major cities, like New York, but the mysterious noise changed very little. They took apart the maser amplifier, found it blameless, and put it back together again. (Penzias had built masers for his Ph.D. and knew them backward and forward.) They estimated the faint hum in the connecting cables. Could the noise possibly arise from the riveted joints in the antenna, which could slightly heat up and emit feeble radio waves? The astronomers put aluminum tape over the suspicious joints. The noise hardly budged.

A pair of pigeons was found roosting in the small part of the horn and had generously used it as their toilet. As Wilson later recounted in his Nobel address: "We evicted the pigeons and cleaned up their mess, but obtained only a small reduction in the [noise]." (The pigeons were promptly sent by company post to Whippany, New Jersey, the main offices of Bell Telephone. As the birds happened to be homing pigeons, they returned to their hi-tech nest within two days. But the pigeons had already been ruled out.) The residual noise, like a disgruntled mumbling at a party, would not go away.

These two men were skilled. Penzias had received his Ph.D. in experimental physics from Columbia University, working with Nobel Prize winners Isidor Rabi and Charles Townes, the inventor of the maser. Wilson had repaired radios and televisions in high school and gotten his doctorate in physics at the California Institute of Technology. Like other teams we have encountered, their temperaments supported each other. Arno Penzias, who had fled Germany with his family in 1939, was ambitious, bold, ebullient, wide-ranging in thought, with less interest in details. Wilson was quiet, painstaking, and precise, in love with the fine print, able to tinker with an instrument for hours until it gave its optimum performance. Taken together, the two scientists were confident in their abilities. And they believed that they had identified all known sources of noise in the Crawford radio telescope. Yet something unknown remained. A bit of hiss. A fuzz in the background, constant in time, constant in all directions of space. As if their radio were submerged in a radio bath at three degrees above absolute zero. What could that be? Arno Penzias and Bob Wilson were concerned.

In December 1964, on an airplane returning from an astronomical conference, Penzias mentioned his "noise problem" to Bernie Burke, a fellow radio astronomer from the Carnegie Institution in Washington, D.C. A short time later, Burke called Penzias back with some interesting news. He had just seen a draft of a paper by a bright twenty-nine-year-old theoretician named James Peebles, who worked at Princeton University under the guidance of physicist Robert Dicke. Peebles was a Big Bang cosmologist. In his paper, he predicted that as a result of the intense heat at the birth of the universe, there should exist today a relict background of radio waves, filling all space uniformly and constantly, like tepid bath water. The temperature of this cosmic bath should be roughly 10 degrees above absolute zero. And it should appear as a constant hiss in a good radio telescope.

At that very moment, in fact, two of Dicke's other protégés, Peter Roll and David Wilkinson, were finishing construction of a radio receiver to search for the predicted cosmic background radiation. Soon they would be on the air.

After the phone call from Burke, Penzias called Dicke. They discussed the matter. Dicke got in his car and drove the thirty-odd miles to Holmdel to inspect the horn and Penzias's data. Both a superb experimentalist and a theorist himself, Dicke was pretty sure he knew what he was looking at. One can only imagine his amazement, excitement, and silent grief. Having made the cosmic prediction some time earlier, he was only months away from discovering it himself. Now he had been scooped. Penzias and Wilson were more cautious in interpreting their results, even after the Princeton revelation. "It never exactly hit us all in one day," Penzias commented later. Said Wilson: "We thought that our measurement was independent of the [Big Bang] theory and might outlive it." The "theory" was alive and well in 1978 when Penzias and Wilson received the Nobel Prize.

By accident, Penzias and Wilson had discovered what most scientists agree is the radiation left over from the origin of the universe, some 15 billion years in the past, "the primeval fireball," to use the language of Princeton physicist John Wheeler. That radiation, called the cosmic background radiation, provides a major confirmation of the Big Bang theory. While Edwin Hubble's discovery in 1929 of the ongoing expansion of the universe (see Chapter 12) had given the first experimental support of the theory, that finding actually probed only the recent past. But by all calculations, the cosmic background radiation resulted from events that took place when the universe was only one second old! The cosmic radio waves offered the first glimmerings of the universe in its infancy. Never before had human beings been so close to the primordial creation. "Hold infinity in the palm of your hand," wrote the mystical poet William Blake.

Just as with many discoveries in science, the story of the cosmic background radiation is an ironical concoction of ignorance and brilliance, missed opportunities, good and bad public relations, the overspecialization of science, and serendipity. Robert Dicke first predicted the cosmic background radiation on a scorchingly hot day in the summer of 1964. Arno Penzias and Robert Wilson accidentally discovered the cosmic background radiation in the fall of 1964. But in fact, the thing had been both predicted and discovered several times earlier.

The first prediction of the cosmic background radiation was based on the calculations of physicists George Gamow, Ralph Alpher, and Robert Herman in the late 1940s, working at the Applied Physics Laboratory of Johns Hopkins University. These scientists were attempting to explain the creation of the chemical elements by nuclear reactions in the young universe. One of the by-products of their investigation was the prediction of a cosmic bath of background radiation, created when the universe was only seconds old. In 1949, with more careful calculations, Alpher and Herman argued that the temperature of this cosmic background radiation today would be about 5 degrees Celsius above absolute zero, placing it in the radio region of the electromagnetic spectrum.

By 1956, Gamow, Alpher, and Herman had discussed their predicted radiation in a half-dozen papers published in leading journals of physics. Furthermore, the technology was available to measure the phenomenon. Why wasn't it done? For one thing, the scientists had doubts that the radiation would have survived the billions of years until now. And they wondered whether other sources of radio emission might obscure the cosmic background radiation. Gamow, Alpher, and Herman were theorists, not experimentalists, and not really sure what could be measured and what not. Furthermore, they were physicists, not astronomers. The two communities were not in good communication with each other, although the situation is much better today. And even within the field of astronomy, the radio astronomers who could actually make the measurements were a somewhat isolated group. So this extremely important prediction slipped away into obscurity. The three physicists moved on to other projects.

Also perplexing is that Dicke's group, which included theoreticians, did not know of the earlier published predictions of Gamow, Alpher, and Herman. Indeed, they believed that they were the first to predict the cosmic radiation. As Peebles later told me somewhat sheepishly, "we hadn't done our homework." In 1975, Dicke wrote in his memoir that "I must take the major blame for this, for the others in our group were too young to know these old papers."

Now for the discovery. The cosmic background radiation was first discovered *indirectly* by American astronomers W. S. Adams and T. Dunham, Jr., in the late 1930s and early 1940s. These scientists discovered faint signs of absorption of light by molecules of cyanogen (carbon bonded to nitrogen) in outer space. In 1941, Andrew McKellar of the Dominican Astrophysical Observatory in Canada, using the data of Adams and Dunham, calculated that these molecules were being pumped up to higher energy states by some background thermal bath at a temperature of 2.3 K, or 2.3 degrees above absolute zero. About the same time, the distinguished molecular spectroscopist Gerhard Herzberg also measured the emissions from interstellar cyanogen and came to the same conclusion. However, in the early 1940s no theory existed for relating a background radiation to the cosmological Big Bang. Thus, the larger meaning of the discovery could not have been guessed. The 2.3-degree background radiation in space was only one of thousands of facts and details that experimental scientists had to carry around in their heads. Without a theoretical framework and explanation, such facts had limited significance.

Unfortunately, a decade later, when Gamow and company produced the first theory, they were ignorant of the earlier experimental findings of Adams, Dunham, and McKellar, published in another field. Thus, again, theory could not be joined with experiment. In turn, Dicke and his group were unaware of both Gamow et al. and Adams et al. As were Penzias and Wilson. In 1964, two Russian physicists, A. G. Doroshkevich and I. D. Novikov, published a paper in English in the journal *Soviet Physics Doklady* claiming not only that the predicted cosmic background radiation could be measured, but also that the best telescope in the world for doing the job was the Crawford radio telescope in Holmdel, New Jersey! No one knew of their work for several years. What an irony that the various groups were, in a sense, receiving cosmic communications from billions of light-years away, and yet not in communication with each other, only hundreds of miles apart.

After their surreal meeting on Crawford Hill in early 1965, Dicke and Penzias agreed to publish their papers back-to-back in the same issue of the *Astrophysical Journal.*

Big Bang cosmology is based upon Einstein's theory of gravity, called general relativity and formulated in 1915 (see Chapter 12). When applied to the cosmos as a whole, that theory makes the simplifying assumption that the mass of the universe is not clustered in planets and stars and galaxies but instead spread out evenly like grains of sand on a beach. One then works with an average density of matter and energy. One can also talk about an average temperature. The equations of the theory show how the average density of matter and temperature change with time, through the force of gravity.

Like financial data given to a tax accountant, three parameters must be provided to the theory: the density of matter today, determined by measuring the total mass in a large volume of space and then dividing by the volume; the rate of expansion of the universe today, determined by how fast the galaxies are moving away from each other and their distances apart; and the "acceleration" of the universe today, which, analogous to the rate of change of velocity, is the rate of change of the expansion speed. Once these three parameters have been measured directly or indirectly, the theory determines everything, from the distant past to the distant future.

A key feature of the Big Bang cosmology is that the universe was hotter in the past. As discussed in Chapter 12, evidence that the universe is expanding—with all the galaxies flying away from each other—means that it was smaller and more compressed in the past. And just as air in a car tire heats up when it is compressed, the temperature of matter and energy in space was higher in the past. The further back in the past, the denser the universe and the higher its temperature. Indeed, the Big Bang model decrees that the universe began in a state of *infinite* density and temperature, expanding and cooling from there.

As far as the cosmic background radiation is concerned, much of the important action occurred when the universe was one second old and younger. According to Einstein's equations of gravity, when the universe was about one second old, its temperature was about ten billion degrees Celsius. At that temperature, galaxies could not exist, stars and planets could not exist, even individual atoms could not exist. The universe at ten billion degrees consisted only of a thick roiling soup of subatomic particles careening this way and that, and electromagnetic radiation at extremely high frequencies, in the gamma-ray region of the spectrum. (Visible light is electromagnetic radiation within a certain narrow range of frequencies. Radio waves have lower frequencies, X-rays and gamma rays higher.)

It is, of course, nearly impossible to grasp such extreme conditions. These are forms of matter and temperatures far higher than anything we have experienced on earth, even in our laboratories. But such a reality is required by our theories and equations. We must extend our imaginations.

Now, we arrive at the origin of the cosmic background radiation, which is a special form of electromagnetic radiation. Electromagnetic radiation is emitted and absorbed by all electrically charged particles. When there are sufficient numbers of such particles, the electromagnetic radiation inevitably becomes black-body radiation. As discussed in Chapter 1, black-body radiation has a particular amount of energy at each frequency range and is completely fixed by the temperature. At a temperature of ten billion degrees, electrons and positrons (the antiparticles of electrons) would be created in huge numbers out of the energy of other moving particles. As a result, the electromagnetic radiation would be converted to black-body radiation.

What happens to this radiation after one second? A theoretical result of Big Bang physics is that if the number of radiation particles, called photons, is much larger than the number of matter particles, then the electromagnetic radiation filling the universe would *remain* black-body radiation even as the universe expanded and cooled. The only change with expansion would be a decrease of the temperature of the radiation. Thus, even today, some fifteen billion years later, space should be filled with black-body radiation. And that is the predicted cosmic background radiation.

Finally, one can estimate the temperature of the cosmic black-body radiation today by how much the universe has expanded since it was a few seconds old. That estimate was performed by Alpher and Herman in 1949 and then repeated independently by Peebles in 1964. The calculation hinges on the observation that about 25 percent of the mass in the universe is helium, believed to have been formed out of protons and neutrons when the universe was about one hundred seconds old, at a temperature of a billion degrees. To explain the fraction of helium, certain conditions in the infant universe are required. By comparing the theoretically required density of matter at one hundred seconds to the roughly measured density of matter today, one infers that the universe has expanded by a factor of approximately one hundred million. Correspondingly, the temperature should have decreased by a factor of one hundred million. The result is that the cosmic radiation should have cooled off from a billion degrees then to about three degrees today.

Robert Dicke was born in May 1916 in St. Louis, Missouri, the son of a patent examiner originally trained as an electrical engineer. Young Dicke was unusually precocious and imaginative in science. In high school, he conceived of and performed a cosmological experiment to test the density of matter in outer space. Dicke put a flashlight in a Wheatstone bridge, which is a device that measures electrical resistance, and alternately shined the flashlight into the heavens and then at the floor. The boy reasoned that if the light pointing upward were not absorbed by matter in distant space, as it was by the floor, the electrical resistance would change. Such an ingenious experiment, although naive, was a herald of Dicke's ability to excel both as a theorist and as an experimentalist, a combination rarely found in physicists. Eventually, Dicke would have fifty patents to his name, ranging from clothes dryers to lasers.

Dicke obtained his A.B. degree from Princeton in 1939 and his Ph.D. in physics from the University of Rochester in 1946. He was a genius at electronic devices. During the war years Dicke invented various microwave-circuit devices and radar systems at the famous Radiation Lab at MIT. In 1944 he invented the Dicke radiometer. That device can distinguish very weak radio signals from background noise by switching the amplifier rapidly between one detector pointed toward the signal-producing region and another pointed at a cold bath of liquid helium. The true signal varies with the switch timing, while the noise does not.

Thereafter, the Dicke radiometer became standard equipment for all radio telescopes, including the one used by Penzias and Wilson.

In the early 1960s, Dicke did the most accurate measurement proving the equivalence of gravitational and inertial mass, the cornerstone of Einstein's theory of gravity. As a theorist, in the 1950s Dicke conceived of and worked out the first quantum theory of the laser, laying the foundations for that instrument. In the early 1960s he proposed his own theory of gravity, for some time a rival to Einstein's theory.

By the mid-1960s Dicke was a legend. Practically worshiped by the young members of his research group, whom he called his "boys" (as did Rutherford, see Chapter 5), he was a quiet, modest, good-humored man with a large head and large ears and the prescience of a Greek oracle.

Jim Peebles remembers the moment when Dicke first conceived of the cosmic background radiation.

> It would be around 1964. It was in the summer, I do remember that, a very hot day. We met in his usual evening group but with a small number of people. For some reason, we met in the attic in Palmer lab. He explained to us first why one might want to think that the universe was hot in its early phases. His thought at that time—and one that he does keep returning to—is that the universe might oscillate, and an oscillating universe does require something to destroy heavy elements so one could start again with hydrogen. The way you destroy heavy elements is to thermally decompose them in black-body radiation. So he explained to us then why you would like a universe that's filled with black-body radiation.

Dicke's chain of reasoning is fascinating. To begin with, he believed in an "oscillating universe." Such a conception, a variation of the standard Big Bang, was first extensively discussed by Richard Tolman of the California Institute of Technology in the early 1930s, although the idea harks back to ancient Buddhist and Hindu cosmologies. In an oscillating universe, the universe expands, then contracts, then expands again, endlessly repeating the cycles of expansion and contraction like the breathing of a great cosmic lung. An oscillating universe is eternal.

For various reasons, Robert Dicke did not want to confront the "initial conditions" of the universe at $t = 0$, did not want to postulate how matter and energy were created from nothing, why matter so dominated antimatter, and other such primordial questions. In an oscillating universe, one does not have to grapple with these questions. One can say that the universe is as it is because it was always that way. The universe always contained matter and energy. Matter always dominated antimatter. And so on. As Dicke writes in his paper, "This [hypothesis of an oscillating universe] relieves us of the necessity of understanding the origin of matter at any finite time in the past."

Dicke's argument then goes something like this: we believe that the heavier chemical elements, like carbon and oxygen, are continuously synthesized from lighter elements in the nuclear reactions of stars. But if those elements are not destroyed at some point in each cycle of an oscillating universe (which goes on forever), then eventually the universe would consist *entirely* of heavy elements, contradicting the observation that most of the material in the universe is hydrogen and helium, the two lightest elements. Therefore, at each cosmic contraction, when the universe returns to its hottest and densest state, the temperature must reach at least ten billion degrees, high enough to destroy all the heavy elements and begin over with nascent hydrogen. And at ten billion degrees, as we have seen earlier, there will be so many electrons that space will be filled with black-body radiation, that is, the cosmic background radiation.

It is important to point out that Dicke's prediction of a cosmic background radiation does not really depend on his adopted model of an oscillating universe. The same prediction results for any universe that was once ten billion degrees or hotter. In particular, the prediction holds for the standard Big Bang model.

Dicke, Peebles, Roll, and Wilkinson begin their paper by expressing dismay over the conventional version of the Big Bang model, which starts in a "singularity" of infinite density at the beginning of time. In particular, they do not see any explanation for why there exists much more matter than antimatter in the universe. Each kind of subatomic particle is paired with an antiparticle, identical to it in most respects but with opposite electrical charge. One would think that there should be equal numbers of particles and their corresponding antiparticles. "In the framework of conventional theory," Dicke writes, "we cannot understand the origin of matter [in excess over antimatter] or of the universe." For these scientists, the essential problem lies in the creation of the universe from nothing—requiring that so much be postulated from the beginning, including the amount of matter versus antimatter.

They then consider several attempts to deal with the problem of the beginning at $t = 0$, also known as the singularity. Number 3, which they prefer, proposes that the singularity of the conventional Big Bang theory never happened. It is a fiction caused by unrealistic assumptions, and such a sharp beginning does not actually occur "in the real world." The scientists then move quickly to Dicke's pet theory, the oscillating universe. As mentioned before, an oscillating universe has the apparent advantage that it had no beginning, so that nothing must be postulated. An oscillating universe lasts forever, continually expanding and contracting.

The oscillating universe is also "closed." A closed universe has enough matter and energy to eventually overwhelm—by the attractive force of its gravity—the outward expansion and cause the universe to begin collapsing. By contrast, an "open" universe does not have sufficient gravity to halt its expansion, and it keeps expanding forever.

The baryons that Dicke et al. refer to are a particular class of subatomic particles. According to theories of particle physics current in 1965, the total number of baryons cannot change. Thus rather than attempt to explain that number from scratch (at $t = 0$), Dicke defers to the oscillating universe model. The number of baryons is simply what it has been from time immemorial.

Dicke and his colleagues then go through the remaining steps of the argument outlined above, leading to a prediction of black-body radiation. They clearly understand that their argument is valid for any universe that was once ten billion (10^{10}) degrees or hotter, oscillating or not. The temperature scale used here and in the paper by Penzias and Wilson is the absolute temperature scale, where the degrees are denoted by K. A K is equal to a degree C, with the origin of the scale shifted. Absolute zero, the coldest possible temperature, is at 0 on the K scale and –273 on the C scale.

Dicke quotes the well-known result that for a universe that is "adiabatically" cooling, meaning that its energy remains constant, its radiation temperature varies inversely with its radius. For example, when the universe expands by a factor of ten, its temperature decreases by a factor of ten.

After considering the production of electrons and their antiparticles, the positrons, Dicke and his colleagues consider the production of neutrinos, another type of subatomic particle with very small mass, and their antiparticles.

In this paper, Dicke and his collaborators do not attempt to estimate the temperature of the cosmic background radiation today, although, as mentioned earlier, Peebles did so in another paper. However, the scientists do argue that the temperature must be less than 40 K. Otherwise, there would be so much gravity associated with the energy in the radiation that the universe would already have begun contracting. Dicke states this latter consideration in other terms, referring to the "Hubble constant," which is the rate of expansion of the universe, and the "acceleration parameter," which is the rate of change of the Hubble constant. (Twenty years earlier, Dicke himself had pointed one of his radiometers at the sky and experimentally concluded that the temperature of space had to be less than 20 K, but he has forgotten his own measurement.)

Dicke et al. then signal us that they are taking their prediction seriously: "Evidently, it would be of considerable interest to attempt to detect this primeval thermal radiation directly." Words are backed by action. They mention, in fact, that two of the group, Roll and Wilkinson, have already constructed a radio receiver to search for the cosmic radiation. Their receiver is tuned to a wavelength of 3.2 centimeters, corresponding to a frequency of 9,370 megacyles per second. Dicke then discusses why this is a good frequency to monitor outer space for the predicted radiation. At much higher frequencies (shorter wavelengths), the cosmic radiation would be absorbed by the atmosphere before it reached the detector, and at much lower frequencies (longer wavelengths) the strong radio emission of gas in the galaxy would mask the cosmic background radiation, effectively shouting it down. The scientists can subtract out the radio emission from the atmosphere by "tipping" their antenna. Such tipping points the antenna through different thicknesses of air and causes a variation in the intensity of atmospheric radio emission, whereas the cosmic background radiation should not vary with direction. It is ubiquitous and constant.

The scientists then acknowledge that, while poised to begin their own measurements, they have learned of the results of Penzias and Wilson. They point out that it is necessary to measure the cosmic background radiation at many wavelengths, in order to see if its spectrum has the predicted shape of black-body radiation. Penzias and Wilson have measured the radiation at only a single wavelength and frequency.

Peebles's calculations on element synthesis are mentioned. Ironically, Dicke refers to the element synthesis calculations of Gamow, Alpher, and Herman, but not to the particular papers where they predicted the cosmic background radiation.

In the "Conclusion" section, Dicke reveals his strong prejudice for an oscillating universe. As Peebles had shown, there is a relation between the cosmic abundance of helium (measured to be about 25 percent), the density of matter today, and the temperature of the observed cosmic background radiation. Given the measurements of Penzias and Wilson, pinning down the latter, that leaves a relation between the helium abundance and the density of matter. Taking a maximum value of the helium abundance to be 25 percent gives a maximum matter density of 3×10^{-32} grams per cubic centimeter, far below the density required to "close" the universe and allow oscillations. A higher density of matter would produce too much helium. Thus, it would seem that an oscillating universe is ruled out by the data.

To save the oscillating universe, Dicke makes a number of unusual and fairly unorthodox proposals: (1) If the universe were expanding much faster when it was at a temperature of a billion degrees, then as little as 25 percent helium could be created with a higher density of matter. A hypothetical form of energy, called a zero-mass scalar, would cause the more rapid expansion. (Dicke's own theory of gravity, called the Brans-Dicke theory, contains such a zero-mass scalar.) (2) Another form of energy, such as gravitational radiation, might serve to make the universe expand faster. (3) Neutrinos in extremely high numbers, technically in a "degenerate" state, could allow higher mass density without raising the helium abundance above 25 percent.

Like many great scientists, Dicke had definite philosophical preferences. At the same time, he was open and unconstrained when it came to experiments. As he wrote in his unpublished autobiography, "I have long believed that an experimentalist should not be unduly inhibited by theoretical untidiness. If he insists on having every last theoretical T crossed before he starts his research the chances are that he will never do a significant experiment."

The first thing one notices about the landmark paper by Penzias and Wilson is the understated title: "A Measurement of Excess Antenna Temperature at 4080 Mc/s." As described earlier, the two scientists were cautious about attaching any grand cosmological significance to their results. They would leave that extrapolation to Dicke and his colleagues. Indeed, they are careful not to mention any theoretical interpretation whatsoever in their paper.

By "excess antenna temperature," the scientists mean the intensity of radio waves in their horn that cannot be attributed to known sources. At each frequency, the intensity of energy can be expressed in terms of an effective temperature, and the temperatures can be added and subtracted like money. The bookkeeping thus goes like this: the total intensity of energy received (at 4,080 megacycles per second) corresponds to a temperature of 6.7 K. Of that total, the radio emission of the atmosphere contributes 2.3 K. The emission by the inevitable heating up of the antenna (called the ohmic losses) and by the warm ground below the antenna (called the back-lobe response) sum to 0.9 K. Subtracting 2.3 K and 0.9 K from 6.7 K leaves 3.5 K unaccounted for. That 3.5 K, with an uncertainty of 1 K in either direction, is the excess antenna temperature. That 3.5 K is the cosmic background radiation, the fading whisper of the origin of the universe. Penzias and Wilson, "cautiously optimistic" to use Wilson's words later, never refer to it as the cosmic background radiation but only as an "excess antenna temperature." In retrospect, their "excess" is like the last paragraph of Lincoln's Gettysburg Address.

The investigators then go on to describe their equipment. The various errors and losses are discussed in terms of their effective temperatures. Then comes a more detailed discussion of contributions from the atmosphere, from the connections between the antenna and the maser amplifier, from the antenna (the horn), and from the ground emission. Finally, Penzias and Wilson refer to previous measurements of noise in the Crawford radio telescope by other scientists at other frequencies. Their results are consistent with the more accurate results reported here.

In the acknowledgments, Penzias and Wilson thank Dicke and his colleagues for "fruitful discussions," just as Dicke thanked them at the end of his paper. It is almost certain that Penzias and Wilson would have reported their discovery of an "excess antenna temperature" even if they had never spoken to Dicke, but they would most likely have buried it in a paper on other subjects.

Subsequent experiments have measured the cosmic background radiation at many other frequencies. By December 1965, Roll and Wilkinson of Dicke's group had completed their own experiment at 9,370 megacycles per second. Within a year, other scientists had measured the radiation at frequencies ranging from 1,430 to 115,340 megacycles per second. To get to the higher frequencies, which don't penetrate the earth's atmosphere, measurements have been undertaken from mountain tops, airplanes, balloons, and rockets. All of these experiments have confirmed that the cosmic background radiation indeed has the special black-body form, as predicted, with a temperature of about 2.7 K.

The discovery of the cosmic background radiation has provided a powerful confirmation of the Big Bang theory of the universe. In recent years, measurements of this radiation have even gone beyond the Big Bang theory and tested new theories of elementary particle physics. Some of those theories predict in quantitative detail how a completely smooth sea of matter and energy in the very young universe would develop the slight clumpiness necessary for the eventual formation of galaxies and stars. That slight clumpiness, in turn, translates into slight variations in the intensity of the cosmic background radiation from different directions in the sky. Those predicted variations are only a few parts in a million. Yet they have been measured by the sensitive instruments aboard the *Cosmic Background Explorer* satellite, launched by NASA in 1989, and the even more sensitive *Wilkinson Microwave Anisotropy Probe* satellite, launched in 2001. The cosmic background radiation is like a cosmic version of Assyrian cuneiform, a bit of history from the early moments of our universe.

One cold day in early 1988, I visited Bob Dicke in his office at Princeton. He was seventy-one at the time, officially retired but still working every day. Beneath the equations scrawled on his blackboard was a printed message to the janitorial staff: PLEASE DO NOT ERASE. By now, Dicke's full head of hair was mostly gray and white. He wore a tie, a herringbone jacket, and large black-rimmed eyeglasses. I remember his hands especially, delicate with slender fingers, not the hands I would have expected from someone who had built electrical circuits and radiometers.

Dicke was gracious and polite, answering my questions with a kind of childlike simplicity and honesty. Sometimes I had to strain to hear him, for his voice was almost a whisper, like the faint radio hiss of the cosmic background radiation. Toward the end of the interview he drifted to a crucial aspect of cosmology that has continued to trouble human beings from the ancient Babylonians to late-twentieth-century physicists. "There's still one point in cosmology that I find very disagreeable," he said, "and that's the idea of time and space having no meaning up to a certain point and then suddenly appearing . . . a universe that is suddenly switched on . . . I guess what bothers me is a sudden barrier, a discontinuity, whether it's in time or space, because I'm used to continuity."

A MEASUREMENT OF EXCESS ANTENNA TEMPERATURE AT 4080 MC/S

Arno A. Penzias and Robert W. Wilson

The Astrophysical Journal (1965)

MEASUREMENTS OF THE EFFECTIVE zenith noise temperature of the 20-foot horn-reflector antenna (Crawford, Hogg, and Hunt 1961) at the Crawford Hill Laboratory, Holmdel, New Jersey, at 4080 Mc/s have yielded a value about 3.5° K higher than expected. This excess temperature is, within the limits of our observations, isotropic, unpolarized, and free from seasonal variations (July, 1964–April, 1965). A possible explanation for the observed excess noise temperature is the one given by Dicke, Peebles, Roll, and Wilkinson (1965) in a companion letter in this issue.

The total antenna temperature measured at the zenith is 6.7° K of which 2.3° K is due to atmospheric absorption. The calculated contribution due to ohmic losses in the antenna and back-lobe response is 0.9° K.

The radiometer used in this investigation has been described elsewhere (Penzias and Wilson 1965). It employs a traveling-wave maser, a low-loss (0.027-db) comparison switch, and a liquid helium–cooled reference termination (Penzias 1965). Measurements were made by switching manually between the antenna input and the reference termination. The antenna, reference termination, and radiometer were well matched so that a round-trip return loss of more than 55 db existed throughout the measurement; thus errors in the measurement of the effective temperature due to impedance mismatch can be neglected. The estimated error in the measured value of the total antenna temperature is 0.3° K and comes largely from uncertainty in the absolute calibration of the reference termination.

The contribution to the antenna temperature due to atmospheric absorption was obtained by recording the variation in antenna temperature with elevation angle and employing the secant law. The result, 2.3° ± 0.3° K, is in good agreement with published values (Hogg 1959; DeGrasse, Hogg, Ohm, and Scovil 1959; Ohm 1961).

The contribution to the antenna temperature from ohmic losses is computed to be 0.8° ± 0.4° K. In this calculation we have divided the antenna into three parts: (1) two non-uniform tapers approximately 1 m in total length which transform between the 2⅛-inch round output waveguide and the 6-inch-square antenna throat opening; (2) a double-choke rotary joint located between these two tapers; (3) the antenna itself. Care was taken to clean and align joints between these parts so that they would not significantly increase the loss in the structure. Appropriate tests were made for leakage and loss in the rotary joint with negative results.

The possibility of losses in the antenna horn due to imperfections in its seams was eliminated by means of a taping test. Taping all the seams in the section near the throat and most of the others with aluminum tape caused no observable change in antenna temperature.

The backlobe response to ground radiation is taken to be less than 0.1° K for two reasons: (1) Measurements of the response of the antenna to a small transmitter located on the ground in its vicinity indicate that the average back-lobe level is more than 30 db below isotropic response. The horn-reflector antenna was pointed to the zenith for these measurements, and complete rotations in azimuth were made with the transmitter in each of ten locations using horizontal and vertical transmitted polarization from each position. (2) Measurements on smaller horn-reflector antennas at these laboratories, using pulsed measuring sets on flat antenna ranges, have consistently shown a back-lobe level of 30 db below isotropic response. Our larger antenna would be expected to have an even lower back-lobe level.

From a combination of the above, we compute the remaining unaccounted-for antenna temperature to be 3.5° ± 1.0° K at 4080 Mc/s. In connection with this result it should be noted that DeGrasse *et al.* (1959) and Ohm (1961) give total system temperatures at 5650 Mc/s and 2390 Mc/s, respectively. From these it is possible to infer upper limits to the background temperatures at these frequencies. These limits are, in both cases, of the same general magnitude as our value.

We are grateful to R. H. Dicke and his associates for fruitful discussions of their results prior to publication. We also wish to acknowledge with thanks the useful comments and advice of A. B. Crawford, D. C. Hogg, and E. A. Ohm in connection with the problems associated with this measurement.

Note added in proof.—The highest frequency at which the background temperature of the sky had been measured previously was 404 Mc/s (Pauliny-Toth and Shakeshaft 1962), where a minimum temperature of 16° K was observed. Combining this value with our result, we find that the average spectrum of the background radiation over this frequency range can be no steeper than $\lambda^{0.7}$. This clearly eliminates the possibility that the radiation we observe is due to radio sources of types known to exist, since in this event, the spectrum would have to be very much steeper.

A. A. Penzias
R. W. Wilson

May 13, 1965
Bell Telephone Laboratories, Inc.
Crawford Hill, Holmdel, New Jersey

REFERENCES

Crawford, A. B., Hogg, D. C., and Hunt, L. E. 1961, *Bell System Tech. J.*, **40,** 1095.
DeGrasse, R. W., Hogg, D. C., Ohm, E. A., and Scovil, H. E. D. 1959, "Ultra-low Noise Receiving System for Satellite or Space Communication," *Proceedings of the National Electronics Conference*, **15,** 370.
Dicke, R. H., Peebles, P. J. E., Roll, P. G., and Wilkinson, D. T. 1965, *Ap. J.*, **142,** 414.
Hogg, D. C. 1959, *J. Appl. Phys.*, **30,** 1417.
Ohm, E. A. 1961, *Bell System Tech. J.*, **40,** 1065.
Pauliny-Toth, I. I. K., and Shakeshaft, J. R. 1962, *M.N.*, **124,** 61.
Penzias, A. A. 1965, *Rev. Sci. Instr.*, **36,** 68.
Penzias, A. A., and Wilson, R. W. 1965, *Ap. J.* (in press).

COSMIC BLACK-BODY RADIATION[1]

Robert H. Dicke, P. James E. Peebles, Peter G. Roll, and David T. Wilkinson

The Astrophysical Journal (1965)

ONE OF THE BASIC PROBLEMS OF COSMOLOGY is the singularity characteristic of the familiar cosmological solutions of Einstein's field equations. Also puzzling is the presence of matter in excess over antimatter in the universe, for baryons and leptons are thought to be conserved. Thus, in the framework of conventional theory we cannot understand the origin of matter or of the universe. We can distinguish three main attempts to deal with these problems.

1. The assumption of continuous creation (Bondi and Gold 1948; Hoyle 1948), which avoids the singularity by postulating a universe expanding for all time and a continuous but slow creation of new matter in the universe.
2. The assumption (Wheeler 1964) that the creation of new matter is intimately related to the existence of the singularity, and that the resolution of both paradoxes may be found in a proper quantum mechanical treatment of Einstein's field equations.
3. The assumption that the singularity results from a mathematical over-idealization, the requirement of strict isotropy or uniformity, and that it would not occur in the real world (Wheeler 1958; Lifshitz and Khalatnikov 1963).

If this third premise is accepted tentatively as a working hypothesis, it carries with it a possible resolution of the second paradox, for the matter we see about us now may represent the same baryon content of the previous expansion of a closed universe, oscillating for all time. This relieves us of the necessity of understanding the origin of matter at any finite time in the past. In this picture it is essential to suppose that at the time of maximum collapse the temperature of the universe would exceed 10^{10} ° K, in order that the ashes of the previous cycle would have been reprocessed back to the hydrogen required for the stars in the next cycle.

Even without this hypothesis it is of interest to inquire about the temperature of the universe in these earlier times. From this broader viewpoint we need not limit the discussion to closed oscillating models. Even if the universe had a singular origin it might have been extremely hot in the early stages.

Could the universe have been filled with black-body radiation from this possible high-temperature state? If so, it is important to notice that as the universe expands the cosmological redshift would serve to adiabatically cool the radiation, while preserving the thermal character. The radiation temperature would vary inversely as the expansion parameter (radius) of the universe.

The presence of thermal radiation remaining from the fireball is to be expected if we can trace the expansion of the universe back to a time when the temperature was of the order of 10^{10} ° K ($\sim m_e c^2$). In this state, we would expect to find that the electron abundance had increased very substantially, due to thermal electron-pair production, to a density characteristic of the temperature only. One readily verifies that, whatever the previous history of the universe, the photon absorption length would have been short with this high electron density, and the radiation content of the universe would have promptly adjusted to a thermal equilibrium distribution due to pair-creation and annihilation processes. This adjustment requires a time interval short compared with the characteristic expansion time of the universe, whether the cosmology is general relativity or the more rapidly evolving Brans-Dicke theory (Brans and Dicke 1961).

The above equilibrium argument may be applied also to the neutrino abundance. In the epoch where $T > 10^{10}$ ° K, the very high thermal electron and photon abundance would be sufficient to assure an equilibrium thermal abundance of electron-type neutrinos, assuming the presence of neutrino-antineutrino pair-production processes. This means that a strictly thermal neutrino and antineutrino distribution, in thermal equilibrium with the radiation, would have issued from the highly contracted phase. Conceivably, even gravitational radiation could be in thermal equilibrium.

Without some knowledge of the density of matter in the primordial fireball we cannot predict the present radiation temperature. However, a rough upper limit is provided by the observation that black-body radiation at a temperature of 40° K provides an energy density of 2×10^{-29} gm cm^3, very roughly the maximum total energy density compatible with the observed Hubble constant and acceleration parameter. Evidently, it would be of considerable interest to attempt to detect this primeval thermal radiation directly.

Two of us (P. G. R. and D. T. W.) have constructed a radiometer and receiving horn capable of an absolute measure of thermal radiation at a wavelength of 3 cm. The choice of wavelength was dictated by two considerations, that at much shorter wavelengths atmospheric absorption would be troublesome, while at longer wavelengths galactic and extragalactic emission would be appreciable. Extrapolating from the observed background radiation at longer wavelengths (~100 cm) according to the power-law spectra characteristic of synchrotron radiation or bremsstrahlung, we can conclude that the total background at 3 cm due to the Galaxy and the extragalactic sources should not exceed 5×10^{-3} ° K when averaged over all directions. Radiation from stars at 3 cm is $<10^{-9}$ ° K. The contribution to the background due to the atmosphere is expected to be approximately 3.5° K, and this can be accurately measured by tipping the antenna (Dicke, Beringer, Kyhl, and Vane 1946).

While we have not yet obtained results with our instrument, we recently learned that Penzias and Wilson (1965) of the Bell Telephone Laboratories have observed background radiation at 7.3-cm wavelength. In attempting to eliminate (or account for) every contribution to the noise seen at the output of their receiver, they ended with a residual of 3.5° ± 1° K. Apparently this could only be due to radiation of unknown origin entering the antenna.

It is evident that more measurements are needed to determine a spectrum, and we expect to continue our work at 3 cm. We also expect to go to a wavelength of 1 cm. We understand that measurements at wavelengths greater than 7 cm may be filled in by Penzias and Wilson.

A temperature in excess of 10^{10} ° K during the highly contracted phase of the universe is strongly implied by a present temperature of 3.5° K for black-body radiation. There are two reasonable cases to consider. Assuming a singularity-free oscillating cosmology, we believe that the temperature must have been high enough to decompose the heavy elements from the previous cycle, for there is no observational evidence for significant amounts of heavy elements in the outer parts of the oldest stars in our Galaxy. If the cosmological solution has a singularity, the temperature would rise much higher than 10^{10} ° K in approaching the singularity (see, e.g., Fig. 1).

It has been pointed out by one of us (P. J. E. P.) that the observation of a temperature as low as 3.5° K, together with the estimated abundance of helium in the protogalaxy, provides some important evidence on possible cosmologies (Peebles 1965). This comes about in the following way. Considering again the epoch $T \gg 10^{10}$ ° K, we see that the presence of the thermal electrons and neutrinos would have assured nearly equal abundances of neutrons and protons. Once the temperature has fallen so low that photodissociation of deuterium is not too great, the neutrons and protons can combine to form deuterium, which in turn readily burns to helium. This was the type of process envisioned by Gamow, Alpher, Herman, and others (Alpher, Bethe, and Gamow 1948; Alpher, Follin, and Herman 1953; Hoyle and Tayler 1964). Evidently the amount of helium produced depends on the density of matter at the time helium formation became possible. If at this time the nucleon density were great enough, an appreciable amount of helium would have been produced before the density fell too low for reactions to occur. Thus, from an upper limit on the possible helium abundance in the protogalaxy we can place an upper limit on the matter density at the time of helium formation (which occurs at a fairly definite temperature, almost independent of density) and hence, given the density of matter in the present universe, we have a lower limit on the present radiation temperature. This limit varies as the cube root of the assumed present mean density of matter.

While little is reliably known about the possible helium content of the protogalaxy, a reasonable upper bound consistent with present abundance observations is 25 per cent helium by mass. With this limit, and assuming that general relativity is valid, then if the present radiation temperature were 3.5° K, we conclude that the matter density in the universe could not exceed 3×10^{-32} gm cm^3. (See Peebles 1965 for a detailed development of the factors determining this value.) This is a factor of 20 below the estimated average density from matter in galaxies (Oort 1958), but the estimate probably is not reliable enough to rule out this low density.

CONCLUSIONS

While all the data are not yet in hand we propose to present here the possible conclusions to be drawn if we tentatively assume that the measurements of Penzias and Wilson (1965) do indicate black-body radiation at 3.5° K. We also assume that the universe can be considered to be isotropic and uniform, and that the present energy density in gravitational radiation is a small part of the whole. Wheeler (1958) has remarked that gravitational radiation could be important.

For the purpose of obtaining definite numerical results we take the present Hubble redshift age to be 10^{10} years.

Assuming the validity of Einstein's field equations, the above discus-

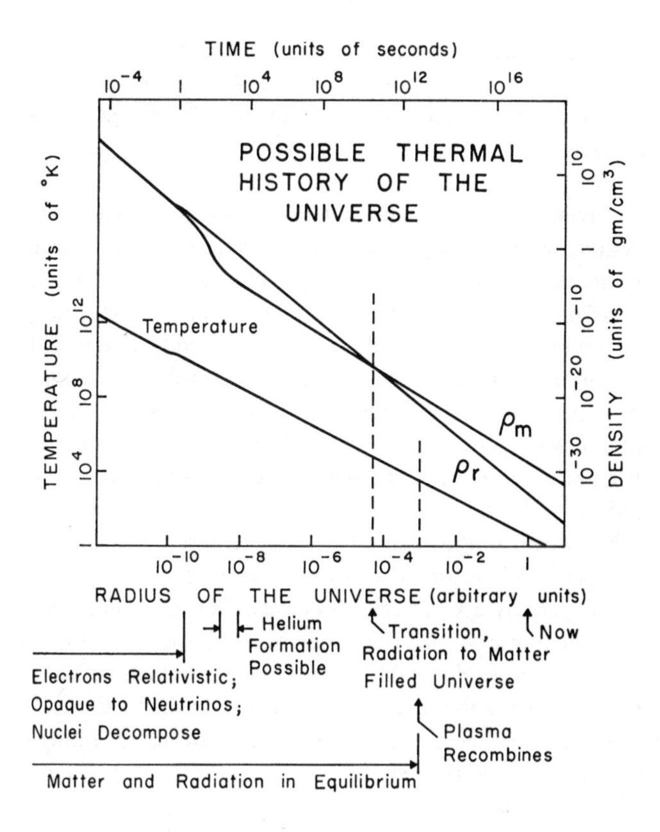

Figure 1. Possible thermal history of the Universe. The figure shows the previous thermal history of the Universe assuming a homogeneous isotropic general-relativity cosmological model (no scalar field) with present matter density 2×10^{-29} gm/cm^3 and present thermal radiation temperature 3.5° K. The bottom horizontal scale may be considered simply the proper distance between two chosen fiducial co-moving galaxies (*points*). The top horizontal scale is the proper world time. The line marked "temperature" refers to the temperature of the thermal radiation. Matter remains in thermal equilibrium with the radiation until the plasma recombines, at the time indicated. Thereafter further expansion cools matter not gravitationally bound faster than the radiation. The mass density in radiation is ρ_r. At present ρ_r is substantially below the mass density in matter, ρ_m, but, in the early Universe ρ_r exceeded ρ_m. We have indicated the time when the Universe exhibited a transition from the characteristics of a radiation-filled model to those of a matter-filled model.

Looking back in time, as the temperature approaches 10^{10} ° K the electrons become relativistic, and thermal electron-pair creation sharply increases the matter density. At temperatures somewhat greater than 10^{10} ° K these electrons should be so abundant as to assure a thermal neutrino abundance and a thermal neutron-proton abundance ratio. A temperature of this order would be required also to decompose the nuclei from the previous cycle in an oscillating Universe. Notice that the nucleons are nonrelativistic here.

The thermal neutrons decay at the right-hand limit of the indicated region of helium formation. There is a left-hand limit on this region because at higher temperatures photodissociation removes the deuterium necessary to form helium. The difficulty with this model is that most of the matter would end up in helium.

sion and numerical values impose severe restrictions on the cosmological problem. The possible conclusions are conveniently discussed under two headings, the assumption of a universe with either an open or a closed space.

OPEN UNIVERSE. From the present observations we cannot exclude the possibility that the total density of matter in the universe is substantially below the minimum value 2×10^{-29} gm cm^3 required for a closed universe. Assuming general relativity is valid, we have concluded from the discussion of the connection between helium production and the present radiation temperature that the present density of material in the universe must be $\gtrsim 3 \times 10^{-32}$ gm cm^3, a factor of 600 smaller than the limit for a closed universe. The thermal-radiation energy density is even smaller, and from the above arguments we expect the same to be true of neutrinos.

Apparently, with the assumption of general relativity and a primordial temperature consistent with the present 3.5° K, we are forced to adopt an open space, with very low density. This rules out the possibility of an oscillating universe. Furthermore, as Einstein (1950) remarked, this result is distinctly non-Machian, in the sense that, with such a low mass density, we cannot reasonably assume that the local inertial properties of space are determined by the presence of matter, rather than by some absolute property of space.

CLOSED UNIVERSE. This could be the type of oscillating universe visualized in the introductory remarks, or it could be a universe expanding from a singular state. In the framework of the present discussion the required mass density in excess of 2×10^{-29} gm cm^3 could not be due to thermal radiation, or to neutrinos, and it must be presumed that it is due to ordinary matter, perhaps intergalactic gas uniformly distributed or else in large clouds (small protogalaxies) that have not yet generated stars (see Fig. 1).

With this large matter content, the limit placed on the radiation temperature by the low helium content of the solar system is very severe. The present black-body temperature would be expected to exceed 30° K (Peebles 1965). One way that we have found reasonably capable of decreasing this lower bound to 3.5° K is to introduce a zero-mass scalar field into the cosmology. It is convenient to do this without invalidating the Einstein field equation, and the form of the theory for which the scalar interaction appears as an ordinary matter interaction (Dicke 1962) has been employed. The cosmological equation (Brans and Dicke 1961) was originally integrated for a cold universe only, but a recent investigation of the solutions for a hot universe indicates that with the scalar field the universe would have expanded through the temperature range $T \sim 10^9$ °K so fast that essentially no helium would have been formed. The reason for this is that the static part of the scalar field contributes a pressure just equal to the scalar-field energy density. By contrast, the pressure due to incoherent electromagnetic radiation or to relativistic particles is one third of the energy density. Thus, if we traced back to a highly contracted universe, we would find that the scalar-field energy density exceeded all other contributions, and that this fast increasing scalar-field energy caused the universe to expand through the highly contracted phase much more rapidly than would be the case if the scalar field vanished. The essential element is that the pressure approaches the energy density, rather than one third of the energy density. Any other interaction which would cause this, such as the model given by Zel'dovich (1962), would also prevent appreciable helium production in the highly contracted universe.

Returning to the problem stated in the first paragraph, we conclude that it is possible to save baryon conservation in a reasonable way if the universe is closed and oscillating. To avoid a catastrophic helium production, either the present matter density should be $< 3 \times 10^{-32}$ gm/cm^3, or there should exist some form of energy content with very high pressure, such as the zero-mass scalar, capable of speeding the universe through the period of helium formation. To have a closed space, an energy density of 2×10^{-29} gm/cm^3 is needed. Without a zero-mass scalar, or some other "hard" interaction, the energy could not be in the form of ordinary matter and may be presumed to be gravitational radiation (Wheeler 1958).

One other possibility for closing the universe, with matter providing the energy content of the universe, is the assumption that the universe contains a net electron-type neutrino abundance (in excess of antineutrinos) greatly larger than the nucleon abundance. In this case, if the neutrino abundance were so great that these neutrinos are degenerate, the degeneracy would have forced a negligible equilibrium neutron abundance in the early, highly contracted universe, thus removing the possibility of nuclear reactions leading to helium formation. However, the required ratio of lepton to baryon number must be $> 10^9$.

We deeply appreciate the helpfulness of Drs. Penzias and Wilson of the Bell Telephone Laboratories, Crawford Hill, Holmdel, New Jersey, in discussing with us the result of their measurements and in showing us their receiving system. We are also grateful for several helpful suggestions of Professor J. A. Wheeler.

R. H. Dicke
P. J. E. Peebles
P. G. Roll
D. T. Wilkinson
May 7, 1965
Palmer Physical Laboratory
Princeton, New Jersey

1. This research was supported in part by the National Science Foundation and the Office of Naval Research of the U.S. Navy.

REFERENCES

Alpher, R. A., Bethe, H. A., and Gamow, G. 1948, *Phys. Rev.*, **73,** 803.
Alpher, R. A., Follin, J. W., and Herman, R. C. 1953, *Phys. Rev.*, **92,** 1347.
Bondi, H., and Gold, T. 1948, *M.N.*, **108,** 252.
Brans, C., and Dicke, R. H. 1961, *Phys. Rev.*, **124,** 925.
Dicke, R. H. 1962, *Phys. Rev.*, **125,** 2163.
Dicke, R. H., Beringer, R., Kyhl, R. L., and Vane, A. B. 1946, *Phys. Rev.*, **70,** 340.
Einstein, A., 1950, *The Meaning of Relativity* (3d ed.; Princeton, N.J.: Princeton University Press), p. 107.
Hoyle, F. 1948, *M.N.*, **108,** 372.
Hoyle, F., and Tayler, R. J. 1964, *Nature*, **203,** 1108.
Liftshitz, E. M., and Khalatnikov, I. M. 1963, *Adv. in Phys.*, **12,** 185.
Oort, J. H. 1958, *La Structure et l'évolution de l'universe* (11th Solvay Conf. [Brussels: Éditions Stoops]), p. 163.
Peebles, P. J. E. 1965, *Phys. Rev.* (in press).
Penzias, A. A., and Wilson, R. W. 1965, private communication.
Wheeler, J. A., 1958, *La Structure et l'évolution de l'universe* (11th Solvay Conf. [Brussels: Éditions Stoops]), p. 112.
———. 1964, in *Relativity, Groups and Topology*, ed. C. DeWitt and B. DeWitt (New York: Gordon & Breach).
Zel'dovich, Ya. B. 1962, *Soviet Phys.—J.E.T.P.*, **14,** 1143.

20

A UNIFIED THEORY OF FORCES

As he remembers it, Steve Weinberg was driving to his office at MIT one morning in the early fall of 1967 when he had a sudden revelation about the forces of nature. At the time, Weinberg was thirty-four years old, a visiting professor at MIT on leave from Berkeley. Fleshy-faced, with blond-and-red curly hair and a generally serious manner, he had been enamored of theoretical physics since his adolescent years at the famous Bronx High School of Science in New York. Subsequently, his training took him to Cornell, the Institute for Theoretical Physics in Copenhagen, and Princeton, where he received his doctorate in 1957. Now he'd been out of graduate school for a decade, nearing the end of prime time for a theoretical physicist.

The neighborhood in East Cambridge near MIT is not pretty. Yellow brick smokestacks belch yellowish smoke, railroad tracks cross the cracked pavement like metallic Band-Aids, and the streets are littered with rundown apartments and ramshackle storefronts like Advance Tire on Broadway or the dark and deserted Necco warehouse on Mass Avenue and Sidney. In short, the physical setting would not seem to inspire brilliant thoughts. But then theoretical physics lives largely in the mind.

"It occurred to me that I had been applying the right ideas to the wrong problem," Weinberg recalls. For some time, Weinberg had been trying to develop a theory of the "strong nuclear force," modeled on the highly successful theory of electricity and magnetism called quantum electrodynamics (QED). He had failed. The "right ideas" were the mathematical concepts of QED. The "wrong problem" was the strong nuclear force. Weinberg's insight while driving through seedy East Cambridge was that the right problem for the right ideas was the "weak force." The weak force, along with the strong nuclear force, the electromagnetic force, and the force of gravity, constitute the four fundamental forces of nature. From the pulls and tugs of everyday human existence, we are familiar with the electromagnetic and gravitational forces. The other two forces are less evident. It is the strong nuclear force that holds neutrons and protons locked together within the tiny prison of the nucleus of the atom. The weak force causes neutrons to change into protons and other bizarre phenomena at the subatomic level. Among other things, the weak force is critical for stars to produce energy and thus for all life as we know it.

Previous theorists of the stature of Enrico Fermi and Richard Feynman had labored to construct partial theories of the weak force, but all such theories suffered from a terrible flaw: they yielded an answer of *infinity* for certain reasonable questions. Mathematicians and poets love infinity, but physicists do not. A single speck of infinity could demolish all of physics.

The good idea that Weinberg realized he could apply to the weak force was *symmetry*, which was discussed briefly in Chapter 4. In general terms, a principle of symmetry says that you can look at something from certain different points of view and it still looks the same. A square has a four-sided symmetry. Rotate it by ninety degrees and its appearance is identical. Many objects in nature have symmetry. A human face has left-right symmetry. Starfish grow a five-armed symmetry. Snowflakes have six identical sectors. Small hailstones have spherical symmetry—they look the same from any direction.

But the most important symmetries in physics are not the symmetries of material objects, but the symmetries of laws—that is, laws that say the same thing from different points of view. For example, one symmetry of the laws of physics is that all directions of space are equivalent. As a consequence, the force between two magnets on a table is the same no matter how you orient the table. The acceleration of a billiard ball upon being struck by a cue stick is the same no matter which direction the strike comes from.

Nature does not obey all symmetry principles we can imagine. For example, it is not true that the mirror image of every physical process is identical to the original. The violation of that particular symmetry principle, called parity, was found in the mid-1950s. But when nature does follow a symmetry principle, the laws are greatly simplified.

The first person to imagine a symmetry principle in physics was a young patent clerk in Switzerland named Albert Einstein, in 1905. In the opening pages of his theory of relativity, Einstein proposed a principle of "motional" symmetry: the laws of physics should appear the same for all observers moving past each other at constant speed. In all experiments to date, that symmetry principle has been verified. One consequence of Einstein's motional symmetry was that electricity and magnetism were no longer considered separate forces but unified into the "electromagnetic force."

Steven Weinberg loves principles of symmetry. In fact, one might say that he is obsessed with symmetry principles, in the way that Plato was obsessed with the Platonic forms. The ancient Greek philosopher believed that the world was based on a small number of eternal and perfect forms, including the four-sided pyramid, the six-sided cube, the eight-sided octahedron. These forms were not material stuff but foundational ideas. Symmetry principles in physics are foundational ideas. Weinberg believes that symmetry principles come before everything else, before the four fundamental forces and even before matter itself. In his book *Dreams of a Final Theory* (1992), he writes:

> Matter thus loses its central role in physics: all that is left is principles of symmetry . . . Symmetry principles have moved to a new level of importance in this [twentieth] century and especially in the last few decades: there are symmetry principles that dictate the very existence of all the known forces of nature . . . We believe that, if we ask why the world is the way it is and then ask why that answer is the way it is, at the end of this chain of explanations we shall find a few simple principles of compelling beauty.

Weinberg often uses the word "beauty" when he speaks of theories in physics, and symmetry principles in particular. Other words he commands to express beauty in science are "simplicity" and "inevitability." When he talks and writes about physics, Weinberg often sounds like a poet. Indeed, like the late biochemist Max Perutz, whom we have encountered earlier, Steven Weinberg is a gifted writer whose articles appear frequently in the literary *New York Review of Books* and whose books have received popular acclaim.

But Weinberg's devotion to principles of symmetry is more than an artist's devotion to beauty. Weinberg, and other physicists as well, hope that certain guiding principles, like symmetry, will so rigidly limit possible theories that only one description of nature is possible. Weinberg does not want many possible theories of gravity, many possible theories of the weak force. He dreams of a universe in which only a single theory is possible. We have all experienced the shaping power of principles. Certain principles of democracy, like "checks and balances," are so influential and constraining that nearly every democratic government on earth is constructed in a similar manner, with a legislative branch, a judicial branch, and an executive branch. In architecture, the simple principle that a building unit should be small enough to be easily lifted but big enough that the minimum are needed is so constraining a principle that nearly every brick through the centuries has been made the same size. The German philosopher and mathematician Leibniz is famous for his statement "This is the best of all possible worlds." Steven Weinberg hopes that this is the *only* possible world.

On that early fall day in 1967, the idea that suddenly occurred to Weinberg was that the weak force might have far more symmetry than had appeared in experiments, what physicists sometimes called hidden symmetry or broken symmetry. In particular, Weinberg imagined that the weak force acts symmetrically on certain pairs of subatomic particles, such as electrons and neutrinos. Although the particles in each pair appear different, they might be identical as far as the weak force is concerned, just as yellow and white tennis balls are identical in the game of tennis. The equivalence between electrons and neutrinos, in turn, would lead to other symmetries similar to those found in QED, the successful theory of the electromagnetic force.

Weinberg rapidly cast his idea in the mathematical language of quantum physics. To his surprise, he found that his new theory of the weak force, with the symmetry between electrons and neutrinos, was two theories in one. It naturally included the electromagnetic force along with the weak force. Indeed, Weinberg's new theory *required* that the two forces be joined at the hip. As he later recalled, "I found in doing this, although it had not been my idea at all to start with, that it turned out to be a theory not only of the weak force, based on an analogy with electromagnetism; it turned out to be a unified theory of the weak and electromagnetic forces . . ."

We have seen how experimental scientists, like Ernest Starling and Alexander Fleming in biology or Ernest Rutherford and Otto Hahn in physics, sometimes discover what they didn't expect. Weinberg's experience shows that the same can happen even in theoretical physics, where the surprise happens with pencil and paper.

Weinberg's unified theory is called the electroweak theory. Just as electricity and magnetism can generate each other and are part of a unifying electromagnetic force, the electroweak theory proposes that the weak force and the electromagnetic force are different aspects of the same force, the electroweak force. The electroweak theory was the first theory of the twentieth century to unify any of the fundamental forces of nature. The electroweak theory does not give infinite answers to any questions asked of it. Its predictions of new subatomic particles and new kinds of reactions have been largely confirmed by experiment. And it is beautiful.

The weak force was first witnessed in 1896, by Antoine-Henri Becquerel. Although he had no way to fully grasp the phenomenon—except that it clearly produced a great deal of penetrating energy—the French physicist had in fact observed a kind of radioactivity in which a neutron converts into a proton in the nucleus of an atom. An electron, called a beta ray, is produced in the conversion and goes flying off at great speed. As a whole, the process is called beta decay. In 1896, neutrons and protons had not been discovered, nor the atomic nucleus for that matter. Beta decay is caused by the weak force. Among the four fundamental forces of nature, only the weak force can change a neutron into a proton.

By the early 1930s the neutron had been discovered, and beta decay was understood as a process in which the electrically neutral neutron changed into three other subatomic particles: a positively charged proton, a negatively charged electron, and an electrically neutral "antineutrino." The reaction can be represented as

$$n \rightarrow p + e + \overline{\nu}$$

where n stands for neutron, p for proton, e for electron, and $\overline{\nu}$ for antineutrino. The "neutrino" was a new kind of subatomic particle, neutral and practically massless, and the antineutrino was the "antiparticle" of the neutrino. Since the early 1930s, physicists had realized that every subatomic particle has a twin, called an antiparticle. Particle and antiparticle twins are almost identical but have certain features that are opposite to each other, such as electrical charge. For example, the electron e has a negative electrical charge, while the antielectron $\overline{e}$ (also called the positron) has an identical mass but a positive electrical charge. Particles and their antiparticle twins can annihilate each other completely to produce pure energy. In some science fiction stories, writers have imagined entire antipeople, or even antiplanets, which can annihilate their twins.

The weak force is called weak because it is far weaker than the strong nuclear force, about a hundred thousand times weaker at ordinary energies. The weak force also acts very slowly compared to the strong nuclear force. Yet, as in the race between the tortoise and the hare, the weak force can patiently accomplish things the strong force cannot, such as convert a neutron into a proton.

The weak force causes other reactions similar to the beta decay above, for example, $n + \bar{e} \rightarrow p + \bar{\nu}$, where an antielectron collides with a neutron and produces a proton and an antineutrino; or the reaction $p + e \rightarrow n + \nu$, where an electron collides with a proton and produces a neutron and a neutrino; or $n + \bar{p} \rightarrow e + \bar{\nu}$, where a neutron collides with an antiproton and produces an electron and an antineutrino.

There are three similarities about all these reactions. First, they all involve four particles. Second, the total electrical charge of particles going into the reaction (before the arrow) equals the total electrical charge of particles coming out of the reaction (after the arrow). And third, upon closer scrutiny, we can see that the electrons, antielectrons, neutrinos, and antineutrinos appear together in a certain regular way. Each has something like a charge, but rather than an electrical charge it is called an electron lepton number. (Electrons and neutrinos, both of very small mass, are called leptons, from the Greek word for "light." Certain other subatomic particles are also called leptons and have their own lepton numbers.) If the electron lepton number is 1 for the electron, 1 for the neutrino, −1 for the antielectron, and −1 for the antineutrino, then the total electron lepton number of particles going into each reaction equals the total electron lepton number of particles coming out of the reaction. A physicist would say that the total electron lepton number is conserved, just as the total electric charge is conserved.

The first theory of the weak force was proposed in 1933 by the great Italian-American physicist Enrico Fermi. A short-legged man who loved hiking, Fermi was almost unique among physicists in being superb at both experiment and theory. (Another example was Robert Dicke.) The weak force was already known to act over only extremely short distances, much smaller than the distance across an atomic nucleus. To account for this experimentally observed fact, Fermi suggested that the four particles involved in a weak nuclear reaction had to meet at a point. Furthermore, they had to have fractional "spins," like the electron.

A word about spin. As we discussed in Chapter 11 on Linus Pauling's work, subatomic particles behave as if they are tiny gyroscopes, spinning about an axis through their centers. The rate of spin, like the mass and the electrical charge, is an internal and fixed property of a subatomic particle. The spin of the electron is used as the standard and set equal to ½. Particles with fractional spins, like ½ or ³⁄₂, are called fermions, named after you know who. Particles with whole-number spins, like 1 or 2, are called bosons, named after the Indian physicist Satyandranath Bose. Neutrons, protons, electrons, and neutrinos are all fermions. Photons, the particles that convey the electromagnetic force, are bosons, with spin 1. (Some other bosons are pions and kaons.) In this language, Fermi's theory of the weak force said that the force acts between four fermions meeting at a point.

Fermi's theory worked well to explain the details of beta decay, but it had problems. First, it seemed designed only to explain that single reaction and thus lacked the kind of compelling "inevitability" that Weinberg speaks of. More seriously, when applied to other types of weak interactions—for example, when used to compute the change in mass of a proton due to particle creation in its vicinity—Fermi's theory gave infinite results.

In 1958, Richard Feynman and Murray Gell-Mann of the California Institute of Technology, and independently E. C. G. Sudarshan and R. E. Marshak, substantially revised Fermi's theory. They figured out how the weak force depended on the spins of the four interacting fermions. These physicists and others also proposed that the four fermions did not actually meet at a point, but were separated by a small distance, a gap. Hypothetically, the weak force would be conveyed across this gap by the exchange of a new kind of particle called an intermediate boson. This new particle was called *W*, for weak. For the spins and other motions to add up properly, the *W* would necessarily require a spin of 1, so it was a boson. It would also necessarily come in two varieties, a positively charged W^+ and a negatively charged W^-.

To see the difference in the two pictures, consider the beta decay reaction $n \rightarrow p + e + \bar{\nu}$. Figure 20.1a illustrates the process as envisioned by Fermi, while Figure 20.1b shows it as envisioned by Feynman, Gell-Mann, Sudarshan, and Marshak. The W^- is represented by the wavy line. In Figure 20.1b, we can see why the *W* particle must be negatively

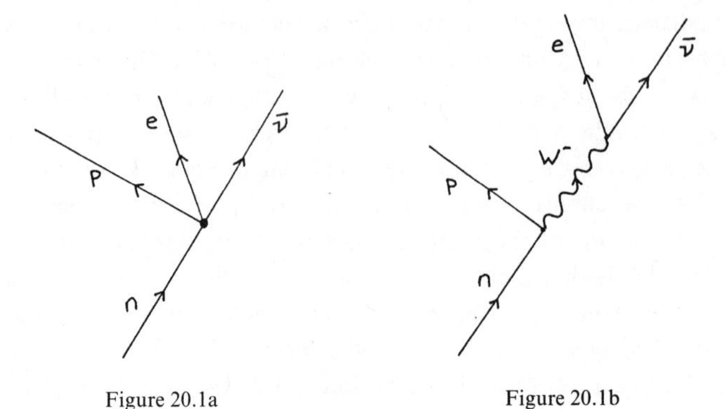

Figure 20.1a

Figure 20.1b

charged in this reaction. Following the arrows, the first thing that happens is that a neutron changes into a proton and creates a W^-. Since total electrical charge cannot be created or destroyed, and the neutron is electrically neutral, the positive charge of the proton must be canceled by the negative charge of the W^-. Next, the W^- travels a bit, disintegrates, and produces an electron and an antineutrino. The negative charge of the W^- is given to the negatively charged electron. Once again, electrical charge is conserved. What could be more simple!

The Feynman–Gell-Mann/Sudarshan-Marshak theory was an advance over Fermi's theory. Yet it also predicted that the energy in certain reasonable reactions would be infinite.

By this time, physicists had found a way of dealing with such infinities in the quantum theory of electromagnetism, called quantum electrodynamics (QED). In all quantum theories of the fundamental forces, the infinities were understood to arise from a peculiar feature of quantum physics that allows some subatomic particles to briefly materialize out of nothing and then disappear. Each subatomic particle, like the electron, is surrounded by a haze of these fleeting ghosts, and they sometimes make an infinite contribution to its energy and mass. In QED, a clever method was found to redefine each particle to *include* its haze. After all, when we measure a particle, we are also measuring the haze of fleeting particles attached to it. Such a redefinition is called renormalization. In a sense, renormalization is like recognizing that in evaluating the balance sheet of the United States, the huge national debt should be offset against the huge wealth in production capacity, even if that capacity is partly hidden from view.

When renormalization occurs, the infinities go away. But only certain theories have the right mathematical properties to permit them to be renormalized. QED could be renormalized. The Fermi/Feynman–Gell-Mann/Sudarshan-Marshak theory of the weak force could not.

Steven Weinberg is quite familiar with the history of work on the weak force. In fact, he has a special interest in history, not only the history of science but also history in general. In an article he wrote for *Daedalus* titled "Physics and History," he sees a similarity between the historical perspective of particle physicists, like himself, and the view of Western religions. By contrast, according to Weinberg, the historical view of other branches of science is more similar to Eastern religious traditions. "Christianity and Judaism teach that history is moving towards a climax—the day of judgment; similarly many elementary particle physicists think that our work . . . will come to an end in a final theory . . . An opposing perception of history is held by those faiths that believe that history will go on forever, that we are bound to the wheel of endless reincarnation . . . Other scientists [non-particle physicists] look forward to an endless future of finding interesting problems."

Weinberg goes on to say that he and other particle physicists must leave behind merely interesting problems on the way to a "final theory" of the forces of nature. "It is not necessary to mop up all the islands of unsolved problems in order to make progress toward a final theory," he writes. "Our situation is a little like that of the United States Navy in World War II: bypassing Japanese strong points like Truk or Rabaul, the Navy instead moved on to take Saipan, which was closer to its goal of the Japanese home islands."

For two decades, following the late 1940s, quantum electrodynamics reigned supreme among all theories in physics. QED was beautiful, its infinities had been cured, and it made exquisitely accurate predictions about the behavior of electrons and photons. Physicists everywhere on the planet wanted to invent theories like QED.

A particular symmetry principle allows QED to be renormalized. That principle says that at each point in space the electron waves that appear in QED can be moved forward or backward by any amount, and if the photon waves (which convey the electromagnetic force between electrons) are altered appropriately, the laws will be unchanged. This symmetry is called a gauge symmetry.

It was also a kind of gauge symmetry that Steven Weinberg imagined when he was driving to work in the early fall of 1967. He pushed his idea about the equivalence of electrons and neutrinos to the more extreme form that a *single particle* could be any combination of electron and neutrino, say 30 percent electron and 70 percent neutrino, and still respond to the weak force in the same way. Furthermore, as with all gauge symmetries, the percentages could vary from one point in space to another. In the strange world of quantum physics, where a particle acts like a wave and can be in several places at once, the very *identity* of a particle can also be blurred. Thus it actually makes sense to speak of a particle that is 30 percent electron and 70 percent neutrino, or 52 percent electron and 48 percent neutrino, and so on.

A fruitful way of expressing this idea is to think of each particle as having a small arrow attached to it and pointing in a certain direction. One particular direction, say the vertical direction, corresponds to pure electron or pure neutrino: If the arrow points straight up, the particle is pure electron. If the arrow points straight down, the particle is pure neutrino. (The arrows of antiparticles point in the opposite direction from those of their particle twins.) Arrows pointing in any direction but vertical represent particles that are part electron and part neutrino, hybrids. *Complete symmetry between electrons and neutrinos means that the laws of the weak interaction look the same when the arrows are rotated in any direction.* So we are back to a kind of rotational symmetry, as with squares and snowflakes, but now the rotations don't occur in the ordinary space we are familiar with, but in an abstract space of electron-neutrino self-identity. These are the kinds of spaces that theoretical physicists adore.

Gradually, we are approaching the full scope of Weinberg's ideas. A key feature of the weak interaction is that the electrons and neutrinos work in pairs, as can be seen in the reactions already considered. (The neutron and proton also work in pairs, but we will disregard them in this discussion.) Furthermore, because of the observation that the total electron lepton number is the same before and after the interaction, it can be shown that every weak interaction is equivalent to an interaction where one member of the pair is a particle and the other an antiparticle. In summary, the weak interaction involves a pair of particles, one of which is either an electron or a neutrino and the other of which is either an antielectron or an antineutrino. There are only four such possible pairs:

$(e\bar{e})$, $(\nu\bar{\nu})$, $(e\bar{\nu})$, and $(\bar{e}\nu)$. These are the "pure pairs," the thoroughbreds of the weak interaction. As rotations occur in "electron-neutrino identity space," each of these four pure pairs becomes a hybrid, a combination of the others. In turn, any general pair of particles in this blurred identity space can be expressed as a specific combination of these four thoroughbreds. For example, a general pair, which we might call (a b), could be 13 percent $(e\bar{e})$, 28 percent $(\nu\bar{\nu})$, 17 percent $(e\bar{\nu})$, and 42 percent $(\bar{e}\nu)$.

Now we come to the intermediate bosons, which are analogous to the photon in QED because they are the particles that convey the force. If we look at Figure 20.1b, we see that a W^- is needed to produce the pure pair $(e\bar{\nu})$. In fact, a different kind of intermediate boson is required to produce each of the pure pairs. Thus, the theory, with its electron-neutrino symmetry, *requires* the existence of four intermediate bosons. Two of the pure pairs have a net electrical charge, the $(e\bar{\nu})$ and the $(\nu\bar{e})$, and thus they will require negative and positive intermediate bosons, respectively, to produce them. These bosons are what we have called before the W^- and the W^+. But there are two more pairs that are electrically neutral, $(e\bar{e})$ and $(\nu\bar{\nu})$. Evidently, the theory requires two more intermediate bosons, and they must be electrically neutral.

So Weinberg's first prediction was that there should be two electrically neutral intermediate bosons. Along with these new particles was the prediction that reactions like $n \rightarrow n + \bar{\nu} + \nu$ should occur. These are called neutral current weak interactions because there is no exchange of electrical charge. Neutral current weak reactions had never been seen before, and would not be seen for a number of years.

Finally, the surprise. We *already know* of an electrically neutral boson that can produce the $(e\bar{e})$ pair, namely the photon, carrier of the electromagnetic force. The theory of QED is full of reactions in which photons produce electron-antielectron pairs, denoted by $\gamma \rightarrow e\bar{e}$. Thus, one of the two neutral bosons predicted by the electroweak theory could be identified with the well-known photon! The other neutral boson Weinberg named the Z. Apparently, the theory contains not only the carriers of the weak force but the carrier of the electromagnetic force as well.

Furthermore, as rotations occur in the strange "electron-neutrino identity space" and the electron-neutrino pairs get mixed together, the intermediate bosons must mix together as well, since hybrid bosons are now needed to disintegrate into hybrid electron-neutrino pairs. (Mixtures of the intermediate bosons are also needed to keep the bookkeeping correct for total electric charge.) In particular, the photon gets mixed together with the Z and the Ws. Thus, a general carrier of the electroweak force will be a combination of the W^+, the W^-, the Z, and the photon, just as a general particle-antiparticle pair will be a combination of the four "pure pairs" discussed above. This kind of mixing is the deep sense in which the electroweak theory is a unified theory. And since all mixtures must be equivalent under the assumed symmetry principle, the individual identities of the electromagnetic force and the weak nuclear force dissolve into the single electroweak force.

Unbeknownst to him, Weinberg was not completely alone in his important ideas. In one of those coincidences that happen frequently in science, a leading Pakistani theoretical physicist named Abdus Salam worked out an essentially identical electroweak theory, independently and simultaneously. Eventually, the two of them would share the Nobel Prize, along with American physicist Sheldon Glashow, who had proposed some of the key concepts in the early 1960s. Indeed, elements of the electroweak theory had been appearing since the work of American physicist Julian Schwinger in 1957. Schwinger first suggested a unification between the photon and the charged Ws. Then, in 1958, S. A. Bludman proposed a Z-like particle, rather than the photon, to mix with the charged Ws. In 1961, Glashow made the important proposal that the theory had to be enlarged to include four intermediate bosons, but the electron and neutrino were never completely equivalent in Glashow's version. It was not until the Weinberg and Salam theory that all the pieces fit together.

For several years, the electroweak theory of Weinberg and Salam received scant attention. First, the new theory did not explain any experimental results that were not already explained by the existing Feynman–Gell-Mann/Sudarshan-Marshak theory. And its predictions of new phenomena, such as the Z particle and the neutral current reactions, had not yet been tested. Finally, and most profoundly, no one knew if the theory could be renormalized.

Then, in 1971, a lowly graduate student at the University of Utrecht named Gerard 't Hooft mathematically proved that the electroweak theory was renormalizable. Suddenly, physicists began to take note of the theory. No fresh experimental evidence had arrived, but a major theoretical obstacle had been overcome. According to the Institute of Scientific Information, Weinberg's 1967 paper on the electroweak theory received a total of four citations in the period 1967–1971. In 1972, after 't Hooft's work, there were sixty-five citations. And the number mounted from there. By 1988, Weinberg's three-page paper in the *Physical Review* was the most frequently cited paper in elementary particle physics since the end of World War II.

Weinberg regrets that it was not he himself who proved renormalization. During the period 1967–1971, in fact, he spent much of his time on a textbook on general relativity and cosmology. As he later lamented in an interview with me in the late 1980s, "I'm sorry I wrote the damn thing because . . . those are the years I should have dropped everything I was doing and worked on proving that the spontaneously broken gauge theories were renormalizable . . . It's wonderful to write a book that has some impact, but it's even more wonderful to make discoveries that have an impact . . ."

In 1973, the first of the electroweak theory's predictions, neutral current reactions involving neutrinos, was experimentally confirmed in the giant particle accelerator at CERN in Geneva and soon after at Fermilab near Chicago. In fact, neutral current reactions had been suggested as far back as 1937. However, such predictions were always made with theories both incomplete and suffering with infinities. After 1973, most physicists *assumed* that the electroweak theory was correct. Twenty years later Weinberg offered an interesting analysis of the psychology, in a type of capitalistic view of the scientific endeavor: "a theory had come along that had the kind of compelling quality, the internal consistency and rigidity, that made it reasonable for physicists to believe they would make more progress in their own scientific work by believing the theory to be true than by waiting for it to go away."

The Nobel Committee in Stockholm did not require further confirmation of the electroweak theory and awarded Weinberg, Salam, and Glashow the Nobel Prize in physics in 1979. In 1983, in another experiment at CERN led by Carlo Rubbia, the W particles were discovered. The Z was found the following year. The measured masses of the W and Z were in good agreement with the predictions of the electroweak theory, each roughly a hundred times as massive as the proton or neutron.

At last we come to Weinberg's paper itself. Of all the landmark papers we have considered, this one is by far the most mathematically technical, and we will content ourselves with a broad-brush discussion.

It is fascinating that in the first paragraph Weinberg says that it is "natural" to unite photons and intermediate bosons—as if that were the germinal notion leading to the theory. Yet, as we have seen earlier, that idea of union was not at all the starting point of Weinberg's ruminations on the weak force. It is sometimes the case that scientists cover the tracks leading to their final accomplishment, especially in their formal papers.

Even without knowledge of any of the symbols or their meanings, one must be impressed by the economy and power of the master equation, equation (4). Here is the "Lagrangian" of the theory. The Lagrangian contains within its tangled mathematical hedges all the laws of the electroweak theory, all the symmetry principles, all the conservation laws (such as conservation of energy). The Lagrangian also harbors, in almost invisible form, all the laws of special relativity and quantum field theory. Einstein's relativistic ideas of time and space are distilled into the Greek subscripts for some of the symbols, such as $\vec{A}_\mu$. The ideas of quantum physics are embodied in the fact that electrons and neutrinos are represented not as definite particles but as waves of probability, hidden in the notation of the "left-handed doublet" L and "right-handed singlet" R. To a physicist, the Lagrangian of equation (4) is a work of art. Aside from the uncertainty of the energy field called ϕ, everything is just as it must be, given the assumed symmetry between electrons and neutrinos.

There are two kinds of quantities in the Lagrangian, the electron and neutrino particles, and the intermediate bosons. We will consider them in turn.

First, the electrons and neutrinos, both members of the electron lepton family. A peculiar aspect of the weak interaction is that it distinguishes between particles whose spin is clockwise about their direction of motion, called right-handed particles, and particles of counterclockwise spin, called left-handed. (This distinction, called parity violation, has been mentioned earlier.) The weak force acts only on the left-handed neutrino, but it can act on both left-handed and right-handed electrons. Thus, the left-handed electron and left-handed neutrino are grouped together in one "left-handed doublet," represented by L in equation (1). The right-handed electron lives by its lonesome self in a "right-handed singlet," represented by R in equation (2).

Next, the four intermediate bosons required by the theory. These are also called "gauge fields." The word "field" here means a packet of energy that occupies a region of space. In modern quantum theory, where everything is represented by a probability wave, there is little distinction between particles and pure energy. Both are considered packets of energy traveling through space, that is, "fields."

Three of the gauge fields are represented by the symbol $\vec{A}_\mu$, a shorthand for A^1_μ, A^2_μ, and A^3_μ. The fourth gauge field, which is separate for technical reasons, is denoted by B_μ. The A^3_μ and B_μ gauge fields are the ones that are electrically neutral. The photon (represented by A_μ without any superscripts) and Z particles are combinations of these, as shown in equations (10) and (11).

Now, we come to the all-important symmetry ideas. Recall that we can think of each particle as having an arrow attached to it, the direction of pointing indicating the particular mixture of electron and neutrino. The arrow is called electronic isospin. The symbol $\vec{T}$ represents the total electronic isospin of the particles. In equation (4), $\vec{t}$ produces rotations of isospin, and the way that $\vec{t}$ enters that equation expresses the symmetry between electrons and neutrinos.

Swiftly going through the various terms of the Lagrangian, the first two terms represent the energies of the $\vec{A}$ and B gauge fields, respectively. The third and fourth terms represent the gauge fields interacting with the right-handed particles and left-handed particles, respectively.

We now arrive at the last three terms of the Lagrangian, all containing the symbol ϕ. The ϕ field is another kind of energy field, called a Higgs field, which Weinberg refers to as a spin-zero doublet. The role of the Higgs field is to "break the symmetry" of the theory. Such theories are called spontaneously broken gauge theories.

Why should we want to "break the symmetry" of such a beautiful theory, a theory giving complete equivalence to electrons and neutrinos, photons and Zs? Because we know from reality that the electron is not *completely* equivalent to the neutrino. The photon cannot be *completely* equivalent to the Z. For example, the electron's mass is far larger than that of the neutrino. Likewise, the photon has no mass, whereas the Z has mass. Something must produce a slight asymmetry between these particles. And that is the job of the Higgs field ϕ.

The Higgs field is the least understood part of the theory. To date, the Higgs field, and the Higgs particles associated with the field, have never been found. Fortunately, very little of the electroweak theory depends on the details of the Higgs field.

Every theory has a number of unknown or adjustable parameters. If there are too many such adjustable parameters, the theory can fit any set of observational results by appropriately adjusting the parameters. A good theory has a small number of adjustable parameters and a much larger number of experimental results that it must fit. What are the unknown parameters in Weinberg's electroweak theory? There are five in all: (1) the strength, g, of the gauge fields $\vec{A}$, (2) The strength, g', of the gauge field B, (3) the strength, G_e, of the ϕ field interaction with electrons, (4) the mass, M_1, of the ϕ field particle, and (5) the lowest energy level, λ, of the ϕ field. Because the mass M_1 does not affect any of the results, we will drop it from consideration and say that there are only four significant adjustable parameters in the theory.

The electron charge, e, comes out of the theory as a specific combination of g and g' given in equation (15). Since the electron charge is well known, the number of adjustable parameters of the theory is now reduced from four to three. Likewise, the electron mass and the rate of beta decay, both experimentally well known, are derived by the theory, as seen in equation (16) and the following equation. Now we are down to only one adjustable parameter. Such economy makes a good theory.

On the other side of the balance sheet, the theory explicitly *predicts* the mass of the W intermediate boson, given in equation (9), and the mass of the Z intermediate boson, given in equation (12), in terms of the parameters of the theory. The theory also implicitly predicts the rates and probabilities of numerous reactions involving electrons and neutrinos in addition to the beta decay reaction.

To summarize, Weinberg's and Salam's electroweak theory has far fewer adjustable parameters than predicted results, making it a good candidate for a powerful and compelling theory. (Of course, its predictions must still be confirmed by experiment.)

The "neutral spin-1 fields" are the Z_μ and the A_μ. The A field has zero mass, equation (13), and Weinberg states that it "is to be identified with the photon."

Toward the end of the short paper, Weinberg seems to discount some of his own calculations by saying, "Of course our model has too many arbitrary features for these predictions to be taken very seriously." Given the proposed symmetries of the theory, the only arbitrary feature, in fact, is the nature of the symmetry-breaking ϕ field. Perhaps Weinberg's uncertainty over the ϕ field—in other words, the precise mechanism of how the symmetry between electrons and neutrinos is slightly broken—is why he expresses some reservations about his theory and calls it a model instead of a theory. Indeed, the word "Model" occurs in his paper's title.

At the end of the paper, Weinberg asks the crucial question: "Is this model renormalizable?" A few years later, he would have the answer to his question.

Steven Weinberg may be the twentieth century's most eloquent advocate of symmetry principles in nature. For him, symmetry principles are both compelling and beautiful, and he often talks about the intellectual pleasure of recognizing such principles. Yet one does not get the impression that Weinberg does physics primarily for pleasure. Rather, he seems on a relentless march to find the most fundamental laws of nature, the "final theory" as he calls it—detouring around all lesser scientific problems as digressions from his goal. It is an urgent goal, one that this physicist burns to reach during his own lifetime. In this regard, Weinberg resembles Einstein. The father of relativity was also single-mindedly bent on discovering the fundamental truths of nature. Even as a young man, as Einstein wrote in his autobiography, he had "learned to scent out that which was able to lead to fundamentals and to turn aside from everything else."

A MODEL OF LEPTONS[1]

Steven Weinberg[2]

Laboratory for Nuclear Science and Physics Department,
Massachusetts Institute of Technology, Cambridge, Massachusetts

Received 17 October 1967

Physical Review Letters (1967)

LEPTONS INTERACT ONLY WITH PHOTONS, and with the intermediate bosons that presumably mediate weak interactions. What could be more natural than to unite[3] these spin-one bosons into a multiplet of gauge fields? Standing in the way of this synthesis are the obvious differences in the masses of the photon and intermediate meson, and in their couplings. We might hope to understand these differences by imagining that the symmetries relating the weak and electromagnetic interactions are exact symmetries of the Lagrangian but are broken by the vacuum. However, this raises the specter of unwanted massless Goldstone bosons.[4] This note will describe a model in which the symmetry between the electromagnetic and weak interactions is spontaneously broken, but in which the Goldstone bosons are avoided by introducing the photon and the intermediate-boson fields as gauge fields.[5] The model may be renormalizable.

We will restrict our attention to symmetry groups that connect the observed electron-type leptons only with each other, i.e., not with muon-type leptons or other unobserved leptons or hadrons. The symmetries then act on a left-handed doublet

$$L \equiv [\tfrac{1}{2}(1+\gamma_5)]\begin{pmatrix}\nu_e \\ e\end{pmatrix} \tag{1}$$

and on a right-handed singlet

$$R \equiv [\tfrac{1}{2}(1-\gamma_5)]e. \tag{2}$$

The largest group that leaves invariant the kinematic terms $-\overline{L}\gamma^\mu\partial_\mu L - \overline{R}\gamma^\mu\partial_\mu R$ of the Lagrangian consists of the electronic isospin $\vec{T}$ acting on L, plus the numbers N_L, N_R of left- and right-handed electron-type leptons. As far as we know, two of these symmetries are entirely unbroken:

the charge $Q = T_3 - N_R - \frac{1}{2}N_L$, and the electron number $N = N_R + N_L$. But the gauge field corresponding to an unbroken symmetry will have zero mass,[6] and there is no massless particle coupled to N,[7] so we must form our gauge group out of the electronic isospin $\vec{T}$ and the electronic hyperchange $Y \equiv N_R + \frac{1}{2}N_L$.

Therefore, we shall construct our Lagrangian out of L and R, plus gauge fields $\vec{A}_\mu$ and B_μ coupled to $\vec{T}$ and Y, plus a spin-zero doublet

$$\varphi = \begin{pmatrix} \varphi^0 \\ \varphi^- \end{pmatrix} \tag{3}$$

whose vacuum expectation value will break $\vec{T}$ and Y and give the electron its mass. The only renormalizable Lagrangian which is invariant under $\vec{T}$ and Y gauge transformations is

$$\begin{aligned} \mathscr{L} = -\tfrac{1}{4}(\partial_\mu \vec{A}_\nu - \partial_\nu \vec{A}_\mu + g\vec{A}_\mu \times \vec{A}_\nu)^2 - \tfrac{1}{4}(\partial_\mu B_\nu - \partial_\nu B_\mu)^2 \\ - \overline{R}\gamma^\mu(\partial_\mu - ig'B_\mu)R - L\gamma^\mu(\partial_\mu ig\vec{t} \cdot \vec{A}_\mu - i\tfrac{1}{2}g'B_\mu)L - \tfrac{1}{2}\partial_\mu\varphi - ig\vec{A}_\mu \cdot \vec{t}\varphi \\ + i\tfrac{1}{2}g'B_\mu\varphi^2 - G_e(\overline{L}\varphi R + \overline{R}\varphi^\dagger L) - M_1{}^2\varphi^\dagger\varphi + h(\varphi^\dagger\varphi)^2. \end{aligned} \tag{4}$$

We have chosen the phase of the R field to make G_e real, and can also adjust the phase of the L and Q fields to make the vacuum expectation value $\lambda \equiv \langle\varphi^0\rangle$ real. The "physical" φ fields are then φ^- and

$$\varphi_1 \equiv (\varphi^0 + \varphi^{0\dagger} - 2\lambda)/\sqrt{2} \qquad \varphi_2 \equiv (\varphi^0 - \varphi^{0\dagger})/i\sqrt{2}. \tag{5}$$

The condition that φ_1 have zero vacuum expectation value to all orders of perturbation theory tells us that $\lambda^2 \cong M_1{}^2/2h$, and therefore the field φ_1 has mass M_1 while φ_2 and φ_- have mass zero. But we can easily see that the Goldstone bosons represented by φ_2 and φ^- have no physical coupling. The Lagrangian is gauge invariant, so we can perform a combined isospin and hypercharge gauge transformation which eliminates φ^- and φ_2 everywhere[8] without changing anything else. We will see that G_e is very small, and in any case M_1 might be very large,[9] so the φ_1 couplings will also be disregarded in the following.

The effect of all this is just to replace φ everywhere by its vacuum expectation value

$$\langle\varphi\rangle = \lambda\begin{pmatrix} 1 \\ 0 \end{pmatrix}. \tag{6}$$

The first four terms in $\mathscr{L}$ remain intact, while the rest of the Lagrangian becomes

$$-\tfrac{1}{8}\lambda^2 g^2[(A_\mu{}^1)^2 + (A_\mu{}^2)^2] - \tfrac{1}{8}\lambda^2(gA_\mu{}^3 + g'B_\mu)^2 - \lambda G_e \bar{e}e. \tag{7}$$

We see immediately that the electron mass is λG_e. The charged spin-1 field is

$$W_\mu \equiv 2^{-\frac{1}{2}}(A_\mu{}^1 + iA_\mu{}^2) \tag{8}$$

and has mass

$$M_W = \tfrac{1}{2}\lambda g. \tag{9}$$

The neutral spin-1 fields of definite mass are

$$Z_\mu = (g^2 + g'^2)^{-\frac{1}{2}}(gA_\mu{}^3 + g'B_\mu), \tag{10}$$

$$A_\mu = (g^2 + g'^2)^{-\frac{1}{2}}(-g'A_\mu{}^3 + gB_\mu). \tag{11}$$

Their masses are

$$M_Z = \tfrac{1}{2}\lambda(g^2 + g'^2)^{\frac{1}{2}}, \tag{12}$$

$$M_A = 0, \tag{13}$$

so A_μ is to be identified as the photon field. The interaction between leptons and spin-1 mesons is

$$\frac{ig}{2\sqrt{2}}\,\bar{e}\gamma^\mu(1+\gamma_5)\nu W_\mu + \text{H.c.} + \frac{igg'}{(g^2+g'^2)^{\frac{1}{2}}}\,\bar{e}\gamma^\mu e A_\mu$$

$$+\frac{i(g^2+g'^2)^{\frac{1}{2}}}{4}\left[\left(\frac{3g'^2-g^2}{g'^2+g^2}\right)\bar{e}\gamma^\mu e - \bar{e}\gamma^\mu\gamma_5 e + \bar{\nu}\gamma^\mu(1+\gamma_5)\nu\right]Z_\mu. \tag{14}$$

We see that the rationalized electric charge is

$$e = gg'/(g^2 + g'^2)^{\frac{1}{2}} \tag{15}$$

and, assuming that W_μ couples as usual to hadrons and muons, the usual coupling constant of weak interactions is given by

$$G_W/\sqrt{2} = g^2/8M_W{}^2 = \tfrac{1}{2}\lambda^2. \tag{16}$$

Note that then the $e - \varphi$ coupling constant is

$$G_e = M_e/\lambda = 2^{\frac{1}{4}} M_e G_W^{\frac{1}{2}} = 2.07 \times 10^{-6}.$$

The coupling of φ_1 to muons is stronger by a factor M_μ/M_e, but still very weak. Note also that (14) gives g and g' larger than e, so (16) tells us that $M_W > 40$ BeV, while (12) gives $M_Z > M_W$ and $M_Z > 80$ BeV.

The only unequivocal new predictions made by this model have to do with the couplings of the neutral intermediate meson Z_μ. If Z_μ does not couple to hadrons then the best place to look for effects of Z_μ is in electron-neutron scattering. Applying a Fierz transformation to the W-exchange terms, the total effective $e - \nu$ interaction is

$$\frac{G_W}{\sqrt{2}} \bar{\nu}\gamma_\mu (1 + \gamma_5)\nu \left\{ \frac{(3g^2 - g'^2)}{2(g^2 + g'^2)} \bar{e}\gamma^\mu e + \tfrac{3}{2}\bar{e}\gamma^\mu \gamma_5 e \right\}.$$

If $g \gg e$ then $g \gg g'$, and this is just the usual $e - \nu$ scattering matrix element times an extra factor $\frac{3}{2}$. If $g \simeq e$ then $g \ll g'$, and the vector interaction is multiplied by a factor $-\frac{1}{2}$ rather than $\frac{3}{2}$. Of course our model has too many arbitrary features for these predictions to be taken very seriously, but it is worth keeping in mind that the standard calculation[10] of the electron-neutrino cross section may well be wrong.

Is this model renormalizable? We usually do not expect non-Abelian gauge theories to be renormalizable if the vector-meson mass is not zero, but our Z_μ and W_μ mesons get their mass from the spontaneous breaking of the symmetry, not from a mass term put in at the beginning. Indeed, the model Lagrangian we start from is probably renormalizable, so the question is whether this renormalizability is lost in the reordering of the perturbation theory implied by our redefinition of the fields. And if this model is renormalizable, then what happens when we extend it to include the couplings of $\vec{A}_\mu$ and B_μ to the hadrons?

I am grateful to the Physics Department of MIT for their hospitality, and to K. A. Johnson for a valuable discussion.

1. This work is supported in part through funds provided by the U. S. Atomic Energy Commission under Contract No. AT(30-1)2098.

2. On leave from the University of California, Berkeley, California.

3. The history of attempts to unify weak and electromagnetic interactions is very long, and will not be reviewed here. Possibly the earliest reference is E. Fermi,

Z. Physik *88*, 161 (1934). A model similar to ours was discussed by S. Glashow, Nucl. Phys. *22*, 579 (1961); the chief difference is that Glashow introduces symmetry-breaking terms into the Lagrangian, and therefore gets less definite predictions.

4. J. Goldstone, Nuovo Cimento *19*, 154 (1961); J. Goldstone, A. Salam, and S. Weinberg, Phys. Rev. *127*, 965 (1962).

5. P. W. Higgs, Phys. Letters *12*, 132 (1964), Phys. Rev. Letters *13*, 508 (1964), and Phys. Rev. *145*, 1156 (1966); F. Englert and R. Brout, Phys. Rev. Letters *13*, 321 (1964); G. S. Guralnik, C. R. Hagen, and T. W. B. Kibble, Phys. Rev. Letters *13*, 585 (1964).

6. See particularly T. W. B. Kibble, Phys. Rev. *155*, 1554 (1967). A similar phenomenon occurs in the strong interactions; the ρ-meson mass in zeroth-order perturbation theory is just the bare mass, while the *A*1 meson picks up an extra contribution from the spontaneous breaking of chiral symmetry. See S. Weinberg, Phys. Rev. Letters *18*, 507 (1967), especially footnote 7; J. Schwinger, Phys. Letters *24B*, 473 (1967); S. Glashow, H. Schnitzer, and S. Weinberg, Phys. Rev. Letters *19*, 139 (1967), Eq. (13) *et seq.*

7. T. D. Lee and C. N. Yang, Phys. Rev. *98*, 101 (1955).

8. This is the same sort of transformation as that which eliminates the nonderivative $\vec{\pi}$ couplings in the σ model; see S. Weinberg, Phys. Rev. Letters *18*, 188 (1967). The $\vec{\pi}$ reappears with derivative coupling because the strong-interaction Lagrangian is not invariant under chiral gauge transformation.

9. For a similar argument applied to the σ meson, see Weinberg, Ref. 8.

10. R. P. Feynman and M. Gell-Mann, Phys. Rev. *109*, 193 (1957).

21

QUARKS: A TINIEST ESSENCE OF MATTER

When my eldest daughter was five years old, I gave her a set of nested Russian dolls, without explanation. She soon found that the outer doll could be opened, to reveal another doll lady enclosed within. To her delight, she discovered that the second doll could also be taken apart, with yet a third inside it. On and on she went, to smaller and smaller ladies, until she got to a tiny wooden figure that wouldn't open.

"Where are the other ladies?" she asked, a glum look on her face.

"There aren't any more," I said. "That's the smallest one."

Evidently she didn't believe me. That night she went to our oil-splattered garage, got a hammer from the tool box, and clobbered the smallest doll. It split into two solid fragments, but no smaller dolls. One of the pieces she carried around in her pocket for a week.

My daughter was not just playing a game. She was haunted. There must be something deep in our natures, even as children, that compels us to seek out the smallest thing from which all else is made, the primordial seed, the unbreakable unit, the thing that endures when all else has crumbled away. We all know the impulse. We want to hold in our hand the innermost thing. The search for that elemental substance has driven scientists and humanists alike since our species could first ponder and feel.

Over two thousand years ago, Aristotle tried to organize the cosmos in terms of five pure ingredients: earth, water, fire, air, and ether. Other ancient thinkers, Democritus and Lucretius, proposed that all matter was constructed of tiny atoms, invisible but indestructible. In this view, material things could be neither created nor destroyed. Every effect had an atomic cause, and thus human beings were freed from the vagaries of the gods.

For the great physicist Isaac Newton, atoms were not so much a liberation as a primordial "oneness," created by God: "It seems probable to me," Newton wrote, "that God in the beginning formed matter in solid, massy, hard, impenetrable, moveable particles . . . and that these primitive particles, being solids, are incomparably harder than porous bodies compounded of them; even so hard as never to wear or break in pieces; no ordinary power being able to divide what God himself made one in the first creation."

Or consider the words of the French novelist Marcel Proust, searching for what remains of the past: "When from a long distant past nothing subsists, after the people are dead, after things are broken and scattered, taste and smell alone . . . more persistent, more faithful, remain poised a long time, like souls . . . and bear unflinchingly, in the tiny and almost impalpable drop of their essence, the vast structure of recollection."

What, then, is the soul and the essence of matter?

One day in May of 2004, I visited physicist Jerome Friedman in his office at MIT. In 1990, Friedman shared the Nobel Prize with Henry Kendall and Richard Taylor for discovering quarks, which along with leptons are the smallest known building blocks of matter. The atom was once thought to be the elemental unit, until scientists discovered the electron and the nucleus of the atom, a hundred thousand times smaller than the atom itself. The atomic nucleus, in turn, was found to be built out of still smaller particles called protons and neutrons. Around 1970, Friedman and his colleagues found that protons and neutrons are made out of three quarks apiece. The quark, it seems, may be the smallest doll in the set.

It was a Friday, late afternoon, and the hallway on the fifth floor of Building 24 was deserted and silent. However, Jerry's door was open. When I walked in, I found him hunched over his computer screen, surrounded by an ocean of books and journals and papers piled several feet high on nearly every horizontal surface in the room. A single chair was not buried. Jerry Friedman, seventy-five years old, is the grandfather you want to have—a big hulking man with a white fringe of hair and a sweet, genuine smile. He smiles when he talks about physics or about art. In fact, he almost became a painter. He continues to paint in his spare time and avidly goes with his wife to art shows, musical concerts, theater, and ballet. On his bookshelves, fighting for space among the textbooks on nuclear physics and quantum electrodynamics, are dozens of Asian ceramics, small pots and vases and sculptures. "Reproductions," he says softly. "I keep the good stuff at home."

Jerry Friedman seems to love life. He is a gentle and good-natured man who never makes an unkind remark, never makes a person feel dumb. On the contrary, he often praises the accomplishments of others. "I have been fortunate to have had outstanding students and colleagues," he once wrote. His doctoral thesis adviser at the University of Chicago, the legendary Enrico Fermi, he admires most of all—not just for Fermi's brilliance but for his human qualities. "Fermi was a quiet man," Friedman says, "a kind man, who went out of his way to help us understand things. Fermi displayed immense modesty in the way that he carried himself and the way that he addressed people." The same might be said of Jerome Friedman.

Thirty-five years ago, Friedman and his colleagues excavated the insides of protons by bombarding them with ultra-high-energy electrons, which acted like exquisitely sharp scalpels. The scientists were surprised to find quarks. "When the experiment was planned," Friedman wrote in his Nobel address, "there was no clear theoretical picture of what to expect."

During our conversation, Jerry goes to the blackboard and draws diagrams, explaining complex physics with simple pictures. I ask for a copy of one of his papers published some years ago. Rising from his impossibly cluttered desk, he goes straight to a particular stack on the floor, sticks his hand in about two feet up, and produces the document. Later, he says to me, "Most of science is having a wonderful time."

The discovery of smaller and smaller elements of matter began in the nineteenth century. In the early years of that century, the British chemist John Dalton found that chemical elements combined to make compounds in very specific relative weights. The scientist conjectured that such definite ratios were evidence for the ancient idea of the atom. No one had seen an atom, nor would they until the advanced electron microscopes of the 1950s and 1960s. However, by the end of the nineteenth century, various indirect methods of measurement, including the scattering of sunlight by molecules of air, indicated that the atom was about a hundred millionth (10^{-8}) of a centimeter in diameter. For comparison, the period at the end of this sentence is about a twentieth of a centimeter.

Soon, it was apparent that the atom was not the smallest unit of matter. In 1897, the stern-faced Joseph John Thomson, director of the famous Cavendish Laboratory in Cambridge, discovered a particle far more diminutive than the atom, the electron. In 1911, Ernest Rutherford and his assistants, also working in England, were astonished to find that the interior of an atom is mostly empty space, with a hard nugget at its center. That nugget was the atomic nucleus (see Chapter 5). The atomic nucleus, containing all of the positive charge of the atom and more than 99 percent of its mass, is confined to a region a hundred thousand times smaller than the atom as a whole. The electrical charge of the atomic nucleus arises from still smaller particles jammed within it, the protons. (The nucleus of a hydrogen atom has a single proton, that of a carbon atom has six, uranium ninety-two.) Eventually, it was found that other, electrically neutral, particles shared the cramped living quarters with the protons. These particles, the neutrons, were discovered by James Chadwick in 1932.

Neutrons and protons were called nucleons. They were tightly bound together within the atomic nucleus by the so-called strong nuclear force. (The strong nuclear force acts equally between any two nucleons, be they two neutrons, two protons, or a neutron and proton.) At the time, two other fundamental forces were known (and had been known since antiquity): the electromagnetic force, acting on electrically charged particles, and the gravitational force, acting on every particle, charged or not.

Within a short time after the discovery of the neutron, it was found that neutrons could change their stripes and become protons by emitting negatively charged electrons and other subatomic particles called neutrinos. The transformation of a neutron into a proton was evidence for a new type of force, the so-called weak force, which is discussed at length in Chapter 20. Thus, by the mid-1930s, it was understood that there were now four fundamental forces of nature.

The gravitational force is the weakest of all the forces, next comes the weak force, then the electromagnetic force, and finally the strong nuclear force. The strength of the strong nuclear force could be gauged by what is required to keep the positively charged protons from flying out of the nucleus under their mutual electrical repulsion. That force is strong indeed—roughly a hundred times stronger than the electromagnetic force and 10^{38} times stronger than the gravitational force! The strong nuclear force is the most powerful force known in the universe. But it has very limited reach. Two nucleons cannot feel each other's strong nuclear force unless they are closer than about 3×10^{-13} centimeters.

Within a few years of the discovery of the neutron, hints began popping up that even nucleons were not "elementary." For one thing, the observed interaction between neutrons and protons was not simple like the electromagnetic interaction between electrons. In fact, the force between neutrons and protons was actually repulsive at very short distances, and attractive at larger distances, until the two particles were out of range altogether. Such a complexity suggested some internal structure of the neutron and proton.

Another problem was the peculiar magnetic moments of the nucleons. The magnetic moment of an elementary particle determines the force exerted on it by a nonuniform magnetic field. It was thought to depend on only three things: the electrical charge of the particle, its mass, and its spin. (Spin is discussed in Chapter 20.) The magnetic moment of the "elementary" electron had been measured and agreed perfectly with the theoretical predictions of relativistic quantum physics. However, the proton had a measured magnetic moment about two and a half times what it should have been according to the theory. The neutron, being electrically neutral, should have had zero magnetic moment but did not. Evidently, something unknown lurked in the innards of neutrons and protons. They were not simple particles like electrons.

Finally, from the 1940s on, dozens of new subatomic particles were discovered in the giant human-made particle accelerators. These machines hurled subatomic particles against each other at ferocious energies and created new zoological species in the process. Names could not be invented fast enough. There were delta particles and lambda particles, sigmas and xis, omegas, pions, kaons, and rhos. When the Greek alphabet was exhausted, physicists resorted to Latin letters. Some of these particles had total lifetimes, from the moment they were created to the moment they disappeared, of a mere 10^{-23} seconds, or 0.00000000000000000000001 seconds!

Physicists attempted to bring some kind of order to the subatomic zoo. First, particles could be classified according to which forces they responded to. All particles subject to the strong nuclear force were called hadrons, named after the Greek word for "strong." Nucleons were hadrons. Electrons and neutrinos were not. The new particles could be further classified according to their masses, electrical charges, and spins.

Taking the electron spin as half a unit, particles with half units of spin were called fermions, while particles with whole units of spin were called bosons. The protons and neutrons were fermions. Pions and kaons were bosons.

By 1960 physicists had discovered over a hundred new subatomic particles. Clearly, these could not all be fundamental. There were simply too many. Furthermore, many of these subatomic particles could transform into their cousins, as the neutron did to the proton. Perhaps a primitive essence did not exist.

As one moved forward in the 1960s, the theory of hadrons gyrated in a turmoil as great as the social and political rebellion taking place in America. There were not a few elementary hadrons, but hundreds. There was not a single theory of the strong force, but many. None of the theories was satisfactory. By contrast, there existed a beautiful and highly accurate theory of the electromagnetic force, called quantum electrodynamics, formulated by Richard Feynman, Julian Schwinger, and Shinichiro Tomonaga in the late 1940s. In quantum electrodynamics, there were two kinds of elementary particles, the electron and the photon. Electrons created the force. Photons conveyed the force from one electron to another, or propagated freely through space. As we have discussed in Chapter 20, theories of the weak force improved during the 1950s and 1960s, culminating in the beautiful electroweak theory of Steven Weinberg, Abdus Salam, and Sheldon Glashow in 1967.

Yet the strong nuclear force remained untamed. Besides the apparent absence of elementary hadrons, it was almost impossible to calculate anything involving the strong force. For the other, far weaker forces, theoretical computations could be made by assuming that the energy associated with the force was smaller than the initial energy of the particle before the force acted. At least the probability for a high-energy event was small. A well-defined mathematical procedure existed for such calculations. But the strong force routinely wielded huge energies, as large as the energy that would be released if a hadron converted its mass completely into energy according to the famous Einsteinian formula $E = mc^2$.

Without a theory of the strong nuclear force, physicists attempted to deduce as much as possible from general principles of physics and from extrapolations of the electromagnetic and weak forces, which were understood. One such attempt, called Regge theory, in vogue in the 1960s, proposed that the observed families of hadron particles were created by starting with a basic, slow-spinning hadron and spinning it faster and faster. Another theory was called current algebra. Invented by Murray Gell-Mann in 1961, current algebra tried to get at hadrons through the side door by considering their interactions with other particles via the electromagnetic and weak forces.

In 1964, two American theoretical physicists, Murray Gell-Mann and George Zweig, independently proposed the idea of quarks. According to their theory, fermion hadrons, like the proton and neutron, carried three quarks, while boson hadrons, like pions and kaons, carried two. Quarks, however, were not envisioned to be material particles. Rather they were mathematical abstractions, whose mathematical properties offered a simple scheme to organize the hundreds of hadrons that had been observed.

Gell-Mann and Zweig began by carefully noting which strong-force interactions occurred and which did not. For example, when negatively charged pions and protons collide, they frequently change into positively charged kaons and negative sigmas, in a reaction written as $\pi^- + p \rightarrow K^+ + \Sigma^-$. However, the very similar reaction $\pi^- + p \rightarrow K^- + \Sigma^+$ never occurs. To explain these and similar findings, the physicists assumed that each hadron had certain intrinsic properties, in addition to spin and electrical charge, called baryon number, isospin, and strangeness. (We have already encountered isospin for electrons and neutrinos in Chapter 20.) These intrinsic properties were called quantum numbers. A given kind of subatomic particle, like the proton or the sigma, was uniquely defined by its quantum numbers. For example, the proton has electrical charge of 1, spin of ½, isospin of ½, baryon number of 1, and strangeness of 0. The lambda has electrical charge 0, spin ½, isospin 0, baryon number 1, and strangeness –1.

The rule for quantum numbers was similar to the rule for electrical charge conservation: although individual subatomic particles could be created or destroyed in interactions, the total value of each quantum number had to be constant, like the total money put into and taken out of a coin-changing machine. (The constancy of some of these quantum numbers, such as isospin, holds only for strong-force interactions.) For example, if a particle of strangeness –1 and another particle of strangeness –2 collided, the total strangeness going into the reaction was –3. The total strangeness of all particles coming out of the reaction had to be −3 as well. No one understood exactly what produced these quantum numbers, but they were intrinsic properties of the particles.

Now we come to quarks. To put the above ideas on a mathematical basis, the quantum numbers were divided among carriers called quarks. In the original theory, there were three types of quarks, called "up," "down," and "strange." All three quarks were to have spin ½ and baryon number ⅓. The up quark and down quark both had strangeness 0. The up quark had isospin ½ and the down quark had isospin −½. The strange quark was to have isospin 0 and strangeness −1. (For the physics cognoscenti: what we are calling isospin is the vertical component of the isospin vector.) The most peculiar aspect of the Gell-Mann–Zweig system was the electrical charges of the quarks. In units where the electrical charge of the electron is −1, the up quark would have a charge of ⅔, the down quark a charge of −⅓, and the strange quark a charge of −⅓. In this plan, the proton consisted of two up quarks and a down quark, for a total electrical charge of 1, and the neutron contained an up and two downs, for a total charge of zero. As another example, the lambda hadron contained a strange quark, an up quark, and a down quark.

By assuming that quarks could be shuffled around between particles in strong interactions but not change their types, Gell-Mann and Zweig could neatly account for the observed family groupings of particles and the constancy of quantum numbers, which in turn explained which reactions occurred and which did not.

Few people believed that quarks were real particles. Despite many searches in accelerators and in cosmic rays, no one had found an isolated quark. Also, physicists bristled at the idea of fractional electrical charges, violating decades of observations showing that all electrical charges came in whole multiples of the electron's charge. As Gell-Mann himself said: "Such particles [quarks] presumably are not real but we may use them in our field theory anyway." Henry Kendall, a collaborator with Friedman on the landmark Friedman-Kendall-Taylor experiment, expressed his team's view about quarks while designing their experiment: "Quark constituent models . . . had serious problems, then unsolved, which made them widely unpopular as models for high energy interactions of hadrons."

In conclusion, at the time of the Friedman-Kendall-Taylor experiment in the late 1960s, the theory of the strong nuclear force and its associated elementary particles was in a terrible mess. A leading theoretician, James Bjorken of Stanford University, complained of "the profound state of theoretical ignorance."

Jerome I. Friedman was born in Chicago in March 1930, the son of Jewish immigrants from Russia. His father had come to the United States in 1913, worked for the Singer Sewing Machine Company, and later set up his own business selling and repairing sewing machines. Friedman's parents struggled financially. "The education of my brother and myself was of paramount importance to my parents," Friedman recalled, "and in addition to their encouragement they were prepared to make any sacrifice to further our intellectual development."

At the end of high school, young Friedman almost accepted a place at the Art Institute of Chicago Museum School, but instead enrolled on full scholarship at the University of Chicago. After attending the Great Books program developed by R. M. Hutchins, he majored in physics. Friedman remained at Chicago for his Ph.D., which he received in 1956. He was Fermi's last doctoral student. (Fermi died of cancer at the age of fifty-three.) "Although some of his major achievements involved formal calculations, Fermi had simple ways of looking at the world," Friedman recalled. "He was a very intuitive physicist, who could look at a complex problem, extract the major elements, and calculate an effect to within ten or fifteen percent. I have always tried to understand physics like that. I try to construct a simple picture of things."

In 1956, the twenty-six-year-old Friedman and Valentine Telegdi were among the first scientists to demonstrate that some subatomic processes in nature do not occur as often as their mirror images—a startling result called parity violation.

The next year, Friedman moved to Stanford University, where he learned the techniques of using high-energy electrons as projectiles to probe heavier subatomic particles. It was at Stanford that the young experimentalist met his two most important collaborators: Henry Kendall and Richard Taylor. While Taylor remained at Stanford, Friedman and Kendall relocated to the Massachusetts Institute of Technology. In 1963, the three physicists began planning new experiments for the Stanford Linear Accelerator (SLAC), a behemoth machine then under construction. SLAC ultimately cost $114 million to build ($460 million in 2005 dollars). When it began operations in 1966, SLAC was the most powerful subatomic particle accelerator on the planet.

The first particle accelerators were built in the 1940s. If you wanted to find out how a small watch worked—the mechanism of its tiny gears and wheels—you would use a tiny and sharp probe. That's the idea of a particle accelerator. A principle of quantum physics relates size and momentum (see Chapter 10). In brief, the smaller the size you want to explore, the higher the required momentum (and energy) of the probe. To go deep inside a proton, which is only 10^{-13} centimeters in size, demands electron probes of ten billion electron volts and higher. Only a particle accelerator can create and control such high energies.

The most stunning architectural feature of the Stanford Linear Accelerator is a straight (linear) tunnel, two miles long. Electromagnetic waves traveling down the tunnel accelerate electrons to higher and higher speeds. At some 260 positions along the tunnel the waves are reinforced, and everything must be precisely timed so that the electrons are given a boost just over the crest of each passing wave. The electrons are like tiny surfers on an electromagnetic sea. They are also projectiles. They are fed in at one end of the tunnel and smash their target at the opposite end, two miles away. For exploring the insides of hadrons, electrons make especially good projectiles because they interact with the mysterious hadrons via a force that is well understood, the electromagnetic force. Using electrons to probe hadrons is like conversing with a foreigner through pictures.

As the electrons are accelerated to higher and higher speeds, they gain more and more energy. So powerful was SLAC that it could accelerate an electron up to an energy of about 20 billion electron volts, or about 40,000 times the energy that would be released if an electron at rest could be converted into pure energy. Such an energy corresponds to an electron speed of 99.99999997 percent of the speed of light!

Various detectors surrounding the target record the vital statistics of electrons rebounding after collisions with the target. Other instruments, called magnetic spectrometers, guide the electrons from the target to these detectors. Magnetic spectrometers contain strong magnetic fields and operate on the principle that an electrically charged particle in motion is deflected by magnetism, the amount of deflection depending on the momentum, or energy, of the particle. Thus, the precise positions at which the electrons strike the detectors, after deflection by a magnetic spectrometer, indicate their energies.

Here is what Friedman, Kendall, Taylor, and their collaborators actually did. They fired electrons of various energies at a target of liquid hydrogen, which is mostly protons. The electrons interacted with the protons and then rebounded, or "scattered," at various angles. The situation is illustrated in Figure 21.1. In this picture, an electron of energy E comes in, strikes the target of protons, and emerges with energy E' at a "scattering angle" of θ. In general, an incoming beam of electrons all of the same energy E will emerge with a range of energies and angles. Downstream of the target, the scientists placed detectors at various angles to the incoming beam. Each of those detectors could measure how many electrons scattered in that direction and their energies. Combining such information with the numbers of incoming electrons, the physicists could then compute the *fraction* of incoming electrons that emerged from the target at each angle θ and at each energy E'. That fraction is called the differential cross section.

The differential cross section is the holy grail of the experiment. Denoted by the ungainly expression $d^2\sigma/d\Omega dE'$, the differential cross section contains all the information about the interactions of the electrons with the protons. All physicists know and love the differential cross section. Different theories make different predictions of what the differential cross section should be. After millions of incident electrons have struck the proton target, scattered, and been detected on the rebound, the scientists may tabulate the differential cross section for each incoming energy E, outgoing energy E', and scattering angle θ.

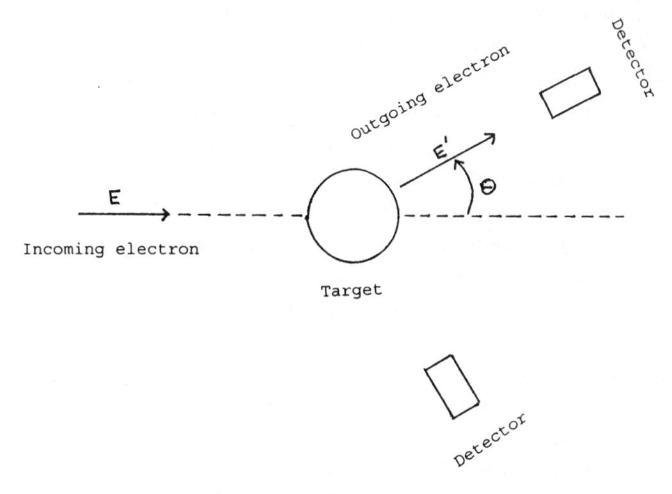

Figure 21.1

The physicists found two surprising results. The first was that a substantial fraction of electrons scattered at much larger angles than expected, as if they had struck something hard and small deep inside the proton. Just as Ernest Rutherford had assumed that his high-speed alpha particles would travel through an atom with little deflection, so did Friedman and his collaborators with their electron projectiles and proton targets. The reasons for such expectations were the same. Many scientists in the early 1900s believed that the positive charge of the atom was spread out evenly throughout its interior, following the so-called plum pudding model of the atom. A high-speed alpha particle passing through such thin material would meet little resistance and thus change its path hardly at all. Likewise for the proton. Experiments with lower-energy electron projectiles by physicist Richard Hofstadter in the mid-1950s had suggested that the proton was a sphere of about 10^{-13} centimeters in diameter, with its positive charge smoothly distributed throughout that sphere. In that case the ultra-high-energy electrons of the Friedman-Kendall-Taylor experiment would have barely budged from their initial path as they passed through their proton targets. In fact, the higher the energy of the electron, the smaller the expected scattering angle. Friedman, Kendall, and Taylor found otherwise. Evidently, the positive charge of the proton was not uniformly spread out, but concentrated in one or more dense and small objects inside the proton.

The second surprise was more subtle. At each scattering angle, the differential cross section did not depend on both E and E' separately, but only on a certain combination of them. In other words, at each angle the differential cross section depended on only a single parameter, not two. Such a result is called scaling. For a more familiar example of scaling, suppose that you wanted to figure out how big an air conditioner was needed to cool a house under construction. You might guess that the required size would depend on the average outside temperature, the square footage of the house, the height of the house, and perhaps other parameters. At each outside temperature, you might then test various air conditioners on houses of different square footages and heights—building up a massive amount of data in the process. What you would find is that the required air conditioner depended only on a single parameter: the square footage of the house multiplied by its height. In other words, the volume of the house. Houses of the same volume require the same size air conditioner, even if they have different square footages and heights. Scaling is a sign that the problem is simpler than what had been imagined. And that simplification is a clue to understanding.

The Stanford theoretician James Bjorken, using "current algebra," had actually predicted scaling for electron scattering from hadrons. As scientists later realized, scaling necessarily follows from the assumption that the fundamental particles have an extremely small, or zero, size. Current algebra, indirectly, makes such an assumption. However, at the time, Bjorken's predictions of scaling were based on abstract and mathematical calculations, and most physicists didn't understand their physical meaning.

Richard Feynman provided that missing physical meaning, with his parton model. In the parton model, each hadron is assumed to consist of more elementary constituents called partons. Partons have zero size. They are point particles. According to Feynman's idea, when a high-energy subatomic particle hits a proton, it actually collides with only one of the proton's partons. There is some interaction, and the struck parton then interacts with the other partons via the strong nuclear force. Thus, the overall collision happens in two steps. The first step can often be calculated, while the second cannot, without a knowledge of the strong force and a very complex computation.

In the summer of 1968, Feynman heard the first report of the Friedman-Kendall-Taylor experimental results, which were about to be presented at the Fourteenth International Conference on High Energy Physics in Vienna. Overnight, Feynman took out his pencil and notepad and applied his new parton model to electron-nucleon scattering. Scaling automatically comes out of the calculation. Feynman's parton model, however, did not specify a theory of the strong nuclear force between partons. Nor did it say what the partons were.

To understand Friedman and collaborators' landmark 1969 paper in more detail, we must consider some of their terminology. For various technical reasons, instead of the two parameters E and E', the scientists use two other parameters: the "electron energy loss," $\nu = E - E'$, and the "square of the four-momentum transfer," $q^2 = 2EE'(1 - cos\theta)$, which is related to the "kick" given to a proton by collision with an incoming electron. The measured cross section is expressed relative to a standard cross section, the Mott cross section, which is the cross section for an electron interacting with a single point of electrical charge.

A common unit of energy is denoted by GeV, standing for 1 billion electron volts. One GeV is approximately 2000 times the energy that would be released if an electron at rest were converted into pure energy.

The dashed curve in Figure 1 of their paper, labeled "elastic scattering," is what would be expected if the electrical charge of the proton were evenly distributed throughout its interior. As can be seen, the dashed curve decreases rapidly with increasing q^2. That behavior is equivalent to the expected result that at each scattering angle, fewer and fewer electrons should be scattered to that angle as the energy of the incoming electrons increases.

In contrast to these expectations are the actual results, indicated by the three curves at the top of Figure 1. Each of these three curves is for a different value of ν at each q^2 (or equivalently, a different value of W for each q^2, where $W = 2M\nu + M^2 - q^2$ and M is the mass of the proton). The important point is that all of these curves show that the measured cross section decreases only slowly with increasing q^2. This crucial and defining result—a result that electrified the scientific community and sent the likes of Feynman to his pad of paper—is stated blandly in the first sentence of the second paragraph: "One of the interesting features of the measurements is the weak momentum-transfer dependence of the inelastic cross sections . . ." The etiquette of modern science clearly demands restrained language in official publications.

The physicists then express the differential cross section in terms of two "form factors," W_1 and W_2. The form factors are determined by the distribution of charge and magnetic moment within the proton. This version of the cross section with the form factors, in fact, is what theoreticians calculate from various theoretical models.

In order to compare their results with the theoretical models, the scientists must produce numbers for W_1 and W_2 separately. To do so, they must measure the differential cross section at many angles, including large angles. In this first experiment, however, they have data only for $\theta = 6°$ and $\theta = 10°$, in which case the cross section mainly measures W_2.

In the next section, the physicists show that they are aware of Bjorken's scaling prediction. In particular, Bjorken has predicted that νW_2 should depend only on the single parameter ν/q^2, rather than on the two parameters ν and q^2 separately. That single parameter is written as $\omega = 2M\nu/q^2$. In Figure 2, we see graphs of νW_2 plotted against ω for various energies E and for various assumptions about the unknown ratio W_2/W_1. That ratio is written in terms of yet another parameter called R. For small R, given in Figures 2a and 2b, the various data points lie on a single "universal" curve, demonstrating scaling. For large R, given in Figures 2c and 2d, the data points diverge at high ω, violating a scaling behavior.

At this point, because they cannot measure W_1, the scientists cannot measure R and thus do not know which sets of curves are closer to reality. Thus, they cannot say for sure whether they have proved or disproved Bjorken's prediction of scaling. Instead, the physicists discuss the various results that follow either the assumption that R is small, which they write as $\sigma_T \gg \sigma_S$, or the assumption that R is large, which they write as $\sigma_S \gg \sigma_T$.

Finally, as good experimentalists do, the scientists compare their results to the various theoretical models. They mention the parton model, Regge theory, the vector dominance model, current algebra, and the quark model. As we have discussed earlier, although these various models make predictions for electrons scattering from protons, none of them contains theories of the difficult and elusive strong nuclear force.

The vector dominance model is ruled out. The parton model seems to agree fairly well with the results. The Regge model is indecisive and does not require the two most interesting results: scaling and the weak dependence of the cross section on q^2.

The theory of current algebra is compared to the experimental results through certain predictions called sum rules. Sum rules relate sums (technically, integrals) of the cross section at all possible values of ω to the structure of the subatomic particle. The sum rules for the quark model, in particular, relate integrals of the cross sections to the *quark charges* carried by the proton or neutron. For example, one sum rule equates an integral over W_2 to the sum of the squares of the quark charges. For the proton, consisting of two up quarks and a down quark, this sum would be $(2/3)^2 + (2/3)^2 + (-1/3)^2 = 1$. For the neutron, consisting of one up quark and two downs, it would be $(2/3)^2 + (-1/3)^2 + (-1/3)^2 = 2/3$. As stated in the last paragraph of the paper, the experimental value of several sum rules is about a half of what the simple quark model predicts.

In conclusion, this first experimental paper by Friedman, Kendall, Taylor, and their collaborators strongly suggests that the proton is made up of smaller constituents, but the paper by itself is unable to confirm the quark model in particular. More work would be needed for that.

That additional work was done in the next few years. By 1970, Friedman and his collaborators had bombarded neutrons, as well as protons, and had measured the cross section at greater angles, so that W_1 and W_2 could both be determined. The R parameter was found to be small, confirming the scaling phenomenon. According to earlier theoretical calculations by other physicists, small R further meant that the parton constituents of the nucleons had to have a spin of ½. Boson partons were thus ruled out.

Over the next two years, a critical complementary set of experiments was performed at the CERN particle accelerator in Geneva, under the direction of D. H. Perkins. Perkins's group bombarded nucleons with neutrinos, which interact with nucleons via the weak force. Those experiments provided complementary sum rules for neutrinos, which, when combined with Friedman, Kendall, and Taylor's earlier sum rule results for electron scattering, showed that only half of the momentum of a proton was carried by its partons. The rest was carried by massless particles called gluons, surrounding the partons. (Gluons conveyed the strong force, as photons did the electromagnetic force.) With that adjustment, the sum rules based on the quark model agreed almost perfectly with all experimental results.

At this point, the array of different experimental results confirmed the details of the quark model. Each nucleon was made of three quarks. The quarks were real particles! What remained to be finished was a theory of the strong nuclear force, the force between quarks.

By 1973, that theory, called quantum chromodynamics, was finally complete. The creation and development of quantum chromodynamics involved contributions from many physicists all over the world, beginning with the work of Japanese-American physicist Yoichiro Nambu in 1966. Quantum chromodynamics incorporates the quark model. A strange and nonintuitive feature of quantum chromodynamics is that the strong nuclear force between any two quarks actually gets stronger the farther apart the quarks are. Such a result, worked out in 1973 by the American theoretical physicists David Gross, David Politzer, and Frank Wilczek, explains why a single, isolated quark has never been observed. A staggeringly high amount of energy would be needed to separate a quark from its quark companions by a macroscopic distance. After the Seventeenth International Conference on High Energy Physics in London in 1974, most physicists considered the quark model, and quantum chromodynamics, to have been confirmed. (Gross, Politzer, and Wilczek shared the 2004 Nobel Prize in physics for their work.)

At the present time, three more types of quarks have been discovered, for a total of six. These six quarks; plus the leptons, which include the electron, the muon, the tau, and their associated neutrinos; plus the force conveyers, which include the photon, the gluon, and the *Z* and *W* bosons, are believed to form the complete list of "elementary particles." That is all. Each of these particles is believed to have no interior structure. Each is believed to be essentially a point, aside from the wave character required by quantum mechanics. These particles are considered the smallest dolls.

Of course, no one knows for sure whether these are the smallest dolls. An untested new theory called string theory proclaims that the smallest entities in nature may not be particles at all, but tiny, stringlike vibrations of energy—at the unimaginably small size of 10^{-33} centimeters. No particle accelerator on earth, nor any conceivable particle accelerator of the remote future, could produce enough energy to probe such minuscule sizes. But practical considerations have never discouraged theoretical physicists.

Knowing that Jerry Friedman was a painter as well as a physicist, I had a few more questions at the end of my interview. I asked him what similarities, if any, he had experienced between science and art. "When a scientist comes out with an idea that jells," he said, "it gives the same pleasure as when a poet finds the right word." Earlier, for a symposium titled "The Humanities and the Sciences," he had written: "A common aspect of all creativity is to give us some sense and meaning of the various observations, impressions, and emotions that fill our lives."

Finally, I asked this codiscoverer of quarks whether scientists might continue to find smaller and smaller particles. After all, that had been the arrow of history. "We have probably found the limit," he said. "My prediction is that it is very likely these [quarks] are the smallest particles." He gave some persuasive reasons for his belief, hesitated, and then grinned. "However, I could be surprised. There are always surprises in science."

OBSERVED BEHAVIOR OF HIGHLY INELASTIC ELECTRON-PROTON SCATTERING

M. Breidenbach, J. I. Friedman, and H. W. Kendall

Department of Physics and Laboratory for Nuclear Science,[1] *Massachusetts Institute of Technology, Cambridge, Massachusetts 02139*

E. D. Bloom, D. H. Coward, H. DeStaebler, J. Drees, L. W. Mo, and R. E. Taylor

Stanford Linear Accelerator Center,[2] *Stanford, California 94305*

Received 22 August 1969

Physical Review Letters (1969)

Results of electron-proton inelastic scattering at 6° and 10° are discussed, and values of the structure function W_2 are estimated. If the interaction is dominated by transverse virtual photons, νW_2 can be expressed as a function of $\omega = 2M\nu/q^2$ within experimental errors for $q^2 > 1$ $(\text{GeV}/c)^2$ and $\omega > 4$, where ν is the invariant energy transfer and q^2 is the invariant momentum transfer of the electron. Various theoretical models and sum rules are briefly discussed.

IN A PREVIOUS LETTER,[3] we have reported experimental results from a Stanford Linear Accelerator Center–Massachusetts Institute of Technology study of high-energy inelastic electron-proton scattering. Measurements of inelastic spectra, in which only the scattered electrons were detected, were made at scattering angles of 6° and 10° and with incident energies between 7 and 17 GeV. In this communication, we discuss some of the salient features of inelastic spectra in the deep continuum region.

One of the interesting features of the measurements is the weak momentum-transfer dependence of the inelastic cross sections for excitations well beyond the resonance region. This weak dependence is illustrated in Fig. 1. Here we have plotted the differential cross section divided by the Mott cross section, $(d^2\sigma/d\Omega dE')/(d\sigma/d\Omega)_{\text{Mott}}$, as a function of the square of the four-momentum transfer, $q^2 = 2EE'(1 - \cos\theta)$, for constant values of the invariant mass of the recoiling target system, W, where $W^2 = 2M(E - E') + M^2 - q^2$. E is the energy of the incident electron, E' is the energy of the final electron, and θ is the scattering angle, all defined in the laboratory system; M is the mass of the proton. The cross section is divided by the Mott cross section

$$\left(\frac{d\sigma}{d\Omega}\right)_{\text{Mott}} = \frac{e^4}{4E^2}\frac{\cos^2\frac{1}{2}\theta}{\sin^4\frac{1}{2}\theta}$$

in order to remove the major part of the well-known four-momentum transfer dependence arising from the photon propagator. Results from both 6° and 10° are included in the figure for each value of W. As W increases, the q^2 dependence appears to decrease. The striking difference between the behavior of the inelastic and elastic cross sections is also illustrated in Fig. 1, where the elastic cross section, divided by the Mott cross section for $\theta = 10°$, is included. The q^2 dependence of the deep continuum is also considerably weaker than that of the electroexcitation of the resonances,[4] which have a q^2 dependence similar to that of elastic scattering for $q^2 > 1$ (GeV/c)^2.

On the basis of general considerations, the differential cross section for inelastic electron scattering in which only the electron is detected can be represented by the following expression[5]:

$$\frac{d^2\sigma}{d\Omega dE'} = \left(\frac{d\sigma}{d\Omega}\right)_{\text{Mott}} (W_2 + 2W_1 \tan^2\tfrac{1}{2}\theta).$$

The form factors W_2 and W_1 depend on the properties of the target system, and can be represented as functions of q^2 and $\nu = E - E'$, the electron energy loss. The ratio W_2/W_1 is given by

$$\frac{W_2}{W_1} = \left(\frac{q^2}{\nu^2 + q^2}\right)(1 + R),\ R \geq 0,$$

where R is the ratio of the photoabsorption cross sections of longitudinal and transverse virtual photons, $R = \sigma_S/\sigma_T$.[6]

The objective of our investigations is to study the behavior of W_1 and W_2 to obtain information about the structure of the proton and its electromagnetic interactions at high energies. Since at present only cross-section measurements at small angles are available, we are unable to make separate determinations of W_2 and W_1. However, we can place limits on W_2 and study the behavior of these limits as a function of the invariants ν and q^2.

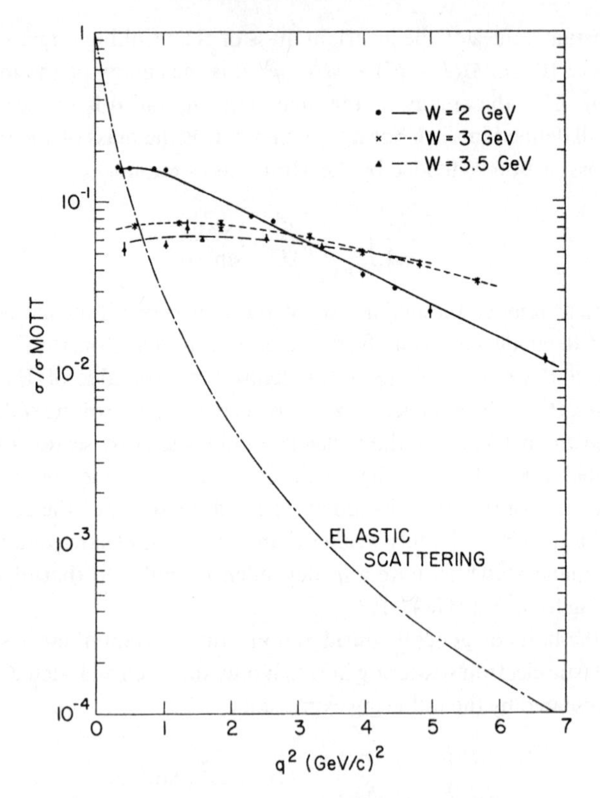

Figure 1. $(d^2\sigma/d\Omega dE')/\sigma_{\text{Mott}}$, in GeV^{-1}, vs q^2 for $W = 2$, 3, and 3.5 GeV. The lines drawn through the data are meant to guide the eye. Also shown is the cross section for elastic *e-p* scattering divided by σ_{Mott}, $(d\sigma/d\Omega)/\sigma_{\text{Mott}}$, calculated for $\theta = 10°$, using the dipole form factor. The relatively slow variation with q^2 of the inelastic cross section compared with the elastic cross section is clearly shown.

Bjorken[7] originally suggested that W_2 could have the form

$$W_2 = (1/\nu)F(\omega),$$

where

$$\omega = 2M\nu/q^2.$$

$F(\omega)$ is a universal function that is conjectured to be valid for large values of ν and q^2. This function is universal in the sense that it manifests scale invariance, that is, it depends only on the ratio ν/q^2. Since

$$\nu W_2 = \frac{\nu d^2\sigma/d\Omega dE'}{(d\sigma/d\Omega)_{\text{Mott}}}\left[1 + 2\,\frac{1}{1+R}\left(1+\frac{\nu^2}{q^2}\right)\tan^2\tfrac{1}{2}\theta\right]^{-1},$$

the value of νW_2 for any given measurement clearly depends on the presently unknown value of R. It should be noted that the sensitivity to R is small when $2(1 + \nu^2/q^2)\tan^2\frac{1}{2}\theta \ll 1$. Experimental limits on W_2 can be calculated on the basis of the extreme assumptions $R = 0$ and $R = \infty$. In Figs. 2(a) and 2(b) the experimental values of νW_2 from the 6° and 10° data for $q^2 > 0.5$ (GeV/c)^2 are shown as a function of ω for the assumption that $R = 0$. Figures 2(c) and 2(d) show the experimental values of νW_2 calculated from the 6° and 10° data with $q^2 > 0.5$ (GeV/c)^2 under the assumption $R = \infty$. The 6°, 7-GeV results for νW_2, all of which have values of $q^2 \leq 0.5$ (GeV/c)^2, are shown for both assumptions in Fig. 2(e). The elastic peaks are not displayed in Fig. 2.

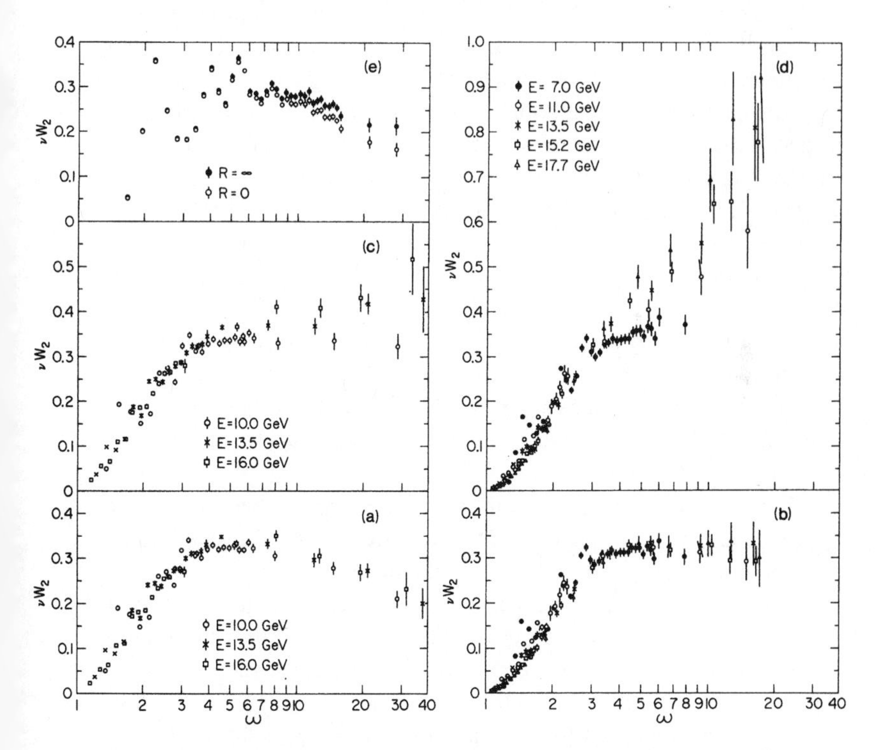

Figure 2. νW_2 vs $\omega = 2M\nu/q^2$ is shown for various assumptions about $R = \sigma_S/\sigma_T$. (a) 6° data except for 7-GeV spectrum for $R = 0$. (b) 10° data for $R = 0$. (c) 6° data except for 7-GeV spectrum for $R = \infty$. (d) 10° data for $R = \infty$. (e) 6°, 7-GeV spectrum for $R = 0$ and $R = \infty$.

The results shown in these figures indicate the following:

(1) If $\sigma_T \gg \sigma_S$, the experimental results are consistent with a universal curve for $\omega \lesssim 4$ and $q^2 \lesssim 0.5$ $(\text{GeV}/c)^2$. Above these values, the measurements at 6° and 10° give the same results within the errors of measurements. The 6°, 7-GeV measurements of νW_2, all of which have values of $q^2 \leq 0.5$ $(\text{GeV}/c)^2$, are somewhat smaller than the results from the other spectra in the continuum region.

The values of νW_2 for $\omega \lesssim 5$ show a gradual decrease as ω increases. In order to test the statistical significance of the observed slope, we have made linear least-squares fits to the values of νW_2 in the region $6 \leq \omega \lesssim 25$. These fits give $\nu W_2 = (0.351 \pm 0.023) - (0.003\,86 \pm 0.000\,88)\omega$ for data with $q^2 > 0.5$ $(\text{GeV}/c)^2$ and $\nu W_2 = (0.366 \pm 0.024) - (0.0045 \pm 0.0019)\omega$ for $q^2 > 1$ $(\text{GeV}/c)^2$. The quoted errors consist of the errors from the fit added in quadrature with estimates of systematic errors.

Since $\sigma_T + \sigma_S \simeq 4\pi^2\alpha\nu W_2/q^2$ for $\omega \gg 1$, our results can provide information about the behavior of σ_T if $\sigma_T \gg \sigma_S$. The scale invariance found in the measurements of νW_2 indicates that the q^2 dependence of σ_T is approximately $1/q^2$. The gradual decrease exhibited in νW_2 for large ω suggests that the photoabsorption cross section for virtual photons falls slowly at constant q^2 as the photon energy ν increases.

The measurements indicate that νW_2 has a broad maximum in the neighborhood of $\omega = 5$. The question of whether this maximum has any correspondence to a possible quasielastic peak[8] requires further investigation.

It should be emphasized that all of the above conclusions are based on the assumption that $\sigma_T \gg \sigma_S$.

(2) If $\sigma_S \gg \sigma_T$, the measurements of νW_2 do not follow a universal curve and have the general feature that at constant $2M\nu/q^2$, the value of νW_2 increases with q^2.

(3) For either assumption, νW_2 shows a threshold behavior in the range $1 \leq \omega \lesssim 4$. W_2 is constrained to be zero at inelastic threshold which corresponds to $\omega \simeq 1$ for large q^2. In the threshold region of νW_2, W_2 falls rapidly as q^2 increases at constant ν. This is qualitatively different from the weak q^2 behavior for $\omega > 4$. For $q^2 \approx 1$ $(\text{GeV}/c)^2$, the threshold region contains the resonances excited in electroproduction. As q^2 increases, the variations due to these resonances damp out and the values of νW_2 do not appear to vary rapidly with q^2 at constant ω.

It can be seen from a comparison of Figs. 2(a) and 2(c) that the 6° data provide a measurement of νW_2 to within 10% up to a value of $\omega \approx 6$, irrespective of the values of R.

There have been a number of different theoretical approaches in the interpretation of the high-energy inelastic electron-scattering results. One class of models,[8–11] referred to as parton models, describes the electron as scattering incoherently from pointlike constituents within the proton. Such models lead to a universal form for νW_2, and the point charges assumed in specific models give the magnitude of νW_2 for $\omega > 2$ to within a factor of 2.[6] Another approach[12,13] relates the inelastic scattering to off-the-mass-shell Compton scattering which is described in terms of Regge exchange using the Pomeranchuk trajectory. Such models lead to a flat behavior of νW_2 as a function of ν but do not require the weak q^2 dependence observed and do not make any numerical predictions at this time. Perhaps the most detailed predictions made at present come from a vector-dominance model which primarily utilizes the ρ meson.[14] This model reproduces the gross behavior of the data and has the feature that νW_2 asymptotically approaches a function of ω as $q^2 \to \infty$. However, a comparison of this model with the data leads to statistically significant discrepancies. This can be seen by noting that the prediction for $d^2\sigma/d\Omega dE'$ contains a parameter ξ, the ratio of the cross sections for longitudinally and transversely polarized ρ mesons on protons, which is expected to be a function of W but which should be independent of q^2. For values of $W \geq 2$ GeV, the experimental values of ξ increase by about (50 ± 5)% as q^2 increases from 1 to 4 (GeV/c)^2. This model predicts that

$$\sigma_S/\sigma_T = \xi(W)(q^2/m_\rho{}^2)[1 - q^2/2m\nu],$$

which will provide the most stringent test of this approach when a separation of W_1 and W_2 can be made.

The application of current algebra[15–19] and the use of current commutators leading to sum rules and sum-rule inequalities provide another way of comparing the measurements with theory. There have been some recent theoretical considerations[20–22] which have pointed to possible ambiguity in these calculations; however, it is still of considerable interest to compare them with experiment.

In general, W_2 and W_1 can be related to commutators of electromagnetic current densities.[8, 18] The experimental value of the energy-weighted sum $\int_1^\infty (d\omega/\omega^2)(\nu W_2)$, which is related to the equal-time commutator of the current and its time derivative, is 0.16 ± 0.01 for $R = 0$ and 0.20 ± 0.03 for $R = \infty$. The integral has been evaluated with an upper limit $\omega = 20$. This integral is also important in parton theories where its value is the mean square charge per parton.

Gottfried[23] has calculated a constant $-q^2$ sum rule for inelastic electron-proton scattering based on a nonrelativistic quark model involving point-like quarks. The resulting sum rule is

$$\int_1^\infty \frac{d\omega}{\omega}(\nu W_2) = \int_{q^2/2M}^\infty d\nu W_2 = 1 - \frac{G_{Ep}{}^2 + (q^2/4M^2)G_{Mp}}{1 + q^2/4M^2},$$

where G_{Ep} and G_{Mp} are the electric and magnetic form factors of the proton. The experimental evaluation of this integral from our data is much more dependent on the assumption about R than the previous integral. We will thus use the 6° measurements of W_2 which are relatively insensitive to R. Our data for a value of $q^2 \simeq 1\ (\mathrm{GeV}/c)^2$, which extend to a value of ν of about 10 GeV, give a sum that is 0.72 ± 0.05 with the assumption that $R = 0$. For $R = \infty$, its value is 0.81 ± 0.06. An extrapolation of our measurements of νW_2 for each assumption suggests that the sum is saturated in the region $\nu \simeq 20$–40 GeV. Bjorken[15] has proposed a constant-q^2 sum-rule inequality for high-energy scattering from the proton and neutron derived on the basis of current algebra. His result states that

$$\int_1^\infty \frac{d\omega}{\omega}\nu(W_{2p} + W_{2n}) = \int_{q^2/2M}^\infty d\nu(W_{2p} + W_{2n}) \geq \tfrac{1}{2},$$

where the subscripts p and n refer to the proton and neutron, respectively. Since there are presently no electron-neutron inelastic scattering results available, we estimate W_{2n} in a model-dependent way. For a quark model[24] of the proton, $W_{2n} \simeq 0.8 W_{2p}$ whereas in the model[8] of Drell and co-workers, W_{2n} rapidly approaches W_{2p} as ν increases. Using our results, this inequality is just satisfied at $\omega \simeq 4.5$ for the quark model and at $\omega \simeq 4.0$ for the other model for either assumption about R. For example, this corresponds to a value of $\nu \simeq 4.5$ GeV for $q^2 = 2\ (\mathrm{GeV}/c)^2$. Bjorken[25] estimates that the experimental value of the sum is too small by about a factor of 2 for either model, but it should be noted that the q^2 dependence found in the data is consistent with the predictions of this calculation.

1. Work supported in part through funds provided by the U. S. Atomic Energy Commission under Contract No. AT(30-1)2098.

2. Work supported by the U. S. Atomic Energy Commission.

3. E. Bloom *et al.*, preceding Letter [*Phys. Rev. Letters 23*, 930 (1969)].

4. Preliminary results from the present experimental program are given in the report by W. K. H. Panofsky, in *Proceedings of the Fourteenth International Conference on High Energy Physics, Vienna, Austria, 1968* (CERN Scientific Information Service, Geneva, Switzerland, 1968), p. 23.

5. R. von Gehlen, *Phys. Rev. 118*, 1455 (1960); J. D. Bjorken, 1960 (unpublished); M. Gourdin, *Nuovo Cimento 21*, 1094 (1961).

6. See L. Hand, in *Proceedings of the Third International Symposium on Electron and Photon Interactions at High Energies, Stanford Linear Accelerator Center, Stanford, California, 1967* (Clearing House of Federal Scientific and Technical Information, Washington, D. C., 1968), or F. J. Gilman, *Phys. Rev. 167*, 1365 (1968).

7. J. D. Bjorken, *Phys. Rev. 179*, 1547 (1969).

8. J. D. Bjorken and E. A. Paschos, Stanford Linear Accelerator Center, Report No. SLAC-PUB-572, 1969 (to be published).

9. R. P. Feynman, private communication.

10. S. J. Drell, D. J. Levy, and T. M. Yan, *Phys. Rev. Letters 22*, 744 (1969).

11. K. Huang, in Argonne National Laboratory Report No. ANL-HEP 6909, 1968 (unpublished), p. 150.

12. H. D. Abarbanel and M. L. Goldberger, *Phys. Rev. Letters 22*, 500 (1969).

13. H. Harari, *Phys. Rev. Letters 22*, 1078 (1969).

14. J. J. Sakurai, *Phys. Rev. Letters 22*, 981 (1969).

15. J. D. Bjorken, *Phys. Rev. Letters 16*, 408 (1966).

16. J. D. Bjorken, in *Selected Topics in Particle Physics, Proceedings of the International School of Physics "Enrico Fermi," Course XLI*, edited by J. Steinberger (Academic Press, Inc., New York, 1968).

17. J. M. Cornwall and R. E. Norton, *Phys. Rev. 177*, 2584 (1969).

18. C. G. Callan, Jr., and D. J. Gross, *Phys. Rev. Letters 21*, 311 (1968).

19. C. G. Callan, Jr., and D. J. Gross, *Phys. Rev. Letters 22*, 156 (1969).

20. R. Jackiw and G. Preparata, *Phys. Rev. Letters 22*, 975 (1969).

21. S. L. Adler and W.-K. Tung, *Phys. Rev. Letters 22*, 978 (1969).

22. H. Cheng and T. T. Wu, *Phys. Rev. Letters 22*, 1409 (1969).

23. K. Gottfried, *Phys. Rev. Letters 18*, 1174 (1967).

24. J. D. Bjorken, Stanford Linear Accelerator Center, Report No. SLAC-PUB-571, 1969 (unpublished).

25. J. D. Bjorken, private communication.

22

THE CREATION OF ALTERED FORMS OF LIFE

THE HISTORY OF SCIENCE IS, in part, a history of human beings trying to gain control over their world. In this baffling cosmos of ours, in this daily explosion of light and sound and tactile sensation, trees, mountains, waves on the ocean, rain and wind, turns of the seasons, heat and cold—in all these diverse phenomena burning through our own mysterious consciousness, we recoil from the idea that we are helpless and ignorant spectators of nature, bits of debris tossed about in the sea of existence. We can accept our own deaths more readily than we can accept a life of accidents and forces beyond understanding. We desire meaning. We desire order. And we desire control.

Knowledge grants one form of control. The ancient Roman poet Lucretius believed that the idea of the conservation of matter—that matter could not be created or destroyed—would free humankind from the capricious interference of the gods. There are more active forms of control. In Sumatra, women who sow rice let their hair hang long down their backs, so that the rice will have long stalks and grow well. The ancient Egyptians crossbred horses, cattle, wheat, and grapes, to produce animals and food of higher quality. The early Romans built massive stone aqueducts to convey water from one place to another.

Of all aspects of nature, the phenomenon of life is the most complex. And the control of life, perhaps, satisfies most deeply our desire for control over our physical world. Indeed, one can view the subject of biology through the centuries as a deepening understanding of the mechanisms and controls within living substance. Just in the twentieth century, following the chapters of this book, one might point to the discovery of hormones, which comprise a chemical command and control system; the discovery of neurotransmitters, the mechanism by which nerves communicate with each other; the discovery of penicillin, the first antibiotic, which gave human beings much more control over infectious disease; the discovery of the structure of DNA and the mechanism by which genetic information is encoded in each living cell; and the discovery of the design of the hemoglobin molecule and the mechanism by which oxygen, that most vital of all gases, is held and released in the body.

In the long list of endeavors to control living substance, the ability to reprogram DNA, to alter the instructions for life within each living cell, is the most profound. Now, we have become architects of life. By splicing genes together, we have created living organisms that never existed before. We have redesigned the lowly bacterium *E. coli* so that it produces insulin for ailing diabetics. We have combined the genes of bacteria with the genes of cotton to create a substance that breathes like cotton but warms like polyester. We have altered the DNA of maize and soybeans to make them resistant to insects and disease. In a flight of fancy, we might even imagine re-creating ourselves—as in M. C. Escher's eerie picture of a hand drawing itself. In which case, we could be the first substance in the universe to design itself. Such power, perhaps the ultimate power, raises more ethical, philosophical, and theological questions than any previous development in science.

The history of gene splicing, also called recombinant DNA or genetic engineering, is recent. It began with a paper by biochemist Paul Berg of Stanford University and his collaborators in 1972. In his goal to insert new genes into living cells, Berg was the first scientist to splice together segments of DNA from different organisms. He was forty-six years old at the time. Soon, Berg became aware that he had set into motion a new biology of unimaginable consequences. Eight years later, on the occasion of his Nobel address, he thanked his students and colleagues for sharing with him "the elation and disappointment of venturing into the unknown."

Paul Berg was born in Brooklyn, New York, in June 1926. By the age of thirteen, he recalls, he had already developed a "strong ambition" to be a scientist. Of particular influence were two books about medical scientists, *Arrowsmith*, by Sinclair Lewis, and *Microbe Hunters*, by Paul de Kruif. Berg also fondly remembers an inspiring high school teacher, Miss Sophie Wolfe, who supervised the stocking of the science laboratories at his school. Wolfe also organized an after-school science program where ambitious students could undertake independent research projects. Over the years Miss Wolfe's encouragement influenced three future Nobel laureates: Berg, Arthur Kornberg, and Jerome Karl. Berg gratefully recalls this first blush of research: "The satisfaction derived from solving a problem with an experiment was a heady experience, almost addicting. Looking back, I realize that nurturing curiosity and the instinct to seek solutions are perhaps the most important contributions education can make."

After graduating from high school, Berg began a course in chemical engineering at the City College of New York but quickly realized that he was much more interested in the chemistry of living systems. He transferred to Pennsylvania State University. After an interruption to serve in the Navy during World War II, he received his B.S. degree in 1948 and then went on to Western Reserve University for his Ph.D. in biochemistry in 1952.

A couple of years later, Berg began working in the laboratory of the great American biologist Arthur Kornberg at Washington University in St. Louis. Kornberg, also a native of Brooklyn, was soon to win a Nobel for his discovery of the enzymes that allow DNA molecules to replicate themselves.

In Kornberg's lab, in the mid-1950s, Berg discovered classes of biological compounds called acyl-adenylates and aminoacyl-adenylates, and their associated enzymes, which help in the translation of the DNA code to assemble proteins. Here, just a couple of years after Watson, Crick, and Franklin's discovery of the structure of DNA, Berg was making a transition from traditional biochemistry to the new field of molecular biology. Gradually, his interests turned to what would be his lifelong obsession: the organization and function of genes. Among the questions that Berg and other biologists wanted answered were: How are genes organized on the chromosome? Precisely how do genes convey their instructions in mammalian cells? (A great deal was already known about the process in bacteria.) In complex organisms like mammals, how do genes tell some cells to be liver cells, other cells to be brain cells, and so on? What is the detailed process by which genes promote communication between different cells?

Berg remained at Washington University until 1959. Then, at the age of thirty-four, he became a professor of biochemistry at Stanford.

The beginnings of Berg's interest in recombinant DNA might be traced to the mid-1960s, when he learned of the work of Renato Dulbecco at the Salk Institute in southern California. Dulbecco was studying the life cycles of the polyoma mouse virus, an organism with a paltry amount of DNA, about five genes' worth. Viruses are the smallest known forms of life. (See Chapter 13.) A virus typically invades the "host" cell of a larger organism and uses the DNA of that cell to reproduce. Some viruses, like polyoma, can transform normal cells into cancerous cells and cause tumors. Dulbecco's research suggested that the polyoma virus, which causes cancer in rodents, somehow *integrated its DNA into the DNA of rodent cells.*

It occurred to Berg that tumor viruses like polyoma might serve as probes into the machinery and operation of DNA in higher animals. Tumor cells would be ideal because they had a small number of identifiable genes that could be monitored much more easily than the genes in a normal cell. Since the 1950s, biologists had achieved much success in using bacterial viruses, called phages, to study the genes of single-celled bacteria. Phages infected bacteria and commandeered their DNA. Perhaps viruses like polyoma could serve the same function with mammalian cells as phages did with bacteria.

Berg's idea, in particular, was to use small tumor viruses as *vehicles* into the genetic landscape of mammalian cells. If those vehicles, or vectors, as they are called, could carry with them specific genes of known functions, then Berg could understand a great deal about mammalian DNA by observing how it responded to these new genes. In an analogous fashion, one might study the creative activity of an architectural firm by slipping under its door the plans for a new kind of window and then watching to see how those windows were incorporated into new buildings designed by the firm.

Berg needed two things for his program: a vehicle organism and a bit of passenger DNA. After working in Dulbecco's lab during his sabbatical year of 1967–1968, Berg decided to use as his vehicle a tumor virus called SV40, which reproduces easily in monkey cells, causes cancer in rodent cells, and, like polyoma, integrates its DNA into the chromosomes of the cells it infects. SV40, like polyoma, is one of the smallest and simplest viruses known. Its DNA comes in a small circular loop and contains a mere 5,243 base pairs, coding for five genes. Recall that each base pair may be thought of as a single letter in the alphabet of the DNA molecule. (See Chapter 17 for a review of DNA.) By comparison, the DNA of a typical bacterium has several million base pairs, and the DNA of a mammal may have several billion. SV40 is to biology what a quark or electron is to physics. It is an elemental life form.

Not having available any pure mammalian genes at the time, for his passenger DNA Berg chose a small loop of DNA from the bacterium

Escherichia coli, a common organism that resides in the human intestines and helps with digestion. The *E. coli* DNA loop contains three genes responsible for ordering the enzymes that metabolize sugar. (These three genes are called the galactose operon, named after the sugar they metabolize, galactose.)

Berg clearly lays out his plan in the first two sentences of his landmark paper of 1972:

> Our goal is to develop a method by which new, functionally defined segments of genetic information can be introduced into mammalian cells. It is known that the DNA of the transforming virus SV40 can enter into a stable . . . association with the genomes of various mammalian cells . . . it seemed possible that SV40 DNA molecules, into which a segment of functionally defined, nonviral DNA had been covalently integrated, could serve as vectors to transport and stabilize these nonviral DNA sequences in the cell genome.

The first step in Berg's program would be to figure out how to "integrate" a piece of nonviral DNA—that is, the *E. coli* genes—into the DNA of the SV40 vehicle. In other words, he had to develop a method to join together two different DNA molecules. That had never been done.

A younger colleague of Paul Berg at Stanford says that he "plays a terrific game of tennis." Photographs of Berg show an athletic man with short wavy hair and a broad smile, given to wearing aviator eyeglasses and short-sleeved shirts with buttoned-down collars. Superficially, such casual dress would seem to match the childlike simplicity of his work. In his laboratory, Berg uses special "scissors" to cut DNA fragments, then rejoins the fragments and "glues" them together; then he uses more glue to fill in the cracks and seal the joints—almost a kindergarten crafts project. However, these cuts and glues operate in the tiny terrain of single molecules, a hundred million times smaller than a thumbtack or sliver of tape.

Figure 22.1 illustrates Berg's procedure in creating the first recombinant DNA. As shown in (a), he started with a DNA loop of SV40 and a DNA loop of *E. coli* called *λdvgal*. These DNA loops are both short segments of DNA wrapped around and joined to themselves to form circles.

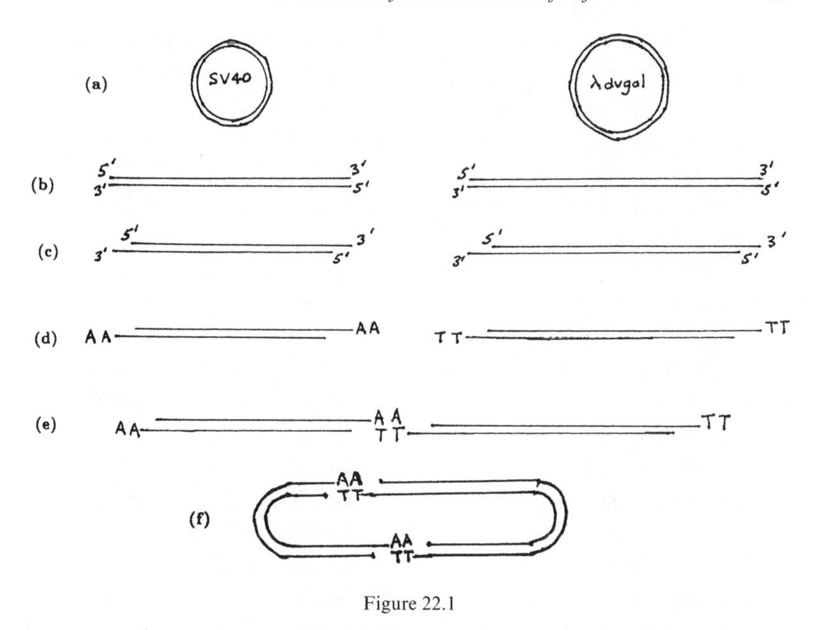

Figure 22.1

In the next step, (b), Berg used a special chemical called a restriction enzyme to cut each of the loops, forming linear segments of DNA. Restriction enzymes, the immensely important "scissors" of DNA research, had been discovered only two years earlier, by Hamilton Smith of Johns Hopkins University. There are now hundreds of different restriction enzymes. Each cuts DNA only at a specific place. For example, the restriction enzyme called HaeIII cuts a strand of DNA wherever the sequence of bases GGCC occurs (and CCGG on the complementary strand). The cut occurs between the G (guanine) and C (cytosine) on each strand. Eco RI, which Berg used in this experiment, cuts a piece of DNA wherever the sequence of bases GAATTC occurs (and CTTAAG on the complementary strand). The cut occurs between the G (guanine) and A (adenine) on each strand. The 5,243 base pairs of the SV40 loop have exactly one occurrence of GAATTC and are cut once. The approximately ten thousand base pairs of the *λdvgal* loop are cut twice, although only one of the two resulting segments is shown in Figure 22.1(b).

Next, Berg used another kind of enzyme called λ exonuclease to snip off about fifty bases from the 5′ end of each DNA strand. One end of each strand of DNA is called the 5′ end and the other is called the 3′ end. (The 5′ end is so named because the final deoxyribose sugar of that end has a phosphate group attached to its fifth carbon atom, counting clockwise from the oxygen atom of the sugar. The final deoxyribose sugar of the 3′ end has a hydroxyl group attached to its third carbon atom. These distinctions are discussed further in Chapter 17.) Because of the complementary nature of the two strands in a double-stranded DNA molecule, the 5′ end of one strand is normally side by side with the 3′ end of the complementary strand. After the snipping shown in (c), the 3′ ends protrude beyond the 5′ ends.

In the next step, shown in (d), Berg and his colleagues employed terminal transferase, an enzyme from the calf thymus, to add adenine bases to the 3′ ends of SV40 DNA and thymine bases to the 3′ ends of the *λdvgal* DNA. In the figure, only two of the bases are shown in each case, denoted by AA or TT. Recall that adenine and thymine are complementary base pairs. They bond together. Thus, by putting adenine tails on one segment of DNA and thymine tails on another, Berg created what molecular biologists call sticky ends. An adenine tail will automatically stick to a thymine tail. Adenine and thymine tails are the "glue."

In (e) and (f) the segments of SV40 DNA and *λdvgal* DNA are mixed together and the sticky ends stick together. Such a process may be visualized as happening in two stages. First an AA sticky end of a SV40 DNA segment binds with a TT sticky end of a *λdvgal* segment, as shown in (e). Then the opposite sticky ends bend around and stick together, forming a loop, like the initial loops in (a).

As seen in (f), there are gaps at the two junctures, missing sections of bases and phosphate-sugar backbones. To fill in these gaps, one adds an enzyme called DNA polymerase I, discovered in 1958, which copies single strands of DNA by transporting and putting into place the missing complementary bases. Finally, the outer phosphate-sugar backbones at the junctures are bonded together by yet another enzyme called DNA ligase, discovered in 1967. DNA ligase helps form covalent chemical bonds joining the phosphate group molecule to the sugar molecule above and below it in the backbone legs of the DNA molecule. (For pictures of the phosphate and sugar molecules in DNA, see Chapter 17.) DNA ligase is a kind of solder at the molecular level.

The final product, shown in (g), is also a DNA loop. But it is a hybrid loop. It is a circular loop of DNA that contains DNA from two different organisms, SV40 and *λdvgal.*

Berg was working in a highly active and rapidly developing field. (Most of his chemical technology had been discovered only in the previous few years.) Berg, Jackson, and Symons's method of splicing fragments of DNA together by adding complementary base "tails" was, in fact, developed simultaneously by another group at Stanford, Peter Lobban and A. D. Kaiser, splicing together two pieces of the phage P22 instead of *λdvgal* and SV40. Within a few months, that method was simplified and improved. In late 1972, three other Stanford scientists, Janet Mertz, Ronald Davis, and Vittorio Sgaramella, discovered that some restriction enzymes automatically leave sticky ends because they produce "staggered" cuts. Figure 22.2 illustrates the idea for Eco RI. The arrows indicate where the cuts are made. As can be seen, the cuts of one segment of DNA leave two ends that will naturally stick together because they have complementary bases. If such cuts are made in two *different* segments of DNA, from two different organisms, then the appropriate ends from each segment will stick together, joining the two segments, without the need to graft on A and T tails as Berg did in his first program. Thus, steps (c) and (d) of Figure 22.1 can be eliminated.

In early 1973, Stanley Cohen, Herbert Boyer, and their collaborators at Stanford and at the University of California at San Francisco produced another hybrid DNA loop. Instead of SV40, these scientists enlisted as their vehicle a DNA loop from *E. coli* named pSC101. Instead of *λdvgal* as the passenger DNA, they used a gene that conferred resistance to penicillin. Such a hybrid DNA loop, reinserted into *E. coli*, would create a bacterium that could swim in penicillin as if it were mother's milk.

Berg, too, was poised to insert his hybrid DNA into the bacterium *E. coli*. Such a procedure would be an intermediate step, with an intermediate host organism, before inserting the hybrid into the cell of a mammal. However, Berg suspended his plan. As he recounted in his Nobel address, "because many colleagues expressed concern about the potential risks

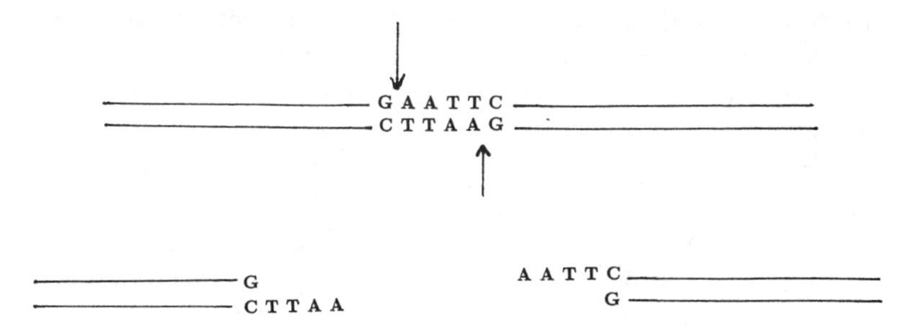

Figure 22.2

of disseminating *E. coli* containing SV40 [cancer-causing genes], the experiments with this recombinant DNA were discontinued."

Paul Berg's voluntary postponement of his work with recombinant DNA, and the subsequent worldwide moratorium on certain kinds of DNA research, was a landmark in the history of science. Scientists, and especially pure scientists, have always plunged ahead with any research they thought useful or interesting. Scientists have always welcomed the unknown. In this case, however, many scientists felt that the unknown held unacceptable risks.

The chain of events leading to Berg's decision began in the summer of 1971. At that time, Janet Mertz, a graduate student of Berg's, attended a course on tumor viruses at the Cold Spring Harbor Laboratory in New York. Mertz told one of the cancer researchers at Cold Spring, Robert Pollack, about Berg's plans to inject his hybrid DNA loop, including the cancer-causing virus SV40, into *E. coli*, producing a "modified" bacterium. Pollack was horrified. The problem, as he saw it, was this: Berg's hybrid DNA could easily live in *E. coli* because it contained *E. coli* DNA. But it also contained a tumor-causing virus. And *E. coli* lived in people. When the very first "modified" *E. coli* reproduced, you would have two modified *E. coli*, since the hybrid DNA would be passed on to the next generation. Then, twenty minutes later, at the next reproduction, you would have four modified *E. coli*. And so on. In one day, there would be 4×10^{21} genetically modified *E. coli*, or four trillion billion of them. Thus, Berg's experiment had the possibility of creating a new kind of cancer-causing organism never before seen in nature, an organism that lived well in human beings and multiplied like crazy.

Pollack called Berg to voice his concerns. Berg called his friends to see if anybody else was worried. They were. News of Berg's proposed experiment had spread almost as quickly as *E. coli* multiplies. At the time, virologist Wallace Rowe of the National Institutes of Health said, "The Berg experiment scares the pants off a lot of people, including [Berg]." Robert Pollack commented to journalist Nicholas Wade, "We're in a pre-Hiroshima situation." After weeks of reflection and discussion with other scientists, Berg decided not to go forward with his experiment until the safety issues could be evaluated.

But what about Boyer and Cohen and all the other biologists who were ready to inject hybrid DNA into living organisms? In June 1973, at the annual Gordon Research Conference on Nucleic Acids, after Boyer described his own plans with pSC101, a number of scientists suggested that the National Academy of Sciences immediately appoint a committee to study the safety issues of recombinant DNA research.

That committee was formed and chaired by Berg. Other members included David Baltimore, Herbert Boyer, Stanley Cohen, Ronald Davis, David Hogness, Daniel Nathans, Richard Roblin, James Watson, Sherman Weissman, and Norton Zinder—all major players in the field. In the July 26, 1974, issues of the widely read science journals *Science* and *Nature*, the committee published a one-page report titled "Potential Biohazards of Recombinant DNA Molecules." There, the scientists recommended a worldwide deferment on certain kinds of recombinant DNA research until the risks were better understood. The article states that "although such [recombinant DNA] experiments are likely to facilitate the solution of important theoretical and practical biological problems, they would also result in the creation of novel types of infectious DNA elements whose biological properties cannot be completely predicted in advance. There is a serious concern that some of the artificial recombinant DNA molecules could prove biologically hazardous."

The Berg letter, as it came to be called, was perhaps only the second time that scientists had attempted to restrain their own research. The first instance occurred during work on the proposed hydrogen bomb, a weapon that, its designers knew, would have a power far exceeding that of the atomic bomb. The Atomic Energy Commission set up a General Advisory Committee to study whether it was worthwhile to pursue such a weapon. In 1949, Robert Oppenheimer and other scientists submitted the majority report of the committee: "We base our recommendation [not to support development of the hydrogen bomb] on our belief that the extreme dangers to mankind in the proposal wholly outweigh any military advantage that would come from this development . . . a super bomb [hydrogen bomb] might become a weapon of genocide." Oppenheimer's recommendation was ignored, research continued on the hydrogen bomb, and it was built by the United States and the Soviet Union.

By contrast, Berg's recommended deferment of recombinant DNA experiments was accepted worldwide, from July 1974 until February 1975, when an international group of 150 biologists and lawyers gathered at the Asilomar conference in Pacific Grove, California. Biologists in attendance expressed a great range of views, from serious concern about the risks of recombinant DNA, to feelings that the risks could not be evaluated, to resentment for having any institutional restraint on basic research, to beliefs that the challenge was mostly a public relations problem. In the end, the attendees voted to replace the existing deferment with a new set of guidelines created by the National Institute of Health. Henceforth, recombinant DNA experiments were to require various degrees of containment appropriate to the estimated risks.

Today, thirty years later, most biologists feel that the fears were not unjustified, but that they were larger than the actual risks. Still, research using certain kinds of recombinant DNA is carried out with caution.

Much of Berg's landmark paper of 1972 deals with the various kinds of chemicals used to manipulate DNA and the methods of labeling, isolating, and identifying DNA.

Before splicing the DNA of *λdvgal* with the DNA of SV40, Berg and his collaborators practice their technique by splicing two pieces of SV40 DNA with themselves. Berg deals with two kinds of SV40 DNA, called SV40(I) and SV40(II). The first is complete DNA, while the second has a break or "nick" in one of the two strands. Both the *λdvgal* and the SV40 DNA are "labeled" with a substance called [^{3}H]dT, a radioactive form of hydrogen that can take the place of normal hydrogen in the DNA molecules. By counting the rate of disintegrations of this radioactive label, the scientists can tell how much [^{3}H]dT has been incorporated into the DNA and thus how much DNA is present in each stage of the process of cutting and gluing. The disintegrations of radioactive hydrogen are "counted" with a device called a scintillation spectrometer. (For a review of radioactivity, see Chapter 15.) Another radioactive label used is [^{32}P], which is a radioactive form of phosphorus that replaces normal phosphorus in the phosphorus-sugar backbone of the DNA molecule.

The DNA is often isolated from other materials by placing it in a centrifuge, a device that spins around at thousands of revolutions per minute. A tube in the centrifuge is filled with a sugar solution whose density increases from the top of the tube to the bottom. When various substances are placed in the tube and spun at high speed, the largest molecules end up near the bottom of the tube and the smallest at the top. The chemical ethidium bromide stains DNA a distinctive color and is often added to the material in the test tube. In another method of centrifugation, the association of different amounts of ethidium bromide with different kinds of DNA is used to separate DNA molecules.

Another, simple device for separating DNA from other substances is Whatman filter paper, or a Whatman disk. When a solution containing

DNA and liquids is poured through the filter paper, the DNA fragments remain behind while the liquids pass through the paper.

While the DNA is being manipulated, it is kept in a liquid bath of chemicals including EDTA and Tris. These provide a "natural" environment for the DNA and a buffer that prevents the DNA from unwanted reactions with other chemicals. When the DNA is mixed with restriction enzyme RI to cut it or with terminal transferase to add on the sticky tails, the solutions are kept at 37°C, which is exactly 98.6°F, body temperature. In general, as much as possible is done to keep the DNA in its natural habitat while it is being manipulated in an unnatural way.

The most sophisticated piece of equipment that Berg and his collaborators use is an electron microscope, which is a device that creates images with electrons rather than with light. Because high-energy electrons have wavelengths much smaller than visible light, an electron microscope can "see" objects much smaller than those that can be seen with an ordinary light microscope, including individual atoms and molecules. The scientists use an electron microscope to measure the shapes of the DNA fragments, linear versus circular, and to measure the lengths of the fragments. In particular, if the SV40 DNA has a length of 1 unit and a section of λ*dvgal* has a length of 2 units, then the length of a hybrid λ*dvgal*-SV40 loop should have a length of 3 units. As seen in Table 1 of the paper, the electron microscope confirms that result. Note also from Table 1 the relatively small number of molecules that the scientists are counting and measuring. They are truly working at the level of individual molecules.

Other terminology helpful for reading Berg's paper: a dimer is a molecule made by joining two smaller molecules; duplex is simply two-stranded DNA; covalent integration is bonding by covalent chemical bonds (see Chapter 11); endonuclease is a restriction enzyme that cuts in the interior of a strand of DNA, while exonuclease cuts off an end of a strand of DNA; dATP and dTTP are deoxyadenine triphosphate and deoxythymine triphosphate, respectively, the forms of adenine and thymine that make the sticky tails on the ends of the DNA strands.

Near the end of their paper, just before the discussion section, the scientists proclaim that all of their tests have indicated success: "We conclude from the experiments described above that λ*dvgal* DNA containing the intact galactose operon from *E. coli* . . . has been covalently inserted into an SV40 genome. These molecules should be useful for testing whether these bacterial genes can be introduced into a mammalian cell genome and whether they can be expressed there."

Finally, the scientists are clearly aware of the general application of their technical advance: "The methods described in this report . . . are general and offer an approach for covalently joining any two DNA molecules together."

As Berg and other scientists hoped, recombinant DNA techniques have shed light on the fundamental nature and operation of genes. For example, such techniques have revealed that almost all mammalian DNA has long stretches of base pairs, called introns, that do not carry any discernible instructions. As of 2005, the function of entrons is still not fully understood.

As another example, recombinant DNA techniques have allowed scientists to create genetic mutations and thus to study the origin and effect of these altered genes. Yet another example is the use of recombinant DNA to study cancer and the mechanism by which tumor viruses act. By January 1979, restrictions on recombinant DNA had been sufficiently softened that scientists could begin inserting tumor viruses into other cells to study how tumors replicate and transform normal cells. As a result, scientists now believe that some cancers begin when there are too many copies of certain regulatory genes or overproduction of certain cellular activities. Barbara McClintock's discovery of "movable genes," genes that change their location on the chromosome, was confirmed and clarified by using recombinant DNA techniques to insert genes into various locations of the *E. coli* DNA.

On the more applied side, the first medical product using a genetically modified organism was human insulin, marketed by Eli Lilly in 1982. Since then, dozens of other precious biological substances have been artificially produced with recombinant DNA technology, including interferon, hepatitis B vaccine, and human growth hormone. Essentially, the DNA of various organisms is modified so that they will produce these substances. In agriculture, recombinant DNA technology has allowed the creation of transgenic crops—including such human staples as maize, soybeans, tomatoes, cotton, potatoes, rice—with tolerance to herbicides and resistance to insects and disease.

But the most important applications and results of recombinant DNA surely lie in the future. The future of gene splicing is unknown. Its full impact is inconceivable. Never before have we wielded such power over the inner workings of life. And, like the Sorcerer's Apprentice, we cannot fathom that power.

BIOCHEMICAL METHOD FOR INSERTING NEW GENETIC INFORMATION INTO DNA OF SIMIAN VIRUS 40: CIRCULAR SV40 DNA MOLECULES CONTAINING LAMBDA PHAGE GENES AND THE GALACTOSE OPERON OF *ESCHERICHIA COLI*

(MOLECULAR HYBRIDS/DNA JOINING/VIRAL TRANSFORMATION/GENETIC TRANSFER)

David A. Jackson,[1] Robert H. Symons,[2] and Paul Berg

Department of Biochemistry, Stanford University Medical Center, Stanford, California 94305

Contributed by Paul Berg, July 31, 1972

Proceedings of the National Academy of Sciences (1972)

ABSTRACT We have developed methods for covalently joining duplex DNA molecules to one another and have used these techniques to construct circular dimers of SV40 DNA and to insert a DNA segment containing lambda phage genes and the galactose operon of *E. coli* into SV40 DNA. The method involves: (*a*) converting circular SV40 DNA to a linear form, (*b*) adding single-stranded homodeoxypolymeric extensions of defined composition and length to the 3′ ends of one of the DNA strands with the enzyme terminal deoxynucleotidyl transferase (*c*) adding complementary homodeoxypolymeric extensions to the other DNA strand, (*d*) annealing the two DNA molecules to form a circular duplex structure, and (*e*) filling the gaps and sealing nicks in this structure with *E. coli* DNA polymerase and DNA ligase to form a covalently closed-circular DNA molecule.

Our goal is to develop a method by which new, functionally defined segments of genetic information can be introduced into mammalian cells. It is known that the DNA of the transforming virus SV40 can enter into a stable, heritable, and presumably covalent association with the genomes of various mammalian cells.[4,5] Since purified SV40 DNA can also transform cells (although with reduced efficiency), it seemed possible that SV40 DNA molecules, into which a segment of functionally defined, nonviral DNA had been covalently integrated, could serve as vectors to transport and stabilize these nonviral DNA sequences in the cell genome. Accordingly, we have developed biochemical techniques that are generally applicable for joining covalently any two DNA molecules.[3] Using these techniques, we have constructed circular dimers of SV40 DNA; moreover, a DNA segment containing λ phage genes and the galactose operon of *Escherichia coli* has been covalently integrated into the circular SV40 DNA molecule. Such hybrid DNA molecules and others like them can be tested for their capacity to transduce foreign DNA sequences into mammalian cells, and can be used to determine whether these new nonviral genes can be expressed in a novel environment.

MATERIALS AND METHODS

DNA. (*a*) Covalently closed-circular duplex SV40 DNA [SV40(I)] (labeled with [^{3}H]dT, 5×10^4 cpm/μg), free from SV40 linear or oligomeric molecules [but containing 3–5% of nicked double-stranded circles—SV40(II)] was purified from SV40-infected CV-1 cells (Jackson, D., & Berg, P., in preparation). (*b*) Closed-circular duplex λ*dvgal* DNA labeled with [^{3}H]dT (2.5×10^4 cpm/μg), was isolated from an *E. coli* strain containing this DNA as an autonomously replicating plasmid[6] by equilibrium sedimentation in CsCl—ethidium bromide gradients[7] after lysis of the cells with detergent. A more detailed characterization of this DNA will be published later. Present information indicates that the λ*dvgal* (λ*dv*-120) DNA is a circular dimer containing tandem duplications of a sequence of several λ phage genes (including C_I, O, and P) joined to the entire galactose operon of *E. coli* (Berg, D., Mertz, J., & Jackson, D., in preparation). DNA concentrations are given as molecular concentrations.

ENZYMES. The circular SV40 and λ*dvgal* DNA molecules were cleaved with the bacterial restriction endonuclease RI (Yoshimori and Boyer, unpublished; the enzyme was generously made available to us by these workers). Phage λ-exonuclease (given to us by Peter Lobban) was prepared according to Little *et al.*,[8] calf-thymus deoxynucleotidyl terminal transferase (terminal transferase), prepared according to Kato *et al.*,[9] was generously sent to us by F. N. Hayes; *E. coli* DNA polymerase I Fraction VII[10] was a gift of Douglas Brutlag; and *E. coli* DNA ligase[11] and exonuclease III[12] were kindly supplied by Paul Modrich.

SUBSTRATES. [α-^{32}P]deoxynucleoside triphosphates (specific activities 5–10 Ci/μmol) were synthesized by the method of Symons.[13] All other reagents were obtained from commercial sources.

CENTRIFUGATIONS. *Alkaline sucrose gradients* were formed by diffusion from equal volumes of 5, 10, 15, and 20% sucrose solutions with 2 mM EDTA containing, respectively, 0.2, 0.4, 0.6, and 0.8 M NaOH, and 0.8, 0.6, 0.4, 0.2 M NaCl. 100-μl samples were run on 3.8-ml gradients in a Beckman SW56 Ti rotor in a Beckman L2-65B ultracentrifuge at 4° and 55,000 rpm for the indicated times. 2- to 10-drop fractions were collected onto 2.5-cm diameter Whatman 3MM discs, dried without washing, and counted in PPO-dimethyl POPOP-toluene scintillator in a Nuclear Chicago Mark II scintillation spectrometer. An overlap of 0.4% of ^{32}P into the ^{3}H channel was not corrected for.

CsCl-ethidium bromide equilibrium centrifugation was performed in a Beckman Type 50 rotor at 4° and 37,000 rpm for 48 hr. SV40 DNA in 10 mM Tris · HCl (pH 8.1)-1 mM Na EDTA-10 mM NaCl was adjusted to 1.566 g/ml of CsCl and 350 μg/ml of ethidium bromide. 30-Drop fractions were collected and aliquots were precipitated on Whatman GF/C filters with cold 2 N HCl; the filters were washed and counted.

ELECTRON MICROSCOPY. DNA was spread for electron microscopy by the aqueous method of Davis *et al.*[14] and photographed in a Phillips EM 300. Projections of the molecules were traced on paper and measured with a Keuffel and Esser map measurer. Plaque-purified SV40(II) DNA was used as an internal length standard.

CONVERSION OF SV40(I) DNA TO UNIT LENGTH LINEAR DNA [SV40(L_{RI})] WITH R_I ENDONUCLEASE. [^{3}H]SV40(I) DNA (18.7 nM) in 100 mM Tris · HCl buffer (pH 7.5)—10 mM $MgCl_2$—2 mM 2-mercaptoethanol was incubated for 30 min at 37° with an amount of R_I previously determined to convert 1.5 times this amount of SV40(I) to linear molecules [SV40(L_{RI})]; Na EDTA (30 mM) was added to stop the reaction, and the DNA was precipitated in 67% ethanol.

REMOVAL OF 5′-TERMINAL REGIONS FROM SV40(L_{RI}) WITH λ EXONUCLEASE. [^{3}H]SV40(L_{RI}) (15 nM) in 67 mM K-glycinate (pH 9.5), 4 mM $MgCl_2$, 0.1 mM EDTA was incubated at 0° with λ-exonuclease (20 μg/ml) to yield [^{3}H]SV40(L_{RI}exo) DNA. Release of [^{3}H]dTMP was measured by chromatographing aliquots of the reaction on polyethyleneimine thin-layer sheets (Brinkmann) in 0.6 *M* NH_4HCO_3 and counting the dTMP spot and the origin (undegraded DNA).

ADDITION OF HOMOPOLYMERIC EXTENSIONS TO SV40(L_{RI}EXO) WITH TERMINAL TRANSFERASE. [^{3}H]SV40(L_{RI}exo) (50 nM) in 100 mM K-cacodylate (pH 7.0), 8 mM $MgCl_2$, 2 mM 2-mercaptoethanol, 150 μg/ml of bovine serum albumin, [α-^{32}P]dNTP (0.2 mM for dATP, 0.4 mM for dTTP) was incubated with terminal transferase (30–60 μg/ml) at 37°. Addition of [^{32}P]dNMP residues to SV40 DNA was measured by spotting aliquots of the reaction mixture on DEAE-paper discs (Whatman DE-81), washing each disc by suction with 50 ml (each) of 0.3 M NH_4-formate (pH 7.8) and 0.25 M NH_4HCO_3, and then with 20 ml of ethanol. To determine the proportion of SV40 linear DNA molecules that had acquired at least one "functional" $(dA)_n$ tail, we measured the amount of SV40 DNA (^{3}H counts) that could be bound to a Whatman GF/C filter (2.4-cm diameter) to which 150 μg of polyuridylic acid had been fixed.[16] 15-μl Aliquots of the reaction mixture were mixed with 5 ml of 0.70 M NaCl–0.07 M Na citrate (pH 7.0)–2% Sarkosyl, and filtered at room temperature through the poly(U) filters, at a flow rate of 3–5 ml/min. Each filter was washed by rapid suction with 50 ml of the same buffer at 0°, dried, and counted. Control experiments showed that 98–100% of [^{3}H]oligo$(dA)_{125}$ bound to the filters under these conditions. When the ratio of [^{32}P]dNMP to [^{3}H]DNA reached the value equivalent to the desired length of the extension, the reaction was stopped with EDTA (30 mM) and 2% Sarkosyl. The [^{3}H]SV40(L_{RI}exo)–[^{32}P]dA or –dT DNA was purified by neutral sucrose gradient zone sedimentation to remove unincorporated dNTP, as well as any traces of SV40(I) or SV40(II) DNA.

FORMATION OF HYDROGEN-BONDED CIRCULAR DNA MOLECULES. [^{32}P]dA and -dT DNAs were mixed at concentrations of 0.15 nM each in 0.1 M NaCl–10 mM Tris · HCl (pH 8.1)–1 mM EDTA. The mixture was kept at 51° for 30 min, then cooled slowly to room temperature.

FORMATION OF COVALENTLY CLOSED-CIRCULAR DNA MOLECULES. After annealing of the DNA, a mixture of the enzymes, substrates, and cofactors needed for closure was added to the DNA solution and the mixture was incubated at 20° for 3–5 hr. The final concentrations in the reaction mixture were: 20 mM Tris · HCl (pH 8.1), 1 mM EDTA, 6 mM $MgCl_2$, 50 μg/ml bovine-serum albumin, 10 mM NH_4Cl, 80 mM NaCl, 0.052 mM DPN, 0.08 mM (each), dATP, dGTP, dCTP, and dTTP, (0.4 μg/ml) *E. coli* DNA polymerase I, (15 units/ml) *E. coli* ligase, and (0.4 unit/ml) *E. coli* exonuclease III.

RESULTS

General approach

Fig. 1 outlines the general approach used to generate circular, covalently-closed DNA molecules from two separate DNAs. Since, in the present case, the units to be joined are themselves circular, the first step requires conversion of the circular structures to linear duplexes. This could be achieved by a double-strand scission at random locations (see *Discussion*) or, as we describe in this paper, at a unique site with R_I restriction endonuclease. Relatively short (50–100 nucleotides) poly(dA) or poly(dT) extensions are added on the 3′-hydroxyl termini of the linear duplexes with terminal transferase; prior removal of a short sequence (30–50 nucleotides) from the 5′-phosphoryl termini by digestion with λ exonuclease facilitates the terminal transferase reaction. Linear duplexes containing $(dA)_n$ extensions are annealed to the DNA to be joined containing $(dT)_n$ extensions at relatively low concentrations. The circular structure formed contains the two DNAs, held together by two hydrogen-bonded homopolymeric regions (Fig. 1). Repair of the four gaps is mediated by *E. coli* DNA polymerase with the four deoxynucleosidetriphosphates, and covalent closure of the ring structure is effected by *E. coli* DNA ligase; *E. coli* exonuclease III removes 3′-phosphoryl residues at any nicks inadvertently introduced during the manipulations (nicks with 3′-phosphoryl ends cannot be sealed by ligase).

Principal steps in the procedure

CIRCULAR SV40 DNA CAN BE OPENED TO LINEAR DUPLEXES BY R_I ENDONUCLEASE. Digestion of SV40(I) DNA with excess R_I endonuclease yields a product that sediments at 14.5 S in neutral sucrose gradients and appears as a linear duplex with the same contour length as SV40(II) DNA when examined by electron microscopy[21] [Jackson and Berg, in preparation; see Table 1]. The point of cleavage is at a unique site on the SV40 DNA, and few if any single-strand breaks are introduced else-

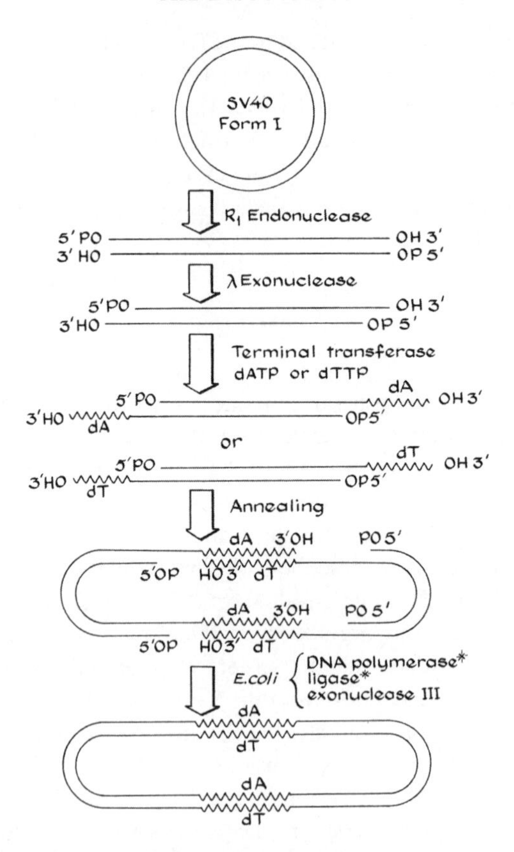

Figure 1. General protocol for producing covalently closed SV40 dimer circles from SV40(I) DNA.

*The four deoxynucleoside triphosphates and DPN are also present for the DNA polymerase and ligase reactions, respectively.

where in the molecule[21]; moreover, the termini at each end are 5′-phosphoryl, 3′-hydroxyl (Mertz, J., Davis, R., in preparation). Digestion of plaque-purified SV40 DNA under our conditions yields about 87% linear molecules, 10% nicked circles, and 3% residual supercoiled circles.

ADDITION OF OLIGO(dA) OR -(dT) EXTENSIONS TO THE 3′-HYDROXYL TERMINI OF SV40 (L_{RI}). Terminal transferase has been used to generate deoxyhomopolymeric extensions on the 3′-hydroxyl termini of DNA[10]; once the chain is initiated, chain propagation is statistical in that each chain grows at about the same rate.[15] Although the length of the extensions can be controlled by variation of either the time of incubation or the amount of substrate, we have varied the time of incubation to minimize spurious nicking of the DNA by trace amounts of endonuclease activity in the enzyme preparation; we have so far been unable to remove or selectively inhibit these nucleases (Jackson and Berg, in preparation).

Incubation of SV40(L_{RI}) with terminal transferase and either dATP or dTTP resulted in appreciable addition of mononucleotidyl units to the DNA. But, for example, after addition of 100 residues of dA per end, only a small proportion of the modified SV40 DNA would bind to filter discs containing poly(U).[16] This result indicated that initiation of terminal nucleotidyl addition was infrequent with SV40(L_{RI}), but that once initiated those termini served as preferential primers for extensive homopolymer synthesis.

Lobban and Kaiser (unpublished) found that P22 phage DNA became a better primer for homopolymer synthesis after incubation of the DNA with λ exonuclease. This enzyme removes, successively, deoxymononucleotides from 5′-phosphoryl termini of double-stranded DNA,[18] thereby rendering the 3′-hydroxyl termini single-stranded. We confirmed their finding with SV40(L_{RI}) DNA; after removal of 30–50 nucleotides per 5′-end (see *Methods*), the number of SV40(L_{RI}) molecules that could be bound to poly(U) filters after incubation with terminal transferase and dATP increased 5- to 6-fold. Even after separation of the strands of the SV40(L_{RI}exo)-dA, a substantial proportion of the ^{3}H-label in the DNA was still bound by the poly(U) filter, indicating that both 3′-hydroxy termini in the duplex DNA can serve as primers.

The weight-average length of the homopolymer extensions was 50–100 residues per end. Zone sedimentation of [^{3}H]-SV40(L_{RI}exo)-[^{32}P]$(dA)_{80}$ (this particular preparation, which is described in *Methods*, had on the average, 80 dA residues per end) in an alkaline sucrose gradient showed that (*i*) 60–70% of the SV40 DNA strands are intact, (*ii*) the [^{32}P]$(dA)_{80}$ is covalently attached to the [^{3}H]SV40 DNA, and (*iii*) the distribution of oligo(dA) chain lengths attached to the SV40 DNA is narrow, indicating that the deviation from the calculated mean length of 80 is small (Fig. 2) [omitted here]. SV40(L_{RI}exo), having $(dT)_{80}$ extensions, was prepared with [^{32}P]dTTP and gave essentially the same results when analyzed as described above.

HYDROGEN-BONDED CIRCULAR MOLECULES ARE FORMED BY ANNEALING SV40(L_{RI}exo)-$(dA)_{80}$ AND SV40(L_{RI}exo)-$(dT)_{80}$ TOGETHER. When SV40 (L_{RI}exo)-$(dA)_{80}$ and SV40(L_{RI}exo)-$(dT)_{80}$ were annealed together,

30–60% of the molecules seen by electron microscopy were circular dimers; linear monomers, linear dimers, and more complex branched forms were also seen. If SV40(L_{RI}exo)-$(dA)_{80}$ or -$(dT)_{80}$ alone was annealed, no circles were found. Centrifugation of annealed preparations in neutral sucrose gradients showed that the bulk of the SV40 DNA sedimented faster than modified unit-length linears (as would be expected for circular and linear dimers, as well as for higher oligomers). Sedimentation in alkaline gradients, however, showed only unit-length single strands containing the oligonucleotide tails (as seen in Fig. 2) [omitted here].

COVALENTLY CLOSED-CIRCULAR DNA MOLECULES ARE FORMED BY INCUBATION OF HYDROGEN-BONDED COMPLEXES WITH DNA POLYMERASE, LIGASE, AND EXONUCLEASE III. The hydrogen-bonded complexes described above can be sealed by incubation with the *E. coli* enzymes DNA polymerase I, ligase, and exonuclease III, plus their substrates and cofactors. Zone sedimentation in alkaline sucrose gradients (Fig. 3) [omitted here] shows that 20% of the input ^{32}P label derived from the oligo(dA) and -(dT) tails sediments with the ^{3}H label present in the SV40 DNA, in the position expected of a covalently closed-circular SV40 dimer (70–75 S). About the same amount of labeled DNA bands in a CsCl-ethidium bromide gradient at a buoyant density characteristic of covalently closed-circular DNA (Fig. 4) [omitted here].

DNA isolated from the heavy band of the CsCl-ethidium bromide gradient contains primarily circular molecules, with a contour length twice that of SV40(II) DNA (Table 1) when viewed by electron microscopy. No covalently closed DNA is formed if either one of the linear precursors is omitted from the annealing step or if the enzymes are left out of the closure reaction. We conclude, therefore, that two unit-length linear SV40 molecules have been joined to form a covalently closed-circular dimer.

Covalent closure of the hydrogen-bonded SV40 DNA dimers is dependent on Mg^{2+}, all four deoxynucleoside triphosphates, *E. coli* DNA polymerase I, and ligase, and is inhibited by 98% if exonuclease III is omitted (Lobban and Kaiser first observed the need for exonuclease III in the joining of P22 molecules; we confirmed their finding with this system). Exonuclease III is probably needed to remove 3′-phosphate groups from 3′-phosphoryl, 5′-hydroxyl nicks introduced by the endonuclease contaminating the terminal transferase preparation. 3′-phosphoryl groups are potent inhibitors of *E. coli* DNA polymerase 1[17] and termini

TABLE 1.

RELATIVE LENGTHS OF SV40 AND λDVGAL-120 DNA MOLECULES

DNA species	Length ± standard deviation in SV40 units*	Number of molecules in sample
SV40(II)	1.00	224
SV40(L_{RI})†	1.00 ± 0.03	108
$(SV40\text{-}dA\ dT)_2$	2.06 ± 0.19	23
λ*dvgal*-120(I)	4.09 ± 0.14	65
λ*dvgal*-120(L_{RI})	2.00 ± 0.04	163
λ*dvgal*-SV40	2.95 ± 0.04	76
λ*dv*-1	2.78 ± 0.05	13

* The contour length of plaque-purified SV40(II) DNA is defined as 1.00 unit.

† Data supplied by J. Morrow.

having 5′-hydroxyl groups cannot be sealed by *E. coli* ligase.[11] The 5′-hydroxyl group can be removed and replaced by a 5′-phosphoryl group by the 5′- to 3′-exonuclease activity of *E. coli* DNA polymerase I.[10]

PREPARATION OF THE GALACTOSE OPERON FOR INSERTION INTO SV40 DNA. The galactose operon of *E. coli* was obtained from a λ*dvgal* DNA; λ*dvgal* is a covalently closed, supercoiled DNA molecule four times as long as SV40(II) DNA (Table 1). After complete digestion of λ*dvgal* DNA with the R_I endonuclease, linear molecules two times the length of SV40(II) DNA are virtually the exclusive product (Table 1). This population has a unimodal length distribution by electron microscopy and appears to be homogeneous by ultracentrifugal criteria (Jackson and Berg, in preparation). The R_I endonuclease seems, therefore, to cut λ*dvgal* circular DNA into two equal length linear molecules. Since one R_I endonuclease cleavage per λ*dv* monomeric unit occurs in the closely related λ*dv*-204 (Jackson and Berg, in preparation), it is likely that λ*dvgal* is cleaved at the same sites and, therefore, that each linear piece contains an intact galactose operon.

The purified λ*dvgal* (L_{RI}) DNA was prepared for joining to SV40 DNA by treatment with λ-exonuclease, followed by terminal transferase and [^{32}P]dTTP, as described for SV40-(L_{RI}).

FORMATION OF COVALENTLY CLOSED-CIRCULAR DNA MOLECULES CONTAINING BOTH SV40 AND λDVGAL DNA. Annealing of [^{3}H]-SV40(L_{RI}exo)-[^{32}P](dA)$_{80}$ with [^{3}H]λ*dvgal*(L_{RI}exo)-[^{32}P]-(dT)$_{80}$, followed by incubation with the enzymes, substrates, and cofactors needed for closure, produced a species of DNA (in about 15% yield) that sedimented rapidly in alkaline sucrose gradients (Fig. 5) [omitted here] and that formed a band in a CsCl-ethidium bromide gradient at the position expected for covalently closed DNA (Fig. 6) [omitted here]. The putative λ*dvgal*-SV40 circular DNA sediments just ahead of λ*dv*-1, a supercoiled circular DNA marker [2.8 times the length of SV40(II)DNA], and behind λ*dvgal* supercoiled circles [4.1 times SV40(II)DNA] in the alkaline sucrose gradient. Electron microscopic measurements of the DNA recovered from the dense band of the CsCl-ethidium bromide gradient showed a mean contour length for the major species of 2.95 ± 0.04 times that of SV40(II) DNA (Table 1). Each of these measurements supports the conclusion that the newly formed, covalently closed-circular DNA contains one SV40 DNA segment and one λ*dvgal* DNA monomeric segment. . . .

Omission of the enzymes from the reaction mixture prevents λ*dvgal*-SV40 DNA formation (Figs. 5 and 6) [omitted here]. No covalently closed product is detectable (Fig. 5) [omitted here] if λ*dvgal* and SV40 linear molecules with identical, rather than complementary, tails are annealed and incubated with the enzymes. This result demonstrates directly that the formation of covalently closed DNA depends on complementarity of the homopolymeric tails.

We conclude from the experiments described above that λ*dvgal* DNA containing the intact galactose operon from *E. coli*, together with some phage λ genes, has been covalently inserted into an SV40 genome. These molecules should be useful for testing whether these bacterial genes can be introduced into a mammalian cell genome and whether they can be expressed there.

DISCUSSION

The methods described in this report for the covalent joining of two SV40 molecules and for the insertion of a segment of DNA containing the galactose operon of *E. coli* into SV40 are general and offer an approach for covalently joining any two DNA molecules together. With the exception of the fortuitous property of the R_I endonuclease, which creates convenient linear DNA precursors, none of the techniques used depends upon any unique property of SV40 and/or the λ*dvgal* DNA. By the use of known enzymes and only minor modifications of the methods described here, it should be possible to join DNA molecules even if they have the wrong combination of hydroxyl and phosphoryl groups at their termini. By judicious use of generally available enzymes, even DNA duplexes with protruding 5′- or 3′-ends can be modified to become suitable substrates for the joining reaction.

One important feature of this method, which is different from some other techniques that can be used to join unrelated DNA molecules to one another,[19,22] is that here the joining is directed by the homopolymeric tails on the DNA. In our protocol, molecule A and molecule B can only be joined to each other; all AA and BB intermolecular joinings and all A and B intramolecular joinings (circularizations) are prevented. The yield of the desired product is thus increased, and subsequent purification problems are greatly reduced.

For some purposes, however, it may be desirable to insert *λdvgal* or other DNA molecules at other specific, or even random, locations in the SV40 genome. Other specific placements could be accomplished if other endonucleases could be found that cleave the SV40 circular DNA specifically. Since pancreatic DNase in the presence of Mn^{2+} produces randomly located, double-strand scissions (17) of SV40 circular DNA (Jackson and Berg, in preparation), it should be possible to insert a DNA segment at a large number of positions in the SV40 genome.

Although the *λdvgal* DNA segment is integrated at the same location in each SV40 DNA molecule, it should be emphasized that the orientation of the two DNA segments to each other is probably not identical. This follows from the fact that each of the two strands of a duplex can be joined to *either* of the two strands of the other duplex

$$\left(\text{e.g., } \begin{smallmatrix} W \frown W \\ \vdots \quad \vdots \\ C \smile C \end{smallmatrix} \text{ or } \begin{smallmatrix} W \frown C \\ \vdots \quad \vdots \\ C \smile W \end{smallmatrix}\right)^{23}$$

What possible consequences this fact has on the genetic expression of these segments remains to be seen.

We have no information concerning the biological activities of the SV40 dimer or the *λdvgal*-SV40 DNAs, but appropriate experiments are in progress. It is clear, however, that the location of the R_I break in the SV40 genome will be crucial in determining the biological potential of these molecules; preliminary evidence suggests that the break occurs in the late genes of SV40 (Morrow, Kelly, Berg, and Lewis, in preparation).

A further feature of these molecules that may bear on their usefulness is the $(dA \cdot dT)_n$ tracts that join the two DNA segments. They could be helpful (as a physical or genetic marker) or a hindrance (by making the molecule more sensitive to degradation) for their potential use as a transducer.

The λ*dvgal*-SV40 DNA produced in these experiments is, in effect, a trivalent biological reagent. It contains the genetic information to code for most of the functions of SV40, all of the functions of the *E. coli* galactose operon, and those functions of the λ bacteriophage required for autonomous replication of circular DNA molecules in *E. coli*. Each of these sets of functions has a wide range of potential uses in studying the molecular biology of SV40 and the mammalian cells with which this virus interacts.

We are grateful to Peter Lobban for many helpful discussions. D.A.J. was a Basic Science Fellow of the National Cystic Fibrosis Research Foundation; R.H.S. was on study leave from the Department of Biochemistry, University of Adelaide, Australia, and was supported in part by a grant from the USPHS. This research was supported by Grant GM-13235 from the USPHS and Grant VC-23A from the American Cancer Society.

1. Present address: Department of Microbiology, University of Michigan Medical Center, Ann Arbor, Mich. 48104.

2. Present address: Department of Biochemistry, University of Adelaide, Adelaide, South Australia, 5001 Australia.

3. Drs. Peter Lobban and A. D. Kaiser of this department have performed experiments similar to ours and have obtained similar results using bacteriophage P22 DNA (Lobban, P. and Kaiser, A. D., in preparation).

4. Sambrook, J., Westphal, H., Srinivasan, P. R. & Dulbecco, R. (1968) *Proc. Nat. Acad. Sci. USA* **60,** 1288–1295.

5. Dulbecco, R. (1969) *Science* **166,** 962–968.

6. Matsubara, K., & Kaiser, A. D. (1968) *Cold Spring Harbor Symp. Quant. Biol.* **33,** 27–34.

7. Radloff, R., Bauer, W., Vinograd, J. (1967) *Proc. Nat. Acad. Sci. USA* **57,** 1514–1521.

8. Little, J. W., Lehman, I. R. & Kaiser, A. D. (1967) *J. Biol. Chem.* **242,** 672–678.

9. Kato, K., Goncalves, J. M., Houts, G. E., & Bollum, F. J. (1967) *J. Biol. Chem.* **242,** 2780–2789.

10. Jovin, T. M., Englund, P. T. & Kornberg, A. (1969) *J. Biol. Chem.* **244,** 2996–3008.

11. Olivera, B. M., Hall, Z. W., Anraku, Y., Chien, J. R. & Lehman, I. R. (1968) *Cold Spring Harbor Symp. Quant. Biol.* **33,** 27–34.

12. Richardson, C. C., Lehman, I. R. & Kornberg, A. (1964) *J. Biol. Chem.* **239,** 251–258.

13. Symons, R. H. (1969) *Biochim. Biophys. Acta* **190,** 548–550.

14. Davis, R., Simon, M. & Davidson, N. (1971) in *Methods in Enzymology*, eds. Grossman, L. & Moldave, K. (Academic Press, New York), Vol. 21, pp. 413–428.

15. Chang, L. M. S. & Bollum, F. J. (1971) *Biochemistry* **10,** 536–542.

16. Sheldon, R., Jurale, C. & Kates, J. (1972) *Proc. Nat. Acad. Sci. USA* 69, 417–421.

17. Richardson, C. C., Schildkraut, C. L. & Kornberg, A. (1963) *Cold Spring Harbor Symp. Quant. Biol.* **28,** 9–19.

18. Little, J. W. (1967) *J. Biol. Chem.* **242,** 679–686.

19. Sgaramella, V., van de Sande, J. H. & Khorana, H. G. (1970) *Proc. Nat. Acad. Sci. USA* 67, 1468–1475.

20. Melgar, E. & Goldthwait, D. A. (1968) *J. Biol. Chem.* **243,** 4409–4416.

21. Morrow, J. F. & Berg, P. (1972) *Proc. Nat. Acad. Sci. USA* **69,** in press.

22. Sgaramella, V. & Lobban, P. (1972) *Nature*, in press.

23. The symbols W and C refer to one or the other complementary strands of a DNA duplex, and the "connectors" indicate how the strands can be joined in the closed-circular duplex.

EPILOGUE

THE ANCIENT ORACLE OF Apollo at Delphi warned petitioners, "Know thyself." Although the wise oracle was probably referring to the psychological and moral self, we might extend the recommendation to the physical self as well. Or even to all life on the planet, the mission of biology. And perhaps, since living substance is a part of nature, we might say, "Know nature." Know trees, know rocks, know raindrops, know elephants, know amoebae, know atoms, know stars.

Measuring in powers of ten, we human beings are almost exactly midway between the largest material objects in the universe, the galaxies, and the smallest that we have explored in our particle accelerators, the electrons and quarks. We stand in the middle. From our thin sliver of existence, we want to know everything. The intricacies. The sweeping principles and the details, the forces, the patterns and cycles, the movements and mechanisms, the secrets of life, the nature of time and of space. We want to know it all, we are driven to know. We discover, we invent, we create, we question. The deeper we go, the more beauty we find, and mystery. Ultimately, the universe is stranger than we can conceive.

Many years ago, I traveled to Font-de-Gaume, a prehistoric cave in France. The walls of Font-de-Gaume are adorned with paintings made 15,000 years ago—graceful drawings of horses and bison and reindeer. One painting I still remember vividly. Two reindeer face each other, antlers touching. The two figures are perfect, and a single, loose-flowing line forms both of their upper bodies, joining them, blending them into one. The artificial light in the cave was quite dim, and the colors were faded, but I was spellbound. Here were my ancestors, struggling to fathom their world. Across the centuries, I tried to imagine their thoughts as they crouched in this dark place. What did they know of the reindeer and bison? How had they so carefully observed these animals to render such accurate drawings? What protections and powers did the paintings bestow? What needs did they satisfy? My eye followed the curving lines on the wall, across the stone and dirt floor, and then out to the mouth of the cave. An oval of light glowed in the distance, perhaps the sun in late afternoon, or the moon risen above the trees.

NOTES

PREFATORY NOTE: Many, but not all, of the landmark discoveries considered in this book resulted in Nobel Prizes. All of the Nobel Prize lectures (given by the prize-winning scientists) together with some biographical information about the scientists can be found at the Web site www.nobel.se.

INTRODUCTION

"ancient rumors about this Cape . . ." Gomes Eanes de Zurara, *The Chronicles of the Discovery and Conquest of Guinea*, edited and translated by C. Raymond Beazley and Edgar Prestage (Hakluyt Society Publications, 1896); also quoted in Daniel J. Boorstin, *The Discoverers* (New York: Random House, 1983), 166.

"he had a wish to know the land" Zurara, *Chronicles*; also quoted in Boorstin, *The Discoverers*, 165.

"it was almost three o'clock in the morning . . ." Werner Heisenberg, *Physics and Beyond*, translated from the German by Arnold J. Pomerans (New York: Harper and Row, 1971), 60–61.

"I would solve some of the problems in ways that weren't the answers the instructor expected . . ." Barbara McClintock, interview by Evelyn Fox Keller, archived at the American Philosophical Society, Philadelphia; quoted in Evelyn Fox Keller, *A Feeling for the Organism* (New York: Freeman, 1983), 26.

"Please forgive me for presenting . . ." Max Perutz, Nobel Prize lecture, December 11, 1962, 669, www.nobel.se.

A NOTE ON NUMBERS

"There are some, King Gelon, who think that the number . . ." Archimedes, "The Sand Reckoner," in *The World of Mathematics*, edited by James R. Newman (New York: Simon and Schuster, 1956), vol. 1, p. 420.

1. THE QUANTUM

Full citation for historical paper: Max Planck, "Zur Theorie des Gesetzes der Energieverteilung im Normalspectrum," *Verhandlungen der Deutschen Physikalis-*

chen Gesellschaft 2 (1900): 237–245. English translation: "On the Theory of the Energy Distribution Law of the Normal Spectrum," translated by D. ter Haar, in *The Old Quantum Theory* (Oxford: Oxford University Press, 1967).

Planck guessed the formula for black-body radiation. More precisely, Planck guessed a formula for the energy dependence of the entropy of his idealized "resonators" in equilibrium with black-body radiation. That formula led to the formula for black-body radiation.

"retain their significance for all times . . ." Max Planck, *Sitzungberichte der Königlich Preussischen Akademie der Wissenschaften* (1899), 440; translated and referred to in M. J. Klein, *Physics Today* 19 (November 1966): 26.

"the search for the absolute" was "the loftiest goal of all scientific activity." Max Planck, *Scientific Autobiography and Other Papers*, translated by F. Gaynor (New York: Philosophical Library, 1949), 35.

"he desired only to understand . . ." Philipp von Jolly is quoted in J. L. Heilbron, *The Dilemmas of an Upright Man* (Berkeley: University of California Press, 1986), 10.

For Marga Planck's comments about her husband being reserved, see Marga Planck to Ehrenfest, April 26, 1933, Ehrenfest Scientific Correspondence, Museum Boerhaave, Leyden; translated and quoted in Heilbron, *Dilemmas*, 33.

"How wonderful it is to set everything else aside . . ." Planck to Runge, July 31, 1877, Carl Rung Papers, Staatsbibliothek Preussischerkulturbesitz, Berlin, translated and quoted in Heilbron, *Dilemmas*, 33.

"He only showed himself fully in all his human qualities in the family." Marga Planck to Einstein, February 1, 1948, Albert Einstein Papers, Jerusalem, translated and quoted in Heilbron, *Dilemmas*, 33.

"only domain in life in which [Planck] gave his spirit free rein." Hans Hartmann, *Max Planck als Mensch und Denker* (Basel, Thun, and Düsseldorf: Ott, 1953), 11–12, translated and quoted in Heilbron, *Dilemmas*, 34.

"he showed convincingly that in addition to the atomistic structure of matter . . ." Einstein, "Max Planck Memorial Services" (1948), *Ideas and Opinions* (New York: Modern Library, 1994), 85.

"The introduction of the quantum . . ." Max Planck, "Physikalische Abhandlungen und Vorträge," *Braunschweig Vieweg* (1910), vol. 2, p. 247, translated and quoted in Heilbron, *Dilemmas*, 21.

2. HORMONES

Full citation for historical paper: William Bayliss and Ernest Starling, "The Mechanism of Pancreatic Secretion," *Journal of Physiology* 28 (September 12, 1902): 325–353.

Characterization of Bayliss's lab in Charles Lovatt Evans, *Reminiscences of Bayliss and Starling* (Cambridge, U.K.: Cambridge University Press, 1964), 3.

Personal characterizations of Bayliss and Starling in ibid., 2–4.

Quote by Sir Charles Martin in ibid., 14.

"In living nature the elements seem to obey . . ." Berzelius in *Lärbok i kemien*

(1808), translated and quoted in Henry M. Leicester, "Berzelius," *Dictionary of Scientific Biography* (*DSB*) (New York: Scribners, 1981) vol. 2, p. 96a.

"a vital phenomenon has . . ." Bernard in *Introduction to the Study of Experimental Medicine* (1865), quoted in M. D. Grmek, "Bernard," *DSB*, vol. 2, p. 32b.

Descartes's idea about the soul and the pineal gland is expressed in *De l'homme* (written in 1630s, published in 1660s), translated and quoted in Theodore M. Brown, "Descartes," *DSB*, vol. 4, p. 63a.

Starling's quote on education comes from "Science in Education," *Science Progress* 13 (1918–1919): 466–475, quoted in *DSB*, vol. 12, p. 618.

3. THE PARTICLE NATURE OF LIGHT

Full citation for historical paper: Albert Einstein, "Über einen die Erzeugung und Verwandlung des Lichtes betreffenden heuristischen Gesichtspunkt," *Annalen der Physik* 17, 4th series (June 9, 1905): 132–148. English translation: "On a Heuristic Point of View Concerning the Production and Transformation of Light," translated by John Stachel, Trevor Lipscombe, Alice Calaprice, and Sam Elworthy, in *Einstein's Miraculous Year* (Princeton, N.J.: Princeton University Press, 1998).

"the aim of science . . ." Albert Einstein, *Journal of the Franklin Institute* 221, no. 3 (March 1936), in Albert Einstein, *Ideas and Opinions* (New York: Modern Library, 1994), 318.

"the artist should compose . . ." Françoise Gilot, *Life with Picasso* (New York: Avon Books, 1981), 51–52.

"she is a book like you . . ." Einstein to Maric, July 29, 1900, in *Collected Papers of Albert Einstein*, translated by Anna Beck (Princeton, N.J.: Princeton University Press, 1987), vol. 1, p. 142.

"Since that bore Kleiner . . ." Einstein to Maric, December 17, 1901, in *Collected Papers*, vol. 1, p. 187.

"very revolutionary" Einstein to Conrad Habicht, May 18 or 25, 1905, in *Collected Papers*, vol. 5, p. 19.

"Einstein was motivated not by logic . . ." Banesh Hoffman in *Some Strangeness in the Proportion: A Centennial Symposium to Celebrate the Achievements of Albert Einstein*, ed. Harry Woolf (Reading, MA: Addison Wesley, 1980), 476.

"Einstein is tall, with broad shoulders . . ." Charles Nordmann, in *L'Illustration*, Paris, April 15, 1922, quoted in Albrecht Fölsing, *Albert Einstein*, translated into English by Ewald Osers (New York: Viking, 1997), 547–548.

"There aren't always subjects that are ripe . . ." Einstein to Conrad Habicht, June 30–September 22, 1905, in *Collected Papers*, vol. 5, p. 20.

4. SPECIAL RELATIVITY

Full citation for historical paper: Albert Einstein, "Zur Elektrodynamik bewegter Körper," *Annalen der Physik* 17 (1905): 891–921. English Translation: "On the Electrodynamics of Moving Bodies," translated by W. Perrett and G. R. Jeffery, in *The Principle of Relativity* (New York: Dover and Methuen, 1952).

For Einstein's having thought about light since age sixteen, see Albert Einstein, "Autobiographical Notes," in *Albert Einstein: Philosopher-Scientist*, edited by P. A. Schilpp (Evanston, Ill.: Open Court, 1949), 53.

Marcus Aurelius's conception of time: *The Meditations of Marcus Aurelius*, *Harvard Classics*, vol. 2, p. 221.

Kant's conception of time: Immanuel Kant, *Critique of Pure Reason* (1781), translated by J.M.D. Meiklejohn, in Encyclopedia Britannica's *Great Books of the Western World* (Chicago: University of Chicago Press), vol. 42, pp. 26–27.

Shelley's "leaden footed time," slower than thought: "The Cenci," Act IV, Scene II, *Harvard Classics*, vol. 18, p. 324.

"One had to understand clearly . . ." Einstein, "Autobiographical Notes," 55.

"We now know that science . . ." Albert Einstein in *Emanuel Libman Anniversary Volumes* (New York: International, 1932), vol. 1, p. 363.

"We would read a page . . ." Maurice Solovine in Albert Einstein and Maurice Solovine, *Letters to Solovine* (New York: Carol Publishing Group, 1993), 9.

"My passionate sense of social justice . . ." Albert Einstein, *Century and Forum* 84 (1931): 193–194; also in Albert Einstein, *Ideas and Opinions* (New York: Modern Library, 1994), 10.

"In your short active existence . . ." Albert Einstein in Einstein and Solovine, *Letters to Solovine*, 143.

5. THE NUCLEUS OF THE ATOM

Full citation for historical paper: Ernest Rutherford, "The Scattering of Alpha and Beta Particles by Matter and the Structure of the Atom," *London, Edinburgh and Dublin Philosophical Magazine and Journal of Science* 21, 6th series (May 1911): 669–688.

"exuberant, outgoing . . ." C. P. Snow, *The Physicists* (Boston: Little, Brown, 1981), 35.

"damn fool experiment" Rutherford quoted in Lawrence Badash, "Rutherford," *Dictionary of Scientific Biography* (New York: Scribners, 1981), vol. 12, p. 31a.

"It was almost as incredible . . ." Ibid.

"atomic bombs" H. G. Wells, *The World Set Free* (New York: Dutton, 1914), 109.

"The Professor is a deceptive . . ." Kapitza quoted in Snow, *The Physicists,* 35.

6. THE SIZE OF THE COSMOS

Full citation for historical paper: Henrietta Leavitt [article signed by Edward C. Pickering], "Periods of 25 Variable Stars in the Small Magellanic Cloud," *Circular of the Astronomical Observatory of Harvard College*, no. 173 (March 3, 1912).

"heavenly writing" and "river of heaven" *Dictionary of Scientific Biography*, (New York: Scribners, 1981) vol. 15, pp. 639–640.

Newton's estimate of the distance to nearest stars in *Principia*, vol. 2, *The System of the World*, section 57.

Miss Leavitt's Stars, by George Johnson (New York: W. W. Norton, 2005).

"Miss Leavitt inherited . . ." Solon Bailey's obituary of Henrietta Leavitt, *Popular Astronomy* 30, no. 4 (April 1922): 197–199.

"with the clarity of her mind and the sweet reasonableness of her nature" Lucy A. Patton, short biography of Henrietta Leavitt, Radcliffe archives.

"I am more sorry that I can tell you . . ." Henrietta Leavitt to Pickering, May 13, 1902, Harvard Archives, HCO Correspondence.

"At last I find myself free . . ." HSL to Pickering, August 25, 1902, Harvard Archives, HCO correspondence.

"While we cannot maintain that in everything woman is man's equal . . ." Williamina Fleming, "A Field for Woman's Work in Astronomy," *Astronomy and Astrophysics* 12 (1893): 683.

For statistics on women hired in astronomy, see Pamela Mack, "Women in Astronomy in the United States, 1875–1920" (B.A. Honors Thesis, Harvard University, 1977), chapter 4.

"Many of the assistants [almost all women] are skilful only in their own particular work . . ." Edward C. Pickering, "Fifty-Third Annual Report of the Director of the Astronomical Observatory of Harvard College, for the year ending September 30, 1898," 4.

"Pickering chose his staff to work, not to think." Celia Payne-Gaposchkin, *An Autobiography and Other Recollections*, edited by Katherine Haramundanis (Cambridge, Mass.: Harvard University Press, 1984), 149 and 147.

"What a variable star 'fiend' Miss Leavitt is . . ." letter from Professor Charles Young to Pickering, March 1, 1905, quoted by Jones and Boyd, *Harvard College Observatory*, 367.

"Not the least of my trial in being ill . . ." HSL to Pickering, mid-December 1909, Harvard Archives, HCO correspondence.

"Her discovery of the relation of period to brightness is destined . . ." Harlow Shapley to Pickering, September 24, 1917, Harvard Archives, Shapley correspondence.

"It may have been a wise decision to assign the problems of photographic photometry to Miss Leavitt . . ." Celia Payne-Gaposchkin, *Autobiography,* 147.

"We shall never understand it until we find a way to send up a net and fetch the thing down!" Ibid., 140.

HSL will and estate in probate court records for Middlesex County, Massachusetts; quoted in Johnson, *Miss Leavitt's Stars*, 88–89.

7. THE ARRANGEMENT OF ATOMS IN SOLID MATTER

Full citation for historical paper: W. Friedrich, P. Knipping, and M. von Laue, "Interferenz-Erscheinungen bei Rontgenstrahlen," *Sitzungsberichte der Königlich Bayerischen Akademie der Wissenschaften*, June 1912: 303–322. Also published in *Annalen der Physik* 41 (August 5, 1913): 971. English translation: "Interference Phenomena with Röntgen Rays," translated for this book by Dagmar Ringe.

"suddenly struck" Von Laue in his Nobel Prize lecture, November 12, 1915, *Nobel Lectures*, 351, www.nobel.se.

"the acknowledged masters of our science . . ." Ibid., 351–352.

"your experiment is one of the finest . . ." Albert Einstein to Max von Laue, June 10, 1912, quoted in Albrecht Fölsing, *Albert Einstein* (New York: Viking, 1997), 323.

"It is the most wonderful thing I have ever seen . . ." Albert Einstein to Ludwig Hopf, June 12, 1912, quoted in ibid., 323.

"I had finally been able to cultivate . . ." Von Laue, Nobel Prize lecture, 350.

"You have already honoured . . ." William Lawrence Bragg in his Nobel Prize lecture, September 6, 1922, *Nobel Lectures*, 370.

8. THE QUANTUM ATOM

Full citation for historical paper: Niels Bohr, "On the Constitution of Atoms and Molecules," *Philosophical Magazine* 26 (1913): 1–25.

"Speak as clearly as you think . . ." Told to me by John Archibald Wheeler, one of Bohr's collaborators.

"But Einstein was delighted . . . he proclaimed that this was one of the great discoveries of science." C. P. Snow, *The Physicists* (Boston: Little, Brown, 1981), 58.

"I am afraid I must hurry . . ." *Collected Works of Niels Bohr*, edited by Leon Rosenfeld (Amsterdam: North-Holland, 1972), also quoted in John L. Heilbron, "Bohr's First Theories of the Atom," in *Niels Bohr*, edited by A. P. French and P. J. Kennedy (Cambridge, Mass.: Harvard University Press, 1985), 43.

"As soon as I saw Balmer's formula . . ." Ibid.

"Indeed, we find ourselves here on the very path taken by Einstein . . ." Bohr, *Nature* (Supplement), April 14, 1928: 580.

For Hevesy's recollection of Einstein's saying that he "had no pluck to develop it," see Hevesy to Bohr, September 23, 1913, in *Collected Works of Niels Bohr*, edited by Ulrich Hoyer (Amsterdam: North Holland, 1982), vol. 2, p. 532. Quoted in John Stachel, *Einstein from B to Z* (Boston: Birkhäuser, 2002), 369.

Wheeler's recollections of his student days with Bohr are taken from *Nuclear Physics in Retrospect*, edited by R. H. Stuewer (Minneapolis: University of Minnesota Press, 1979).

9. THE MEANS OF COMMUNICATION BETWEEN NERVES

Full citation for historical paper: Otto Loewi, "Über humorale Übertragbarkeit der Herznervenwirkung," *Pflügers Archiv* 189 (1921): 239–242, English translation: "On the Humoral Transmission of the Action of the Cardiac Nerve," translated for this book by Alison Abbott.

"The night before Easter Sunday . . ." Otto Loewi, "Autobiographic Sketch," *Perspectives in Biology and Medicine* 4 (1925): 17.

"shows that an idea may sleep for decades . . ." Ibid., 18.

"a ready and uninhibited talker" whose "enthusiasms and dislikes . . ." Henry

H. Dale, "Otto Loewi," *Biographical Memoirs of Fellows of the Royal Society* 8 (1962): 80.

"I have not time . . ." Ibid., 71.

"a great scientist and person." Loewi, "Autobiographic Sketch," 8.

"For a long time I encountered great difficulties . . ." Ibid., 9.

"charmed by Starling's appearance, his expressive features, his shining eyes." Ibid., 10.

"to step suddenly beyond the normal limits of the research programme . . ." Dale, "Otto Loewi," 76.

"suffered from a kind of stage fright." Loewi, "Autobiographic Sketch," 14.

"so much for the field of activity and the importance . . ." Otto Loewi, Nobel Prize lecture, December 12, 1936, p. 5, www.nobel.se.

"When I was awakened that night and saw the pistols directed at me . . ." Loewi, "Autobiographic Sketch," 21.

10. THE UNCERTAINTY PRINCIPLE

Full citation for historical paper: Werner Heisenberg, "Über den anschaulichen Inhalt der quantentheoretishcen Kinematik und Mechanik," *Zeitschrift fur Physik* 43 (May 31, 1927): 172–198. English translation: "On the Physical Content of Quantum Kinematics and Mechanics," translated by John Archibald Wheeler and Wojciek Hubert Zurek, in *Quantum Theory and Measurement* (Princeton, N.J.: Princeton University Press, 1983).

"half serious, half joking" Edward Teller, *Memoirs* (Cambridge, Mass.: Perseus, 2001), 57.

"looked like a simple farm boy . . ." Max Born, *My Life* (New York: Scribner, 1978), 212.

"the great Architect did wisely . . ." John Milton, *Paradise Lost*, Book VIII, lines 72–75.

"I made straight for Heligoland . . ." Werner Heisenberg, *Physics and Beyond*, translated from the German by Arnold J. Pomerans (New York: Harper and Row, 1971), 60–61.

"the natural phenomena . . ." Werner Heisenberg, "The Development of Quantum Mechanics," Nobel Prize lecture, December 11, 1933, p. 1, www.nobel.se.

"I learned physics, along with a dash . . ." Heisenberg quoted in Elisabeth Heisenberg, *Inner Exile: Recollections of a Life with Werner Heisenberg*, translated by S. Cappellarii and C. Morris (Boston: Birkhauser, 1984), 32.

"It must have driven him to utter despair . . ." Victor Weisskopf, introduction to E. Heisenberg, *Inner Exile*, xiii.

"he would have abandoned his friends . . ." E. Heisenberg, *Inner Exile*, 67.

11. THE CHEMICAL BOND

Full citation for historical paper: Linus Pauling, "The Shared-Electron Chemical Bond," *Proceedings of the National Academy of Sciences* 14 (1928): 359–362.

"I am going to be a chemist!" Linus Pauling, "Starting Out," in *Linus Pauling in His Own Words*, edited by Barbara Marinacci (New York: Simon and Schuster, 1995), 31.

"I had the feeling that I could understand everything . . ." Ibid., 28.

"Today I am beginning to write the history of my life . . ." Ava Helen and Linus Pauling Papers at Oregon State University, quoted in *Linus Pauling, Scientist and Peacemaker*, edited by Cliff Mead and Tom Hager (Corvallis: Oregon State University Press, 2001), 25.

"A surprising result of the calculation, of great chemical significance . . ." Linus Pauling, *The Nature of the Chemical Bond*, 3d ed. (Ithaca, N.Y.: Cornell University Press, 1939), 113–114.

"I tried to fit knowledge that I acquired . . ." Linus Pauling to Dan Campbell, 1980, quoted in Mead and Hager, eds., *Linus Pauling*, 81.

"The way to have good ideas is to have lots of ideas." Recollected in May 2003 by Robert Silbey, chemist at MIT.

"How do people of different beliefs . . ." Linus Pauling, "The Ultimate Decision," quoted in Mead and Hager, eds., *Linus Pauling*, 198–199.

12. THE EXPANSION OF THE UNIVERSE

Full citation for historical paper: Edwin Hubble, "A Relation Between Distance and Radial Velocity Among Extra-Galactic Nebulae," *Proceedings of the National Academy of Sciences* 15 (March 15, 1929): 168–173.

"a gasp of astonishment swept through the library." Walter B. Clausen, Associated Press release, February 4, 1931, quoted in Gale E. Christianson, *Edwin Hubble, Mariner of the Nebulae* (New York: Farrar, Straus and Giroux, 1995), 210.

"a way of acting as though he had all the answers . . ." Albert A. Colvin to Charles Whitney, June 14, 1971, quoted in Christianson, *Edwin Hubble*, 25.

"tried to do things to prove he was capable of doing them" Elizabeth Hubble, quoted in Christianson, *Edwin Hubble*, 50.

"Throughout all past time, according to the records handed down from generation to generation . . ." Aristotle, *On the Heavens*, Book I, Chapter III, translated by W. K. C. Guthrie, *Loeb Classical Library* (Cambridge, Mass.: Harvard University Press, 1971), 25.

"The state of immobility is regarded as . . ." Copernicus, *On the Revolutions*, translated by Charles Glenn Wallis in Encyclopedia Britannica's *Great Books of the Western World* (Chicago: University of Chicago, 1987), vol. 16, p. 520.

"But I am constant as the northern star . . ." Shakespeare, *Julius Caesar*, III, i, 60–62.

"That [lambda] term is necessary only for the purpose . . ." Albert Einstein, "Cosmological Considerations of the General Theory of Relativity," *Sitzungsberichte der Preussischen Akademie der Wissenschaften*, I (1917): 142–152, translated by W. Perrett and G. B. Jeffery, in *The Principle of Relativity* (New York: Dover, 1952), 188.

"One of the most perplexing problems of cosmogony . . ." Arthur Eddington, *The Mathematical Theory of Relativity* (Cambridge, U.K.: Cambridge University Press, 1923), 161.

"The lines in the spectra of very distant stars or nebulae . . ." Wilhelm de Sitter, *Monthly Notices of the Royal Astronomical Society* 78 (1917): 26.

"The receding velocities of extragalactic nebulae are a cosmical effect . . ." Georges Lemaître, "A Homogeneous Universe of Constant Mass and Increasing Radius Accounting for the Radial Velocity of Extra-Galactic Nebulae," *Annales de la Société scientifique de Bruzelles* 47A (1927): 49; translated into English and reprinted in *Monthly Notices of the Royal Astronomical Society* 91 (1931): 483, quote from p. 489.

"Your husband's work is beautiful." Grace Hubble's diaries, "E.P.H.: Some People," 2, quoted in Christianson, *Edwin Hubble*, 211.

"Science is the one human activity . . ." Edwin Hubble, *Realm of the Nebulae* (New Haven, Conn.: Yale University Press, 1936), 1.

13. ANTIBIOTICS

Full citation for historical paper: Alexander Fleming, "On the Antibacterial Action of Cultures of a Penicillium, with Special Reference to Their Use in the Isolation of B. Influenzae," *British Journal of Experimental Pathology* 10, no. 3 (1929): 226–236.

Opening passage from Thucydides, *The History of the Peloponnesian War*, Book II, Chapter VII, sections [47], [49], [52], translated by Richard Crawley in Encyclopedia Britannica's *Great Books of the Western World* (Chicago: University of Chicago, 1987), vol. 6, pp. 399–400.

"could be more eloquently silent . . ." Manuscripts of John Freeman, deposited in the Sir Alexander Fleming Museum in the Wright-Fleming Institute of Microbiology, London (as are all manuscripts referred to in Maurois's book), quoted in André Maurois, *The Life of Sir Alexander Fleming*, translated from the French by Gerard Hopkin (New York: Dutton, 1959), 54.

"never liked talking, but when he did . . ." Manuscripts of C. A. Pannett, quoted in Maurois, *Life of Sir Alexander Fleming*, 57.

"delighted in making difficulties for himself . . ." Manuscripts of C. A. Pannett, quoted in Maurois, *Life of Sir Alexander Fleming*, 32.

"Things fall out of the air." Manuscript of D. M. Pryce, quoted in Maurois, *Life of Sir Alexander Fleming*, 125.

"What struck me was . . ." Ibid.

"In the inferior organisms . . ." Pasteur, *Works*, vol. VI, p. 178, quoted in Maurois, *Life of Sir Alexander Fleming*, 129.

"I had no knowledge of any of these three . . ." Fleming's diaries, quoted in Maurois, *Life of Sir Alexander Fleming*, 30–31.

"of great use to me" Fleming, Nobel Prize lecture, December 11, 1945, *Nobel Lectures*, 84, www.nobel.se.

"was very shy . . ." Maurois, *Life of Sir Alexander Fleming*, 136.

"completely changed the medical mind in . . ." Fleming, Nobel lecture, p. 92.

"In a time when annihilation and destruction . . ." Professor G. Liljestrand of the Royal Caroline Institute, in his presentation of the 1945 Nobel Prize in physiology or medicine, www.nobel.se.

14. THE MEANS OF PRODUCTION OF ENERGY IN LIVING ORGANISMS

Full citation for historical paper: Hans Krebs and W. A. Johnson, "The Role of Citric Acid in Intermediate Metabolism in Animal Tissues," *Enzymologia* 4 (1937): 148–156.

"self conscious, timid, and solitary . . ." Hans Krebs, *Reminiscences and Reflections* (Oxford, U.K.: Clarendon Press, 1981), 9.

"He set an example of high standards . . ." Ibid., 27.

"he did not think I had sufficient ability . . ." Ibid., 40.

"I came to the conclusion . . ." Ibid., 42.

"British friendliness and human warmth . . ." Ibid., 38.

"In visualizing the cycle mechanism . . ." Ibid., 118.

"there can be one pleasure still greater . . ." Ibid., 229.

15. NUCLEAR FISSION

Full citations for historical papers: O. Hahn and F. Strassmann, "Über den Nachweis und das Verhalten der bei der Bestrahlung des Urans mittels Neutronen entstehended Erdalkalimetalle," *Die Naturwissenschaften* 27 (1939): 11; partial English translation: "Concerning the Existence of Alkaline Earth Metals Resulting from Neutron Irradiation of Uranium," translated by Hans G. Graetzer, in Hans G. Graetzer and David L. Anderson, *The Discovery of Nuclear Fission* (New York: Van Nostrand Reinhold, 1971), 44–47. Lise Meitner and O. R. Frisch, "Disintegration of Uranium by Neutrons: A New Type of Nuclear Reaction," *Nature* 143 (February 11, 1939): 239–240.

"Perhaps you can come up with some sort of fantastic explanation . . ." Hahn to Meitner, December 19, 1938, Meitner Collection, Churchill College Archives Centre, Cambridge, U.K., quoted in Ruth Lewin Sime, *Lise Meitner: A Life in Physics* (Berkeley: University of California Press, 1996), 233.

"Thinking back to the time of my youth . . ." Meitner to Frl. Hitzenberger, March 29/April 10, 1951, Meitner Collection, quoted in Sime, *Lise Meitner*, 7.

Anecdote about Meitner and sewing on the Sabbath: Lilli Eppstein, personal communication to Ruth Sime, Stocksund, September 12, 1987, referred to in Sime, *Lise Meitner*, 5.

"Oh, I thought you were a man!" Meitner, "Looking Back," *Bulletin of the Atomic Scientists* 20 (November 1964): 5.

"Dear Herr Hahn! The pitchblende experiment . . ." Meitner to Hahn, February 22, 1917, Otto Hahn Nachlass, Archiv zur Geschichte der Max-Planck Gesellschaft, Berlin, quoted in Sime, *Lise Meitner*, 63–64.

"[Meitner] urgently requested . . ." Strassmann, *Kernspaltung: Berlin Dezember 1938*, 18, 20, translated and referred to in Sime, *Lise Meitner*, 229.

"Oh, what fools we have been! . . ." Frisch's recollection in "How It All Began," *Physics Today*, November 1967, p. 47.

"trusted for his human qualities, simplicity of manner, transparent honesty, common sense, and loyalty." Rod Spence, *Biographical Memoirs of the Royal Society* 16 (1970): 302.

"I have no self confidence . . ." Meitner to Walter Meitner, February 6, 1939, Meitner Collection, quoted in Sime, *Lise Meitner*, 255.

"Surely Hahn fully deserved the Nobel Prize in chemistry . . ." Lise Meitner to Birgit Broomé Aminoff, November 20, 1945, Meitner Collection, quoted in Sime, *Lise Meitner*, 327.

"Hahn was completely shattered by the news . . ." T. H. Rittner to M. Perrin, Lt. Comdr. Welsh, and Capt. Davis for Gen. [Leslie] Groves, Top Secret Report 4, Operation "Epsilon" (August 6–7, 1945), reprinted in *Hitler's Uranium Club: The Secret Recordings at Farm Hall*, annotated by Jeremy Bernstein (New York: Copernicus Books, 2001), 115.

"one could love one's work . . ." Meitner to James Franck, March 1958, James Franck Papers, Joseph Regenstein Library, University of Chicago, quoted in Sime, *Lise Meitner*, 375.

16. THE MOVABILITY OF GENES

Full citation for historical paper: Barbara McClintock, "Mutable Loci in Maize," *Carnegie Institution of Washington Yearbook* 47 (1948): 155–169.

"I was just so interested in what I was doing . . ." Barbara McClintock, interview by Evelyn Fox Keller, American Philosophical Society, Philadelphia, reported in Evelyn Fox Keller, *A Feeling for the Organism* (New York: Freeman, 1983), 70.

"had a greenhouse man but he is not too bright." Barbara McClintock to George Beadle, January 28, 1951, Caltech Archives, quoted in Nathaniel Comfort, *The Tangled Field* (Cambridge, Mass.: Harvard University Press, 2001), 99.

too "delicate" and "feminine" McClintock, interview by Evelyn Fox Keller, September 24, 1978, American Philosophical Society, quoted in Comfort, *Tangled Field*, 19.

"used to love to be alone . . . just thinking about things." McClintock quoted in Keller, *Feeling for the Organism*, 22.

"Here was a dividing line . . ." Ibid., 33.

"There was not that strong necessity for a personal attachment to anybody . . ." Ibid., 34.

"At this stage, in the mid thirties, a career for women did not receive very much approbation." Ibid., 72.

"I would solve some of the problems in ways that weren't the answers the instructor expected . . ." Ibid., 26.

"[McClintock] was just beside herself with excitement . . ." Evelyn Witkin, interview by Nathaniel Comfort, March 19, 1996, quoted in Comfort, *The Tangled Field*, 113.

"was so striking that I dropped everything . . ." McClintock quoted in Keller, *Feeling for the Organism*, 124.

"Continued study . . . has revealed a type of event involving the Ds locus . . ." McClintock, "Mutable Loci in Maize," *Carnegie Institution of Washington Yearbook*, vol. 48 (1949) 142–143.

"It seems to me that we might as well stop dealing with genes in the old sense . . ." McClintock to George Beadle, January 28, 1951, Caltech Archives, quoted in Comfort, *Tangled Field*, 99.

"When you suddenly see the problem, something happens . . ." McClintock quoted in Keller, *Feeling for the Organism*, 103.

17. THE STRUCTURE OF DNA

Full citation for historical papers: J. D. Watson and F.H.C. Crick, "Molecular Structure of Nucleic Acids," *Nature* 171 (April 25, 1953): 737–738. Rosalind E. Franklin and R. G. Gosling, "Molecular Configuration in Sodium Thymonucleate," *Nature* 171 (April 25, 1953): 740–741.

"Jim and I hit it off immediately . . ." Francis Crick, *What Mad Pursuit* (New York: Basic Books, 1968), 64.

"talked louder and faster than anyone else . . ." James D. Watson, *Double Helix* (New York: New American Library, 1969), 16.

"very private person with very high personal and scientific standards . . ." Frederick Dainton to Anne Sayre, in archives at University of Maryland, quoted in *Physics Today*, March 2003, p. 45.

"stubborn and difficult to supervise." Norrish to Anne Sayre, September 22, 1970, quoted in Anne Sayre, *Rosalind Franklin and DNA* (New York: Norton, 1975), 58.

"for all practical purposes the personal property . . ." Watson, *Double Helix*, 19.

"the facts will speak for themselves." Frederick Dainton to Anne Sayre, in archives at University of Maryland, quoted in *Physics Today*, March 2003, p. 45.

"Our lunch conversations quickly centered on how genes were put together . . ." Watson, *Double Helix*, 37.

"I was soon taught [by Crick] . . ." Ibid., 38.

"quick, nervous . . . without a trace of warmth or frivolity . . ." Ibid., 51.

Crick's detailed calculations about expected X-ray diffraction patterns from helical molecules were done with W. Cochran and V. Vand and reported in *Acta Crystallographica* 5 (1952): 581–586. According to these results, an absent "fourth layer line," as was seen in Franklin's photos of DNA B, would necessarily result from the interference of multiple strands and further indicated a molecule with an even number of strands. However, other diffraction data showed that the molecule was not wide enough to support four strands, thus leaving only the two-stranded possibility. Private conversation with Alexander Rich of MIT, November 25, 2003.

"to drink a toast to the Pauling failure." Watson, *Double Helix*, 104.

"the instant I saw the picture . . ." Ibid., 107.

"important biological objects come in pairs." Ibid., 108.

"When I got to our still empty office the following morning . . ." Ibid., 123.

18. THE STRUCTURE OF PROTEINS

Full citation for historical paper: M. F. Perutz, M. G. Rossmann, Ann F. Cullis, Hilary Muirhead, and Georg Will, "Structure of Haemoglobin," *Nature* 185 (1960): 416–422.

"a mind like a razor and an elegant command of language" Anthony Tucker, *The Guardian*, February 7, 2002.

"outward manner that was reserved and quiet . . ." Alexander Rich, in a private conversation with Alan Lightman, MIT, March 30, 2004.

"I found in Kendrew an outstandingly able . . ." Max Perutz, appreciation of Kendrew, September 30, 1997, in *MRC Newsletter*, Fall 1997, Medical Research Council of the Laboratory of Molecular Biology, Cambridge University.

"tied up in some very interesting research at the time." Perutz quoted by George Rada, chief executive of Britain's Medical Research Council, in the MRC press release upon Perutz's death, February 6, 2002.

"wasted five semesters in an exacting course of inorganic analysis" Perutz, quoted in Nobel Prize biography of Max Perutz, www.nobel.se.

"I took several hundred X-ray diffraction pictures . . ." Max Perutz, *I Wish I'd Made You Angry Earlier* (Cold Spring, New York: Cold Spring Harbor Laboratory Press, 1998), x–xi.

"The wide distances separating the haem groups was perhaps the greatest surprise . . ." Perutz, Nobel Prize lecture, December 11, 1962, p. 665, www.nobel.se.

The quotes from Perutz's commonplace book are in the last section of Perutz, *I Wish I'd Made You Angry Earlier*.

"Please forgive me for presenting . . ." Perutz, Nobel Prize lecture, 669.

"The most beautiful experience we can have is the mysterious . . ." Albert Einstein, "The World as I See It," *Forum and Century* 84 (1931): 193–194, also quoted in Einstein, *Ideas and Opinions* (New York: Modern Library, 1994), 11.

19. RADIO WAVES FROM THE BIG BANG

Full citation for historical papers: A. A. Penzias and R. W. Wilson, "A Measurement of Excess Antenna Temperature at 4080 Mc/s," *Astrophysical Journal* 142 (1965): 419–421. R. H. Dicke, P.J.E. Peebles, P. G. Roll, and D. T. Wilkinson, "Cosmic Black-Body Radiation," *Astrophysical Journal* 142 (1965): 414–419.

"We evicted pigeons and cleaned up their mess, but obtained only a small reduction in the [noise]." Robert Wilson, Nobel Prize lecture, December 8, 1978, p. 475, www.nobel.se.

"It never exactly hit us all in one day" Arno Penzias, quoted in Timothy Ferris, *The Red Limit* (New York: Bantam, 1977), 96–97.

"We thought that our measurement was independent of the theory and might outlive it." Wilson, Nobel lecture, 476.

"Hold infinity in the palm of your hand" William Blake, from *Auguries of Innocence* (1805).

"we hadn't done our homework." P.J.E. Peebles, in a telephone interview with Alan Lightman, April 26, 2004.

"I must take the major blame . . ." Robert Dicke, unpublished scientific autobiography, 1975, stored in the Membership office of the National Academy of Sciences.

"It would be around 1964 . . ." James Peebles, quoted in Alan Lightman and Roberta Brawer, *Origins* (Cambridge, Mass.: Harvard University Press, 1990), 218.

"I have long believed that an experimentalist . . ." Dicke, unpublished scientific autobiography.

"cautiously optimistic" Wilson, Nobel lecture, 476.

"There's still one point in cosmology . . ." Robert Dicke, quoted in Lightman and Brawer, *Origins*, 212.

20. A UNIFIED THEORY OF FORCES

Full citation for historical paper: Steven Weinberg, "A Model of Leptons," *Physical Review Letters* 19 (1967): 1264–1266.

"It occurred to me that I had been applying the right ideas to the wrong problem." Steven Weinberg, Nobel Prize lecture, December 8, 1979, p. 548, www.nobel.se.

"Matter thus loses its central role . . ." Steven Weinberg, *Dreams of a Final Theory* (Pantheon: New York, 1992), 138–139, 142, 165.

"I found in doing this, although it had not been my idea at all to start with . . ." Ibid., 119.

"Christianity and Judaism teach that history is moving towards a climax . . ." Steven Weinberg, "Physics and History," *Daedalus*, 127 (Winter 1998): 152.

"It is not necessary to mop up all the islands of unsolved problems . . ." Ibid., 153.

"I'm sorry I wrote the damn thing because . . ." Steven Weinberg, quoted in Alan Lightman and Roberta Brawer, *Origins*: *The Lives and Worlds of Modern Cosmologists* (Cambridge, Mass.: Harvard University Press, 1990), 456.

"a theory had come along that had the kind of compelling quality . . ." Weinberg, *Dreams of a Final Theory*, 123.

"learned to scent out that which was able to lead to fundamentals . . ." Albert Einstein, "Autobiographical Notes," in *Albert Einstein*: *Philosopher-Scientist*, edited by Paul Arthur Schilpp (Evanston, Ill.: Open Court, 1949), 17.

21. QUARKS: A TINIEST ESSENCE OF MATTER

Full citation for historical paper: M. Breidenbach, J. I. Friedman, H. W. Kendall, E. D. Bloom, D. H. Coward, H. DeStaebler, J. Drees, L. W. Mo, and R. E.

Taylor, "Observed Behavior of Highly Inelastic Electron-Proton Scattering," *Physical Review Letters* 23 (1969): 935–939.

"It seems probable to me that God . . ." Isaac Newton, *Optics*, Book III, Part 1, translated by Andrew Motte and revised by Florian Cajori, in Encyclopedia Britannica's *Great Books of the Western World* (Chicago: University of Chicago, 1987), vol. 34, p. 541.

"When from a long distant past nothing subsists . . ." Marcel Proust, *A la recherche du temps perdu*, vol. 1, *Swann's Way* (1913), translated by C. K. Scott Moncrieff (New York: Modern Library, 2003), 63–64.

"Reproductions . . . I keep the good stuff at home." Jerome Friedman, interview by Alan Lightman May 28, 2004, Cambridge, Mass.

"I have been fortunate to have had outstanding . . ." Jerome Friedman, Nobel autobiography, www.nobel.se.

"Fermi was a quiet man . . ." Friedman, interview.

"When the experiment was planned . . ." Jerome Friedman, Nobel Prize lecture, December 8, 1990, p. 717, www.nobel.se.

"Most of science is having a wonderful time." Friedman, interview.

"Such particles [quarks] presumably are not real but we may use them in our field theory anyway." Murray Gell-Mann, *Physics* 1 (1964): 63.

"Quark constituent models . . . had serious problems . . ." Henry Kendall, Nobel Prize lecture, December 8, 1990, p. 678, www.nobel.se.

"the profound state of theoretical ignorance." James Bjorken, *Physical Review* 179 (1969): 1547.

"The education of my brother and myself was of paramount importance . . ." Friedman, Nobel autobiography.

". . . Fermi had simple ways of looking at the world . . ." Friedman, interview.

"When a scientist comes out with an idea that jells . . ." Friedman, interview.

"A common aspect of all creativity . . ." Jerome Friedman, "The Humanities and the Sciences," symposium of the American Council of Learned Societies, May 1, 1999, Philadelphia, *ACLS Occasional Paper*, No. 47.

"We have probably found the limit . . ." Friedman, interview.

22. THE CREATION OF ALTERED FORMS OF LIFE

Full citation for historical paper: David A. Jackson, Robert H. Symons, and Paul Berg, "Biochemical Method for Inserting New Genetic Information into DNA of Simian Virus 40," *Proceedings of the National Academy of Sciences* 69 (1972): 2904–2909.

"the elation and disappointment of venturing into the unknown." Paul Berg, Nobel Prize lecture, December 8, 1980, p. 385, www.nobel.se.

"The satisfaction derived from solving a problem with an experiment . . ." Paul Berg, Nobel Prize autobiography, www.nobel.se.

"plays a terrific game of tennis." Suzanne Pfeffer, professor of biochemistry at Stanford University, quoted in profile of Paul Berg for the American Society of Cell Biology, 1996, www.ascb.org/profiles.

"because many colleagues expressed concern . . ." Berg, Nobel lecture, 393.

"The Berg experiment scares the pants off a lot of people . . ." Wallace Rowe, quoted in Nicholas Wade, "Microbiology: Hazardous Profession Faces New Uncertainties," *Science* 182 (November 9, 1973): 566, also quoted in Nicholas Wade, *The Ultimate Experiment* (New York: Walker, 1977), 34.

"We're in a pre-Hiroshima situation." Robert Pollack, quoted in Wade, "Microbiology," 567.

"We base our recommendation [not to support development of the hydrogen bomb] on our belief that the extreme dangers . . ." Robert Oppenheimer et al, quoted in Edward Teller, *Memoirs* (Cambridge, Massachusetts: Perseus, 2001), 287.

ABRIDGMENTS OF PAPERS

ALL OF THE SHORTER PAPERS have been included in full. For the longer papers, I have omitted some of the less important or more detailed material.

Paper 1 by Planck: 20 percent has been omitted.
Paper 2 by Bayliss and Starling: 66 percent has been omitted.
Paper 3 by Einstein: 63 percent has been omitted.
Paper 4 by Einstein: 54 percent has been omitted.
Paper 5 by Rutherford: 20 percent has been omitted.
Paper 6 by Leavitt: none has been omitted.
Paper 7 by von Laue et al.: 59 percent has been omitted.
Paper 8 by Bohr: 47 percent has been omitted.
Paper 9 by Loewi: none has been omitted.
Paper 10 by Heisenberg: 68 percent has been omitted.
Paper 11 by Pauling: none has been omitted.
Paper 12 by Hubble: none has been omitted.
Paper 13 by Fleming: 23 percent has been omitted.
Paper 14 by Krebs and Johnson: none has been omitted.
Papers 15: none of Meitner and Frisch's paper has been omitted; 67 percent of Hahn and Strassmann's paper has been omitted.
Paper 16 by McClintock: 40 percent has been omitted.
Papers 17: none of Watson and Crick's paper has been omitted; none of Franklin and Gosling's paper has been omitted.
Paper 18 by Perutz et al.: none has been omitted.
Papers 19: none of Dicke et al.'s paper has been omitted; none of Penzias and Wilson's paper has been omitted.
Paper 20 by Weinberg: none has been omitted.
Paper 21 by Friedman et al.: none has been omitted.
Paper 22 by Berg et al.: 20 percent has been omitted.

ACKNOWLEDGMENTS

I AM GRATEFUL TO A NUMBER of scientists for counsel and critique, including Henry Abarbanel, Paul Berg, Emilio Bizzi, Carolyn Cohen, Nathaniel Comfort, Shane Crotty, Rick Danheiser, Jerome Friedman, Margaret Geller, Robert Jaffe, David Kirsch, Daniel Kleppner, Harvey Lodish, Robert Naumann, James Peebles, Alex Rich, George Rybicki, Paul Schechter, Phillip Sharp, Robert Silbey, Steven Weinberg, and Rainer Weiss. For helping to locate scientific journal articles, I thank the librarians at the Massachusetts Institute of Technology, especially Jennifer Edelman. For her help in securing the various reprinting permissions, I thank Celeste Parker Bates. I am especially grateful to Alison Abbott for her translation of Otto Loewi's paper and to Dagmar Ringe for her translation of the paper by Max von Laue and his colleagues. I thank Brian Barth for his beautiful cover, Walter Havighurst for his fine copyediting work, and Rahel Lerner for her excellent job handling the complex logistics of putting the whole book together. I thank Jessie Shelton for her proofreading. As always, I thank my longtime friend and editor, Dan Frank, and my longtime friend and literary agent, Jane Gelfman, for their continued guidance and encouragement.

PERMISSIONS ACKNOWLEDGMENTS

FOR PREVIOUSLY PUBLISHED SCIENTIFIC PAPERS:

Max Planck, "On the Theory of the Energy Distribution Law of the Normal Spectrum," translated from the German by D. ter Haar, in *The Old Quantum Theory* (Oxford: Pergamon Press, 1967), 38–58. *Original publication*: From Max Planck, "Zur Theorie des Gesetzes der Energieverteilung im Normalspectrum," *Verhandlungen der Deutschen Physikalischen Gesellschaft* 2 (1900): 237–245. Reprinted by permission of Deutschen Physikalischen Gesellschaft E.V.

Albert Einstein, "On a Heuristic Point of View Concerning the Production and Transformation of Light," translated from the German by John Stachel, Trevor Lipscombe, Alice Calaprice, and Sam Elworthy, in *Einstein's Miraculous Year* (Copyright © 1998 by Princeton University Press). Reprinted by permission of Princeton University Press. *Original publication*: From Albert Einstein, "Über einen die Erzeugung und Verwandlung des Lichtes betreffenden heuristischen Gesichtspunkt," *Annalen der Physik* 17, 4th series (June 9, 1905): 132–148. Reprinted by permission of *Annalen der Physik*, Wiley-VCH, STM.

Albert Einstein, "On the Electrodynamics of Moving Bodies," translated from the German by W. Perrett and G. R. Jeffery, in *The Principle of Relativity* (New York: Dover and Methuen 1952). Reprinted by permission of Taylor & Francis, Limited, and the Albert Einstein Archives, The Jewish National University Library, The Hebrew University of Jerusalem, Israel. *Original publication*: From Albert Einstein, "Zur Elektrodynamik bewegter Körper," *Annalen der Physik* 17 (1905): 891–921. Reprinted by permission of *Annalen der Physik*, Wiley-VCH, STM.

Ernest Rutherford, "The Scattering of Alpha and Beta Particles by Matter and the Structure of the Atom," *London, Edinburgh and Dublin Philosophical Magazine and Journal of Science* 21, 6th series (May 1911): 669–688. Reprinted by permission of Tayor & Francis, Limited. Available online at www.tandf.co.uk/journals.

Niels Bohr, "On the Constitution of Atoms and Molecules," *Philosophical Magazine* 26 (1913): 1–25. Reprinted by permission of Taylor & Francis, Limited. Available online at www.tandf.co.uk/journals.

Werner Heisenberg, "On the Physical Content of Quantum Kinematics and Mechanics," translated from the German by John Archibald Wheeler and Wojciek

Hubert Zurek, in *Quantum Theory and Measurement* (Princeton, N.J.: Princeton University Press, 1983). Copyright © 1983 by Princeton University Press. Reprinted by permission of Princeton University Press. *Original publication*: From Werner Heisenberg, "Über den anschaulichen Inhalt der quantentheoretishcen Kinematik und Mechanik," *Zeitschrift fur Physik* 43 (May 31, 1927): 172–198. Copyright © 1927 Springer Verlag GmbH & Co. KG. Reprinted by permission of the publisher.

Linus Pauling, "The Shared-Electron Chemical Bond," *Proceedings of the National Academy of Science USA* 14 (1928): 359–362. Reprinted by permission of the Ava Helen and Linus Pauling Papers, Oregon State University Special Collections.

Edwin Hubble, "A Relation Between Distance and Radial Velocity Among Extra-Galactic Nebulae," *Proceedings of the National Academy of Science USA* 15 (March 15, 1929): 168–173. Reprinted by permission of The Huntington Library, San Marino, California.

Alexander Fleming, "On the Antibacterial Action of Cultures of a Penicillium, with Special Reference to Their Use in the Isolation of B. Influenzae," *British Journal of Experimental Pathology* 10, no. 3 (1929): 226–236. Reprinted by permission of Blackwell Publishing, Limited.

Hans Krebs and W. A. Johnson, "The Role of Citric Acid in Intermediate Metabolism in Animal Tissues," *Enzymologia* 4 (1937): 148–156. Reprinted by permission of Kluwer Academic/Plenum Publishers.

O. Hahn and F. Strassmann "Concerning the Existence of Alkaline Earth Metals Resulting from Neutron Irradiation of Uranium," translated from the German by Hans G. Graetzer, in Hans G. Graetzer and David L. Anderson, *The Discovery of Nuclear Fission* (New York: Van Nostrand Reinhold, 1971), 44–47. Reprinted by permission of the American Institute of Physics. *Original publication*: From O. Hahn and F. Strassmann, "Über den Nachweis und das Verhalten der bei der Bestrahlung des Urans mittels Neutronen entstehended Erdalkalimetalle," *Die Naturwissenschaften* 27 (1939): 11. Copyright © 1939 Springer Verlag GmbH & Co. KG. Reprinted by permission of the publisher.

Lise Meitner and O. R. Frisch, "Disintegration of Uranium by Neutrons: A New Type of Nuclear Reaction," *Nature* 143 (February 11, 1939): 239–240. Copyright © 1939 Macmillan Publishers, Limited. Reprinted by permission of *Nature.*

Barbara McClintock, "Mutable Loci in Maize," *Carnegie Institution of Washington Yearbook* 47 (1948): 155–169. Reprinted by permission of the Carnegie Institution of Washington.

J. D. Watson and F.H.C. Crick, "Molecular Structure of Nucleic Acids: A Structure for Deoxyribose Nucleic Acid," *Nature* 171 (April 25, 1953): 737–738. Copyright © 1953 Macmillan Publishers, Limited. Reprinted by permission of *Nature.*

Rosalind E. Franklin and R. G. Gosling, "Molecular Configuration in Sodium Thymonucleate," *Nature* 171 (April 25, 1953): 740–741. Copyright © 1953 Macmillan Publishers, Limited. Reprinted by permission of *Nature.*

M. F. Perutz, M. G. Rossmann, Ann F. Cullis, Hilary Muirhead, and Georg

FOR ILLUSTRATIONS:

INDEX

ALSO BY ALAN LIGHTMAN

THE DIAGNOSIS

The Diagnosis is a harrowing tale of one man's struggle to cope in a wired world, even as his own biological wiring short-circuits. As Bill Chalmers attempts to find a diagnosis for his sudden, deteriorating illness, he discovers that he is fighting not just for his body but also for his soul.

Fiction/Literature/0-375-72550-4

EINSTEIN'S DREAMS

A modern classic, *Einstein's Dreams* is a fictional collage of stories dreamed by Albert Einstein in 1905, as the young genius is creating his theory of relativity, a new conception of time. *Einstein's Dreams* explores the connections between science and art, the process of creativity, and ultimately the fragility of human existence.

Fiction/1-4000-7780-X

REUNION

Charles, a once-promising poet, relives an intense love affair at his thirtieth college reunion. As Charles remembers contradictory versions of events, reality and identity dissolve into a haze of illusion. *Reunion* explores the pain of self-examination, the clay-like nature of memory, and the fatal power of first love.

Fiction/0-375-71344-1

A SENSE OF THE MYSTERIOUS

A gifted physicist and novelist, Alan Lightman has lived in the dual worlds of science and art for much of his life. In these brilliant essays, the two worlds meet. In *A Sense of the Mysterious*, rather than finding a forbidding gulf between the two cultures, Lightman discovers complementary ways of looking at the world, both part of being human.

Science History/1-4000-7819-9